The Lepton

Edited by Paul F. Kisak

Contents

Chapter 1

Overview

1.1 Lepton

A **lepton** is an elementary, half-integer spin (spin $\frac{1}{2}$) particle that does not undergo strong interactions, but is subject to the Pauli exclusion principle.[*][1] The best known of all leptons is the electron, which is directly tied to all chemical properties. Two main classes of leptons exist: charged leptons (also known as the *electron-like* leptons), and neutral leptons (better known as neutrinos). Charged leptons can combine with other particles to form various composite particles such as atoms and positronium, while neutrinos rarely interact with anything, and are consequently rarely observed.

There are six types of leptons, known as *flavours*, forming three *generations*.[*][2] The first generation is the *electronic leptons*, comprising the electron (e−) and electron neutrino (ν
e); the second is the *muonic leptons*, comprising the muon (μ−) and muon neutrino (ν
μ); and the third is the *tauonic leptons*, comprising the tau (τ−) and the tau neutrino (ν
τ). Electrons have the least mass of all the charged leptons. The heavier muons and taus will rapidly change into electrons through a process of particle decay: the transformation from a higher mass state to a lower mass state. Thus electrons are stable and the most common charged lepton in the universe, whereas muons and taus can only be produced in high energy collisions (such as those involving cosmic rays and those carried out in particle accelerators).

Leptons have various intrinsic properties, including electric charge, spin, and mass. Unlike quarks however, leptons are not subject to the strong interaction, but they are subject to the other three fundamental interactions: gravitation, electromagnetism (excluding neutrinos, which are electrically neutral), and the weak interaction. For every lepton flavor there is a corresponding type of antiparticle, known as antilepton, that differs from the lepton only in that some of its properties have equal magnitude but opposite sign. However, according to certain theories, neutrinos may be their own antiparticle, but it is not currently known whether this is the case or not.

The first charged lepton, the electron, was theorized in the mid-19th century by several scientists[*][3][*][4][*][5] and was discovered in 1897 by J. J. Thomson.[*][6] The next lepton to be observed was the muon, discovered by Carl D. Anderson in 1936, which was classified as a meson at the time.[*][7] After investigation, it was realized that the muon did not have the expected properties of a meson, but rather behaved like an electron, only with higher mass. It took until 1947 for the concept of "leptons" as a family of particle to be proposed.[*][8] The first neutrino, the electron neutrino, was proposed by Wolfgang Pauli in 1930 to explain certain characteristics of beta decay.[*][8] It was first observed in the Cowan–Reines neutrino experiment conducted by Clyde Cowan and Frederick Reines in 1956.[*][8][*][9] The muon neutrino was discovered in 1962 by Leon M. Lederman, Melvin Schwartz and Jack Steinberger,[*][10] and the tau discovered between 1974 and 1977 by Martin Lewis Perl and his colleagues from the Stanford Linear Accelerator Center and Lawrence Berkeley National Laboratory.[*][11] The tau neutrino remained elusive until July 2000, when the DONUT collaboration from Fermilab announced its discovery.[*][12][*][13]

Leptons are an important part of the Standard Model. Electrons are one of the components of atoms, alongside protons and neutrons. Exotic atoms with muons and taus instead of electrons can also be synthesized, as well as lepton–antilepton particles such as positronium.

1.1.1 Etymology

The name *lepton* comes from the Greek λεπτός *leptós*, "fine, small, thin" (neuter form: λεπτόν *leptón*);[*][14][*][15] the earliest attested form of the word is the Mycenaean Greek 𐀩𐀡𐀵, *re-po-to*, written in Linear B syllabic script.[*][16] *Lepton* was first used by physicist Léon

Rosenfeld in 1948:[*][17]

> Following a suggestion of Prof. C. Møller, I adopt —as a pendant to "nucleon" —the denomination "lepton" (from λεπτός, small, thin, delicate) to denote a particle of small mass.

The etymology incorrectly implies that all the leptons are of small mass. When Rosenfeld named them, the only known leptons were electrons and muons, which are in fact of small mass —the mass of an electron $(0.511\ \mathrm{MeV}/c^2)$[*][18] and the mass of a muon (with a value of $105.7\ \mathrm{MeV}/c^2$)[*][19] are fractions of the mass of the "heavy" proton $(938.3\ \mathrm{MeV}/c^2)$.[*][20] However, the mass of the tau (discovered in the mid 1970s) $(1777\ \mathrm{MeV}/c^2)$[*][21] is nearly twice that of the proton, and about 3,500 times that of the electron.

1.1.2   History

See also: Electron § Discovery, Muon § History and Tau (particle) § History

The first lepton identified was the electron, discovered

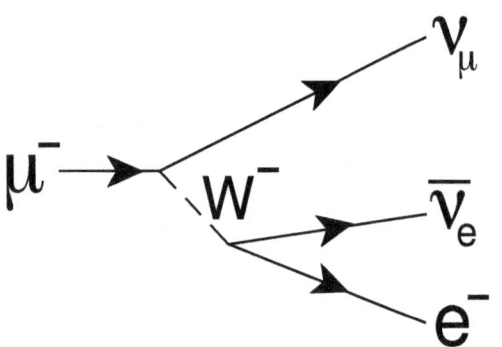

A muon transmutes into a muon neutrino by emitting a W− boson. The W− boson subsequently decays into an electron and an electron antineutrino.

by J.J. Thomson and his team of British physicists in 1897.[*][22][*][23] Then in 1930 Wolfgang Pauli postulated the electron neutrino to preserve conservation of energy, conservation of momentum, and conservation of angular momentum in beta decay.[*][24] Pauli theorized that an undetected particle was carrying away the difference between the energy, momentum, and angular momentum of the initial and observed final particles. The electron neutrino was simply called the neutrino, as it was not yet known that neutrinos came in different flavours (or different "generations").

Nearly 40 years after the discovery of the electron, the muon was discovered by Carl D. Anderson in 1936. Due to its mass, it was initially categorized as a meson rather than a lepton.[*][25] It later became clear that the muon was much more similar to the electron than to mesons, as muons do not undergo the strong interaction, and thus the muon was reclassified: electrons, muons, and the (electron) neutrino were grouped into a new group of particles – the leptons. In 1962 Leon M. Lederman, Melvin Schwartz and Jack Steinberger showed that more than one type of neutrino exists by first detecting interactions of the muon neutrino, which earned them the 1988 Nobel Prize, although by then the different flavours of neutrino had already been theorized.[*][26]

The tau was first detected in a series of experiments between 1974 and 1977 by Martin Lewis Perl with his colleagues at the SLAC LBL group.[*][27] Like the electron and the muon, it too was expected to have an associated neutrino. The first evidence for tau neutrinos came from the observation of "missing" energy and momentum in tau decay, analogous to the "missing" energy and momentum in beta decay leading to the discovery of the electron neutrino. The first detection of tau neutrino interactions was announced in 2000 by the DONUT collaboration at Fermilab, making it the latest particle of the Standard Model to have been directly observed,[*][28] apart from the Higgs boson, which probably has been discovered in 2012.

Although all present data is consistent with three generations of leptons, some particle physicists are searching for a fourth generation. The current lower limit on the mass of such a fourth charged lepton is $100.8\ \mathrm{GeV}/c^2$,[*][29] while its associated neutrino would have a mass of at least $45.0\ \mathrm{GeV}/c^2$.[*][30]

1.1.3   Properties

Spin and chirality

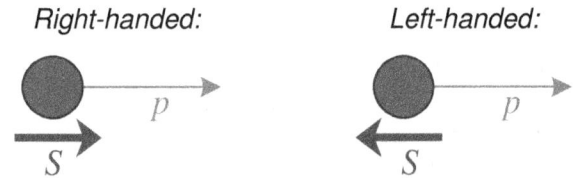

Left-handed and right-handed helicities

Leptons are spin-$\frac{1}{2}$ particles. The spin-statistics theorem thus implies that they are fermions and thus that they are subject to the Pauli exclusion principle; no two leptons of the same species can be in exactly the same state at the same time. Furthermore, it means that a lepton can have only two possible spin states, namely up or down.

A closely related property is chirality, which in turn is closely related to a more easily visualized property called helicity. The helicity of a particle is the direction of its spin relative to its momentum; particles with spin in the same direction as their momentum are called *right-handed* and otherwise they are called *left-handed*. When a particle is mass-less, the direction of its momentum relative to its spin is frame independent, while for massive particles it is possible to 'overtake' the particle by a Lorentz transformation flipping the helicity. Chirality is a technical property (defined through the transformation behaviour under the Poincaré group) that agrees with helicity for (approximately) massless particles and is still well defined for massive particles.

In many quantum field theories—such as quantum electrodynamics and quantum chromodynamics—left and right-handed fermions are identical. However in the Standard Model left-handed and right-handed fermions are treated asymmetrically. Only left-handed fermions participate in the weak interaction, while there are no right-handed neutrinos. This is an example of parity violation. In the literature left-handed fields are often denoted by a capital L subscript (e.g. e$-_L$) and right-handed fields are denoted by a capital R subscript.

Electromagnetic interaction

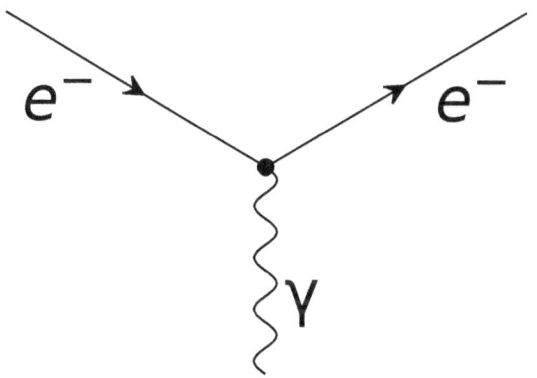

Lepton–photon interaction

One of the most prominent properties of leptons is their electric charge, Q. The electric charge determines the strength of their electromagnetic interactions. It determines the strength of the electric field generated by the particle (see Coulomb's law) and how strongly the particle reacts to an external electric or magnetic field (see Lorentz force). Each generation contains one lepton with $Q = -e$ (conventionally the charge of a particle is expressed in units of the elementary charge) and one lepton with zero electric charge. The lepton with electric charge is commonly simply referred to as a 'charged lepton' while the neutral lepton is called a neutrino. For example the first generation consists of the electron e− with a negative electric charge and

the electrically neutral electron neutrino ν
e.

In the language of quantum field theory the electromagnetic interaction of the charged leptons is expressed by the fact that the particles interact with the quantum of the electromagnetic field, the photon. The Feynman diagram of the electron-photon interaction is shown on the right.

Because leptons possess an intrinsic rotation in the form of their spin, charged leptons generate a magnetic field. The size of their magnetic dipole moment μ is given by,

$$\mu = g\frac{Q\,\hbar}{4m},$$

where m is the mass of the lepton and g is the so-called g-factor for the lepton. First order approximation quantum mechanics predicts that the g-factor is 2 for all leptons. However, higher order quantum effects caused by loops in Feynman diagrams introduce corrections to this value. These corrections, referred to as the anomalous magnetic dipole moment, are very sensitive to the details of a quantum field theory model and thus provide the opportunity for precision tests of the standard model. The theoretical and measured values for the electron anomalous magnetic dipole moment are within agreement within eight significant figures.*[31]

Weak Interaction

In the Standard Model the left-handed charged lepton and the left-handed neutrino are arranged in doublet (ν
e$_L$, e$-_L$) that transforms in the spinor representation ($T = \frac{1}{2}$) of the weak isospin SU(2) gauge symmetry. This means that these particles are eigenstates of the isospin projection T_3 with eigenvalues $\frac{1}{2}$ and $-\frac{1}{2}$ respectively. In the meantime, the right-handed charged lepton transforms as a weak isospin scalar ($T = 0$) and thus does not participate in the weak interaction, while there is no right-handed neutrino at all.

The Higgs mechanism recombines the gauge fields of the weak isospin SU(2) and the weak hypercharge U(1) symmetries to three massive vector bosons (W+, W−, Z0) mediating the weak interaction, and one massless vector boson, the photon, responsible for the electromagnetic interaction. The electric charge Q can be calculated from the isospin projection T_3 and weak hypercharge Y_W through the Gell-Mann–Nishijima formula,

$$Q = T_3 + Y_W/2$$

To recover the observed electric charges for all particles the left-handed weak isospin doublet (ν

e$_L$, e$-_L$) must thus have $Y_W = -1$, while the right-handed isospin scalar e−
R must have $Y_W = -2$. The interaction of the leptons with the massive weak interaction vector bosons is shown in the figure on the left.

Mass

In the Standard Model each lepton starts out with no intrinsic mass. The charged leptons (i.e. the electron, muon, and tau) obtain an effective mass through interaction with the Higgs field, but the neutrinos remain massless. For technical reasons the masslessness of the neutrinos implies that there is no mixing of the different generations of charged leptons as there is for quarks. This is in close agreement with current experimental observations.*[32]

However, it is known from experiments – most prominently from observed neutrino oscillations*[33] – that neutrinos do in fact have some very small mass, probably less than 2 eV/c^2.*[34] This implies the existence of physics beyond the Standard Model. The currently most favoured extension is the so-called seesaw mechanism, which would explain both why the left-handed neutrinos are so light compared to the corresponding charged leptons, and why we have not yet seen any right-handed neutrinos.

Leptonic numbers

Main article: Lepton number

The members of each generation's weak isospin doublet are assigned leptonic numbers that are conserved under the Standard Model.*[35] Electrons and electron neutrinos have an *electronic number* of $L_e = 1$, while muons and muon neutrinos have a *muonic number* of $L_\mu = 1$, while tau particles and tau neutrinos have a *tauonic number* of $L_\tau = 1$. The antileptons have their respective generation's leptonic numbers of −1.

Conservation of the leptonic numbers means that the number of leptons of the same type remains the same, when particles interact. This implies that leptons and antileptons must be created in pairs of a single generation. For example, the following processes are allowed under conservation of leptonic numbers:

$$\begin{pmatrix} \nu_e \\ e^- \end{pmatrix}, \begin{pmatrix} \nu_\mu \\ \mu^- \end{pmatrix}, \begin{pmatrix} \nu_\tau \\ \tau^- \end{pmatrix}$$

Each generation forms a weak isospin doublet.

e− + e+ → γ + γ,

τ− + τ+ → Z0 + Z0,

but not these:

γ → e− + μ+,

W− → e− + ν

τ,

Z0 → μ− + τ+.

However, neutrino oscillations are known to violate the conservation of the individual leptonic numbers. Such a violation is considered to be smoking gun evidence for physics beyond the Standard Model. A much stronger conservation law is the conservation of the total number of leptons (L), conserved even in the case of neutrino oscillations, but even it is still violated by a tiny amount by the chiral anomaly.

1.1.4 Universality

The coupling of the leptons to gauge bosons are flavour-independent (i.e., the interactions between leptons and gauge bosons are the same for all leptons).*[35] This property is called *lepton universality* and has been tested in measurements of the tau and muon lifetimes and of Z boson partial decay widths, particularly at the Stanford Linear Collider (SLC) and Large Electron-Positron Collider (LEP) experiments.*[36]:241–243*[37]:138

The decay rate (Γ) of muons through the process μ− → e−
+ ν
e + ν
μ is approximately given by an expression of the form (see muon decay for more details)*[35]

$$\Gamma\left(\mu^- \to e^- + \bar{\nu}_e + \nu_\mu\right) = K_1 G_F^2 m_\mu^5,$$

where K_1 is some constant, and G_F is the Fermi coupling constant. The decay rate of tau particles through the process τ− → e− + ν
e + ν
τ is given by an expression of the same form*[35]

$$\Gamma\left(\tau^- \to e^- + \bar{\nu}_e + \nu_\tau\right) = K_2 G_F^2 m_\tau^5,$$

where K_2 is some constant. Muon–Tauon universality implies that $K_1 = K_2$. On the other hand, electron–muon universality implies*[35]

$$\Gamma\left(\tau^- \to e^- + \bar{\nu}_e + \nu_\tau\right) = \Gamma\left(\tau^- \to \mu^- + \bar{\nu}_\mu + \nu_\tau\right).$$

This explains why the branching ratios for the electronic mode (17.85%) and muonic (17.36%) mode of tau decay are equal (within error).[*][21]

Universality also accounts for the ratio of muon and tau lifetimes. The lifetime of a lepton (τ_l) is related to the decay rate by[*][35]

$$\tau_l = \frac{B\left(l^- \rightarrow e^- + \bar{\nu}_e + \nu_l\right)}{\Gamma\left(l^- \rightarrow e^- + \bar{\nu}_e + \nu_l\right)},$$

where $B(\text{x} \rightarrow \text{y})$ and $\Gamma(\text{x} \rightarrow \text{y})$ denotes the branching ratios and the resonance width of the process x → y.

The ratio of tau and muon lifetime is thus given by[*][35]

$$\frac{\tau_\tau}{\tau_\mu} = \frac{B\left(\tau^- \rightarrow e^- + \bar{\nu}_e + \nu_\tau\right)}{B\left(\mu^- \rightarrow e^- + \bar{\nu}_e + \nu_\mu\right)} \left(\frac{m_\mu}{m_\tau}\right)^5.$$

Using the values of the 2008 *Review of Particle Physics* for the branching ratios of muons[*][19] and tau[*][21] yields a lifetime ratio of ~$1.29 \times 10^{*-7}$, comparable to the measured lifetime ratio of ~$1.32 \times 10^{*-7}$. The difference is due to K_1 and K_2 not actually being constants; they depend on the mass of leptons.

1.1.5 Table of leptons

1.1.6 See also

- Koide formula

- List of particles

- Preons – hypothetical particles which were once postulated to be subcomponents of quarks and leptons

1.1.7 Notes

[1] "Lepton (physics)". *Encyclopædia Britannica*. Retrieved 2010-09-29.

[2] R. Nave. "Leptons". *HyperPhysics*. Georgia State University, Department of Physics and Astronomy. Retrieved 2010-09-29.

[3] W.V. Farrar (1969). "Richard Laming and the Coal-Gas Industry, with His Views on the Structure of Matter". *Annals of Science* **25** (3): 243–254. doi:10.1080/00033796900200141.

[4] T. Arabatzis (2006). *Representing Electrons: A Biographical Approach to Theoretical Entities*. University of Chicago Press. pp. 70–74. ISBN 0-226-02421-0.

[5] J.Z. Buchwald, A. Warwick (2001). *Histories of the Electron: The Birth of Microphysics*. MIT Press. pp. 195–203. ISBN 0-262-52424-4.

[6] J.J. Thomson (1897). "Cathode Rays". *Philosophical Magazine* **44** (269): 293. doi:10.1080/14786449708621070.

[7] S.H. Neddermeyer, C.D. Anderson; Anderson (1937). "Note on the Nature of Cosmic-Ray Particles". *Physical Review* **51** (10): 884–886. Bibcode:1937PhRv...51..884N. doi:10.1103/PhysRev.51.884.

[8] "The Reines-Cowan Experiments: Detecting the Poltergeist" (PDF). *Los Alamos Science* **25**: 3. 1997. Retrieved 2010-02-10.

[9] F. Reines, C.L. Cowan, Jr.; Cowan (1956). "The Neutrino". *Nature* **178** (4531): 446. Bibcode:1956Natur.178..446R. doi:10.1038/178446a0.

[10] G. Danby; Gaillard, J-M.; Goulianos, K.; Lederman, L.; Mistry, N.; Schwartz, M.; Steinberger, J. et al. (1962). "Observation of high-energy neutrino reactions and the existence of two kinds of neutrinos". *Physical Review Letters* **9**: 36. Bibcode:1962PhRvL...9...36D. doi:10.1103/PhysRevLett.9.36.

[11] M.L. Perl; Abrams, G.; Boyarski, A.; Breidenbach, M.; Briggs, D.; Bulos, F.; Chinowsky, W.; Dakin, J.; Feldman, G.; Friedberg, C.; Fryberger, D.; Goldhaber, G.; Hanson, G.; Heile, F.; Jean-Marie, B.; Kadyk, J.; Larsen, R.; Litke, A.; Lüke, D.; Lulu, B.; Lüth, V.; Lyon, D.; Morehouse, C.; Paterson, J.; Pierre, F.; Pun, T.; Rapidis, P.; Richter, B.; Sadoulet, B. et al. (1975). "Evidence for Anomalous Lepton Production in e+e– Annihilation". *Physical Review Letters* **35** (22): 1489. Bibcode:1975PhRvL..35.1489P. doi:10.1103/PhysRevLett.35.1489.

[12] "Physicists Find First Direct Evidence for Tau Neutrino at Fermilab" (Press release). Fermilab. 20 July 2000.

[13] K. Kodama *et al.* (DONUT Collaboration); Kodama; Ushida; Andreopoulos; Saoulidou; Tzanakos; Yager; Baller; Boehnlein; Freeman; Lundberg; Morfin; Rameika; Yun; Song; Yoon; Chung; Berghaus; Kubantsev; Reay; Sidwell; Stanton; Yoshida; Aoki; Hara; Rhee; Ciampa; Erickson; Graham et al. (2001). "Observation of tau neutrino interactions". *Physics Letters B* **504** (3): 218. arXiv:hep-ex/0012035. Bibcode:2001PhLB..504..218D. doi:10.1016/S0370-2693(01)00307-0.

[14] "lepton". *Online Etymology Dictionary*.

[15] λεπτός. Liddell, Henry George; Scott, Robert; *A Greek–English Lexicon* at the Perseus Project.

[16] Found on the KN L 693 and PY Un 1322 tablets. "The Linear B word re-po-to". *Palaeolexicon. Word study tool of ancient languages*. Raymoure, K.A. "re-po-to". *Minoan Linear A & Mycenaean Linear B*. Deaditerranean. "KN 693 L (103)". "PY 1322 Un + fr. (Cii)". *DĀMOS: Database of Mycenaean at Oslo*. University of Oslo.

[17] L. Rosenfeld (1948)

[18] C. Amsler *et al.* (2008): Particle listings – e–

[19] C. Amsler *et al.* (2008): Particle listings – μ–

[20] C. Amsler *et al.* (2008): Particle listings – p+

[21] C. Amsler *et al.* (2008): Particle listings – τ–

[22] S. Weinberg (2003)

[23] R. Wilson (1997)

[24] K. Riesselmann (2007)

[25] S.H. Neddermeyer, C.D. Anderson (1937)

[26] I.V. Anicin (2005)

[27] M.L. Perl et al. (1975)

[28] K. Kodama (2001)

[29] C. Amsler *et al.* (2008) Heavy Charged Leptons Searches

[30] C. Amsler *et al.* (2008) Searches for Heavy Neutral Leptons

[31] M.E. Peskin, D.V. Schroeder (1995), p. 197

[32] M.E. Peskin, D.V. Schroeder (1995), p. 27

[33] Y. Fukuda *et al.* (1998)

[34] C.Amsler et al. (2008): Particle listings – Neutrino properties

[35] B.R. Martin, G. Shaw (1992)

[36] J. P. Cumalat (1993). *Physics in Collision 12*. Atlantica Séguier Frontières. ISBN 978-2-86332-129-4.

[37] G Fraser (1 January 1998). *The Particle Century*. CRC Press. ISBN 978-1-4200-5033-2.

[38] J. Peltoniemi, J. Sarkamo (2005)

1.1.8 References

- C. Amsler *et al.* (Particle Data Group); Amsler; Doser; Antonelli; Asner; Babu; Baer; Band; Barnett; Bergren; Beringer; Bernardi; Bertl; Bichsel; Biebel; Bloch; Blucher; Blusk; Cahn; Carena; Caso; Ceccucci; Chakraborty; Chen; Chivukula; Cowan; Dahl; d'Ambrosio; Damour et al. (2008). "Review of Particle Physics". *Physics Letters B* **667**: 1. Bibcode:2008PhLB..667....1P. doi:10.1016/j.physletb.2008.07.018.

- I.V. Anicin (2005).   "The Neutrino – Its Past, Present and Future". *SFIN (Institute of Physics, Belgrade) year XV, Series A: Conferences, No. A2 (2002) 3–59*: 3172. arXiv:physics/0503172. Bibcode:2005physics...3172A.

- Y.Fukuda; Hayakawa, T.; Ichihara, E.; Inoue, K.; Ishihara, K.; Ishino, H.; Itow, Y.; Kajita, T. et al.   (1998).   "Evidence for Oscillation of Atmospheric Neutrinos". *Physical Review Letters* **81** (8): 1562–1567. arXiv:hep-ex/9807003.   Bibcode:1998PhRvL..81.1562F. doi:10.1103/PhysRevLett.81.1562.

- K. Kodama; Ushida, N.; Andreopoulos, C.; Saoulidou, N.; Tzanakos, G.; Yager, P.; Baller, B.; Boehnlein, D.; Freeman, W.; Lundberg, B.; Morfin, J.; Rameika, R.; Yun, J.C.; Song, J.S.; Yoon, C.S.; Chung, S.H.; Berghaus, P.; Kubantsev, M.; Reay, N.W.; Sidwell, R.; Stanton, N.; Yoshida, S.; Aoki, S.; Hara, T.; Rhee, J.T.; Ciampa, D.; Erickson, C.; Graham, M.; Heller, K. et al. (2001).   "Observation of tau neutrino interactions". *Physics Letters B* **504** (3): 218.   arXiv:hep-ex/0012035. Bibcode:2001PhLB..504..218D. doi:10.1016/S0370-2693(01)00307-0.

- B.R. Martin, G. Shaw (1992). "Chapter 2 – Leptons, quarks and hadrons". *Particle Physics*. John Wiley & Sons. pp. 23–47. ISBN 0-471-92358-3.

- S.H. Neddermeyer, C.D. Anderson; Anderson (1937).   "Note on the Nature of Cosmic-Ray Particles".   *Physical Review* **51** (10): 884–886.   Bibcode:1937PhRv...51..884N. doi:10.1103/PhysRev.51.884.

- J. Peltoniemi, J. Sarkamo (2005). "Laboratory measurements and limits for neutrino properties". *The Ultimate Neutrino Page*. Retrieved 2008-11-07.

- M.L. Perl; Abrams, G.; Boyarski, A.; Breidenbach, M.; Briggs, D.; Bulos, F.; Chinowsky, W.; Dakin, J. et al. (1975).   "Evidence for Anomalous Lepton Production in e^*+-e^*- Annihilation".   *Physical Review Letters* **35** (22): 1489–1492.   Bibcode:1975PhRvL..35.1489P. doi:10.1103/PhysRevLett.35.1489.

- M.E. Peskin, D.V. Schroeder (1995). *Introduction to Quantum Field Theory*. Westview Press. ISBN 0-201-50397-2.

- K. Riesselmann (2007). "Logbook: Neutrino Invention". *Symmetry Magazine* **4** (2).

- L. Rosenfeld (1948). *Nuclear Forces*. Interscience Publishers. p. xvii.

- R. Shankar (1994). "Chapter 2 – Rotational Invariance and Angular Momentum". *Principles of Quantum Mechanics* (2nd ed.). Springer. pp. 305–352. ISBN 978-0-306-44790-7.

- S. Weinberg (2003). *The Discovery of Subatomic Particles*. Cambridge University Press. ISBN 0-521-82351-X.

- R. Wilson (1997). *Astronomy Through the Ages: The Story of the Human Attempt to Understand the Universe*. CRC Press. p. 138. ISBN 0-7484-0748-0.

1.1.9 External links

- Particle Data Group homepage. The PDG compiles authoritative information on particle properties.

- Leptons, a summary of leptons from *Hyperphysics*.

Chapter 2

Charged leptons

2.1 Electron

For other uses, see Electron (disambiguation).

The **electron** is a subatomic particle, symbol e– or β–, with a negative elementary electric charge.[*][7] Electrons belong to the first generation of the lepton particle family,[*][8] and are generally thought to be elementary particles because they have no known components or substructure.[*][1] The electron has a mass that is approximately 1/1836 that of the proton.[*][9] Quantum mechanical properties of the electron include an intrinsic angular momentum (spin) of a half-integer value in units of $\hbar$, which means that it is a fermion. Being fermions, no two electrons can occupy the same quantum state, in accordance with the Pauli exclusion principle.[*][8] Like all matter, electrons have properties of both particles and waves, and so can collide with other particles and can be diffracted like light. The wave properties of electrons are easier to observe with experiments than those of other particles like neutrons and protons because electrons have a lower mass and hence a higher De Broglie wavelength for typical energies.

Many physical phenomena involve electrons in an essential role, such as electricity, magnetism, and thermal conductivity, and they also participate in gravitational, electromagnetic and weak interactions.[*][10] An electron generates an electric field surrounding it. An electron moving relative to an observer generates a magnetic field. External magnetic fields deflect an electron. Electrons radiate or absorb energy in the form of photons when accelerated. Laboratory instruments are capable of containing and observing individual electrons as well as electron plasma using electromagnetic fields, whereas dedicated telescopes can detect electron plasma in outer space. Electrons have many applications, including electronics, welding, cathode ray tubes, electron microscopes, radiation therapy, lasers, gaseous ionization detectors and particle accelerators.

Interactions involving electrons and other subatomic particles are of interest in fields such as chemistry and nuclear physics. The Coulomb force interaction between positive protons inside atomic nuclei and negative electrons composes atoms. Ionization or changes in the proportions of particles changes the binding energy of the system. The exchange or sharing of the electrons between two or more atoms is the main cause of chemical bonding.[*][11] British natural philosopher Richard Laming first hypothesized the concept of an indivisible quantity of electric charge to explain the chemical properties of atoms in 1838;[*][3] Irish physicist George Johnstone Stoney named this charge 'electron' in 1891, and J. J. Thomson and his team of British physicists identified it as a particle in 1897.[*][5][*][12][*][13] Electrons can also participate in nuclear reactions, such as nucleosynthesis in stars, where they are known as beta particles. Electrons may be created through beta decay of radioactive isotopes and in high-energy collisions, for instance when cosmic rays enter the atmosphere. The antiparticle of the electron is called the positron; it is identical to the electron except that it carries electrical and other charges of the opposite sign. When an electron collides with a positron, both particles may be totally annihilated, producing gamma ray photons.

2.1.1 History

See also: History of electromagnetism

The ancient Greeks noticed that amber attracted small objects when rubbed with fur. Along with lightning, this phenomenon is one of humanity's earliest recorded experiences with electricity. [*][14] In his 1600 treatise *De Magnete*, the English scientist William Gilbert coined the New Latin term *electricus*, to refer to this property of attracting small objects after being rubbed. [*][15] Both *electric* and *electricity* are derived from the Latin *ēlectrum* (also the root of the alloy of the same name), which came from the Greek word for amber, ἤλεκτρον (*ēlektron*).

In the early 1700s, Francis Hauksbee and French chemist Charles François de Fay independently discovered what they believed were two kinds of frictional electricity—one generated from rubbing glass, the other from rubbing resin. From this, Du Fay theorized that electricity consists of two electrical fluids, *vitreous* and *resinous*, that are separated by friction, and that neutralize each other when combined.[16] A decade later Benjamin Franklin proposed that electricity was not from different types of electrical fluid, but the same electrical fluid under different pressures. He gave them the modern charge nomenclature of positive and negative respectively.[17] Franklin thought of the charge carrier as being positive, but he did not correctly identify which situation was a surplus of the charge carrier, and which situation was a deficit.[18]

Between 1838 and 1851, British natural philosopher Richard Laming developed the idea that an atom is composed of a core of matter surrounded by subatomic particles that had unit electric charges.[2] Beginning in 1846, German physicist William Weber theorized that electricity was composed of positively and negatively charged fluids, and their interaction was governed by the inverse square law. After studying the phenomenon of electrolysis in 1874, Irish physicist George Johnstone Stoney suggested that there existed a "single definite quantity of electricity", the charge of a monovalent ion. He was able to estimate the value of this elementary charge *e* by means of Faraday's laws of electrolysis.[19] However, Stoney believed these charges were permanently attached to atoms and could not be removed. In 1881, German physicist Hermann von Helmholtz argued that both positive and negative charges were divided into elementary parts, each of which "behaves like atoms of electricity".[3]

Stoney initially coined the term *electrolion* in 1881. Ten years later, he switched to *electron* to describe these elementary charges, writing in 1894: "... an estimate was made of the actual amount of this most remarkable fundamental unit of electricity, for which I have since ventured to suggest the name *electron*". A 1906 proposal to change to *electrion* failed because Hendrik Lorentz preferred to keep *electron*.[20][21] The word *electron* is a combination of the words *electric* and *ion*.[22] The suffix *-on* which is now used to designate other subatomic particles, such as a proton or neutron, is in turn derived from electron.[23][24]

Discovery

The German physicist Johann Wilhelm Hittorf studied electrical conductivity in rarefied gases: in 1869, he discovered a glow emitted from the cathode that increased in size with decrease in gas pressure. In 1876, the German physicist Eugen Goldstein showed that the rays from this glow cast a shadow, and he dubbed the rays cathode rays.[26] During the 1870s, the English chemist and physicist Sir William

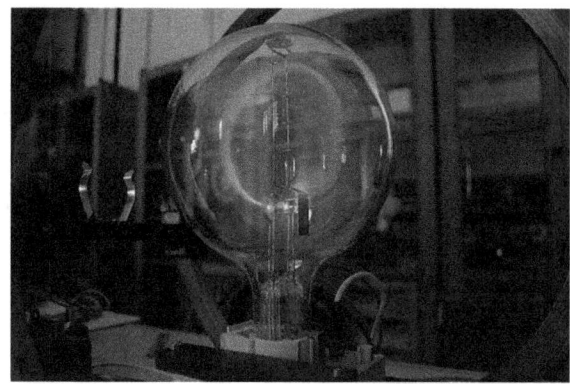

A beam of electrons deflected in a circle by a magnetic field[25]

Crookes developed the first cathode ray tube to have a high vacuum inside.[27] He then showed that the luminescence rays appearing within the tube carried energy and moved from the cathode to the anode. Furthermore, by applying a magnetic field, he was able to deflect the rays, thereby demonstrating that the beam behaved as though it were negatively charged.[28][29] In 1879, he proposed that these properties could be explained by what he termed 'radiant matter'. He suggested that this was a fourth state of matter, consisting of negatively charged molecules that were being projected with high velocity from the cathode.[30]

The German-born British physicist Arthur Schuster expanded upon Crookes' experiments by placing metal plates parallel to the cathode rays and applying an electric potential between the plates. The field deflected the rays toward the positively charged plate, providing further evidence that the rays carried negative charge. By measuring the amount of deflection for a given level of current, in 1890 Schuster was able to estimate the charge-to-mass ratio of the ray components. However, this produced a value that was more than a thousand times greater than what was expected, so little credence was given to his calculations at the time.[28][31]

In 1892 Hendrik Lorentz suggested that the mass of these particles (electrons) could be a consequence of their electric charge.[32]

In 1896, the British physicist J. J. Thomson, with his colleagues John S. Townsend and H. A. Wilson,[12] performed experiments indicating that cathode rays really were unique particles, rather than waves, atoms or molecules as was believed earlier.[5] Thomson made good estimates of both the charge *e* and the mass *m*, finding that cathode ray particles, which he called "corpuscles," had perhaps one thousandth of the mass of the least massive ion known: hydrogen.[5][13] He showed that their charge to mass ratio, *e/m*, was independent of cathode material. He further showed that the negatively charged particles produced by radioactive materials, by heated materials and by illuminated materials were universal.[*][5][*][33] The name electron was again proposed for these particles by the Irish physicist George F. Fitzgerald, and the name has since gained universal acceptance.[*][28]

Robert Millikan

While studying naturally fluorescing minerals in 1896, the French physicist Henri Becquerel discovered that they emitted radiation without any exposure to an external energy source. These radioactive materials became the subject of much interest by scientists, including the New Zealand physicist Ernest Rutherford who discovered they emitted particles. He designated these particles alpha and beta, on the basis of their ability to penetrate matter.[*][34] In 1900, Becquerel showed that the beta rays emitted by radium could be deflected by an electric field, and that their mass-to-charge ratio was the same as for cathode rays.[*][35] This evidence strengthened the view that electrons existed as components of atoms.[*][36][*][37]

The electron's charge was more carefully measured by the American physicists Robert Millikan and Harvey Fletcher in their oil-drop experiment of 1909, the results of which were published in 1911. This experiment used an electric field to prevent a charged droplet of oil from falling as a result of gravity. This device could measure the electric charge from as few as 1–150 ions with an error margin of less than 0.3%. Comparable experiments had been done earlier by Thomson's team,[*][5] using clouds of charged water droplets generated by electrolysis,[*][12] and in 1911 by Abram Ioffe, who independently obtained the same result as Millikan using charged microparticles of metals, then published his results in 1913.[*][38] However, oil drops were more stable than water drops because of their slower evaporation rate, and thus more suited to precise experimentation over longer periods of time.[*][39]

Around the beginning of the twentieth century, it was found that under certain conditions a fast-moving charged particle caused a condensation of supersaturated water vapor along its path. In 1911, Charles Wilson used this principle to devise his cloud chamber so he could photograph the tracks of charged particles, such as fast-moving electrons.[*][40]

Atomic theory

See also: The proton–electron model of the nucleus

By 1914, experiments by physicists Ernest Rutherford,

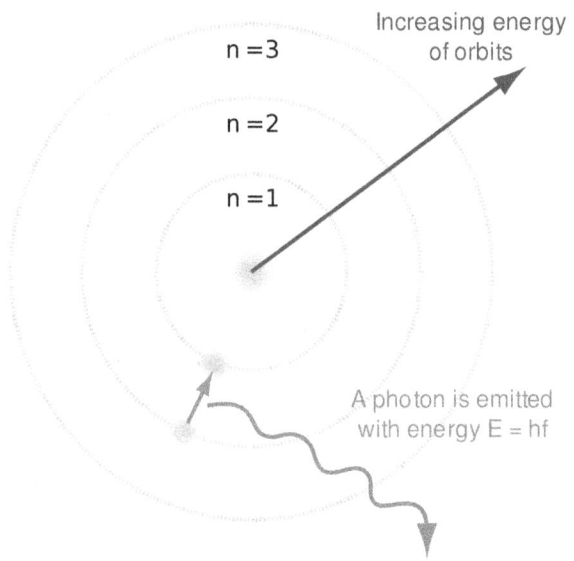

The Bohr model of the atom, showing states of electron with energy quantized by the number n. An electron dropping to a lower orbit emits a photon equal to the energy difference between the orbits.

Henry Moseley, James Franck and Gustav Hertz had largely established the structure of an atom as a dense nucleus of positive charge surrounded by lower-mass electrons.[*][41] In 1913, Danish physicist Niels Bohr postulated that electrons resided in quantized energy states, with the energy determined by the angular momentum of the electron's orbits about the nucleus. The electrons could move between these states, or orbits, by the emission or absorption of photons at specific frequencies. By means of these quantized orbits, he accurately explained the spectral lines of the hydrogen atom.[42] However, Bohr's model failed to account for the relative intensities of the spectral lines and it was unsuccessful in explaining the spectra of more complex atoms.[41]

Chemical bonds between atoms were explained by Gilbert Newton Lewis, who in 1916 proposed that a covalent bond between two atoms is maintained by a pair of electrons shared between them.[43] Later, in 1927, Walter Heitler and Fritz London gave the full explanation of the electron-pair formation and chemical bonding in terms of quantum mechanics.[44] In 1919, the American chemist Irving Langmuir elaborated on the Lewis' static model of the atom and suggested that all electrons were distributed in successive "concentric (nearly) spherical shells, all of equal thickness" .[45] The shells were, in turn, divided by him in a number of cells each containing one pair of electrons. With this model Langmuir was able to qualitatively explain the chemical properties of all elements in the periodic table,[44] which were known to largely repeat themselves according to the periodic law.[46]

In 1924, Austrian physicist Wolfgang Pauli observed that the shell-like structure of the atom could be explained by a set of four parameters that defined every quantum energy state, as long as each state was inhabited by no more than a single electron. (This prohibition against more than one electron occupying the same quantum energy state became known as the Pauli exclusion principle.)[47] The physical mechanism to explain the fourth parameter, which had two distinct possible values, was provided by the Dutch physicists Samuel Goudsmit and George Uhlenbeck. In 1925, Goudsmit and Uhlenbeck suggested that an electron, in addition to the angular momentum of its orbit, possesses an intrinsic angular momentum and magnetic dipole moment.[41][48] The intrinsic angular momentum became known as spin, and explained the previously mysterious splitting of spectral lines observed with a high-resolution spectrograph; this phenomenon is known as fine structure splitting.[49]

Quantum mechanics

See also: History of quantum mechanics

In his 1924 dissertation *Recherches sur la théorie des quanta* (Research on Quantum Theory), French physicist Louis de Broglie hypothesized that all matter possesses a de Broglie wave similar to light.[50] That is, under the appropriate conditions, electrons and other matter would show properties of either particles or waves. The corpuscular properties of a particle are demonstrated when it is shown to have a localized position in space along its trajectory at any given moment.[51] Wave-like nature is observed, for example,

when a beam of light is passed through parallel slits and creates interference patterns. In 1927, the interference effect was found in a beam of electrons by English physicist George Paget Thomson with a thin metal film and by American physicists Clinton Davisson and Lester Germer using a crystal of nickel.[52]

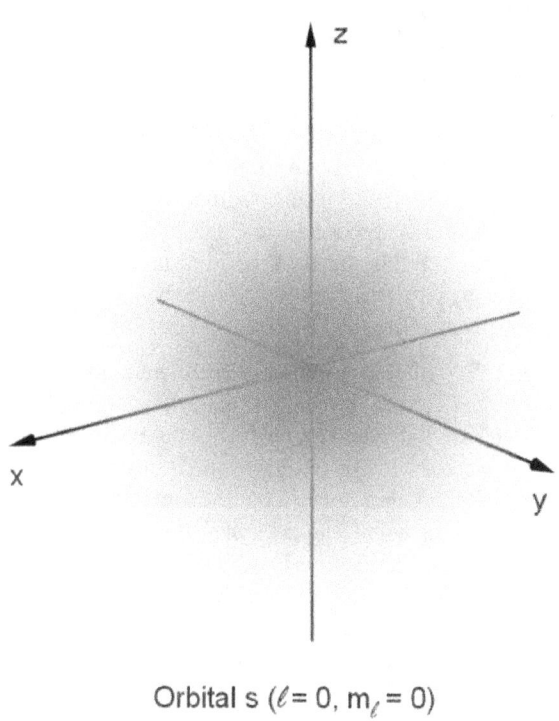

Orbital s ($\ell = 0$, $m_\ell = 0$)

In quantum mechanics, the behavior of an electron in an atom is described by an orbital, which is a probability distribution rather than an orbit. In the figure, the shading indicates the relative probability to "find" the electron, having the energy corresponding to the given quantum numbers, at that point.

De Broglie's prediction of a wave nature for electrons led Erwin Schrödinger to postulate a wave equation for electrons moving under the influence of the nucleus in the atom. In 1926, this equation, the Schrödinger equation, successfully described how electron waves propagated.[53] Rather than yielding a solution that determined the location of an electron over time, this wave equation also could be used to predict the probability of finding an electron near a position, especially a position near where the electron was bound in space, for which the electron wave equations did not change in time. This approach led to a second formulation of quantum mechanics (the first being by Heisenberg in 1925), and solutions of Schrödinger's equation, like Heisenberg's, provided derivations of the energy states of an electron in a hydrogen atom that were equivalent to those that had been derived first by Bohr in 1913, and that were known to reproduce the hydrogen spectrum.[54] Once spin and the interaction between multiple electrons were considered, quantum mechanics later made it possible to predict the config-uration of electrons in atoms with higher atomic numbers than hydrogen.*[55]

In 1928, building on Wolfgang Pauli's work, Paul Dirac produced a model of the electron – the Dirac equation, consistent with relativity theory, by applying relativistic and symmetry considerations to the hamiltonian formulation of the quantum mechanics of the electro-magnetic field.*[56] To resolve some problems within his relativistic equation, in 1930 Dirac developed a model of the vacuum as an infinite sea of particles having negative energy, which was dubbed the Dirac sea. This led him to predict the existence of a positron, the antimatter counterpart of the electron.*[57] This particle was discovered in 1932 by Carl Anderson, who proposed calling standard electrons *negatrons*, and us-ing *electron* as a generic term to describe both the positively and negatively charged variants.

In 1947 Willis Lamb, working in collaboration with grad-uate student Robert Retherford, found that certain quan-tum states of hydrogen atom, which should have the same energy, were shifted in relation to each other, the differ-ence being the Lamb shift. About the same time, Polykarp Kusch, working with Henry M. Foley, discovered the mag-netic moment of the electron is slightly larger than predicted by Dirac's theory. This small difference was later called anomalous magnetic dipole moment of the electron. This difference was later explained by the theory of quantum electrodynamics, developed by Sin-Itiro Tomonaga, Julian Schwinger and Richard Feynman in the late 1940s.*[58]

Particle accelerators

With the development of the particle accelerator during the first half of the twentieth century, physicists began to delve deeper into the properties of subatomic parti-cles.*[59] The first successful attempt to accelerate elec-trons using electromagnetic induction was made in 1942 by Donald Kerst. His initial betatron reached energies of 2.3 MeV, while subsequent betatrons achieved 300 MeV. In 1947, synchrotron radiation was discovered with a 70 MeV electron synchrotron at General Electric. This radi-ation was caused by the acceleration of electrons, moving near the speed of light, through a magnetic field.*[60]

With a beam energy of 1.5 GeV, the first high-energy particle collider was ADONE, which began operations in 1968.*[61] This device accelerated electrons and positrons in opposite directions, effectively doubling the energy of their collision when compared to striking a static target with an electron.*[62] The Large Electron–Positron Col-lider (LEP) at CERN, which was operational from 1989 to 2000, achieved collision energies of 209 GeV and made important measurements for the Standard Model of particle physics.*[63]*[64]

Confinement of individual electrons

Individual electrons can now be easily confined in ultra small ($L = 20$ nm, $W = 20$ nm) CMOS transistors oper-ated at cryogenic temperature over a range of −269 °C (4 K) to about −258 °C (15 K).*[65] The electron wavefunc-tion spreads in a semiconductor lattice and negligibly inter-acts with the valence band electrons, so it can be treated in the single particle formalism, by replacing its mass with the effective mass tensor.

2.1.2 Characteristics

Classification

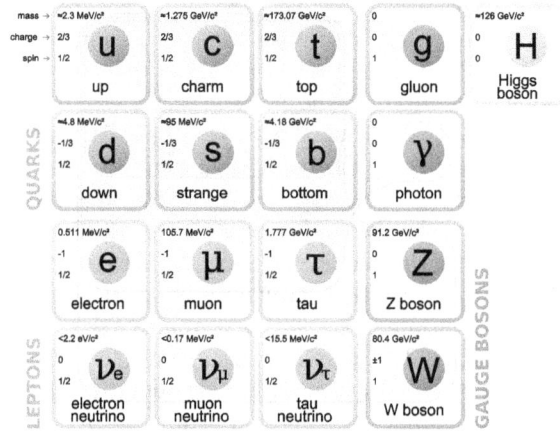

Standard Model of elementary particles. The electron (symbol e) is on the left.

In the Standard Model of particle physics, electrons belong to the group of subatomic particles called leptons, which are believed to be fundamental or elementary particles. Elec-trons have the lowest mass of any charged lepton (or elec-trically charged particle of any type) and belong to the first-generation of fundamental particles.*[66] The second and third generation contain charged leptons, the muon and the tau, which are identical to the electron in charge, spin and interactions, but are more massive. Leptons differ from the other basic constituent of matter, the quarks, by their lack of strong interaction. All members of the lepton group are fermions, because they all have half-odd integer spin; the electron has spin $\frac{1}{2}$.*[67]

Fundamental properties

The invariant mass of an electron is approximately 9.109×10^{-31} kilograms,[68] or 5.489×10^{-4} atomic mass units. On the basis of Einstein's principle of mass–energy equivalence, this mass corresponds to a rest energy of 0.511 MeV. The ratio between the mass of a proton and that of an electron is about 1836.[9][69] Astronomical measurements show that the proton-to-electron mass ratio has held the same value for at least half the age of the universe, as is predicted by the Standard Model.[70]

Electrons have an electric charge of -1.602×10^{-19} coulomb,[68] which is used as a standard unit of charge for subatomic particles, and is also called the elementary charge. This elementary charge has a relative standard uncertainty of 2.2×10^{-8}.[68] Within the limits of experimental accuracy, the electron charge is identical to the charge of a proton, but with the opposite sign.[71] As the symbol e is used for the elementary charge, the electron is commonly symbolized by e−, where the minus sign indicates the negative charge. The positron is symbolized by e+ because it has the same properties as the electron but with a positive rather than negative charge.[67][68]

The electron has an intrinsic angular momentum or spin of $\frac{1}{2}$.[68] This property is usually stated by referring to the electron as a spin-$\frac{1}{2}$ particle.[67] For such particles the spin magnitude is $\sqrt{3}/2\ \hbar$.[note 3] while the result of the measurement of a projection of the spin on any axis can only be $\pm \hbar/2$. In addition to spin, the electron has an intrinsic magnetic moment along its spin axis.[68] It is approximately equal to one Bohr magneton,[72][note 4] which is a physical constant equal to $9.27400915(23) \times 10^{-24}$ joules per tesla.[68] The orientation of the spin with respect to the momentum of the electron defines the property of elementary particles known as helicity.[73]

The electron has no known substructure.[1][74] and it is assumed to be a point particle with a point charge and no spatial extent.[8] In classical physics, the angular momentum and magnetic moment of an object depend upon its physical dimensions. Hence, the concept of a dimensionless electron possessing these properties might seem paradoxical and inconsistent to experimental observations in Penning traps which point to finite non-zero radius of the electron. A possible explanation of this paradoxical situation is given below in the "Virtual particles" subsection by taking into consideration the Foldy-Wouthuysen transformation. The issue of the radius of the electron is a challenging problem of the modern theoretical physics. The admission of the hypothesis of a finite radius of the electron is incompatible to the premises of the theory of relativity. On the other hand, a point-like electron (zero radius) generates serious mathematical difficulties due to the self-energy of the electron tending to infinity.[75] These aspects have been analyzed in detail by Dmitri Ivanenko and Arseny Sokolov.

Observation of a single electron in a Penning trap shows the upper limit of the particle's radius is 10^{-22} meters.[76] There *is* a physical constant called the "classical electron radius", with the much larger value of 2.8179×10^{-15} m, greater than the radius of the proton. However, the terminology comes from a simplistic calculation that ignores the effects of quantum mechanics; in reality, the so-called classical electron radius has little to do with the true fundamental structure of the electron.[77][note 5]

There are elementary particles that spontaneously decay into less massive particles. An example is the muon, which decays into an electron, a neutrino and an antineutrino, with a mean lifetime of 2.2×10^{-6} seconds. However, the electron is thought to be stable on theoretical grounds: the electron is the least massive particle with non-zero electric charge, so its decay would violate charge conservation.[78] The experimental lower bound for the electron's mean lifetime is 4.6×10^{26} years, at a 90% confidence level.[79][80]

Quantum properties

As with all particles, electrons can act as waves. This is called the wave–particle duality and can be demonstrated using the double-slit experiment.

The wave-like nature of the electron allows it to pass through two parallel slits simultaneously, rather than just one slit as would be the case for a classical particle. In quantum mechanics, the wave-like property of one particle can be described mathematically as a complex-valued function, the wave function, commonly denoted by the Greek letter psi (ψ). When the absolute value of this function is squared, it gives the probability that a particle will be observed near a location—a probability density.[81]:162–218

Electrons are identical particles because they cannot be distinguished from each other by their intrinsic physical properties. In quantum mechanics, this means that a pair of interacting electrons must be able to swap positions without an observable change to the state of the system. The wave function of fermions, including electrons, is antisymmetric, meaning that it changes sign when two electrons are swapped; that is, $\psi(r_1, r_2) = -\psi(r_2, r_1)$, where the variables r_1 and r_2 correspond to the first and second electrons, respectively. Since the absolute value is not changed by a sign swap, this corresponds to equal probabilities. Bosons, such as the photon, have symmetric wave functions instead.[81]:162–218

In the case of antisymmetry, solutions of the wave equation for interacting electrons result in a zero probability that each pair will occupy the same location or state. This is respon-

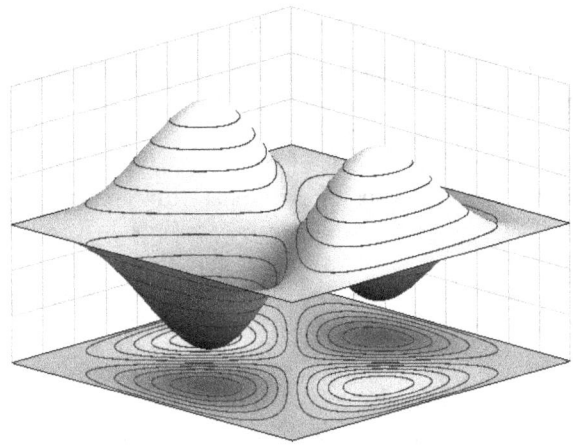

Example of an antisymmetric wave function for a quantum state of two identical fermions in a 1-dimensional box. If the particles swap position, the wave function inverts its sign.

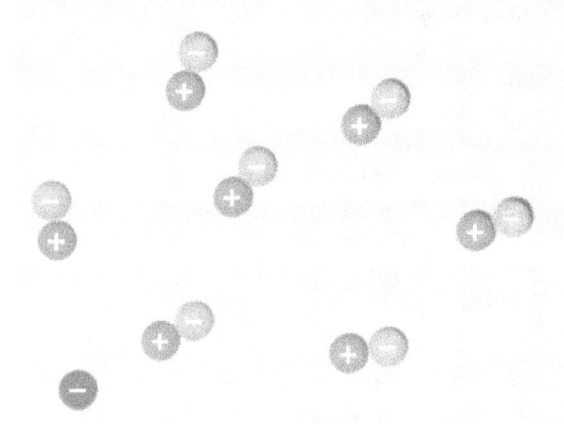

A schematic depiction of virtual electron–positron pairs appearing at random near an electron (at lower left)

sible for the Pauli exclusion principle, which precludes any two electrons from occupying the same quantum state. This principle explains many of the properties of electrons. For example, it causes groups of bound electrons to occupy different orbitals in an atom, rather than all overlapping each other in the same orbit.[81][*]:162–218

Virtual particles

Main article: Virtual particle

In a simplified picture, every photon spends some time as a combination of a virtual electron plus its antiparticle, the virtual positron, which rapidly annihilate each other shortly thereafter.[82] The combination of the energy variation needed to create these particles, and the time during which they exist, fall under the threshold of detectability expressed by the Heisenberg uncertainty relation, $\Delta E \cdot \Delta t \geq \hbar$. In effect, the energy needed to create these virtual particles, ΔE, can be "borrowed" from the vacuum for a period of time, Δt, so that their product is no more than the reduced Planck constant, $\hbar \approx 6.6 \times 10^{*}{-}16$ eV·s. Thus, for a virtual electron, Δt is at most $1.3 \times 10^{*}{-}21$ s.[83]

While an electron–positron virtual pair is in existence, the coulomb force from the ambient electric field surrounding an electron causes a created positron to be attracted to the original electron, while a created electron experiences a repulsion. This causes what is called vacuum polarization. In effect, the vacuum behaves like a medium having a dielectric permittivity more than unity. Thus the effective charge of an electron is actually smaller than its true value, and the charge decreases with increasing distance from the electron.[84][85] This polarization was confirmed ex-

perimentally in 1997 using the Japanese TRISTAN particle accelerator.[86] Virtual particles cause a comparable shielding effect for the mass of the electron.[87]

The interaction with virtual particles also explains the small (about 0.1%) deviation of the intrinsic magnetic moment of the electron from the Bohr magneton (the anomalous magnetic moment).[72][88] The extraordinarily precise agreement of this predicted difference with the experimentally determined value is viewed as one of the great achievements of quantum electrodynamics.[89]

The apparent paradox (mentioned above in the properties subsection) of a point particle electron having intrinsic angular momentum and magnetic moment can be explained by the formation of virtual photons in the electric field generated by the electron. These photons cause the electron to shift about in a jittery fashion (known as zitterbewegung),[90] which results in a net circular motion with precession. This motion produces both the spin and the magnetic moment of the electron.[8][91] In atoms, this creation of virtual photons explains the Lamb shift observed in spectral lines.[84]

Interaction

An electron generates an electric field that exerts an attractive force on a particle with a positive charge, such as the proton, and a repulsive force on a particle with a negative charge. The strength of this force is determined by Coulomb's inverse square law.[92] When an electron is in motion, it generates a magnetic field.[81][*]:140 The Ampère-Maxwell law relates the magnetic field to the mass motion of electrons (the current) with respect to an observer. This property of induction supplies the magnetic field that drives an electric motor.[93] The electromagnetic field of an arbitrary moving charged particle is expressed by the Liénard–Wiechert potentials, which are valid even when the particle's speed is close to that of light (relativistic).

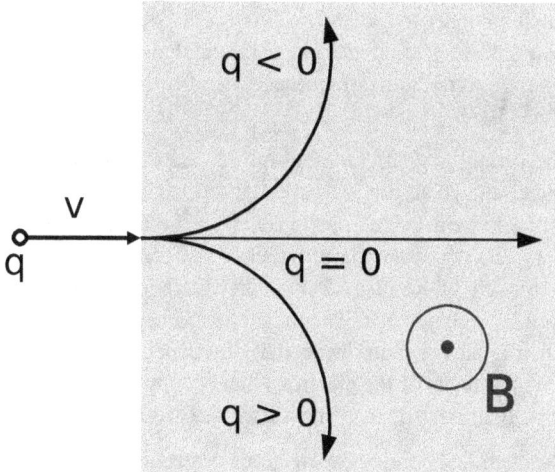

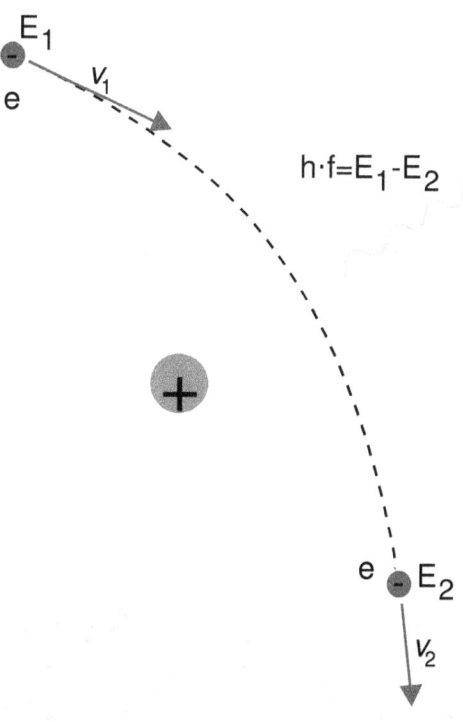

A particle with charge q *(at left) is moving with velocity* v *through a magnetic field* B *that is oriented toward the viewer. For an electron,* q *is negative so it follows a curved trajectory toward the top.*

Here, Bremsstrahlung is produced by an electron e *deflected by the electric field of an atomic nucleus. The energy change* $E_2 - E_1$ *determines the frequency* f *of the emitted photon.*

When an electron is moving through a magnetic field, it is subject to the Lorentz force that acts perpendicularly to the plane defined by the magnetic field and the electron velocity. This centripetal force causes the electron to follow a helical trajectory through the field at a radius called the gyroradius. The acceleration from this curving motion induces the electron to radiate energy in the form of synchrotron radiation.[81][*:160][94][note 6] The energy emission in turn causes a recoil of the electron, known as the Abraham–Lorentz–Dirac Force, which creates a friction that slows the electron. This force is caused by a back-reaction of the electron's own field upon itself.[95]

Photons mediate electromagnetic interactions between particles in quantum electrodynamics. An isolated electron at a constant velocity cannot emit or absorb a real photon; doing so would violate conservation of energy and momentum. Instead, virtual photons can transfer momentum between two charged particles. This exchange of virtual photons, for example, generates the Coulomb force.[96] Energy emission can occur when a moving electron is deflected by a charged particle, such as a proton. The acceleration of the electron results in the emission of Bremsstrahlung radiation.[97]

An inelastic collision between a photon (light) and a solitary (free) electron is called Compton scattering. This collision results in a transfer of momentum and energy between the particles, which modifies the wavelength of the photon by an amount called the Compton shift.[note 7] The maximum magnitude of this wavelength shift is h/m_ec, which is known as the Compton wavelength.[98] For an electron,

it has a value of 2.43×10^{-12} m.[68] When the wavelength of the light is long (for instance, the wavelength of the visible light is 0.4–0.7 μm) the wavelength shift becomes negligible. Such interaction between the light and free electrons is called Thomson scattering or Linear Thomson scattering.[99]

The relative strength of the electromagnetic interaction between two charged particles, such as an electron and a proton, is given by the fine-structure constant. This value is a dimensionless quantity formed by the ratio of two energies: the electrostatic energy of attraction (or repulsion) at a separation of one Compton wavelength, and the rest energy of the charge. It is given by $\alpha \approx 7.297353 \times 10^{-3}$, which is approximately equal to $1/137$.[68]

When electrons and positrons collide, they annihilate each other, giving rise to two or more gamma ray photons. If the electron and positron have negligible momentum, a positronium atom can form before annihilation results in two or three gamma ray photons totalling 1.022 MeV.[100][101] On the other hand, high-energy photons may transform into an electron and a positron by a process called pair production, but only in the presence of a nearby charged particle, such as a nucleus.[102][103]

In the theory of electroweak interaction, the left-handed component of electron's wavefunction forms a weak isospin doublet with the electron neutrino. This means that during weak interactions, electron neutrinos behave like electrons. Either member of this doublet can undergo a charged current interaction by emitting or absorbing a W and be converted into the other member. Charge is conserved during this reaction because the W boson also carries a charge, canceling out any net change during the transmutation. Charged current interactions are responsible for the phenomenon of beta decay in a radioactive atom. Both the electron and electron neutrino can undergo a neutral current interaction via a Z0 exchange, and this is responsible for neutrino-electron elastic scattering.[*][104]

Atoms and molecules

Main article: Atom
An electron can be *bound* to the nucleus of an atom by the

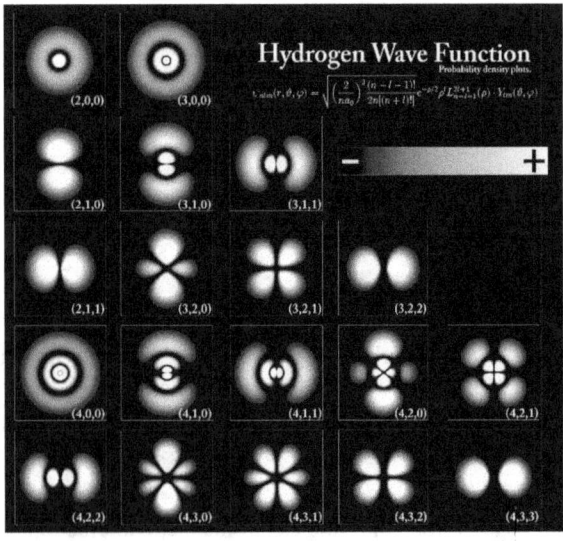

Probability densities for the first few hydrogen atom orbitals, seen in cross-section. The energy level of a bound electron determines the orbital it occupies, and the color reflects the probability of finding the electron at a given position.

attractive Coulomb force. A system of one or more electrons bound to a nucleus is called an atom. If the number of electrons is different from the nucleus' electrical charge, such an atom is called an ion. The wave-like behavior of a bound electron is described by a function called an atomic orbital. Each orbital has its own set of quantum numbers such as energy, angular momentum and projection of angular momentum, and only a discrete set of these orbitals exist around the nucleus. According to the Pauli exclusion principle each orbital can be occupied by up to two electrons, which must differ in their spin quantum number.

Electrons can transfer between different orbitals by the emission or absorption of photons with an energy that matches the difference in potential.[*][105] Other methods of orbital transfer include collisions with particles, such as electrons, and the Auger effect.[*][106] To escape the atom, the energy of the electron must be increased above its binding energy to the atom. This occurs, for example, with the photoelectric effect, where an incident photon exceeding the atom's ionization energy is absorbed by the electron.[*][107]

The orbital angular momentum of electrons is quantized. Because the electron is charged, it produces an orbital magnetic moment that is proportional to the angular momentum. The net magnetic moment of an atom is equal to the vector sum of orbital and spin magnetic moments of all electrons and the nucleus. The magnetic moment of the nucleus is negligible compared with that of the electrons. The magnetic moments of the electrons that occupy the same orbital (so called, paired electrons) cancel each other out.[*][108]

The chemical bond between atoms occurs as a result of electromagnetic interactions, as described by the laws of quantum mechanics.[*][109] The strongest bonds are formed by the sharing or transfer of electrons between atoms, allowing the formation of molecules.[*][11] Within a molecule, electrons move under the influence of several nuclei, and occupy molecular orbitals; much as they can occupy atomic orbitals in isolated atoms.[*][110] A fundamental factor in these molecular structures is the existence of electron pairs. These are electrons with opposed spins, allowing them to occupy the same molecular orbital without violating the Pauli exclusion principle (much like in atoms). Different molecular orbitals have different spatial distribution of the electron density. For instance, in bonded pairs (i.e. in the pairs that actually bind atoms together) electrons can be found with the maximal probability in a relatively small volume between the nuclei. On the contrary, in non-bonded pairs electrons are distributed in a large volume around nuclei.[*][111]

Conductivity

If a body has more or fewer electrons than are required to balance the positive charge of the nuclei, then that object has a net electric charge. When there is an excess of electrons, the object is said to be negatively charged. When there are fewer electrons than the number of protons in nuclei, the object is said to be positively charged. When the number of electrons and the number of protons are equal, their charges cancel each other and the object is said to be electrically neutral. A macroscopic body can develop an electric charge through rubbing, by the triboelectric effect.[*][115]

Independent electrons moving in vacuum are termed *free* electrons. Electrons in metals also behave as if they were

A lightning discharge consists primarily of a flow of electrons.[112] *The electric potential needed for lightning may be generated by a triboelectric effect.*[113]*[114]

free. In reality the particles that are commonly termed electrons in metals and other solids are quasi-electrons—quasiparticles, which have the same electrical charge, spin and magnetic moment as real electrons but may have a different mass.*[116] When free electrons—both in vacuum and metals—move, they produce a net flow of charge called an electric current, which generates a magnetic field. Likewise a current can be created by a changing magnetic field. These interactions are described mathematically by Maxwell's equations.*[117]

At a given temperature, each material has an electrical conductivity that determines the value of electric current when an electric potential is applied. Examples of good conductors include metals such as copper and gold, whereas glass and Teflon are poor conductors. In any dielectric material, the electrons remain bound to their respective atoms and the material behaves as an insulator. Most semiconductors have a variable level of conductivity that lies between the extremes of conduction and insulation.*[118] On the other hand, metals have an electronic band structure containing partially filled electronic bands. The presence of such bands allows electrons in metals to behave as if they were free or delocalized electrons. These electrons are not associated with specific atoms, so when an electric field is applied, they are free to move like a gas (called Fermi gas)*[119] through the material much like free electrons.

Because of collisions between electrons and atoms, the drift velocity of electrons in a conductor is on the order of mil-

limeters per second. However, the speed at which a change of current at one point in the material causes changes in currents in other parts of the material, the velocity of propagation, is typically about 75% of light speed.*[120] This occurs because electrical signals propagate as a wave, with the velocity dependent on the dielectric constant of the material.*[121]

Metals make relatively good conductors of heat, primarily because the delocalized electrons are free to transport thermal energy between atoms. However, unlike electrical conductivity, the thermal conductivity of a metal is nearly independent of temperature. This is expressed mathematically by the Wiedemann–Franz law,*[119] which states that the ratio of thermal conductivity to the electrical conductivity is proportional to the temperature. The thermal disorder in the metallic lattice increases the electrical resistivity of the material, producing a temperature dependence for electric current.*[122]

When cooled below a point called the critical temperature, materials can undergo a phase transition in which they lose all resistivity to electric current, in a process known as superconductivity. In BCS theory, this behavior is modeled by pairs of electrons entering a quantum state known as a Bose–Einstein condensate. These Cooper pairs have their motion coupled to nearby matter via lattice vibrations called phonons, thereby avoiding the collisions with atoms that normally create electrical resistance.*[123] (Cooper pairs have a radius of roughly 100 nm, so they can overlap each other.)*[124] However, the mechanism by which higher temperature superconductors operate remains uncertain.

Electrons inside conducting solids, which are quasiparticles themselves, when tightly confined at temperatures close to absolute zero, behave as though they had split into three other quasiparticles: spinons, Orbitons and holons.*[125]*[126] The former carries spin and magnetic moment, the next carries its orbital location while the latter electrical charge.

Motion and energy

According to Einstein's theory of special relativity, as an electron's speed approaches the speed of light, from an observer's point of view its relativistic mass increases, thereby making it more and more difficult to accelerate it from within the observer's frame of reference. The speed of an electron can approach, but never reach, the speed of light in a vacuum, c. However, when relativistic electrons—that is, electrons moving at a speed close to c—are injected into a dielectric medium such as water, where the local speed of light is significantly less than c, the electrons temporarily travel faster than light in the medium. As they interact with the medium, they generate a faint light called Cherenkov radiation.[127]

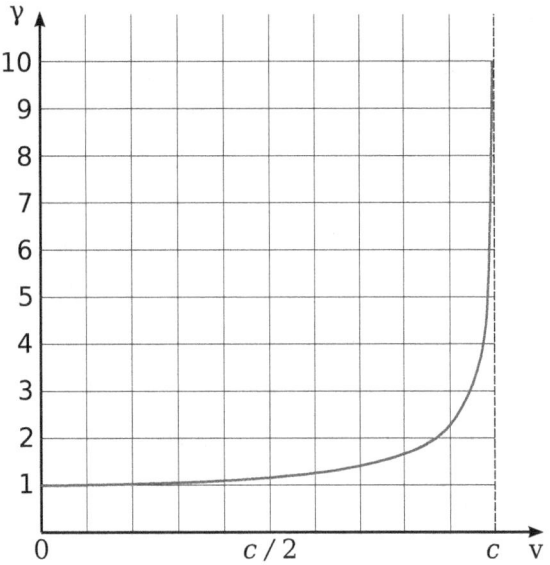

Lorentz factor as a function of velocity. It starts at value 1 and goes to infinity as v *approaches* c.

The effects of special relativity are based on a quantity known as the Lorentz factor, defined as $\gamma = 1/\sqrt{1 - v^2/c^2}$ where v is the speed of the particle. The kinetic energy K_e of an electron moving with velocity v is:

$$K_e = (\gamma - 1)m_e c^2,$$

where m_e is the mass of electron. For example, the Stanford linear accelerator can accelerate an electron to roughly 51 GeV.[128] Since an electron behaves as a wave, at a given velocity it has a characteristic de Broglie wavelength. This is given by $\lambda_e = h/p$ where h is the Planck constant and p is the momentum.[50] For the 51 GeV electron above, the wavelength is about 2.4×10^{-17} m, small enough to explore structures well below the size of an atomic nucleus.[129]

2.1.3 Formation

The Big Bang theory is the most widely accepted scientific theory to explain the early stages in the evolution of the Universe.[130] For the first millisecond of the Big Bang, the temperatures were over 10 billion Kelvin and photons had mean energies over a million electronvolts. These photons were sufficiently energetic that they could react with each other to form pairs of electrons and positrons. Likewise, positron-electron pairs annihilated each other and emitted energetic photons:

$$\gamma + \gamma \leftrightarrow e+ + e-$$

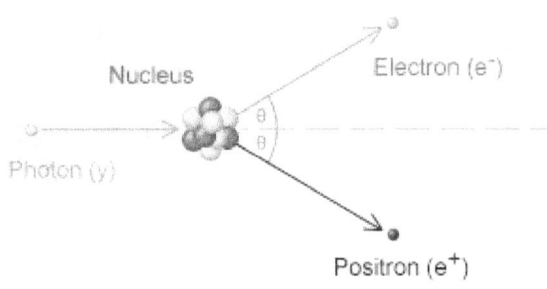

Pair production caused by the collision of a photon with an atomic nucleus

An equilibrium between electrons, positrons and photons was maintained during this phase of the evolution of the Universe. After 15 seconds had passed, however, the temperature of the universe dropped below the threshold where electron-positron formation could occur. Most of the surviving electrons and positrons annihilated each other, releasing gamma radiation that briefly reheated the universe.[131]

For reasons that remain uncertain, during the process of leptogenesis there was an excess in the number of electrons over positrons.[132] Hence, about one electron in every billion survived the annihilation process. This excess matched the excess of protons over antiprotons, in a condition known as baryon asymmetry, resulting in a net charge of zero for the universe.[133][134] The surviving protons and neutrons began to participate in reactions with each other—in the process known as nucleosynthesis, forming isotopes of hydrogen and helium, with trace amounts of lithium. This process peaked after about five minutes.[135] Any leftover neutrons underwent negative beta decay with a half-life of about a thousand seconds, releasing a proton and electron in the process,

$$n \rightarrow p + e^- + \overline{\nu}_e$$

For about the next 300000–400000 years, the excess electrons remained too energetic to bind with atomic nuclei.[136] What followed is a period known as recombination, when neutral atoms were formed and the expanding universe became transparent to radiation.[137]

Roughly one million years after the big bang, the first generation of stars began to form.[137] Within a star, stellar nucleosynthesis results in the production of positrons from the fusion of atomic nuclei. These antimatter particles immediately annihilate with electrons, releasing gamma rays. The net result is a steady reduction in the number of electrons, and a matching increase in the number of neutrons. However, the process of stellar evolution can result in the synthesis of radioactive isotopes. Selected isotopes can subsequently undergo negative beta decay, emitting an electron and antineutrino from the nucleus.[138] An example is the cobalt-60 (^{60}Co) isotope, which decays to form nickel-60 (60Ni).[139]

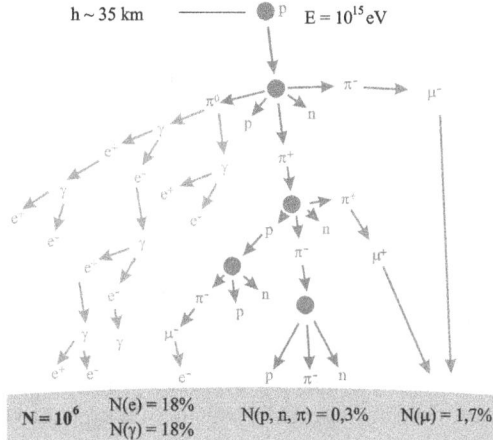

$$N = 10^6 \quad \begin{array}{l} N(e) = 18\% \\ N(\gamma) = 18\% \end{array} \quad N(p, n, \pi) = 0,3\% \quad N(\mu) = 1,7\%$$

An extended air shower generated by an energetic cosmic ray striking the Earth's atmosphere

At the end of its lifetime, a star with more than about 20 solar masses can undergo gravitational collapse to form a black hole.[140] According to classical physics, these massive stellar objects exert a gravitational attraction that is strong enough to prevent anything, even electromagnetic radiation, from escaping past the Schwarzschild radius. However, quantum mechanical effects are believed to potentially allow the emission of Hawking radiation at this distance. Electrons (and positrons) are thought to be created at the event horizon of these stellar remnants.

When pairs of virtual particles (such as an electron and positron) are created in the vicinity of the event horizon, the random spatial distribution of these particles may permit one of them to appear on the exterior; this process is called quantum tunnelling. The gravitational potential of the black hole can then supply the energy that transforms this virtual particle into a real particle, allowing it to radiate away into space.[141] In exchange, the other member of the pair is given negative energy, which results in a net loss of mass-energy by the black hole. The rate of Hawking radiation increases with decreasing mass, eventually causing the black hole to evaporate away until, finally, it explodes.[142]

Cosmic rays are particles traveling through space with high energies. Energy events as high as 3.0×10^{20} eV have been

recorded.[143] When these particles collide with nucleons in the Earth's atmosphere, a shower of particles is generated, including pions.[144] More than half of the cosmic radiation observed from the Earth's surface consists of muons. The particle called a muon is a lepton produced in the upper atmosphere by the decay of a pion.

$$\pi^- \rightarrow \mu^- + \bar{\nu}_\mu$$

A muon, in turn, can decay to form an electron or positron.[145]

$$\mu^- \rightarrow e^- + \bar{\nu}_e + \nu_\mu$$

2.1.4 Observation

Aurorae are mostly caused by energetic electrons precipitating into the atmosphere.[146]

Remote observation of electrons requires detection of their radiated energy. For example, in high-energy environments such as the corona of a star, free electrons form a plasma that radiates energy due to Bremsstrahlung radiation. Electron gas can undergo plasma oscillation, which is waves caused by synchronized variations in electron density, and these produce energy emissions that can be detected by using radio telescopes.[147]

The frequency of a photon is proportional to its energy. As a bound electron transitions between different energy levels of an atom, it absorbs or emits photons at characteristic frequencies. For instance, when atoms are irradiated by a source with a broad spectrum, distinct absorption lines appear in the spectrum of transmitted radiation. Each element or molecule displays a characteristic set of spectral lines, such as the hydrogen spectral series. Spectroscopic measurements of the strength and width of these lines allow the composition and physical properties of a substance to be determined.[*][148][*][149]

In laboratory conditions, the interactions of individual electrons can be observed by means of particle detectors, which allow measurement of specific properties such as energy, spin and charge.[*][107] The development of the Paul trap and Penning trap allows charged particles to be contained within a small region for long durations. This enables precise measurements of the particle properties. For example, in one instance a Penning trap was used to contain a single electron for a period of 10 months.[*][150] The magnetic moment of the electron was measured to a precision of eleven digits, which, in 1980, was a greater accuracy than for any other physical constant.[*][151]

The first video images of an electron's energy distribution were captured by a team at Lund University in Sweden, February 2008. The scientists used extremely short flashes of light, called attosecond pulses, which allowed an electron's motion to be observed for the first time.[*][152][*][153]

The distribution of the electrons in solid materials can be visualized by angle-resolved photoemission spectroscopy (ARPES). This technique employs the photoelectric effect to measure the reciprocal space—a mathematical representation of periodic structures that is used to infer the original structure. ARPES can be used to determine the direction, speed and scattering of electrons within the material.[*][154]

2.1.5 Plasma applications

Particle beams

During a NASA wind tunnel test, a model of the Space Shuttle is targeted by a beam of electrons, simulating the effect of ionizing gases during re-entry.[*][155]

Electron beams are used in welding.[*][156] They allow energy densities up to 10^7 W·cm[*]−2 across a narrow focus diameter of 0.1–1.3 mm and usually require no filler material. This welding technique must be performed in a vacuum to prevent the electrons from interacting with the gas before reaching their target, and it can be used to join conductive materials that would otherwise be considered unsuitable for welding.[*][157][*][158]

Electron-beam lithography (EBL) is a method of etching semiconductors at resolutions smaller than a micrometer.[*][159] This technique is limited by high costs, slow performance, the need to operate the beam in the vacuum and the tendency of the electrons to scatter in solids. The last problem limits the resolution to about 10 nm. For this reason, EBL is primarily used for the production of small numbers of specialized integrated circuits.[*][160]

Electron beam processing is used to irradiate materials in order to change their physical properties or sterilize medical and food products.[*][161] Electron beams fluidise or quasi-melt glasses without significant increase of temperature on intensive irradiation: e.g. intensive electron radiation causes a many orders of magnitude decrease of viscosity and stepwise decrease of its activation energy.[*][162]

Linear particle accelerators generate electron beams for treatment of superficial tumors in radiation therapy. Electron therapy can treat such skin lesions as basal-cell carcinomas because an electron beam only penetrates to a limited depth before being absorbed, typically up to 5 cm for electron energies in the range 5–20 MeV. An electron beam can be used to supplement the treatment of areas that have been irradiated by X-rays.[*][163][*][164]

Particle accelerators use electric fields to propel electrons and their antiparticles to high energies. These particles emit synchrotron radiation as they pass through magnetic fields. The dependency of the intensity of this radiation upon spin polarizes the electron beam—a process known as the Sokolov–Ternov effect.[*][note 8] Polarized electron beams can be useful for various experiments. Synchrotron radiation can also cool the electron beams to reduce the momentum spread of the particles. Electron and positron beams are collided upon the particles' accelerating to the required energies; particle detectors observe the resulting energy emissions, which particle physics studies .[*][165]

Imaging

Low-energy electron diffraction (LEED) is a method of bombarding a crystalline material with a collimated beam of electrons and then observing the resulting diffraction patterns to determine the structure of the material. The required energy of the electrons is typically in the range 20–

200 eV.[166] The reflection high-energy electron diffraction (RHEED) technique uses the reflection of a beam of electrons fired at various low angles to characterize the surface of crystalline materials. The beam energy is typically in the range 8–20 keV and the angle of incidence is 1–4°.[167][168]

The electron microscope directs a focused beam of electrons at a specimen. Some electrons change their properties, such as movement direction, angle, and relative phase and energy as the beam interacts with the material. Microscopists can record these changes in the electron beam to produce atomically resolved images of the material.[169] In blue light, conventional optical microscopes have a diffraction-limited resolution of about 200 nm.[170] By comparison, electron microscopes are limited by the de Broglie wavelength of the electron. This wavelength, for example, is equal to 0.0037 nm for electrons accelerated across a 100,000-volt potential.[171] The Transmission Electron Aberration-Corrected Microscope is capable of sub-0.05 nm resolution, which is more than enough to resolve individual atoms.[172] This capability makes the electron microscope a useful laboratory instrument for high resolution imaging. However, electron microscopes are expensive instruments that are costly to maintain.

Two main types of electron microscopes exist: transmission and scanning. Transmission electron microscopes function like overhead projectors, with a beam of electrons passing through a slice of material then being projected by lenses on a photographic slide or a charge-coupled device. Scanning electron microscopes rasteri a finely focused electron beam, as in a TV set, across the studied sample to produce the image. Magnifications range from 100× to 1,000,000× or higher for both microscope types. The scanning tunneling microscope uses quantum tunneling of electrons from a sharp metal tip into the studied material and can produce atomically resolved images of its surface.[173][174][175]

Other applications

In the free-electron laser (FEL), a relativistic electron beam passes through a pair of undulators that contain arrays of dipole magnets whose fields point in alternating directions. The electrons emit synchrotron radiation that coherently interacts with the same electrons to strongly amplify the radiation field at the resonance frequency. FEL can emit a coherent high-brilliance electromagnetic radiation with a wide range of frequencies, from microwaves to soft X-rays. These devices may find manufacturing, communication and various medical applications, such as soft tissue surgery.[176]

Electrons are important in cathode ray tubes, which have been extensively used as display devices in laboratory instruments, computer monitors and television sets.[177] In a photomultiplier tube, every photon striking the photocathode initiates an avalanche of electrons that produces a detectable current pulse.[178] Vacuum tubes use the flow of electrons to manipulate electrical signals, and they played a critical role in the development of electronics technology. However, they have been largely supplanted by solid-state devices such as the transistor.[179]

2.1.6 See also

- Anyon
- Electride
- Electron bubble
- Exoelectron emission
- *g*-factor
- Periodic systems of small molecules
- Spintronics
- Stern–Gerlach experiment
- Townsend discharge
- Zeeman effect
- List of particles
- Lepton

2.1.7 Notes

[1] The fractional version's denominator is the inverse of the decimal value (along with its relative standard uncertainty of 4.2×10^{-13} u).

[2] The electron's charge is the negative of elementary charge, which has a positive value for the proton.

[3] This magnitude is obtained from the spin quantum number as

$$S = \sqrt{s(s+1)} \cdot \frac{h}{2\pi}$$
$$= \frac{\sqrt{3}}{2}$$

for quantum number $s = 1/2$.
See: Gupta, M.C. (2001). *Atomic and Molecular Spectroscopy*. New Age Publishers. p. 81. ISBN 81-224-1300-5.

[4] Bohr magneton:

$$\mu_B = \frac{e\hbar}{2m_e}.$$

[5] The classical electron radius is derived as follows. Assume that the electron's charge is spread uniformly throughout a spherical volume. Since one part of the sphere would repel the other parts, the sphere contains electrostatic potential energy. This energy is assumed to equal the electron's rest energy, defined by special relativity ($E = mc^2$).

From electrostatics theory, the potential energy of a sphere with radius r and charge e is given by:

$$E_p = \frac{e^2}{8\pi\varepsilon_0 r},$$

where ε_0 is the vacuum permittivity. For an electron with rest mass m_0, the rest energy is equal to:

$$E_p = m_0 c^2,$$

where c is the speed of light in a vacuum. Setting them equal and solving for r gives the classical electron radius.
See: Haken, H.; Wolf, H.C.; Brewer, W.D. (2005). *The Physics of Atoms and Quanta: Introduction to Experiments and Theory*. Springer. p. 70. ISBN 3-540-67274-5.

[6] Radiation from non-relativistic electrons is sometimes termed cyclotron radiation.

[7] The change in wavelength, $\Delta\lambda$, depends on the angle of the recoil, θ, as follows,

$$\Delta\lambda = \frac{h}{m_e c}(1 - \cos\theta),$$

where c is the speed of light in a vacuum and m_e is the electron mass. See Zombeck (2007: 393, 396).

[8] The polarization of an electron beam means that the spins of all electrons point into one direction. In other words, the projections of the spins of all electrons onto their momentum vector have the same sign.

2.1.8   References

[1] Eichten, E.J.; Peskin, M.E.; Peskin, M. (1983). "New Tests for Quark and Lepton Substructure". *Physical Review Letters* **50** (11): 811–814. Bibcode:1983PhRvL..50..811E. doi:10.1103/PhysRevLett.50.811.

[2] Farrar, W.V. (1969). "Richard Laming and the Coal-Gas Industry, with His Views on the Structure of Matter". *Annals of Science* **25** (3): 243–254. doi:10.1080/00033796900200141.

[3] Arabatzis, T. (2006). *Representing Electrons: A Biographical Approach to Theoretical Entities*. University of Chicago Press. pp. 70–74. ISBN 0-226-02421-0.

[4] Buchwald, J.Z.; Warwick, A. (2001). *Histories of the Electron: The Birth of Microphysics*. MIT Press. pp. 195–203. ISBN 0-262-52424-4.

[5] Thomson, J.J. (1897). "Cathode Rays". *Philosophical Magazine* **44** (269): 293. doi:10.1080/14786449708621070.

[6] P.J. Mohr, B.N. Taylor, and D.B. Newell (2011), "The 2010 CODATA Recommended Values of the Fundamental Physical Constants" (Web Version 6.0). This database was developed by J. Baker, M. Douma, and S. Kotochigova. Available: http://physics.nist.gov/constants [Thursday, 02-Jun-2011 21:00:12 EDT]. National Institute of Standards and Technology, Gaithersburg, MD 20899.

[7] "JERRY COFF". Retrieved 10 September 2010.

[8] Curtis, L.J. (2003). *Atomic Structure and Lifetimes: A Conceptual Approach*. Cambridge University Press. p. 74. ISBN 0-521-53635-9.

[9] "CODATA value: proton-electron mass ratio". *2006 CODATA recommended values*. National Institute of Standards and Technology. Retrieved 2009-07-18.

[10] Anastopoulos, C. (2008). *Particle Or Wave: The Evolution of the Concept of Matter in Modern Physics*. Princeton University Press. pp. 236–237. ISBN 0-691-13512-6.

[11] Pauling, L.C. (1960). *The Nature of the Chemical Bond and the Structure of Molecules and Crystals: an introduction to modern structural chemistry* (3rd ed.). Cornell University Press. pp. 4–10. ISBN 0-8014-0333-2.

[12] Dahl (1997:122–185).

[13] Wilson, R. (1997). *Astronomy Through the Ages: The Story of the Human Attempt to Understand the Universe*. CRC Press. p. 138. ISBN 0-7484-0748-0.

[14] Shipley, J.T. (1945). *Dictionary of Word Origins*. The Philosophical Library. p. 133. ISBN 0-88029-751-4.

[15] Baigrie, B. (2006). *Electricity and Magnetism: A Historical Perspective*. Greenwood Press. pp. 7–8. ISBN 0-313-33358-0.

[16] Keithley, J.F. (1999). *The Story of Electrical and Magnetic Measurements: From 500 B.C. to the 1940s*. IEEE Press. pp. 15, 20. ISBN 0-7803-1193-0.

[17] "Benjamin Franklin (1706–1790)". *Eric Weisstein's World of Biography*. Wolfram Research. Retrieved 2010-12-16.

[18] Myers, R.L. (2006). *The Basics of Physics*. Greenwood Publishing Group. p. 242. ISBN 0-313-32857-9.

[19] Barrow, J.D. (1983). "Natural Units Before Planck". *Quarterly Journal of the Royal Astronomical Society* **24**: 24–26. Bibcode:1983QJRAS..24...24B.

[20] Sōgo Okamura (1994). *History of Electron Tubes*. IOS Press. p. 11. ISBN 978-90-5199-145-1. Retrieved 29 May 2015. In 1881, Stoney named this electromagnetic 'electrolion'. It came to be called 'electron' from 1891. [...] In 1906, the suggestion to call cathode ray particles 'electrions' was brought up but through the opinion of Lorentz of Holland 'electrons' came to be widely used.

[21] Stoney, G.J. (1894). "Of the "Electron," or Atom of Electricity". *Philosophical Magazine* **38** (5): 418–420. doi:10.1080/14786449408620653.

[22] "electron, n.2". OED Online. March 2013. Oxford University Press. Accessed 12 April 2013

[23] Soukhanov, A.H. ed. (1986). *Word Mysteries & Histories.* Houghton Mifflin Company. p. 73. ISBN 0-395-40265-4.

[24] Guralnik, D.B. ed. (1970). *Webster's New World Dictionary.* Prentice Hall. p. 450.

[25] Born, M.; Blin-Stoyle, R.J.; Radcliffe, J.M. (1989). *Atomic Physics.* Courier Dover. p. 26. ISBN 0-486-65984-4.

[26] Dahl (1997:55–58).

[27] DeKosky, R.K. (1983). "William Crookes and the quest for absolute vacuum in the 1870s". *Annals of Science* **40** (1): 1–18. doi:10.1080/00033798300200101.

[28] Leicester, H.M. (1971). *The Historical Background of Chemistry.* Courier Dover. pp. 221–222. ISBN 0-486-61053-5.

[29] Dahl (1997:64–78).

[30] Zeeman, P.; Zeeman, P. (1907). "Sir William Crookes, F.R.S". *Nature* **77** (1984): 1–3. Bibcode:1907Natur..77....1C. doi:10.1038/077001a0.

[31] Dahl (1997:99).

[32] Frank Wilczek: "Happy Birthday, Electron" *Scientific American,* June 2012.

[33] Thomson, J.J. (1906). "Nobel Lecture: Carriers of Negative Electricity" (PDF). The Nobel Foundation. Retrieved 2008-08-25.

[34] Trenn, T.J. (1976). "Rutherford on the Alpha-Beta-Gamma Classification of Radioactive Rays". *Isis* **67** (1): 61–75. doi:10.1086/351545. JSTOR 231134.

[35] Becquerel, H. (1900). "Déviation du Rayonnement du Radium dans un Champ Électrique". *Comptes rendus de l'Académie des sciences* (in French) **130**: 809–815.

[36] Buchwald and Warwick (2001:90–91).

[37] Myers, W.G. (1976). "Becquerel's Discovery of Radioactivity in 1896". *Journal of Nuclear Medicine* **17** (7): 579–582. PMID 775027.

[38] Kikoin, I.K.; Sominskiĭ, I.S. (1961). "Abram Fedorovich Ioffe (on his eightieth birthday)". *Soviet Physics Uspekhi* **3** (5): 798–809. Bibcode:1961SvPhU...3..798K. doi:10.1070/PU1961v003n05ABEH005812. Original publication in Russian: Кикоин, И.К.; Соминский, М.С. (1960). "Академик А.Ф. Иоффе" (PDF). *Успехи Физических Наук* **72** (10): 303–321.

[39] Millikan, R.A. (1911). "The Isolation of an Ion, a Precision Measurement of its Charge, and the Correction of Stokes' Law". *Physical Review* **32** (2): 349–397. Bibcode:1911PhRvI..32..349M. doi:10.1103/PhysRevSeriesI.32.349.

[40] Das Gupta, N.N.; Ghosh, S.K. (1999). "A Report on the Wilson Cloud Chamber and Its Applications in Physics". *Reviews of Modern Physics* **18** (2): 225–290. Bibcode:1946RvMP...18..225G. doi:10.1103/RevModPhys.18.225.

[41] Smirnov, B.M. (2003). *Physics of Atoms and Ions.* Springer. pp. 14–21. ISBN 0-387-95550-X.

[42] Bohr, N. (1922). "Nobel Lecture: The Structure of the Atom" (PDF). The Nobel Foundation. Retrieved 2008-12-03.

[43] Lewis, G.N. (1916). "The Atom and the Molecule". *Journal of the American Chemical Society* **38** (4): 762–786. doi:10.1021/ja02261a002.

[44] Arabatzis, T.; Gavroglu, K. (1997). "The chemists' electron". *European Journal of Physics* **18** (3): 150–163. Bibcode:1997EJPh...18..150A. doi:10.1088/0143-0807/18/3/005.

[45] Langmuir, I. (1919). "The Arrangement of Electrons in Atoms and Molecules". *Journal of the American Chemical Society* **41** (6): 868–934. doi:10.1021/ja02227a002.

[46] Scerri, E.R. (2007). *The Periodic Table.* Oxford University Press. pp. 205–226. ISBN 0-19-530573-6.

[47] Massimi, M. (2005). *Pauli's Exclusion Principle, The Origin and Validation of a Scientific Principle.* Cambridge University Press. pp. 7–8. ISBN 0-521-83911-4.

[48] Uhlenbeck, G.E.; Goudsmith, S. (1925). "Ersetzung der Hypothese vom unmechanischen Zwang durch eine Forderung bezüglich des inneren Verhaltens jedes einzelnen Elektrons". *Die Naturwissenschaften* (in German) **13** (47): 953. Bibcode:1925NW.....13..953E. doi:10.1007/BF01558878.

[49] Pauli, W. (1923). "Über die Gesetzmäßigkeiten des anomalen Zeemaneffektes". *Zeitschrift für Physik* (in German) **16** (1): 155–164. Bibcode:1923ZPhy...16..155P. doi:10.1007/BF01327386.

[50] de Broglie, L. (1929). "Nobel Lecture: The Wave Nature of the Electron" (PDF). The Nobel Foundation. Retrieved 2008-08-30.

[51] Falkenburg, B. (2007). *Particle Metaphysics: A Critical Account of Subatomic Reality.* Springer. p. 85. ISBN 3-540-33731-8.

[52] Davisson, C. (1937). "Nobel Lecture: The Discovery of Electron Waves" (PDF). The Nobel Foundation. Retrieved 2008-08-30.

[53] Schrödinger, E. (1926). "Quantisierung als Eigenwertproblem". *Annalen der Physik* (in German) **385** (13): 437–490. Bibcode:1926AnP...385..437S. doi:10.1002/andp.19263851302.

[54] Rigden, J.S. (2003). *Hydrogen*. Harvard University Press. pp. 59–86. ISBN 0-674-01252-6.

[55] Reed, B.C. (2007). *Quantum Mechanics*. Jones & Bartlett Publishers. pp. 275–350. ISBN 0-7637-4451-4.

[56] Dirac, P.A.M. (1928). "The Quantum Theory of the Electron". *Proceedings of the Royal Society A* **117** (778): 610–624. Bibcode:1928RSPSA.117..610D. doi:10.1098/rspa.1928.0023.

[57] Dirac, P.A.M. (1933). "Nobel Lecture: Theory of Electrons and Positrons" (PDF). The Nobel Foundation. Retrieved 2008-11-01.

[58] "The Nobel Prize in Physics 1965". The Nobel Foundation. Retrieved 2008-11-04.

[59] Panofsky, W.K.H. (1997). "The Evolution of Particle Accelerators & Colliders" (PDF). *Beam Line* (Stanford University) **27** (1): 36–44. Retrieved 2008-09-15.

[60] Elder, F.R. et al. (1947). "Radiation from Electrons in a Synchrotron". *Physical Review* **71** (11): 829–830. Bibcode:1947PhRv...71..829E. doi:10.1103/PhysRev.71.829.5.

[61] Hoddeson, L. et al. (1997). *The Rise of the Standard Model: Particle Physics in the 1960s and 1970s*. Cambridge University Press. pp. 25–26. ISBN 0-521-57816-7.

[62] Bernardini, C. (2004). "AdA: The First Electron–Positron Collider". *Physics in Perspective* **6** (2): 156–183. Bibcode:2004PhP.....6..156B. doi:10.1007/s00016-003-0202-y.

[63] "Testing the Standard Model: The LEP experiments". CERN. 2008. Retrieved 2008-09-15.

[64] "LEP reaps a final harvest". *CERN Courier* **40** (10). 2000.

[65] Prati, E.; De Michielis, M.; Belli, M.; Cocco, S.; Fanciulli, M.; Kotekar-Patil, D.; Ruoff, M.; Kern, D. P.; Wharam, D. A.; Verduijn, J.; Tettamanzi, G. C.; Rogge, S.; Roche, B.; Wacquez, R.; Jehl, X.; Vinet, M.; Sanquer, M. (2012). "Few electron limit of n-type metal oxide semiconductor single electron transistors". *Nanotechnology* **23** (21): 215204. doi:10.1088/0957-4484/23/21/215204. PMID 22552118.

[66] Frampton, P.H.; Hung, P.Q.; Sher, Marc (2000). "Quarks and Leptons Beyond the Third Generation". *Physics Reports* **330** (5–6): 263–348. arXiv:hep-ph/9903387. Bibcode:2000PhR...330..263F. doi:10.1016/S0370-1573(99)00095-2.

[67] Raith, W.; Mulvey, T. (2001). *Constituents of Matter: Atoms, Molecules, Nuclei and Particles*. CRC Press. pp. 777–781. ISBN 0-8493-1202-7.

[68] The original source for CODATA is Mohr, P.J.; Taylor, B.N.; Newell, D.B. (2006). "CODATA recommended values of the fundamental physical constants". *Reviews of Modern Physics* **80** (2): 633–730. arXiv:0801.0028. Bibcode:2008RvMP...80..633M. doi:10.1103/RevModPhys.80.633.

Individual physical constants from the CODATA are available at: "The NIST Reference on Constants, Units and Uncertainty". National Institute of Standards and Technology. Retrieved 2009-01-15.

[69] Zombeck, M.V. (2007). *Handbook of Space Astronomy and Astrophysics* (3rd ed.). Cambridge University Press. p. 14. ISBN 0-521-78242-2.

[70] Murphy, M.T. et al. (2008). "Strong Limit on a Variable Proton-to-Electron Mass Ratio from Molecules in the Distant Universe". *Science* **320** (5883): 1611–1613. arXiv:0806.3081. Bibcode:2008Sci...320.1611M. doi:10.1126/science.1156352. PMID 18566280.

[71] Zorn, J.C.; Chamberlain, G.E.; Hughes, V.W. (1963). "Experimental Limits for the Electron-Proton Charge Difference and for the Charge of the Neutron". *Physical Review* **129** (6): 2566–2576. Bibcode:1963PhRv..129.2566Z. doi:10.1103/PhysRev.129.2566.

[72] Odom, B. et al. (2006). "New Measurement of the Electron Magnetic Moment Using a One-Electron Quantum Cyclotron". *Physical Review Letters* **97** (3): 030801. Bibcode:2006PhRvL..97c0801O. doi:10.1103/PhysRevLett.97.030801. PMID 16907490.

[73] Anastopoulos, C. (2008). *Particle Or Wave: The Evolution of the Concept of Matter in Modern Physics*. Princeton University Press. pp. 261–262. ISBN 0-691-13512-6.

[74] Gabrielse, G. et al. (2006). "New Determination of the Fine Structure Constant from the Electron *g* Value and QED". *Physical Review Letters* **97** (3): 030802(1–4). Bibcode:2006PhRvL..97c0802G. doi:10.1103/PhysRevLett.97.030802.

[75] Eduard Shpolsky, Atomic physics (Atomnaia fizika),second edition, 1951

[76] Dehmelt, H. (1988). "A Single Atomic Particle Forever Floating at Rest in Free Space: New Value for Electron Radius". *Physica Scripta* **T22**: 102–10. Bibcode:1988PhST...22..102D. doi:10.1088/0031-8949/1988/T22/016.

[77] Meschede, D. (2004). *Optics, light and lasers: The Practical Approach to Modern Aspects of Photonics and Laser Physics*. Wiley-VCH. p. 168. ISBN 3-527-40364-7.

[78] Steinberg, R.I. et al. (1999). "Experimental test of charge conservation and the stability of the electron". *Physical Review D* **61** (2): 2582–2586. Bibcode:1975PhRvD..12.2582S. doi:10.1103/PhysRevD.12.2582.

[79] J. Beringer (Particle Data Group) et al. (2012). "Review of Particle Physics: [electron properties]" (PDF). *Physical Review D* **86** (1): 010001. Bibcode:2012PhRvD..86a0001B. doi:10.1103/PhysRevD.86.010001.

[80] Back, H. O. et al. (2002). "Search for electron decay mode $e \rightarrow \gamma + \nu$ with prototype of Borexino detector". *Physics Letters B* **525**: 29–40. Bibcode:2002PhLB..525...29B. doi:10.1016/S0370-2693(01)01440-X.

[81] Munowitz, M. (2005). *Knowing, The Nature of Physical Law*. Oxford University Press. ISBN 0-19-516737-6.

[82] Kane, G. (October 9, 2006). "Are virtual particles really constantly popping in and out of existence? Or are they merely a mathematical bookkeeping device for quantum mechanics?". Scientific American. Retrieved 2008-09-19.

[83] Taylor, J. (1989). "Gauge Theories in Particle Physics". In Davies, Paul. *The New Physics*. Cambridge University Press. p. 464. ISBN 0-521-43831-4.

[84] Genz, H. (2001). *Nothingness: The Science of Empty Space*. Da Capo Press. pp. 241–243, 245–247. ISBN 0-7382-0610-5.

[85] Gribbin, J. (January 25, 1997). "More to electrons than meets the eye". *New Scientist*. Retrieved 2008-09-17.

[86] Levine, I. et al. (1997). "Measurement of the Electromagnetic Coupling at Large Momentum Transfer". *Physical Review Letters* **78** (3): 424–427. Bibcode:1997PhRvL..78..424L. doi:10.1103/PhysRevLett.78.424.

[87] Murayama, H. (March 10–17, 2006). *Supersymmetry Breaking Made Easy, Viable and Generic. Proceedings of the XLIInd Rencontres de Moriond on Electroweak Interactions and Unified Theories* (La Thuile, Italy). arXiv:0709.3041. —lists a 9% mass difference for an electron that is the size of the Planck distance.

[88] Schwinger, J. (1948). "On Quantum-Electrodynamics and the Magnetic Moment of the Electron". *Physical Review* **73** (4): 416–417. Bibcode:1948PhRv...73..416S. doi:10.1103/PhysRev.73.416.

[89] Huang, K. (2007). *Fundamental Forces of Nature: The Story of Gauge Fields*. World Scientific. pp. 123–125. ISBN 981-270-645-3.

[90] Foldy, L.L.; Wouthuysen, S. (1950). "On the Dirac Theory of Spin 1/2 Particles and Its Non-Relativistic Limit". *Physical Review* **78**: 29–36. Bibcode:1950PhRv...78...29F. doi:10.1103/PhysRev.78.29.

[91] Sidharth, B.G. (2008). "Revisiting Zitterbewegung". *International Journal of Theoretical Physics* **48** (2): 497–506. arXiv:0806.0985. Bibcode:2009IJTP...48..497S. doi:10.1007/s10773-008-9825-8.

[92] Elliott, R.S. (1978). "The History of Electromagnetics as Hertz Would Have Known It". *IEEE Transactions on Microwave Theory and Techniques* **36** (5): 806–823. Bibcode:1988ITMTT..36..806E. doi:10.1109/22.3600.

[93] Crowell, B. (2000). *Electricity and Magnetism*. Light and Matter. pp. 129–152. ISBN 0-9704670-4-4.

[94] Mahadevan, R.; Narayan, R.; Yi, I. (1996). "Harmony in Electrons: Cyclotron and Synchrotron Emission by Thermal Electrons in a Magnetic Field". *The Astrophysical Journal* **465**: 327–337. arXiv:astro-ph/9601073. Bibcode:1996ApJ...465..327M. doi:10.1086/177422.

[95] Rohrlich, F. (1999). "The Self-Force and Radiation Reaction". *American Journal of Physics* **68** (12): 1109–1112. Bibcode:2000AmJPh..68.1109R. doi:10.1119/1.1286430.

[96] Georgi, H. (1989). "Grand Unified Theories". In Davies, Paul. *The New Physics*. Cambridge University Press. p. 427. ISBN 0-521-43831-4.

[97] Blumenthal, G.J.; Gould, R. (1970). "Bremsstrahlung, Synchrotron Radiation, and Compton Scattering of High-Energy Electrons Traversing Dilute Gases". *Reviews of Modern Physics* **42** (2): 237–270. Bibcode:1970RvMP...42..237B. doi:10.1103/RevModPhys.42.237.

[98] Staff (2008). "The Nobel Prize in Physics 1927". The Nobel Foundation. Retrieved 2008-09-28.

[99] Chen, S.-Y.; Maksimchuk, A.; Umstadter, D. (1998). "Experimental observation of relativistic nonlinear Thomson scattering". *Nature* **396** (6712): 653–655. arXiv:physics/9810036. Bibcode:1998Natur.396..653C. doi:10.1038/25303.

[100] Beringer, R.; Montgomery, C.G. (1942). "The Angular Distribution of Positron Annihilation Radiation". *Physical Review* **61** (5–6): 222–224. Bibcode:1942PhRv...61..222B. doi:10.1103/PhysRev.61.222.

[101] Buffa, A. (2000). *College Physics* (4th ed.). Prentice Hall. p. 888. ISBN 0-13-082444-5.

[102] Eichler, J. (2005). "Electron–positron pair production in relativistic ion–atom collisions". *Physics Letters A* **347** (1–3): 67–72. Bibcode:2005PhLA..347...67E. doi:10.1016/j.physleta.2005.06.105.

[103] Hubbell, J.H. (2006). "Electron positron pair production by photons: A historical overview". *Radiation Physics and Chemistry* **75** (6): 614–623. Bibcode:2006RaPC...75..614H. doi:10.1016/j.radphyschem.2005.10.008.

[104] Quigg, C. (June 4–30, 2000). *The Electroweak Theory. TASI 2000: Flavor Physics for the Millennium* (Boulder, Colorado): 80. arXiv:hep-ph/0204104.

[105] Mulliken, R.S. (1967). "Spectroscopy, Molecular Orbitals, and Chemical Bonding". *Science* **157** (3784): 13–24. Bibcode:1967Sci...157...13M. doi:10.1126/science.157.3784.13. PMID 5338306.

[106] Burhop, E.H.S. (1952). *The Auger Effect and Other Radiationless Transitions*. Cambridge University Press. pp. 2–3. ISBN 0-88275-966-3.

[107] Grupen, C. (2000). "Physics of Particle Detection". *AIP Conference Proceedings* **536**: 3–34. arXiv:physics/9906063. doi:10.1063/1.1361756.

[108] Jiles, D. (1998). *Introduction to Magnetism and Magnetic Materials*. CRC Press. pp. 280–287. ISBN 0-412-79860-3.

[109] Löwdin, P.O.; Erkki Brändas, E.; Kryachko, E.S. (2003). *Fundamental World of Quantum Chemistry: A Tribute to the Memory of Per- Olov Löwdin*. Springer. pp. 393–394. ISBN 1-4020-1290-X.

[110] McQuarrie, D.A.; Simon, J.D. (1997). *Physical Chemistry: A Molecular Approach*. University Science Books. pp. 325–361. ISBN 0-935702-99-7.

[111] Daudel, R. et al. (1973). "The Electron Pair in Chemistry". *Canadian Journal of Chemistry* **52** (8): 1310–1320. doi:10.1139/v74-201.

[112] Rakov, V.A.; Uman, M.A. (2007). *Lightning: Physics and Effects*. Cambridge University Press. p. 4. ISBN 0-521-03541-4.

[113] Freeman, G.R.; March, N.H. (1999). "Triboelectricity and some associated phenomena". *Materials Science and Technology* **15** (12): 1454–1458. doi:10.1179/026708399101505464.

[114] Forward, K.M.; Lacks, D.J.; Sankaran, R.M. (2009). "Methodology for studying particle–particle triboelectrification in granular materials". *Journal of Electrostatics* **67** (2–3): 178–183. doi:10.1016/j.elstat.2008.12.002.

[115] Weinberg, S. (2003). *The Discovery of Subatomic Particles*. Cambridge University Press. pp. 15–16. ISBN 0-521-82351-X.

[116] Lou, L.-F. (2003). *Introduction to phonons and electrons*. World Scientific. pp. 162, 164. ISBN 978-981-238-461-4.

[117] Guru, B.S.; Hızıroğlu, H.R. (2004). *Electromagnetic Field Theory*. Cambridge University Press. pp. 138, 276. ISBN 0-521-83016-8.

[118] Achuthan, M.K.; Bhat, K.N. (2007). *Fundamentals of Semiconductor Devices*. Tata McGraw-Hill. pp. 49–67. ISBN 0-07-061220-X.

[119] Ziman, J.M. (2001). *Electrons and Phonons: The Theory of Transport Phenomena in Solids*. Oxford University Press. p. 260. ISBN 0-19-850779-8.

[120] Main, P. (June 12, 1993). "When electrons go with the flow: Remove the obstacles that create electrical resistance, and you get ballistic electrons and a quantum surprise". *New Scientist* **1887**: 30. Retrieved 2008-10-09.

[121] Blackwell, G.R. (2000). *The Electronic Packaging Handbook*. CRC Press. pp. 6.39–6.40. ISBN 0-8493-8591-1.

[122] Durrant, A. (2000). *Quantum Physics of Matter: The Physical World*. CRC Press. pp. 43, 71–78. ISBN 0-7503-0721-8.

[123] Staff (2008). "The Nobel Prize in Physics 1972". The Nobel Foundation. Retrieved 2008-10-13.

[124] Kadin, A.M. (2007). "Spatial Structure of the Cooper Pair". *Journal of Superconductivity and Novel Magnetism* **20** (4): 285–292. arXiv:cond-mat/0510279. doi:10.1007/s10948-006-0198-z.

[125] "Discovery About Behavior Of Building Block Of Nature Could Lead To Computer Revolution". *ScienceDaily*. July 31, 2009. Retrieved 2009-08-01.

[126] Jompol, Y. et al. (2009). "Probing Spin-Charge Separation in a Tomonaga-Luttinger Liquid". *Science* **325** (5940): 597–601. arXiv:1002.2782. Bibcode:2009Sci...325..597J. doi:10.1126/science.1171769. PMID 19644117.

[127] Staff (2008). "The Nobel Prize in Physics 1958, for the discovery and the interpretation of the Cherenkov effect". The Nobel Foundation. Retrieved 2008-09-25.

[128] Staff (August 26, 2008). "Special Relativity". Stanford Linear Accelerator Center. Retrieved 2008-09-25.

[129] Adams, S. (2000). *Frontiers: Twentieth Century Physics*. CRC Press. p. 215. ISBN 0-7484-0840-1.

[130] Lurquin, P.F. (2003). *The Origins of Life and the Universe*. Columbia University Press. p. 2. ISBN 0-231-12655-7.

[131] Silk, J. (2000). *The Big Bang: The Creation and Evolution of the Universe* (3rd ed.). Macmillan. pp. 110–112, 134–137. ISBN 0-8050-7256-X.

[132] Christianto, V. (2007). "Thirty Unsolved Problems in the Physics of Elementary Particles" (PDF). *Progress in Physics* **4**: 112–114.

[133] Kolb, E.W.; Wolfram, Stephen (1980). "The Development of Baryon Asymmetry in the Early Universe". *Physics Letters B* **91** (2): 217–221. Bibcode:1980PhLB...91..217K. doi:10.1016/0370-2693(80)90435-9.

[134] Sather, E. (Spring–Summer 1996). "The Mystery of Matter Asymmetry" (PDF). *Beam Line*. University of Stanford. Retrieved 2008-11-01.

[135] Burles, S.; Nollett, K.M.; Turner, M.S. (1999). "Big-Bang Nucleosynthesis: Linking Inner Space and Outer Space". arXiv:astro-ph/9903300 [astro-ph].

[136] Boesgaard, A.M.; Steigman, G. (1985). "Big bang nucleosynthesis – Theories and observations" . *Annual Review of Astronomy and Astrophysics* **23** (2): 319–378. Bibcode:1985ARA&A..23..319B. doi:10.1146/annurev.aa.23.090185.001535.

[137] Barkana, R. (2006). "The First Stars in the Universe and Cosmic Reionization" . *Science* **313** (5789): 931–934. arXiv:astro-ph/0608450. Bibcode:2006Sci...313..931B. doi:10.1126/science.1125644. PMID 16917052.

[138] Burbidge, E.M. et al. (1957). "Synthesis of Elements in Stars" . *Reviews of Modern Physics* **29** (4): 548–647. Bibcode:1957RvMP...29..547B. doi:10.1103/RevModPhys.29.547.

[139] Rodberg, L.S.; Weisskopf, V. (1957). "Fall of Parity: Recent Discoveries Related to Symmetry of Laws of Nature" . *Science* **125** (3249): 627–633. Bibcode:1957Sci...125..627R. doi:10.1126/science.125.3249.627. PMID 17810563.

[140] Fryer, C.L. (1999). "Mass Limits For Black Hole Formation" . *The Astrophysical Journal* **522** (1): 413–418. arXiv:astro-ph/9902315. Bibcode:1999ApJ...522..413F. doi:10.1086/307647.

[141] Parikh, M.K.; Wilczek, F. (2000). "Hawking Radiation As Tunneling" . *Physical Review Letters* **85** (24): 5042–5045. arXiv:hep-th/9907001. Bibcode:2000PhRvL..85.5042P. doi:10.1103/PhysRevLett.85.5042. PMID 11102182.

[142] Hawking, S.W. (1974). "Black hole explosions?". *Nature* **248** (5443): 30–31. Bibcode:1974Natur.248...30H. doi:10.1038/248030a0.

[143] Halzen, F.; Hooper, D. (2002). "High-energy neutrino astronomy: the cosmic ray connection" . *Reports on Progress in Physics* **66** (7): 1025–1078. arXiv:astro-ph/0204527. Bibcode:2002astro.ph..4527H. doi:10.1088/0034-4885/65/7/201.

[144] Ziegler, J.F. (1998). "Terrestrial cosmic ray intensities" . *IBM Journal of Research and Development* **42** (1): 117–139. doi:10.1147/rd.421.0117.

[145] Sutton, C. (August 4, 1990). "Muons, pions and other strange particles" . *New Scientist*. Retrieved 2008-08-28.

[146] Wolpert, S. (July 24, 2008). "Scientists solve 30-year-old aurora borealis mystery" . University of California. Retrieved 2008-10-11.

[147] Gurnett, D.A.; Anderson, R. (1976). "Electron Plasma Oscillations Associated with Type III Radio Bursts" . *Science* **194** (4270): 1159–1162. Bibcode:1976Sci...194.1159G. doi:10.1126/science.194.4270.1159. PMID 17790910.

[148] Martin, W.C.; Wiese, W.L. (2007). "Atomic Spectroscopy: A Compendium of Basic Ideas, Notation, Data, and Formulas" . National Institute of Standards and Technology. Retrieved 2007-01-08.

[149] Fowles, G.R. (1989). *Introduction to Modern Optics*. Courier Dover. pp. 227–233. ISBN 0-486-65957-7.

[150] Staff (2008). "The Nobel Prize in Physics 1989" . The Nobel Foundation. Retrieved 2008-09-24.

[151] Ekstrom, P.; Wineland, David (1980). "The isolated Electron" (PDF). *Scientific American* **243** (2): 91–101. doi:10.1038/scientificamerican0880-104. Retrieved 2008-09-24.

[152] Mauritsson, J. "Electron filmed for the first time ever" (PDF). Lund University. Archived from the original (PDF) on March 25, 2009. Retrieved 2008-09-17.

[153] Mauritsson, J. et al. (2008). "Coherent Electron Scattering Captured by an Attosecond Quantum Stroboscope" . *Physical Review Letters* **100** (7): 073003. arXiv:0708.1060. Bibcode:2008PhRvL.100g3003M. doi:10.1103/PhysRevLett.100.073003. PMID 18352546.

[154] Damascelli, A. (2004). "Probing the Electronic Structure of Complex Systems by ARPES" . *Physica Scripta* **T109**: 61–74. arXiv:cond-mat/0307085. Bibcode:2004PhST..109...61D. doi:10.1238/Physica.Topical.109a00061.

[155] Staff (April 4, 1975). "Image # L-1975-02972" . Langley Research Center, NASA. Retrieved 2008-09-20.

[156] Elmer, J. (March 3, 2008). "Standardizing the Art of Electron-Beam Welding" . Lawrence Livermore National Laboratory. Retrieved 2008-10-16.

[157] Schultz, H. (1993). *Electron Beam Welding*. Woodhead Publishing. pp. 2–3. ISBN 1-85573-050-2.

[158] Benedict, G.F. (1987). *Nontraditional Manufacturing Processes*. Manufacturing engineering and materials processing **19**. CRC Press. p. 273. ISBN 0-8247-7352-7.

[159] Ozdemir, F.S. (June 25–27, 1979). *Electron beam lithography*. Proceedings of the 16th Conference on Design automation (San Diego, CA, USA: IEEE Press): 383–391. Retrieved 2008-10-16.

[160] Madou, M.J. (2002). *Fundamentals of Microfabrication: the Science of Miniaturization* (2nd ed.). CRC Press. pp. 53–54. ISBN 0-8493-0826-7.

[161] Jongen, Y.; Herer, A. (May 2–5, 1996). *Electron Beam Scanning in Industrial Applications*. APS/AAPT Joint Meeting (American Physical Society). Bibcode:1996APS..MAY.H9902J.

[162] Mobus G. et al. (2010). Journal of Nuclear Materials, v. 396, 264–271, doi:10.1016/j.jnucmat.2009.11.020

[163] Beddar, A.S.; Domanovic, Mary Ann; Kubu, Mary Lou; Ellis, Rod J.; Sibata, Claudio H.; Kinsella, Timothy J. (2001). "Mobile linear accelerators for intraoperative radiation therapy" . *AORN Journal* **74** (5): 700. doi:10.1016/S0001-2092(06)61769-9.

[164] Gazda, M.J.; Coia, L.R. (June 1, 2007). "Principles of Radiation Therapy" (PDF). Retrieved 2013-10-31.

[165] Chao, A.W.; Tigner, M. (1999). *Handbook of Accelerator Physics and Engineering*. World Scientific. pp. 155, 188. ISBN 981-02-3500-3.

[166] Oura, K. et al. (2003). *Surface Science: An Introduction*. Springer. pp. 1–45. ISBN 3-540-00545-5.

[167] Ichimiya, A.; Cohen, P.I. (2004). *Reflection High-energy Electron Diffraction*. Cambridge University Press. p. 1. ISBN 0-521-45373-9.

[168] Heppell, T.A. (1967). "A combined low energy and reflection high energy electron diffraction apparatus". *Journal of Scientific Instruments* **44** (9): 686–688. Bibcode:1967JScI...44..686H. doi:10.1088/0950-7671/44/9/311.

[169] McMullan, D. (1993). "Scanning Electron Microscopy: 1928–1965". University of Cambridge. Retrieved 2009-03-23.

[170] Slayter, H.S. (1992). *Light and electron microscopy*. Cambridge University Press. p. 1. ISBN 0-521-33948-0.

[171] Cember, H. (1996). *Introduction to Health Physics*. McGraw-Hill Professional. pp. 42–43. ISBN 0-07-105461-8.

[172] Erni, R. et al. (2009). "Atomic-Resolution Imaging with a Sub-50-pm Electron Probe". *Physical Review Letters* **102** (9): 096101. Bibcode:2009PhRvL.102i6101E. doi:10.1103/PhysRevLett.102.096101. PMID 19392535.

[173] Bozzola, J.J.; Russell, L.D. (1999). *Electron Microscopy: Principles and Techniques for Biologists*. Jones & Bartlett Publishers. pp. 12, 197–199. ISBN 0-7637-0192-0.

[174] Flegler, S.L.; Heckman Jr., J.W.; Klomparens, K.L. (1995). *Scanning and Transmission Electron Microscopy: An Introduction* (Reprint ed.). Oxford University Press. pp. 43–45. ISBN 0-19-510751-9.

[175] Bozzola, J.J.; Russell, L.D. (1999). *Electron Microscopy: Principles and Techniques for Biologists* (2nd ed.). Jones & Bartlett Publishers. p. 9. ISBN 0-7637-0192-0.

[176] Freund, H.P.; Antonsen, T. (1996). *Principles of Free-Electron Lasers*. Springer. pp. 1–30. ISBN 0-412-72540-1.

[177] Kitzmiller, J.W. (1995). *Television Picture Tubes and Other Cathode-Ray Tubes: Industry and Trade Summary*. DIANE Publishing. pp. 3–5. ISBN 0-7881-2100-6.

[178] Sclater, N. (1999). *Electronic Technology Handbook*. McGraw-Hill Professional. pp. 227–228. ISBN 0-07-058048-0.

[179] Staff (2008). "The History of the Integrated Circuit". The Nobel Foundation. Retrieved 2008-10-18.

2.1.9   External links

- "The Discovery of the Electron". American Institute of Physics, Center for History of Physics.

- "Particle Data Group". University of California.

- Bock, R.K.; Vasilescu, A. (1998). *The Particle Detector BriefBook* (14th ed.). Springer. ISBN 3-540-64120-3.

- Copeland, Ed. "Spherical Electron". *Sixty Symbols*. Brady Haran for the University of Nottingham.

2.2   Positron (antielectron)

For other uses, see Positron (disambiguation).

The **positron** or **antielectron** is the antiparticle or the antimatter counterpart of the electron. The positron has an electric charge of +1 e, a spin of ½, and has the same mass as an electron. When a low-energy positron collides with a low-energy electron, annihilation occurs, resulting in the production of two or more gamma ray photons (see electron–positron annihilation).

Positrons may be generated by positron emission radioactive decay (through weak interactions), or by pair production from a sufficiently energetic photon which is interacting with an atom in a material.

2.2.1   History

Theory

In 1928, Paul Dirac published a paper[*][2] proposing that electrons can have both a positive charge and negative energy. This paper introduced the Dirac equation, a unification of quantum mechanics, special relativity, and the then-new concept of electron spin to explain the Zeeman effect. The paper did not explicitly predict a new particle, but did allow for electrons having either positive or negative energy as solutions. Hermann Weyl then published "Gravitation and the Electron" (Proceedings of the National Academy of Sciences of the United States of America, Vol. 15, No. 4-Apr. 15, 1929, pp. 323–334) discussing the mathematical implications of the negative energy solution. The positive-energy solution explained experimental results, but Dirac was puzzled by the equally valid negative-energy solution that the mathematical model allowed. Quantum mechanics did not allow the negative energy solution to simply be ignored, as classical mechanics often did in such equations;

the dual solution implied the possibility of an electron spontaneously jumping between positive and negative energy states. However, no such transition had yet been observed experimentally. He referred to the issues raised by this conflict between theory and observation as "difficulties" that were "unresolved".

Dirac wrote a follow-up paper in December 1929[*][3] that attempted to explain the unavoidable negative-energy solution for the relativistic electron. He argued that "... an electron with negative energy moves in an external [electromagnetic] field as though it carries a positive charge." He further asserted that all of space could be regarded as a "sea" of negative energy states that were filled, so as to prevent electrons jumping between positive energy states (negative electric charge) and negative energy states (positive charge). The paper also explored the possibility of the proton being an island in this sea, and that it might actually be a negative-energy electron. Dirac acknowledged that the proton having a much greater mass than the electron was a problem, but expressed "hope" that a future theory would resolve the issue.

Robert Oppenheimer argued strongly against the proton being the negative-energy electron solution to Dirac's equation. He asserted that if it were, the hydrogen atom would rapidly self-destruct.[*][4] Persuaded by Oppenheimer's argument, Dirac published a paper in 1931 that predicted the existence of an as-yet unobserved particle that he called an "anti-electron" that would have the same mass as an electron and that would mutually annihilate upon contact with an electron.[*][5]

Feynman, and earlier Stueckelberg, proposed an interpretation of the positron as an electron moving backward in time,[*][6] reinterpreting the negative-energy solutions of the Dirac equation. Electrons moving backward in time would have a positive electric charge. Wheeler invoked this concept to explain the identical properties shared by all electrons, suggesting that "they are all the same electron" with a complex, self-intersecting worldline.[*][7] Yoichiro Nambu later applied it to all production and annihilation of particle-antiparticle pairs, stating that "the eventual creation and annihilation of pairs that may occur now and then is no creation or annihilation, but only a change of direction of moving particles, from past to future, or from future to past." [*][8] The backwards in time point of view is nowadays accepted as completely equivalent to other pictures, but it does not have anything to do with the macroscopic terms "cause" and "effect", which do not appear in a microscopic physical description.

Experimental clues and discovery

Dmitri Skobeltsyn first observed the positron in

1929.[*][9][*][10] While using a Wilson cloud chamber[*][11] to try to detect gamma radiation in cosmic rays, Skobeltsyn detected particles that acted like electrons but curved in the opposite direction in an applied magnetic field.[*][10]

Likewise, in 1929 Chung-Yao Chao, a graduate student at Caltech, noticed some anomalous results that indicated particles behaving like electrons, but with a positive charge, though the results were inconclusive and the phenomenon was not pursued.[*][12]

Carl David Anderson discovered the positron on August 2, 1932,[*][13] for which he won the Nobel Prize for Physics in 1936.[*][14] Anderson did not coin the term *positron*, but allowed it at the suggestion of the Physical Review journal editor to which he submitted his discovery paper in late 1932. The positron was the first evidence of antimatter and was discovered when Anderson allowed cosmic rays to pass through a cloud chamber and a lead plate. A magnet surrounded this apparatus, causing particles to bend in different directions based on their electric charge. The ion trail left by each positron appeared on the photographic plate with a curvature matching the mass-to-charge ratio of an electron, but in a direction that showed its charge was positive.[*][15]

Anderson wrote in retrospect that the positron could have been discovered earlier based on Chung-Yao Chao's work, if only it had been followed up.[*][12] Frédéric and Irène Joliot-Curie in Paris had evidence of positrons in old photographs when Anderson's results came out, but they had dismissed them as protons.[*][15]

2.2.2 Natural production

Main article: Positron emission

Positrons are produced naturally in β[*]+ decays of naturally occurring radioactive isotopes (for example, potassium-40) and in interactions of gamma quanta (emitted by radioactive nuclei) with matter. Antineutrinos are another kind of antiparticle created by natural radioactivity (β[*] − decay). Many different kinds of antiparticles are also produced by (and contained in) cosmic rays. Recent (as of January 2011) research by the American Astronomical Society has discovered antimatter (positrons) originating above thunderstorm clouds; positrons are produced in gamma-ray flashes created by electrons accelerated by strong electric fields in the clouds.[*][16] Antiprotons have also been found to exist in the Van Allen Belts around the Earth by the PAMELA module.[*][17][*][18]

Antiparticles, of which the most common are positrons due to their low mass, are also produced in any environment with a sufficiently high temperature (mean particle energy greater than the pair production threshold). During the period of baryogenesis, when the universe was extremely hot and dense, matter and antimatter were continually produced and annihilated. The presence of remaining matter, and absence of detectable remaining antimatter,[19] also called baryon asymmetry, is attributed to CP-violation: a violation of the CP-symmetry relating matter to antimatter. The exact mechanism of this violation during baryogenesis remains a mystery.

Positrons production from radioactive β+ decay, can be considered both artificial and natural production, as the generation of the radioisotope can be natural or artificial. Perhaps the best known naturally-occurring radioisotope which produces positrons is potassium-40, a long-lived isotope of potassium which occurs as a primordial isotope of potassium, and even though a small percent of potassium, (0.0117%) is the single most abundant radioisotope in the human body. In a human body of 70 kg mass, about 4,400 nuclei of ^{40}K decay per second.[20] The activity of natural potassium is 31 Bq/g.[21] About 0.001% of these ^{40}K decays produce about 4000 natural positrons per day in the human body.[22] These positrons soon find an electron, undergo annihilation, and produce pairs of 511 keV gamma rays, in a process similar (but much lower intensity) to that which happens during a PET scan nuclear medicine procedure.

Observation in cosmic rays

Main article: Cosmic ray

Satellite experiments have found evidence of positrons (as well as a few antiprotons) in primary cosmic rays, amounting to less than 1% of the particles in primary cosmic rays. These do not appear to be the products of large amounts of antimatter from the Big Bang, or indeed complex antimatter in the universe (evidence for which is lacking, see below). Rather, the antimatter in cosmic rays appear to consist of only these two elementary particles, probably made in energetic processes long after the Big Bang.

Preliminary results from the presently operating Alpha Magnetic Spectrometer (*AMS-02*) on board the International Space Station show that positrons in the cosmic rays arrive with no directionality, and with energies that range from 10 GeV to 250 GeV. In September, 2014, new results with almost twice as much data were presented in a talk at CERN and published in Physical Review Letters.[23][24] A new measurement of positron fraction up to 500 GeV was reported, showing that positron fraction peaks at a maximum of about 16% of total electron+positron events, around an energy of 275 ± 32

GeV. At higher energies, up to 500 GeV, the ratio of positrons to electrons begins to fall again. The absolute flux of positrons also begins to fall before 500 GeV, but peaks at energies far higher than electron energies, which peak about 10 GeV.[25] These results on interpretation have been suggested to be due to positron production in annihilation events of massive dark matter particles.[26]

Positrons, like anti-protons, do not appear to originate from any hypothetical "antimatter" regions of the universe. On the contrary, there is no evidence of complex antimatter atomic nuclei, such as antihelium nuclei (i.e., anti-alpha particles), in cosmic rays. These are actively being searched for. A prototype of the *AMS-02* designated *AMS-01*, was flown into space aboard the Space Shuttle *Discovery* on STS-91 in June 1998. By not detecting any antihelium at all, the *AMS-01* established an upper limit of 1.1×10^{-6} for the antihelium to helium flux ratio.[27]

2.2.3 Artificial production

New research has dramatically increased the quantity of positrons that experimentalists can produce. Physicists at the Lawrence Livermore National Laboratory in California have used a short, ultra-intense laser to irradiate a millimetre-thick gold target and produce more than 100 billion positrons.[28][29]

2.2.4 Applications

Certain kinds of particle accelerator experiments involve colliding positrons and electrons at relativistic speeds. The high impact energy and the mutual annihilation of these matter/antimatter opposites create a fountain of diverse subatomic particles. Physicists study the results of these collisions to test theoretical predictions and to search for new kinds of particles.

Gamma rays, emitted indirectly by a positron-emitting radionuclide (tracer), are detected in positron emission tomography (PET) scanners used in hospitals. PET scanners create detailed three-dimensional images of metabolic activity within the human body.[30]

An experimental tool called positron annihilation spectroscopy (PAS) is used in materials research to detect variations in density, defects, displacements, or even voids, within a solid material.[31]

2.2.5 See also

- Beta particle

- Radioactive decay

- List of particles

- Positron emission tomography

- Positronium

- Proton

- Positronic brain

2.2.6 References

Notes

[1] The fractional version's denominator is the inverse of the decimal value (along with its relative standard uncertainty of $4.2 \times 10^{*}-10$).

Citations

[1] The original source for CODATA is:

> Mohr, P.J.; Taylor, B.N.; Newell, D.B. (2006). "CODATA recommended values of the fundamental physical constants". *Reviews of Modern Physics* **80** (2): 633–730. arXiv:0801.0028. Bibcode:2008RvMP...80..633M. doi:10.1103/RevModPhys.80.633.
> Individual physical constants from the CODATA are available at:
> "The NIST Reference on Constants, Units and Uncertainty". National Institute of Standards and Technology. Retrieved 2013-10-24.

[2] P. A. M. Dirac. "The quantum theory of the electron" (PDF).

[3] P. A. M. Dirac. "A Theory of Electrons and Protons" (PDF).

[4] Frank Close (2009). *Antimatter*. Oxford University Press. p. 46. ISBN 978-0-19-955016-6.

[5] P. A. M. Dirac (1931). "Quantised Singularities in the Quantum Field". *Proc. R. Soc. Lond. A* **133** (821): 2–3. Bibcode:1931RSPSA.133...60D. doi:10.1098/rspa.1931.0130.

[6] Feynman, Richard (1949). "The Theory of Positrons". *Physical Review* **76** (76): 749. Bibcode:1949PhRv...76..749F. doi:10.1103/PhysRev.76.749.

[7] Feynman, Richard (1965-12-11). *The Development of the Space-Time View of Quantum Electrodynamics* (Speech). Nobel Lecture. Retrieved 2007-01-02.

[8] Nambu, Yoichiro (1950). "The Use of the Proper Time in Quantum Electrodynamics I". *Progress in Theoretical Physics* **5** (5): 82. Bibcode:1950PThPh...5...82N. doi:10.1143/PTP.5.82.

[9] Frank Close. *Antimatter*. Oxford University Press. pp. 50–52. ISBN 978-0-19-955016-6.

[10] *general chemistry*. Taylor & Francis. 1943. p. 660. GGKEY:0PYLHBL5D4L. Retrieved 15 June 2011.

[11] Cowan, Eugene (1982). "The Picture That Was Not Reversed". *Engineering & Science* **46** (2): 6–28.

[12] Jagdish Mehra; Helmut Rechenberg (2000). *The Historical Development of Quantum Theory, Volume 6: The Completion of. Quantum Mechanics 1926–1941*. Springer. p. 804. ISBN 978-0-387-95175-1.

[13] Anderson, Carl D. (1933). "The Positive Electron". *Physical Review* **43** (6): 491–494. Bibcode:1933PhRv...43..491A. doi:10.1103/PhysRev.43.491.

[14] "The Nobel Prize in Physics 1936". Retrieved 2010-01-21.

[15] GILMER, PENNY J. (19 July 2011). "IRÈNE JOLIOT-CURIE, A NOBEL LAUREATE IN ARTIFICIAL RADIOACTIVITY" (PDF). p. 8. Retrieved 13 July 2013.

[16] "Antimatter caught streaming from thunderstorms on Earth". BBC. 11 January 2011. Archived from the original on 12 January 2011. Retrieved 11 January 2011.

[17] Adriani, O.; Barbarino, G. C.; Bazilevskaya, G. A.; Bellotti, R. et al. (2011). "The Discovery of Geomagnetically Trapped Cosmic-Ray Antiprotons". *The Astrophysical Journal Letters* **737** (2): L29. arXiv:1107.4882v1. Bibcode:2011ApJ...737L..29A. doi:10.1088/2041-8205/737/2/L29.

[18] Than, Ker (10 August 2011). "Antimatter Found Orbiting Earth—A First". National Geographic Society. Retrieved 12 August 2011.

[19] "What's the Matter with Antimatter?". NASA. 29 May 2000. Archived from the original on 4 June 2008. Retrieved 24 May 2008.

[20] "Radiation and Radioactive Decay. Radioactive Human Body". Harvard Natural Sciences Lecture Demonstrations. Retrieved 2011-05-18.

[21] Winteringham, F. P. W; Effects, F.A.O. Standing Committee on Radiation, Land And Water Development Division, Food and Agriculture Organization of the United Nations (1989). *Radioactive fallout in soils, crops and food: a background review*. Food & Agriculture Org. p. 32. ISBN 978-92-5-102877-3.

[22] Engelkemeir, DW; KF Flynn; LE Glendenin (1962). "Positron Emission in the Decay of K^{40}". *Physical Review* **126** (5): 1818. Bibcode:1962PhRv..126.1818E. doi:10.1103/PhysRev.126.1818.

[23] L. Accardo et al. (AMS Collaboration) (18 September 2014). "High Statistics Measurement of the Positron Fraction in Primary Cosmic Rays of 0.5–500 GeV with the Alpha Magnetic Spectrometer on the International Space Station" (PDF). *Physical Review Letters* **113**: 121101. Bibcode:2014PhRvL.113l1101A. doi:10.1103/PhysRevLett.113.121101.

[24] Schirber, Michael. "Synopsis: More Dark Matter Hints from Cosmic Rays?". American Physical Society. Retrieved 21 September 2014.

[25] "New results from the Alpha Magnetic$Spectrometer on the International Space Station" (PDF). *AMS-02 at NASA*. Retrieved 21 September 2014.

[26] Aguilar, M.; Alberti, G.; Alpat, B.; Alvino, A.; Ambrosi, G.; Andeen, K.; Anderhub, H.; Arruda, L.; Azzarello, P.; Bachlechner, A.; Barao, F.; Baret, B.; Barrau, A.; Barrin, L.; Bartoloni, A.; Basara, L.; Basili, A.; Batalha, L.; Bates, J.; Battiston, R.; Bazo, J.; Becker, R.; Becker, U.; Behlmann, M.; Beischer, B.; Berdugo, J.; Berges, P.; Bertucci, B.; Bigongiari, G. et al. (2013). "First Result from the Alpha Magnetic Spectrometer on the International Space Station: Precision Measurement of the Positron Fraction in Primary Cosmic Rays of 0.5–350 GeV". *Physical Review Letters* **110** (14): 141102. Bibcode:2013PhRvL.110n1102A. doi:10.1103/PhysRevLett.110.141102.

[27] AMS Collaboration; Aguilar, M.; Alcaraz, J.; Allaby, J.; Alpat, B.; Ambrosi, G.; Anderhub, H.; Ao, L. et al. (August 2002). "The Alpha Magnetic Spectrometer (AMS) on the International Space Station: Part I – results from the test flight on the space shuttle". *Physics Reports* **366** (6): 331–405. Bibcode:2002PhR...366..331A. doi:10.1016/S0370-1573(02)00013-3.

[28] Bland, E. (1 December 2008). "Laser technique produces bevy of antimatter". MSNBC. Retrieved 2009-07-16. The LLNL scientists created the positrons by shooting the lab's high-powered Titan laser onto a one-millimeter-thick piece of gold.

[29] "Laser creates billions of antimatter particles". *Cosmos Online*.

[30] Phelps, Michael E. (2006). *PET: physics, instrumentation, and scanners*. Springer. pp. 2–3. ISBN 0-387-32302-3.

[31] "Introduction to Positron Research". *St. Olaf College*.

2.2.7   External links

- What is a Positron? (from the Frequently Asked Questions :: Center for Antimatter-Matter Studies)

- Website about positrons and antimatter

- Positron information search at SLAC

- Positron Annihilation as a method of experimental physics used in materials research.

- New production method to produce large quantities of positrons

- Website about antimatter (positrons, positronium and antihydrogen). Positron Laboratory, Como, Italy

- Website of the AEgIS: Antimatter Experiment: Gravity, Interferometry, Spectroscopy, CERN

- Synopsis: Tabletop Particle Accelerator ... new tabletop method for generating electron-positron streams.

2.3   Muon

The **muon** (/ˈmjuːɒn/; from the Greek letter mu (μ) used to represent it) is an elementary particle similar to the electron, with electric charge of −1 e and a spin of $1/2$, but with a much greater mass (105.7 MeV/c^2). It is classified as a lepton, together with the electron (mass 0.511 MeV/c^2), the tau (mass 1776.82 MeV/c^2), and the three neutrinos (electron neutrino ν

e, muon neutrino ν

μ and tau neutrino ν

τ). As is the case with other leptons, the muon is not believed to have any sub-structure—that is, it is not thought to be composed of any simpler particles.

The muon is an unstable subatomic particle with a mean lifetime of 2.2 μs. Among all known unstable subatomic particles, only the neutron (lasting around 15 minutes) and some atomic nuclei have a longer decay lifetime; others decay significantly faster. The decay of the muon (as well as of the neutron, the longest-lived unstable baryon), is mediated by the weak interaction exclusively. Muon decay always produces at least three particles, which must include an electron of the same charge as the muon and two neutrinos of different types.

Like all elementary particles, the muon has a corresponding antiparticle of opposite charge (+1 e) but equal mass and spin: the **antimuon** (also called a *positive muon*). Muons are denoted by μ− and antimuons by μ+. Muons were previously called **mu mesons**, but are not classified as mesons by modern particle physicists (see § History), and that name is no longer used by the physics community.

Muons have a mass of 105.7 MeV/c^2, which is about 207 times that of the electron. Due to their greater mass, muons are not as sharply accelerated when they encounter electromagnetic fields, and do not emit as much bremsstrahlung (deceleration radiation). This allows muons of a given energy to penetrate far more deeply into matter than electrons, since the deceleration of electrons and muons is primarily due to energy loss by the bremsstrahlung mechanism. As an example, so-called "secondary muons", generated by cosmic rays hitting the atmosphere, can penetrate to the Earth's surface, and even into deep mines.

Because muons have a very large mass and energy compared with the decay energy of radioactivity, they are never produced by radioactive decay. They are, however, produced in copious amounts in high-energy interactions in normal matter, in certain particle accelerator experiments with hadrons, or naturally in cosmic ray interactions with matter. These interactions usually produce pi mesons initially, which most often decay to muons.

As with the case of the other charged leptons, the muon has an associated muon neutrino, denoted by v
μ, which is not the same particle as the electron neutrino, and does not participate in the same nuclear reactions.

2.3.1 History

Muons were discovered by Carl D. Anderson and Seth Neddermeyer at Caltech in 1936, while studying cosmic radiation. Anderson had noticed particles that curved differently from electrons and other known particles when passed through a magnetic field. They were negatively charged but curved less sharply than electrons, but more sharply than protons, for particles of the same velocity. It was assumed that the magnitude of their negative electric charge was equal to that of the electron, and so to account for the difference in curvature, it was supposed that their mass was greater than an electron but smaller than a proton. Thus Anderson initially called the new particle a *mesotron*, adopting the prefix *meso-* from the Greek word for "mid-". The existence of the muon was confirmed in 1937 by J. C. Street and E. C. Stevenson's cloud chamber experiment.[*][2]

A particle with a mass in the meson range had been predicted before the discovery of any mesons, by theorist Hideki Yukawa:[*][3]

> "It seems natural to modify the theory of Heisenberg and Fermi in the following way. The transition of a heavy particle from neutron state to proton state is not always accompanied by the emission of light particles. The transition is sometimes taken up by another heavy particle."

Because of its mass, the mu meson was initially thought to be Yukawa's particle, but it later proved to have the wrong properties. Yukawa's predicted particle, the pi meson, was finally identified in 1947 (again from cosmic ray interactions), and shown to differ from the earlier-discovered mu

meson by having the correct properties to be a particle which mediated the nuclear force.

With two particles now known with the intermediate mass, the more general term *meson* was adopted to refer to any such particle within the correct mass range between electrons and nucleons. Further, in order to differentiate between the two different types of mesons after the second meson was discovered, the initial mesotron particle was renamed the *mu meson* (the Greek letter μ (*mu*) corresponds to *m*), and the new 1947 meson (Yukawa's particle) was named the pi meson.

As more types of mesons were discovered in accelerator experiments later, it was eventually found that the mu meson significantly differed not only from the pi meson (of about the same mass), but also from all other types of mesons. The difference, in part, was that mu mesons did not interact with the nuclear force, as pi mesons did (and were required to do, in Yukawa's theory). Newer mesons also showed evidence of behaving like the pi meson in nuclear interactions, but not like the mu meson. Also, the mu meson's decay products included both a neutrino and an antineutrino, rather than just one or the other, as was observed in the decay of other charged mesons.

In the eventual Standard Model of particle physics codified in the 1970s, all mesons other than the mu meson were understood to be hadrons—that is, particles made of quarks—and thus subject to the nuclear force. In the quark model, a *meson* was no longer defined by mass (for some had been discovered that were very massive—more than nucleons), but instead were particles composed of exactly two quarks (a quark and antiquark), unlike the baryons, which are defined as particles composed of three quarks (protons and neutrons were the lightest baryons). Mu mesons, however, had shown themselves to be fundamental particles (leptons) like electrons, with no quark structure. Thus, mu mesons were not mesons at all, in the new sense and use of the term *meson* used with the quark model of particle structure.

With this change in definition, the term *mu meson* was abandoned, and replaced whenever possible with the modern term *muon*, making the term mu meson only historical. In the new quark model, other types of mesons sometimes continued to be referred to in shorter terminology (e.g., *pion* for pi meson), but in the case of the muon, it retained the shorter name and was never again properly referred to by older "mu meson" terminology.

The eventual recognition of the "mu meson" muon as a simple "heavy electron" with no role at all in the nuclear interaction, seemed so incongruous and surprising at the time, that Nobel laureate I. I. Rabi famously quipped, "Who ordered that?"

In the Rossi–Hall experiment (1941), muons were used to

observe the time dilation (or alternately, length contraction) predicted by special relativity, for the first time.

2.3.2   Muon sources

On Earth, most naturally occurring muons are created by quasars and supernovas, which consist mostly of protons, many arriving from deep space at very high energy[*][4]

> About 10,000 muons reach every square meter of the earth's surface a minute; these charged particles form as by-products of cosmic rays colliding with molecules in the upper atmosphere. Traveling at relativistic speeds, muons can penetrate tens of meters into rocks and other matter before attenuating as a result of absorption or deflection by other atoms.[*][5]

When a cosmic ray proton impacts atomic nuclei in the upper atmosphere, pions are created. These decay within a relatively short distance (meters) into muons (their preferred decay product), and muon neutrinos. The muons from these high energy cosmic rays generally continue in about the same direction as the original proton, at a velocity near the speed of light. Although their lifetime *without* relativistic effects would allow a half-survival distance of only about 456 m (2,197 µs×ln(2) × 0,9997×c) at most (as seen from Earth) the time dilation effect of special relativity (from the viewpoint of the Earth) allows cosmic ray secondary muons to survive the flight to the Earth's surface, since in the Earth frame, the muons have a longer half life due to their velocity. From the viewpoint (inertial frame) of the muon, on the other hand, it is the length contraction effect of special relativity which allows this penetration, since in the muon frame, its lifetime is unaffected, but the length contraction causes distances through the atmosphere and Earth to be far shorter than these distances in the Earth rest-frame. Both effects are equally valid ways of explaining the fast muon's unusual survival over distances.

Since muons are unusually penetrative of ordinary matter, like neutrinos, they are also detectable deep underground (700 meters at the Soudan 2 detector) and underwater, where they form a major part of the natural background ionizing radiation. Like cosmic rays, as noted, this secondary muon radiation is also directional.

The same nuclear reaction described above (i.e. hadron-hadron impacts to produce pion beams, which then quickly decay to muon beams over short distances) is used by particle physicists to produce muon beams, such as the beam used for the muon $g - 2$ experiment.[*][6]

2.3.3   Muon decay

See also: Michel parameters

Muons are unstable elementary particles and are heavier

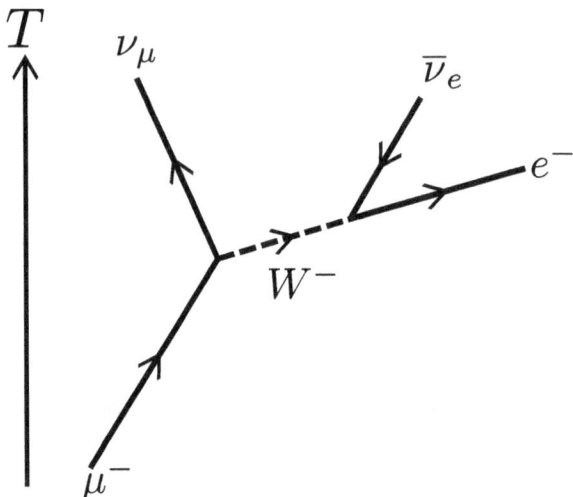

The most common decay of the muon

than electrons and neutrinos but lighter than all other matter particles. They decay via the weak interaction. Because lepton numbers must be conserved, one of the product neutrinos of muon decay must be a muon-type neutrino and the other an electron-type antineutrino (antimuon decay produces the corresponding antiparticles, as detailed below). Because charge must be conserved, one of the products of muon decay is always an electron of the same charge as the muon (a positron if it is a positive muon). Thus all muons decay to at least an electron, and two neutrinos. Sometimes, besides these necessary products, additional other particles that have no net charge and spin of zero (e.g., a pair of photons, or an electron-positron pair), are produced.

The dominant muon decay mode (sometimes called the Michel decay after Louis Michel) is the simplest possible: the muon decays to an electron, an electron antineutrino, and a muon neutrino. Antimuons, in mirror fashion, most often decay to the corresponding antiparticles: a positron, an electron neutrino, and a muon antineutrino. In formulaic terms, these two decays are:

$$\mu- \rightarrow e- + \nu$$
$$e + \nu$$
$$\mu$$
$$\mu+ \rightarrow e+ + \nu$$
$$e + \nu$$
$$\mu$$

The mean lifetime, $\tau = 1/\Gamma$, of the (positive) muon is (2.1969811±0.0000022) μs.[*][1] The equality of the muon and antimuon lifetimes has been established to better than one part in 10^4.

The muon decay width which follows from Fermi's golden rule follows Sargent's law of fifth-power dependence on m_μ,

$$\Gamma = \frac{G_F^2 m_\mu^5}{192\pi^3} I\left(\frac{m_e^2}{m_\mu^2}\right),$$

where $I(x) = 1 - 8x - 12x^2 \ln x + 8x^3 - x^4$, G_F is the Fermi coupling constant and $x = 2E_e/m_\mu c^2$ is the fraction of the maximum energy transmitted to the electron.

The decay distributions of the electron in muon decays have been parameterised using the so-called Michel parameters. The values of these four parameters are predicted unambiguously in the Standard Model of particle physics, thus muon decays represent a good test of the space-time structure of the weak interaction. No deviation from the Standard Model predictions has yet been found.

For the decay of the muon, the expected decay distribution for the Standard Model values of Michel parameters is

$$\frac{d^2\Gamma}{dx\,d\cos\theta} \sim x^2[(3 - 2x) + P_\mu \cos\theta(1 - 2x)]$$

where θ is the angle between the muon's polarization vector $\mathbf{P}_\mu$ and the decay-electron momentum vector, and $P_\mu = |\mathbf{P}_\mu|$ is the fraction of muons that are forward-polarized. Integrating this expression over electron energy gives the angular distribution of the daughter electrons:

$$\frac{d\Gamma}{d\cos\theta} \sim 1 - \frac{1}{3}P_\mu \cos\theta.$$

The electron energy distribution integrated over the polar angle (valid for $x < 1$) is

$$\frac{d\Gamma}{dx} \sim (3x^2 - 2x^3).$$

Due to the muons decaying by the weak interaction, parity conservation is violated. Replacing the $\cos\theta$ term in the expected decay values of the Michel Parameters with a $\cos\omega t$ term, where ω is the Larmor frequency from Larmor precession of the muon in a uniform magnetic field, given by:

$$\omega = \frac{egB}{2m}$$

where m is mass of the muon, e is charge, g is the muon g-factor and B is applied field.

A change in the electron distribution computed using the standard, unprecessional, Michel Parameters can be seen

displaying a periodicity of π radians. This can be shown to physically correspond to a phase change of π, introduced in the electron distribution as the angular momentum is changed by the action of the charge conjugation operator, which is conserved by the weak interaction.

The observation of Parity violation in muon decay can be compared to the concept of violation of parity in weak interactions in general as an extension of The Wu Experiment, as well as the change of angular momentum introduced by a phase change of π corresponding to the charge-parity operator being invariant in this interaction. This fact is true for all lepton interactions in The Standard Model.

Certain neutrino-less decay modes are kinematically allowed but forbidden in the Standard Model. Examples forbidden by lepton flavour conservation are:

μ− → e− + γ and

μ− → e− + e+ + e− .

Observation of such decay modes would constitute clear evidence for theories beyond the Standard Model. Upper limits for the branching fractions of such decay modes were measured in many experiments starting more than 50 years ago. The current upper limit for the μ+ → e+ + γ branching fraction was measured 2013 in the MEG experiment and is $5.7 \times 10^*{-13}.^*$[7]

2.3.4 Muonic atoms

The muon was the first elementary particle discovered that does not appear in ordinary atoms. Negative muons can, however, form muonic atoms (also called mu-mesic atoms), by replacing an electron in ordinary atoms. Muonic hydrogen atoms are much smaller than typical hydrogen atoms because the much larger mass of the muon gives it a much more localized ground-state wavefunction than is observed for the electron. In multi-electron atoms, when only one of the electrons is replaced by a muon, the size of the atom continues to be determined by the other electrons, and the atomic size is nearly unchanged. However, in such cases the orbital of the muon continues to be smaller and far closer to the nucleus than the atomic orbitals of the electrons.

Muonic helium is created by substituting a muon for one of the electrons in helium-4. The muon orbits much closer to the nucleus, so muonic helium can therefore be regarded like an isotope of helium whose nucleus consists of two neutrons, two protons and a muon, with a single electron outside. Colloquially, it could be called "helium 4.1", since the mass of the muon is roughly 0.1 amu. Chemically, muonic helium, possessing an unpaired valence electron, can bond with other atoms, and behaves more like a hydrogen atom than an inert helium atom.*[8]*[9]*[10]

A positive muon, when stopped in ordinary matter, can also bind an electron and form an exotic atom known as muonium (Mu) atom, in which the muon acts as the nucleus. The positive muon, in this context, can be considered a pseudo-isotope of hydrogen with one ninth of the mass of the proton. Because the reduced mass of muonium, and hence its Bohr radius, is very close to that of hydrogen, this short-lived "atom" behaves chemically —to a first approximation —like hydrogen, deuterium and tritium.

2.3.5 Use in measurement of the proton charge radius

The recent culmination of a twelve year experiment at investigating the proton's charge radius involved the use of muonic hydrogen. This form of hydrogen is composed of a muon orbiting a proton.[11] The Lamb shift in muonic hydrogen was measured by driving the muon from its 2s state up to an excited 2p state using a laser. The frequency of the photon required to induce this transition was revealed to be 50 terahertz which, according to present theories of quantum electrodynamics, yields a value of 0.84184 ± 0.00067 femtometres for the charge radius of the proton.[12]

2.3.6 Anomalous magnetic dipole moment

The anomalous magnetic dipole moment is the difference between the experimentally observed value of the magnetic dipole moment and the theoretical value predicted by the Dirac equation. The measurement and prediction of this value is very important in the precision tests of QED (quantum electrodynamics). The E821 experiment[13] at Brookhaven National Laboratory (BNL) studied the precession of muon and anti-muon in a constant external magnetic field as they circulated in a confining storage ring. E821 reported the following average value[14] in 2006:

$$a = \frac{g-2}{2} = 0.00116592080(54)(33)$$

where the first errors are statistical and the second systematic.

The prediction for the value of the muon anomalous magnetic moment includes three parts:

$$\alpha_\mu{}^*\mathrm{SM} = \alpha_\mu{}^*\mathrm{QED} + \alpha_\mu{}^*\mathrm{EW} + \alpha_\mu{}^*\mathrm{had.}$$

The difference between the g-factors of the muon and the electron is due to their difference in mass. Because of the muon's larger mass, contributions to the theoretical calculation of its anomalous magnetic dipole moment from Standard Model weak interactions and from contributions involving hadrons are important at the current level of precision, whereas these effects are not important for the electron. The muon's anomalous magnetic dipole moment is also sensitive to contributions from new physics beyond the Standard Model, such as supersymmetry. For this reason, the muon's anomalous magnetic moment is normally used as a probe for new physics beyond the Standard Model rather than as a test of QED.[15] A new experiment at Fermilab using the E821 magnet will improve the precision of this measurement.[16]

2.3.7 Muon radiography and tomography

Main article: Muon tomography

Since muons are much more deeply penetrating than X-rays or gamma rays, muon imaging can be used with much thicker material or, with cosmic ray sources, larger objects. An important advantage of muon non-ionizing radiation is that it is safe for humans, plants, and animals. One example is commercial muon tomography used to image entire cargo containers to detect shielded nuclear material, as well as explosives or other contraband.[17]

The technique of muon transmission radiography based on cosmic ray sources was first used in the 1950s to measure the depth of the overburden of a tunnel in Australia[18] and in the 1960s to search for possible hidden chambers in the Pyramid of Chephren in Giza.[19]

In 2003, the scientists at Los Alamos National Laboratory developed a new imaging technique: **muon scattering tomography**. With muon scattering tomography, both incoming and outgoing trajectories for each particle are reconstructed, such as with sealed aluminum drift tubes.[20] Since the development of this technique, several companies have started to use it.

In August 2014, Decision Sciences International Corporation announced it had been awarded a contract by Toshiba for use of its muon tracking detectors in reclaiming the Fukushima nuclear complex.[21] The Fukushima Daiichi Tracker (FDT) was proposed to make a few months of muon measurements to show the distribution of the reactor cores.

In December 2014, Tepco reported that they would be using two different muon imaging techniques at Fukushima, "Muon Scanning Method" on Unit 1 (the most badly damaged, where the fuel may have left the reactor vessel) and "Muon Scattering Method" on Unit 2.[22]

The International Research Institute for Nuclear Decommissioning IRID in Japan and the High Energy Accelerator Research Organization KEK call the method they developed for Unit 1 the **muon permeation method**; 1,200 optical fibers for wavelength conversion light up when muons come into contact with them.*[23] After a month of data collection, it is hoped to reveal the location and amount of fuel debris still inside the reactor. The measurements began in February 2015.*[24]

2.3.8 See also

- Muonic atoms
- Muon spin spectroscopy
- Muon-catalyzed fusion
- Muon Tomography
- Mu2e, an experiment to detect neutrinoless conversion of muons to electrons
- List of particles

2.3.9 References

[1] J. Beringer et al. (Particle Data Group) (2012). "PDGLive Particle Summary 'Leptons (e, mu, tau, ... neutrinos ...)'" (PDF). Particle Data Group. Retrieved 2013-01-12.

[2] New Evidence for the Existence of a Particle Intermediate Between the Proton and Electron", Phys. Rev. 52, 1003 (1937).

[3] Yukaya Hideka, On the Interaction of Elementary Particles 1, Proceedings of the Physico-Mathematical Society of Japan (3) 17, 48, pp 139–148 (1935). (Read 17 November 1934)

[4] S. Carroll (2004). *Spacetime and Geometry: An Introduction to General Relativity*. Addison Wesley. p. 204

[5] Mark Wolverton (September 2007). "Muons for Peace: New Way to Spot Hidden Nukes Gets Ready to Debut". *Scientific American* **297** (3): 26–28. doi:10.1038/scientificamerican0907-26.

[6] "Physicists Announce Latest Muon g-2 Measurement" (Press release). Brookhaven National Laboratory. 30 July 2002. Retrieved 2009-11-14.

[7] J. Adam (MEG Collaboration) et al. (2013). "New Constraint on the Existence of the mu+ -> e+ gamma Decay". *Physical Review Letters* **110** (20): 201801. arXiv:1303.0754. Bibcode:2013PhRvL.110t1801A. doi:10.1103/PhysRevLett.110.201801.

[8] Fleming, D. G.; Arseneau, D. J.; Sukhorukov, O.; Brewer, J. H.; Mielke, S. L.; Schatz, G. C.; Garrett, B. C.; Peterson, K. A.; Truhlar, D. G. (28 Jan 2011). "Kinetic Isotope Effects for the Reactions of Muonic Helium and Muonium with H2". *Science* **331** (6016): 448–450. Bibcode:2011Sci...331..448F. doi:10.1126/science.1199421. PMID 21273484.

[9] Moncada, F.; Cruz, D.; Reyes, A. "Muonic alchemy: Transmuting elements with the inclusion of negative muons". *Chemical Physics Letters* **539**: 209–213. Bibcode:2012CPL...539..209M. doi:10.1016/j.cplett.2012.04.062.

[10] Moncada, F.; Cruz, D.; Reyes, A (10 May 2013). "Electronic properties of atoms and molecules containing one and two negative muons". *Chemical Physics Letters* **570**: 16–21. Bibcode:2013CPL...570...16M. doi:10.1016/j.cplett.2013.03.004.

[11] TRIUMF Muonic Hydrogen collaboration. "A brief description of Muonic Hydrogen research". Retrieved 2010-11-7

[12] Pohl, Randolf et al. "*The Size of the Proton*" *Nature* 466, 213–216 (8 July 2010)

[13] "The Muon g-2 Experiment Home Page". G-2.bnl.gov. 2004-01-08. Retrieved 2012-01-06.

[14] "(from the July 2007 review by Particle Data Group)" (PDF). Retrieved 2012-01-06.

[15] Hagiwara, K; Martin, A; Nomura, D; Teubner, T (2007). "Improved predictions for g–2g–2 of the muon and αQED(MZ2)". *Physics Letters B* **649** (2–3): 173. arXiv:hep-ph/0611102. Bibcode:2007PhLB..649..173H. doi:10.1016/j.physletb.2007.04.012.

[16] "Revolutionary muon experiment to begin with 3,200-mile move of 50-foot-wide particle storage ring". May 8, 2013. Retrieved Mar 16, 2015.

[17] "Decision Sciences Corp".

[18] George, E.P. (July 1, 1955). "Cosmic rays measure overburden of tunnel". *Commonwealth Engineer*: 455.

[19] Alvarez, L.W. (1970). "Search for hidden chambers in the pyramids using cosmic rays". *Science* **167**: 832. Bibcode:1970Sci...167..832A. doi:10.1126/science.167.3919.832.

[20] Konstantin N. Borozdin, Gary E. Hogan, Christopher Morris, William C. Priedhorsky, Alexander Saunders, Larry J. Schultz & Margaret E. Teasdale. "Radiographic imaging with cosmic-ray muons". Nature.

[21] http://www.decisionsciencescorp.com/ds-awarded-toshiba-contract-fukushima-daiichi-nuclear-project/

[22] Tepco to start "scanning" inside of Reactor 1 in early February by using muon Fukushima Diary

[23] "Muon measuring instrument production for "muon permeation method" and its review by international experts". IRID.or.jp.

[24] Muon Scans Begin At Fukushima Daiichi - SimplyInfo

- S.H. Neddermeyer, C.D. Anderson; Anderson (1937). "Note on the Nature of Cosmic-Ray Particles". *Physical Review* **51** (10): 884–886. Bibcode:1937PhRv...51..884N. doi:10.1103/PhysRev.51.884.

- J.C. Street, E.C. Stevenson; Stevenson (1937). "New Evidence for the Existence of a Particle of Mass Intermediate Between the Proton and Electron". *Physical Review* **52** (9): 1003–1004. Bibcode:1937PhRv...52.1003S. doi:10.1103/PhysRev.52.1003.

- G. Feinberg, S. Weinberg; Weinberg (1961). "Law of Conservation of Muons". *Physical Review Letters* **6** (7): 381–383. Bibcode:1961PhRvL...6..381F. doi:10.1103/PhysRevLett.6.381.

- Serway & Faughn (1995). *College Physics* (4th ed.). Saunders. p. 841.

- M. Knecht (2003). "The Anomalous Magnetic Moments of the Electron and the Muon". In B. Duplantier, V. Rivasseau. *Poincaré Seminar 2002: Vacuum Energy – Renormalization*. Progress in Mathematical Physics **30**. Birkhäuser Verlag. p. 265. ISBN 3-7643-0579-7.

- E. Derman (2004). *My Life As A Quant*. Wiley. pp. 58–62.

2.3.10   External links

- Muon anomalous magnetic moment and supersymmetry

- g-2 (muon anomalous magnetic moment) experiment

- muLan (Measurement of the Positive Muon Lifetime) experiment

- The Review of Particle Physics

- The TRIUMF Weak Interaction Symmetry Test

- The MEG Experiment (Search for the decay Muon → Positron + Gamma)

- King, Philip. "Making Muons". *Backstage Science*. Brady Haran.

2.4   Tauon

Not to be confused with the τ^*+ of the $\tau{-}\theta$ puzzle, which is now identified as a kaon.

The **tau** (τ), also called the **tau lepton**, **tau particle** or **tauon**, is an elementary particle similar to the electron, with negative electric charge and a spin of $\frac{1}{2}$. Together with the electron, the muon, and the three neutrinos, it is a lepton. Like all elementary particles with half-integral spin, the tau has a corresponding antiparticle of opposite charge but equal mass and spin, which in the tau's case is the **anti-tau** (also called the *positive tau*). Tau particles are denoted by $\tau{-}$ and the antitau by $\tau{+}$.

Tau leptons have a lifetime of $2.9{\times}10^*{-}13$ s and a mass of 1776.82 MeV/c^2 (compared to 105.7 MeV/c^2 for muons and 0.511 MeV/c^2 for electrons). Since their interactions are very similar to those of the electron, a tau can be thought of as a much heavier version of the electron. Because of their greater mass, tau particles do not emit as much bremsstrahlung radiation as electrons; consequently they are potentially highly penetrating, much more so than electrons. However, because of their short lifetime, the range of the tau is mainly set by their decay length, which is too small for bremsstrahlung to be noticeable: their penetrating power appears only at ultra high energy (above PeV energies).[*][4]

As with the case of the other charged leptons, the tau has an associated tau neutrino, denoted by ν
τ.

2.4.1   History

The tau was detected in a series of experiments between 1974 and 1977 by Martin Lewis Perl with his colleagues at the SLAC-LBL group.[*][2] Their equipment consisted of SLAC's then-new e+–e− colliding ring, called SPEAR, and the LBL magnetic detector. They could detect and distinguish between leptons, hadrons and photons. They did not detect the tau directly, but rather discovered anomalous events:

"*We have discovered 64 events of the form*

e+ + e− → e± + μ∓ + at least two undetected particles

for which we have no conventional explanation."

The need for at least two undetected particles was shown by the inability to conserve energy and momentum with only one. However, no other muons, electrons, photons, or hadrons were detected. It was proposed that this event was the production and subsequent decay of a new particle pair:

$$e+ + e- \rightarrow \tau+ + \tau- \rightarrow e\pm + \mu\mp + 4\nu$$

This was difficult to verify, because the energy to produce the $\tau+\tau-$ pair is similar to the threshold for D meson production. Work done at DESY-Hamburg, and with the Direct Electron Counter (DELCO) at SPEAR, subsequently established the mass and spin of the tau.

The symbol τ was derived from the Greek τρίτον (triton, meaning "third" in English), since it was the third charged lepton discovered.[5]

Martin Perl shared the 1995 Nobel Prize in Physics with Frederick Reines. The latter was awarded his share of the prize for experimental discovery of the neutrino.

2.4.2 Tau decay

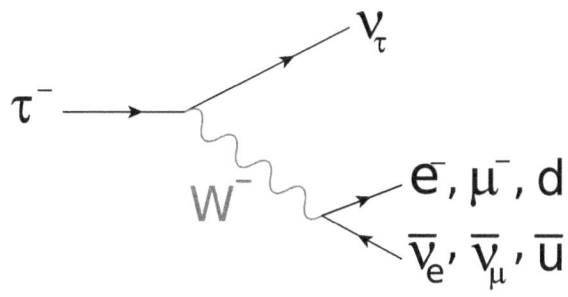

Feynman diagram of the common decays of the tau by emission of a W boson.

The tau is the only lepton that can decay into hadrons – the other leptons do not have the necessary mass. Like the other decay modes of the tau, the hadronic decay is through the weak interaction.[6]

The branching ratio of the dominant hadronic tau decays are:[3]

- 25.52% for decay into a charged pion, a neutral pion, and a tau neutrino;

- 10.83% for decay into a charged pion and a tau neutrino;

- 9.30% for decay into a charged pion, two neutral pions, and a tau neutrino;

- 8.99% for decay into three charged pions (of which two have the same electrical charge) and a tau neutrino;

- 2.70% for decay into three charged pions (of which two have the same electrical charge), a neutral pion, and a tau neutrino;

- 1.05% for decay into three neutral pions, a charged pion, and a tau neutrino.

In total, the tau lepton will decay hadronically approximately 64.79% of the time.

Since the tauonic lepton number is conserved in weak decays, a tau neutrino is always created when a tau decays.[6]

The branching ratio of the common purely leptonic tau decays are:[3]

- 17.82% for decay into a tau neutrino, electron and electron antineutrino;

- 17.39% for decay into a tau neutrino, muon and muon antineutrino.

The similarity of values of the two branching ratios is a consequence of lepton universality.

2.4.3 Exotic atoms

The tau lepton is predicted to form exotic atoms like other charged subatomic particles. One of such, called **tauonium** by the analogy to muonium, consists in antitauon and an electron: $\tau+e-$.[7]

Another one is an onium atom $\tau+\tau-$ called *true tauonium* and is difficult to detect due to tau's extremely short lifetime at low (non-relativistic) energies needed to form this atom. Its detection is important for quantum electrodynamics.[7]

2.4.4 See also

- Koide formula

2.4.5 References

[1] L. B. Okun (1980). *Leptons and Quarks*. V.I. Kisin (trans.). North-Holland Publishing. p. 103. ISBN 978-0444869241.

[2] Perl, M. L.; Abrams, G.; Boyarski, A.; Breidenbach, M.; Briggs, D.; Bulos, F.; Chinowsky, W.; Dakin, J. et al. (1975). "Evidence for Anomalous Lepton Production in e+e– Annihilation". *Physical Review Letters* **35** (22): 1489. Bibcode:1975PhRvL..35.1489P. doi:10.1103/PhysRevLett.35.1489.

[3] J. Beringer *et al.* (Particle Data Group) (2012). "Review of Particle Physics". *Journal of Physics G*

86 (1): 581–651. Bibcode:2012PhRvD..86a0001B. doi:10.1103/PhysRevD.86.010001. |chapter= ignored (help)

[4] D. Fargion, P.G. De Sanctis Lucentini, M. De Santis, M. Grossi (2004). "Tau Air Showers from Earth". *The Astrophysical Journal* **613** (2): 1285. arXiv:hep-ph/0305128. Bibcode:2004ApJ...613.1285F. doi:10.1086/423124.

[5] M.L. Perl (1977). "Evidence for, and properties of, the new charged heavy lepton" (PDF). In T. Thanh Van (ed.). *Proceedings of the XII Rencontre de Moriond*. SLAC-PUB-1923.

[6] Riazuddin (2009). "Non-standard interactions" (PDF). *NCP 5th Particle Physics Sypnoisis* (Islamabad,: Riazuddin, Head of High-Energy Theory Group at National Center for Physics) **1** (1): 1–25.

[7] Brodsky, Stanley J.; Lebed, Richard F. (2009). "Production of the Smallest QED Atom: True Muonium ($\mu^{*}+\mu^{*}-$)". *Physical Review Letters* **102** (21): 213401. arXiv:0904.2225. Bibcode:2009PhRvL.102u3401B. doi:10.1103/PhysRevLett.102.213401.

2.4.6   External links

- Nobel Prize in Physics 1995

- Perl's logbook showing tau discovery

- A Tale of Three Papers gives the covers of the three original papers announcing the discovery.

Chapter 3

Neutrinos

3.1 Neutrino

For other uses, see Neutrino (disambiguation).

A **neutrino** (/nuːˈtriːnoʊ/ or /njuːˈtriːnoʊ/, in Italian [nɛuˈtrino]) is an electrically neutral elementary particle*[4] with half-integer spin. The neutrino (meaning "little neutral one" in Italian) is denoted by the Greek letter ν (*nu*). All evidence suggests that neutrinos have mass but that their masses are tiny, even compared to other subatomic particles. They are the only identified candidate for dark matter, specifically hot dark matter.*[5]

Neutrinos are leptons, along with the charged electrons, muons, and taus, and come in three flavors: electron neutrinos (ν

e), muon neutrinos (ν

μ), and tau neutrinos (ν

τ). Each flavor is also associated with an antiparticle, called an "antineutrino", which also has no electric charge and half-integer spin. Neutrinos are produced in a way that conserves lepton number; i.e., for every electron neutrino produced, a positron (anti-electron) is produced, and for every electron antineutrino produced, an electron is produced as well.

Neutrinos do not carry any electric charge, which means that they are not affected by the electromagnetic force that acts on charged particles, and are leptons, so they are not affected by the strong force that acts on particles inside atomic nuclei. Neutrinos are therefore affected only by the weak subatomic force and by gravity. The weak force is a very short-range interaction, and gravity is extremely weak on the subatomic scale. Thus, neutrinos typically pass through normal matter unimpeded and undetected.

Neutrinos can be created in several ways, including in certain types of radioactive decay, in nuclear reactions such as those that take place in the Sun, in nuclear reactors, when cosmic rays hit atoms and in supernovas. The majority of neutrinos in the vicinity of the earth are from nuclear reactions in the Sun. In fact, about 65 billion (6.5×10^{10}) solar neutrinos per second pass through every square centimeter perpendicular to the direction of the Sun in the region of the Earth.*[6]

Neutrinos are now understood to oscillate between different flavors in flight. That is, an electron neutrino produced in a beta decay reaction may arrive in a detector as a muon or tau neutrino. This oscillation requires that the different neutrino flavors have different masses, although these masses have been shown to be tiny. From cosmological measurements, we know that the sum of the three neutrino masses must be less than one millionth that of the electron.*[7]

3.1.1 History

Pauli's proposal

The neutrino*[nb 1] was postulated first by Wolfgang Pauli in 1930 to explain how beta decay could conserve energy, momentum, and angular momentum (spin). In contrast to Niels Bohr, who proposed a statistical version of the conservation laws to explain the event, Pauli hypothesized an undetected particle that he called a "neutron" in keeping with convention employed for naming both the proton and the electron, which in 1930 were known to be respective products for alpha and beta decay. He considered that the new particle was emitted from the nucleus together with the electron or beta particle in the process of beta decay.*[8]*[nb 2]

James Chadwick discovered a much more massive nuclear particle in 1932 and also named it a neutron, leaving two kinds of particles with the same name. Pauli earlier had used the term "neutron" for both the particle that conserved energy in beta decay, and a presumed neutral particle in the nucleus.*[nb 3] The word "neutrino" entered the international vocabulary through Enrico Fermi, who used it during a conference in Paris in July 1932 and at the Solvay Conference in October 1933, where also Pauli employed it. The name (the Italian equivalent of "little neutral one") was jokingly coined by Edoardo Amaldi during a conversation with Fermi at the Institute of physics of via Panisperna in Rome, in order to distinguish this light neutral particle from Chadwick's neutron. [*][9]

In Fermi's theory of beta decay, Chadwick's large neutral particle could decay to a proton, electron, and the smaller neutral particle (flavored as an electron antineutrino):

$$n0 \rightarrow p+ + e- + \nu e$$

Fermi's paper, written in 1934, unified Pauli's neutrino with Paul Dirac's positron and Werner Heisenberg's neutron–proton model and gave a solid theoretical basis for future experimental work. However, the journal Nature rejected Fermi's paper, saying that the theory was "too remote from reality". He submitted the paper to an Italian journal, which accepted it, but the general lack of interest in his theory at that early date caused him to switch to experimental physics.[*][10][*][11]

Nevertheless, even in 1934 there were hints that Bohr's idea that the energy conservation laws were not followed, was incorrect. At the Solvay conference of 1934, the first measurements of the energy spectra of beta decay were reported, and these spectra were found to impose a strict limit on the energy of electrons from each type of beta decay. Such a limit was not expected if the conservation of energy was not upheld, in which case any amount of energy would be expected to be statistically available in at least a few decays. The natural explanation of the beta decay spectrum as first measured in 1934 was that only a limited (and conserved) amount of energy was available, and a new particle was sometimes taking a varying fraction of this limited energy, leaving the rest for the beta particle. Pauli made use of the occasion to publicly emphasize that the still-undetected "neutrino" must be an actual particle.

Direct detection

In 1942 Wang Ganchang first proposed the use of beta capture to experimentally detect neutrinos.[*][12] In the 20 July 1956 issue of *Science*, Clyde Cowan, Frederick Reines, F. B. Harrison, H. W. Kruse, and A. D. McGuire published confirmation that they had detected the neutrino,[*][13][*][14] a result that was rewarded almost forty years later with the 1995 Nobel Prize.[*][15]

In this experiment, now known as the Cowan–Reines neutrino experiment, antineutrinos created in a nuclear reactor by beta decay reacted with protons to produce neutrons and positrons:

$$\nu\ e + p+ \rightarrow n0 + e+$$

Clyde Cowan conducting the neutrino experiment c. 1956

The positron quickly finds an electron, and they annihilate each other. The two resulting gamma rays (γ) are detectable. The neutron can be detected by its capture on an appropriate nucleus, releasing a gamma ray. The coincidence of both events – positron annihilation and neutron capture – gives a unique signature of an antineutrino interaction.

Neutrino flavor

The antineutrino discovered by Cowan and Reines is the antiparticle of the electron neutrino. In 1962, Leon M. Lederman, Melvin Schwartz and Jack Steinberger showed that more than one type of neutrino exists by first detecting interactions of the muon neutrino (already hypothesised with the name *neutretto*),[*][16] which earned them the 1988 Nobel Prize in Physics. When the third type of lepton, the tau, was discovered in 1975 at the Stanford Linear Accelerator Center, it too was expected to have an associated neutrino (the tau neutrino). First evidence for this third neutrino type came from the observation of missing energy and momentum in tau decays analogous to the beta decay leading to the discovery of the electron neutrino. The first detection of tau neutrino interactions was announced in summer of 2000 by the DONUT collaboration at Fermilab; its existence had already been inferred by both theoretical consistency and experimental data from the Large Electron–Positron Collider.

Solar neutrino problem

Main article: Solar neutrino problem

Starting in the late 1960s, several experiments found that the number of electron neutrinos arriving from the Sun was between one third and one half the number predicted by the Standard Solar Model. This discrepancy, which became known as the solar neutrino problem, remained unresolved for some thirty years. It was resolved by discovery of neutrino oscillation and mass. (The Standard Model of particle physics had assumed that neutrinos are massless and cannot change flavor. However, if neutrinos had mass, they could change flavor, or *oscillate* between flavors).

Oscillation

A practical method for investigating neutrino oscillations was first suggested by Bruno Pontecorvo in 1957 using an analogy with kaon oscillations; over the subsequent 10 years he developed the mathematical formalism and the modern formulation of vacuum oscillations. In 1985 Stanislav Mikheyev and Alexei Smirnov (expanding on 1978 work by Lincoln Wolfenstein) noted that flavor oscillations can be modified when neutrinos propagate through matter. This so-called Mikheyev–Smirnov–Wolfenstein effect (MSW effect) is important to understand because many neutrinos emitted by fusion in the Sun pass through the dense matter in the solar core (where essentially all solar fusion takes place) on their way to detectors on Earth.

Starting in 1998, experiments began to show that solar and atmospheric neutrinos change flavors (see Super-Kamiokande and Sudbury Neutrino Observatory). This resolved the solar neutrino problem: the electron neutrinos produced in the Sun had partly changed into other flavors which the experiments could not detect.

Although individual experiments, such as the set of solar neutrino experiments, are consistent with non-oscillatory mechanisms of neutrino flavor conversion, taken altogether, neutrino experiments imply the existence of neutrino oscillations. Especially relevant in this context are the reactor experiment KamLAND and the accelerator experiments such as MINOS. The KamLAND experiment has indeed identified oscillations as the neutrino flavor conversion mechanism involved in the solar electron neutrinos. Similarly MINOS confirms the oscillation of atmospheric neutrinos and gives a better determination of the mass squared splitting.[17]

Supernova neutrinos

See also: Supernova Early Warning System

Raymond Davis, Jr. and Masatoshi Koshiba were jointly awarded the 2002 Nobel Prize in Physics; Davis for his pioneer work on cosmic neutrinos and Koshiba for the first real time observation of supernova neutrinos. The detection of solar neutrinos, and of neutrinos of the SN 1987A supernova in 1987 marked the beginning of neutrino astronomy. In an average supernova, approximately 10^{57} (an Octodecillion) neutrinos are released.

3.1.2 Properties and reactions

The neutrino has half-integer spin ($\hbar/2$) and is therefore a fermion. Neutrinos interact primarily through the weak force. The discovery of neutrino flavor oscillations implies that neutrinos have mass. The existence of a neutrino mass strongly suggests the existence of a tiny neutrino magnetic moment[18] of the order of $10^{-19} \mu_B$, allowing the possibility that neutrinos may interact electromagnetically as well. An experiment done by C. S. Wu at Columbia University showed that neutrinos always have left-handed chirality.[19] It is very hard to uniquely identify neutrino interactions among the natural background of radioactivity. For this reason, in early experiments a special reaction channel was chosen to facilitate the identification: the interaction of an antineutrino with one of the hydrogen nuclei in the water molecules. A hydrogen nucleus is a single proton, so simultaneous nuclear interactions, which would occur within a heavier nucleus, don't need to be considered for the detection experiment. Within a cubic metre of water placed right outside a nuclear reactor, only relatively few such interactions can be recorded, but the setup is now used for measuring the reactor's plutonium production rate.

Mikheyev–Smirnov–Wolfenstein effect

Main article: Mikheyev–Smirnov–Wolfenstein effect

Neutrinos traveling through matter, in general, undergo a process analogous to light traveling through a transparent material. This process is not directly observable because it does not produce ionizing radiation, but gives rise to the MSW effect. Only a small fraction of the neutrino's energy is transferred to the material.

Nuclear reactions

Neutrinos can interact with a nucleus, changing it to another nucleus. This process is used in radiochemical neutrino detectors. In this case, the energy levels and spin states within the target nucleus have to be taken into account to estimate the probability for an interaction. In general the interaction probability increases with the number of neutrons and protons within a nucleus.

Induced fission

Very much like neutrons do in nuclear reactors, neutrinos can induce fission reactions within heavy nuclei.[20] So far, this reaction has not been measured in a laboratory, but is predicted to happen within stars and supernovae. The process affects the abundance of isotopes seen in the universe.[21] Neutrino fission of deuterium nuclei has been observed in the Sudbury Neutrino Observatory, which uses a heavy water detector.

Types

There are three known types (*flavors*) of neutrinos: electron neutrino ν

e, muon neutrino ν

μ and tau neutrino ν

τ, named after their partner leptons in the Standard Model (see table at right). The current best measurement of the number of neutrino types comes from observing the decay of the Z boson. This particle can decay into any light neutrino and its antineutrino, and the more types of light neutrinos[nb 4] available, the shorter the lifetime of the Z boson. Measurements of the Z lifetime have shown that the number of light neutrino types is 3.[18] The correspondence between the six quarks in the Standard Model and the six leptons, among them the three neutrinos, suggests to physicists' intuition that there should be exactly three types of neutrino. However, actual proof that there are only three kinds of neutrinos remains an elusive goal of particle physics.

The possibility of *sterile* neutrinos—relatively light neutrinos which do not participate in the weak interaction but which could be created through flavor oscillation (see below)—is unaffected by these Z-boson-based measurements, and the existence of such particles is in fact hinted by experimental data from the LSND experiment. However, the currently running MiniBooNE experiment suggested, until recently, that sterile neutrinos are not required to explain the experimental data,[22] although the latest research into this area is on-going and anomalies in the MiniBooNE data may allow for exotic neutrino types, including sterile neutrinos.[23] A recent re-analysis of reference electron spectra

data from the Institut Laue-Langevin[24] has also hinted at a fourth, sterile neutrino.[25]

Recently analyzed data from the Wilkinson Microwave Anisotropy Probe of the cosmic background radiation is compatible with either three or four types of neutrinos. It is hoped that the addition of two more years of data from the probe will resolve this uncertainty.[26]

Antineutrinos

Antineutrinos, the antiparticles of neutrinos, are neutral particles produced in nuclear beta decay. These are emitted during beta particle emissions, in which a neutron decays into a proton, electron, and antineutrino. They have a spin of ½, and are part of the lepton family of particles. All antineutrinos observed thus far possess right-handed helicity (i.e. only one of the two possible spin states has ever been seen), while neutrinos are left-handed. Antineutrinos, like neutrinos, interact with other matter only through the gravitational and weak forces, making them very difficult to detect experimentally. Neutrino oscillation experiments indicate that antineutrinos have mass, but beta decay experiments constrain that mass to be very small. A neutrino–antineutrino interaction has been suggested in attempts to form a composite photon with the neutrino theory of light.

Because antineutrinos and neutrinos are neutral particles, it is possible that they are actually the same particle. Particles that have this property are known as Majorana particles. Majorana neutrinos have the property that the neutrino and antineutrino could be distinguished only by chirality; what experiments observe as a difference between the neutrino and antineutrino could simply be due to one particle with two possible chiralities. If neutrinos are indeed Majorana particles, neutrinoless double beta decay, as well as a range of other lepton number violating phenomena, would be allowed. Several experiments have been and are being conducted to search for this process.

Researchers around the world have begun to investigate the possibility of using antineutrinos for reactor monitoring in the context of preventing the proliferation of nuclear weapons.[27][28][29]

Antineutrinos were first detected as a result of their interaction with protons in a large tank of water. This was installed next to a nuclear reactor as a controllable source of the antineutrinos. (See: Cowan–Reines neutrino experiment)

Only antineutrinos, not neutrinos, take part in the Glashow resonance.

Flavor oscillations

Main article: Neutrino oscillation

Neutrinos are most often created or detected with a well defined flavor (electron, muon, tau). However, in a phenomenon known as neutrino flavor oscillation, neutrinos are able to oscillate among the three available flavors while they propagate through space. Specifically, this occurs because the neutrino flavor eigenstates are not the same as the neutrino mass eigenstates (simply called 1, 2, 3). This allows for a neutrino that was produced as an electron neutrino at a given location to have a calculable probability to be detected as either a muon or tau neutrino after it has traveled to another location. This quantum mechanical effect was first hinted by the discrepancy between the number of electron neutrinos detected from the Sun's core failing to match the expected numbers, dubbed as the "solar neutrino problem". In the Standard Model the existence of flavor oscillations implies nonzero differences between the neutrino masses, because the amount of mixing between neutrino flavors at a given time depends on the differences between their squared masses. There are other possibilities in which neutrino can oscillate even if they are massless. If Lorentz invariance is not an exact symmetry, neutrinos can experience Lorentz-violating oscillations.[*][30]

It is possible that the neutrino and antineutrino are in fact the same particle, a hypothesis first proposed by the Italian physicist Ettore Majorana. The neutrino could transform into an antineutrino (and vice versa) by flipping the orientation of its spin state.[*][31]

This change in spin would require the neutrino and antineutrino to have nonzero mass, and therefore travel slower than light, because such a spin flip, caused only by a change in point of view, can take place only if inertial frames of reference exist that move faster than the particle: such a particle has a spin of one orientation when seen from a frame which moves slower than the particle, but the opposite spin when observed from a frame that moves faster than the particle.

On July 19, 2013 the results from the T2K experiment presented at the European Physical Society Conference on High Energy Physics in Stockholm, Sweden, confirmed neutrino oscillation theory.[*][32][*][33]

Speed

Main article: Measurements of neutrino speed

Before neutrinos were found to oscillate, they were generally assumed to be massless, propagating at the speed of light. According to the theory of special relativity, the question of neutrino velocity is closely related to their mass. If neutrinos are massless, they must travel at the speed of light. However, if they have mass, they cannot reach the speed of light.

Also some Lorentz-violating variants of quantum gravity might allow faster-than-light neutrinos. A comprehensive framework for Lorentz violations is the Standard-Model Extension (SME).

In the early 1980s, first measurements of neutrino speed were done using pulsed pion beams (produced by pulsed proton beams hitting a target). The pions decayed producing neutrinos, and the neutrino interactions observed within a time window in a detector at a distance were consistent with the speed of light. This measurement was repeated in 2007 using the MINOS detectors, which found the speed of 3 GeV neutrinos to be, at the 99% confidence level, in the range between $0.999976\,c$ and $1.000126\,c$. The central value of $1.000051c$ is higher than the speed of light but is also consistent with a velocity of exactly c or even slightly less. This measurement set an upper bound on the mass of the muon neutrino of 50 MeV at 99% confidence.[*][34][*][35] After the detectors for the project were upgraded in 2012, MINOS refined their initial result and found agreement with the speed of light, with the difference in the arrival time of neutrinos and light of -0.0006% ($\pm 0.0012\%$).[*][36]

A similar observation was made, on a much larger scale, with supernova 1987A (SN 1987A). 10-MeV antineutrinos from the supernova were detected within a time window that was consistent with the speed of light for the neutrinos. Currently, the question of whether or not neutrinos have mass cannot be decided; their speed is (as yet) indistinguishable from the speed of light.

In September 2011, the OPERA collaboration released calculations showing velocities of 17-GeV and 28-GeV neutrinos exceeding the speed of light in their experiments (see Faster-than-light neutrino anomaly). In November 2011, OPERA repeated its experiment with changes so that the speed could be determined individually for each detected neutrino. The results showed the same faster-than-light speed. However, in February 2012 reports came out that the results may have been caused by a loose fiber optic cable attached to one of the atomic clocks which measured the departure and arrival times of the neutrinos. An independent recreation of the experiment in the same laboratory by ICARUS found no discernible difference between the speed of a neutrino and the speed of light.[*][37] In June 2012, CERN announced that new measurements conducted by four Gran Sasso experiments (OPERA, ICARUS, Borexino and LVD) found agreement between the speed of light and the speed of neutrinos, finally refuting the initial OPERA result.[*][38]

Mass

The Standard Model of particle physics assumed that neutrinos are massless. However the experimentally established phenomenon of neutrino oscillation, which mixes neutrino flavour states with neutrino mass states (analogously to CKM mixing), requires neutrinos to have nonzero masses.[*][22] Massive neutrinos were originally conceived by Bruno Pontecorvo in the 1950s. Enhancing the basic framework to accommodate their mass is straightforward by adding a right-handed Lagrangian. This can be done in two ways. If, like other fundamental Standard Model particles, mass is generated by the Dirac mechanism, then the framework would require an SU(2) singlet. This particle would have no other Standard Model interactions (apart from the Yukawa interactions with the neutral component of the Higgs doublet), so is called a sterile neutrino. Or, mass can be generated by the Majorana mechanism, which would require the neutrino and antineutrino to be the same particle.

The strongest upper limit on the masses of neutrinos comes from cosmology: the Big Bang model predicts that there is a fixed ratio between the number of neutrinos and the number of photons in the cosmic microwave background. If the total energy of all three types of neutrinos exceeded an average of 50 eV per neutrino, there would be so much mass in the universe that it would collapse.[*][39] This limit can be circumvented by assuming that the neutrino is unstable; however, there are limits within the Standard Model that make this difficult. A much more stringent constraint comes from a careful analysis of cosmological data, such as the cosmic microwave background radiation, galaxy surveys, and the Lyman-alpha forest. These indicate that the summed masses of the three neutrinos must be less than 0.3 eV.[*][40]

In 1998, research results at the Super-Kamiokande neutrino detector determined that neutrinos can oscillate from one flavor to another, which requires that they must have a nonzero mass.[*][41] While this shows that neutrinos have mass, the absolute neutrino mass scale is still not known. This is because neutrino oscillations are sensitive only to the difference in the squares of the masses.[*][42] The best estimate of the difference in the squares of the masses of mass eigenstates 1 and 2 was published by KamLAND in 2005: $\Delta m2$
$21 = 0.000079$ eV^2.[*][43] In 2006, the MINOS experiment measured oscillations from an intense muon neutrino beam, determining the difference in the squares of the masses between neutrino mass eigenstates 2 and 3. The initial results indicate $|\Delta m2$
$32| = 0.0027$ eV^2, consistent with previous results from Super-Kamiokande.[*][44] Since $|\Delta m2$
$32|$ is the difference of two squared masses, at least one of

them has to have a value which is at least the square root of this value. Thus, there exists at least one neutrino mass eigenstate with a mass of at least 0.04 eV.[*][45]

In 2009, lensing data of a galaxy cluster were analyzed to predict a neutrino mass of about 1.5 eV.[*][46] This surprisingly high value requires that the three neutrino masses be nearly equal, with neutrino oscillations of order meV. The masses lie below the Mainz-Troitsk upper bound of 2.2 eV for the electron antineutrino.[*][47] The latter will be tested in 2015 in the KATRIN experiment, that searches for a mass between 0.2 eV and 2 eV.

A number of efforts are under way to directly determine the absolute neutrino mass scale in laboratory experiments. The methods applied involve nuclear beta decay (KATRIN and MARE).

On 31 May 2010, OPERA researchers observed the first tau neutrino candidate event in a muon neutrino beam, the first time this transformation in neutrinos had been observed, providing further evidence that they have mass.[*][48]

In July 2010 the 3-D MegaZ DR7 galaxy survey reported that they had measured a limit of the combined mass of the three neutrino varieties to be less than 0.28 eV.[*][49] A tighter upper bound yet for this sum of masses, 0.23 eV, was reported in March 2013 by the Planck collaboration,[*][50] whereas a February 2014 result estimates the sum as 0.320 ± 0.081 eV based on discrepancies between the cosmological consequences implied by Planck's detailed measurements of the Cosmic Microwave Background and predictions arising from observing other phenomena, combined with the assumption that neutrinos are responsible for the observed weaker gravitational lensing than would be expected from massless neutrinos.[*][51]

If the neutrino is a Majorana particle, the mass may be calculated by finding the half life of neutrinoless double-beta decay of certain nuclei. As of 2015, the lowest upper limit on the Majorana mass of the neutrino has been set by KamLAND-Zen: 0.12–0.25 eV.[*][52]

Size

Standard Model neutrinos are fundamental point-like particles. An effective size can be defined using their electroweak cross section (apparent size in electroweak interaction). The average electroweak characteristic size is r^2 = $n \times 10^{*}$–33 cm^2 ($n \times 1$ nanobarn), where $n = 3.2$ for electron neutrino, $n = 1.7$ for muon neutrino and $n = 1.0$ for tau neutrino; it depends on no other properties than mass.[*][53] However, this is best understood as being relevant only to probability of scattering. Since the neutrino does not interact electromagnetically, and is defined quantum mechanically by a wavefunction, it does not have a size in the same sense as everyday objects.*[54] Furthermore, processes that produce neutrinos impart such high energies to them that they travel at almost the speed of light. Nevertheless, neutrinos are fermions, and thus obey the Pauli exclusion principle, i.e. that increasing their density forces them into progressively higher momentum states.

Chirality

Experimental results show that (nearly) all produced and observed neutrinos have left-handed helicities (spins antiparallel to momenta), and all antineutrinos have right-handed helicities, within the margin of error. In the massless limit, it means that only one of two possible chiralities is observed for either particle. These are the only chiralities included in the Standard Model of particle interactions.

It is possible that their counterparts (right-handed neutrinos and left-handed antineutrinos) simply do not exist. If they do, their properties are substantially different from observable neutrinos and antineutrinos. It is theorized that they are either very heavy (on the order of GUT scale—see *Seesaw mechanism*), do not participate in weak interaction (so-called sterile neutrinos), or both.

The existence of nonzero neutrino masses somewhat complicates the situation. Neutrinos are produced in weak interactions as chirality eigenstates. However, chirality of a massive particle is not a constant of motion; helicity is, but the chirality operator does not share eigenstates with the helicity operator. Free neutrinos propagate as mixtures of left- and right-handed helicity states, with mixing amplitudes on the order of m_ν/E. This does not significantly affect the experiments, because neutrinos involved are nearly always ultrarelativistic, and thus mixing amplitudes are vanishingly small. For example, most solar neutrinos have energies on the order of 100 keV–1 MeV, so the fraction of neutrinos with "wrong" helicity among them cannot exceed 10^*–$10.^*$[55]*[56]

3.1.3 Sources

Artificial

Reactor neutrinos Nuclear reactors are the major source of human-generated neutrinos. Antineutrinos are made in the beta-decay of neutron-rich daughter fragments in the fission process. Generally, the four main isotopes contributing to the antineutrino flux are 235U, 238U, 239Pu and 241Pu (i.e. via the antineutrinos emitted during beta-minus decay of their respective fission fragments). The average nuclear fission releases about 200 MeV of energy, of which roughly 4.5% (or about 9 MeV)*[57] is radiated away as antineutrinos. For a typical nuclear reactor with a thermal power of

4000 MW, meaning that the core produces this much heat, and an electrical power generation of 1300 MW, the total power production from fissioning atoms is actually 4185 MW, of which 185 MW is radiated away as antineutrino radiation and never appears in the engineering. This is to say, 185 MW of fission energy is *lost* from this reactor and does not appear as heat available to run turbines, since antineutrinos penetrate all building materials practically without interaction.*[nb 5]

The antineutrino energy spectrum depends on the degree to which the fuel is burned (plutonium-239 fission antineutrinos on average have slightly more energy than those from uranium-235 fission), but in general, the *detectable* antineutrinos from fission have a peak energy between about 3.5 and 4 MeV, with a maximum energy of about 10 MeV.*[58] There is no established experimental method to measure the flux of low-energy antineutrinos. Only antineutrinos with an energy above threshold of 1.8 MeV can be uniquely identified (see *neutrino detection* below). An estimated 3% of all antineutrinos from a nuclear reactor carry an energy above this threshold. Thus, an average nuclear power plant may generate over 10^{20} antineutrinos per second above this threshold, but also a much larger number (97%/3% = ~30 times this number) below the energy threshold, which cannot be seen with present detector technology.

Accelerator neutrinos Some particle accelerators have been used to make neutrino beams. The technique is to collide protons with a fixed target, producing charged pions or kaons. These unstable particles are then magnetically focused into a long tunnel where they decay while in flight. Because of the relativistic boost of the decaying particle, the neutrinos are produced as a beam rather than isotropically. Efforts to construct an accelerator facility where neutrinos are produced through muon decays are ongoing.*[59] Such a setup is generally known as a neutrino factory.

Nuclear bombs Nuclear bombs also produce very large quantities of neutrinos. Fred Reines and Clyde Cowan considered the detection of neutrinos from a bomb prior to their search for reactor neutrinos; a fission reactor was recommended as a better alternative by Los Alamos physics division leader J.M.B. Kellogg.*[60] Fission bombs produce antineutrinos (from the fission process), and fusion bombs produce both neutrinos (from the fusion process) and antineutrinos (from the initiating fission explosion).

Geologic

Main article: Geoneutrino

Neutrinos are part of the natural background radiation. In particular, the decay chains of 238U and 232Th isotopes, as well as 40K, include beta decays which emit antineutrinos. These so-called geoneutrinos can provide valuable information on the Earth's interior. A first indication for geoneutrinos was found by the KamLAND experiment in 2005. KamLAND's main background in the geoneutrino measurement are the antineutrinos coming from reactors. Several future experiments aim at improving the geoneutrino measurement and these will necessarily have to be far away from reactors.

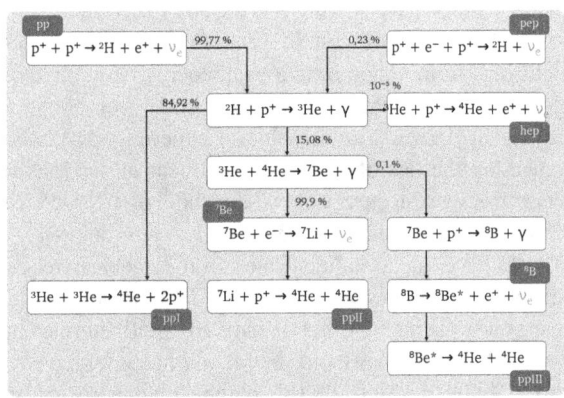

Solar neutrinos (proton–proton chain) in the Standard Solar Model

Atmospheric

Atmospheric neutrinos result from the interaction of cosmic rays with atomic nuclei in the Earth's atmosphere, creating showers of particles, many of which are unstable and produce neutrinos when they decay. A collaboration of particle physicists from Tata Institute of Fundamental Research (India), Osaka City University (Japan) and Durham University (UK) recorded the first cosmic ray neutrino interaction in an underground laboratory in Kolar Gold Fields in India in 1965.

Solar

Solar neutrinos originate from the nuclear fusion powering the Sun and other stars. The details of the operation of the Sun are explained by the Standard Solar Model. In short: when four protons fuse to become one helium nucleus, two of them have to convert into neutrons, and each such conversion releases one electron neutrino.

The Sun sends enormous numbers of neutrinos in all directions. Each second, about 65 billion (6.5×10^{10}) solar neutrinos pass through every square centimeter on the part of the Earth that faces the Sun.[*][6] Since neutrinos are insignificantly absorbed by the mass of the Earth, the surface area on the side of the Earth opposite the Sun receives about the same number of neutrinos as the side facing the Sun.

Supernovae

SN 1987A

In 1966 Colgate and White[*][61] calculated that neutrinos carry away most of the gravitational energy released by the collapse of massive stars, events now categorized as Type Ib and Ic and Type II supernovae. When such stars collapse, matter densities at the core becomes so high (10^{17} kg/m^3) that the degeneracy of electrons is not enough to prevent protons and electrons from combining to form a neutron and an electron neutrino. A second and more important neutrino source is the thermal energy (100 billion kelvins) of the newly formed neutron core, which is dissipated via the formation of neutrino–antineutrino pairs of all flavors.[*][62]

Colgate and White's theory of supernova neutrino production was confirmed in 1987, when neutrinos from supernova 1987A were detected. The water-based detectors Kamiokande II and IMB detected 11 and 8 antineutrinos of thermal origin,[*][62] respectively, while the scintillator-based Baksan detector found 5 neutrinos (lepton number = 1) of either thermal or electron-capture origin, in a burst lasting less than 13 seconds. The neutrino signal from the supernova arrived at earth several hours before the arrival of the first electromagnetic radiation, as expected from the evident fact that the latter emerges along with the shock wave. The exceptionally feeble interaction with normal matter allowed the neutrinos to pass through the churning mass of the exploding star, while the electromagnetic photons were slowed.

Because neutrinos interact so little with matter, it is thought that a supernova's neutrino emissions carry information about the innermost regions of the explosion. Much of the *visible* light comes from the decay of radioactive elements produced by the supernova shock wave, and even light from the explosion itself is scattered by dense and turbulent gases, and thus delayed. The neutrino burst is expected to reach Earth before any electromagnetic waves, including visible light, gamma rays or radio waves. The exact time delay depends on the velocity of the shock wave and on the thickness of the outer layer of the star. For a Type II supernova, astronomers expect the neutrino flood to be released seconds after the stellar core collapse, while the first electromagnetic signal may emerge hours later, after the explosion shock wave has had time to reach the surface of the star. The SNEWS project uses a network of neutrino detectors to monitor the sky for candidate supernova events; the neutrino signal will provide a useful advance warning of a star exploding in the Milky Way.

Although neutrinos pass through the outer gases of a supernova without scattering, they provide information about the deeper supernova core with evidence that here, even neutrinos scatter to a significant extent. In a supernova core the densities are those of a neutron star (which is expected to be formed in this type of supernova),[*][63] becoming large enough to influence the duration of the neutrino signal by delaying some neutrinos. The length of the neutrino signal from SN 1987A, some 13 seconds, was far longer than it would take in theory for neutrinos to pass directly through the neutrino-generating core of a supernova, expected to be only 32 kilometers in diameter SN 1987A. The number of neutrinos counted was also consistent with a total neutrino energy of 2.2 x 10^{46} joules, which was estimated to be nearly all of the total energy of the supernova.[*][64]

Supernova remnants

The energy of supernova neutrinos ranges from a few to several tens of MeV. However, the sites where cosmic rays are accelerated are expected to produce neutrinos that are at least one million times more energetic, produced from turbulent gaseous environments left over by supernova explosions: the supernova remnants. The origin of the cosmic rays was attributed to supernovas by Walter Baade and Fritz Zwicky; this hypothesis was refined by Vitaly L. Ginzburg and Sergei I. Syrovatsky who attributed the origin to supernova remnants, and supported their claim by the crucial remark, that the cosmic ray losses of the Milky Way is compensated, if the efficiency of acceleration in supernova remnants is about 10 percent. Ginzburg and Syrovatskii's hypothesis is supported by the specific mechanism of "shock wave acceleration" happening in supernova remnants, which is consistent with the original theoretical picture drawn by Enrico Fermi, and is receiving support from observational data. The very-high-energy neutrinos are still to be seen, but this branch of neutrino astronomy is just in its infancy. The main existing or forthcoming experiments that aim at observing very-high-energy neutrinos from our galaxy are Baikal, AMANDA, IceCube, ANTARES, NEMO and Nestor. Related information is provided by very-high-energy gamma ray observatories, such as VERITAS, HESS and MAGIC. Indeed, the collisions of cosmic rays are supposed to produce charged pions, whose decay give the neutrinos, and also neutral pions, whose decay give gamma rays: the environment of a supernova remnant is transparent to both types of radiation.

Still-higher-energy neutrinos, resulting from the interactions of extragalactic cosmic rays, could be observed with the Pierre Auger Observatory or with the dedicated experiment named ANITA.

Big Bang

Main article: Cosmic neutrino background

It is thought that, just like the cosmic microwave background radiation left over from the Big Bang, there is a background of low-energy neutrinos in our Universe. In the 1980s it was proposed that these may be the explanation for the dark matter thought to exist in the universe. Neutrinos have one important advantage over most other dark matter candidates: it is known that they exist. However, this idea also has serious problems.

From particle experiments, it is known that neutrinos are very light. This means that they easily move at speeds close to the speed of light. For this reason, dark matter made from neutrinos is termed "hot dark matter". The problem is that being fast moving, the neutrinos would tend to have spread out evenly in the universe before cosmological expansion made them cold enough to congregate in clumps. This would cause the part of dark matter made of neutrinos to be smeared out and unable to cause the large galactic structures that we see.

Further, these same galaxies and groups of galaxies appear to be surrounded by dark matter that is not fast enough to escape from those galaxies. Presumably this matter provided the gravitational nucleus for formation. This implies that neutrinos cannot make up a significant part of the total amount of dark matter.

From cosmological arguments, relic background neutrinos are estimated to have density of 56 of each type per cubic centimeter and temperature 1.9 K ($1.7×10^{*}−4$ eV) if they are massless, much colder if their mass exceeds 0.001 eV. Although their density is quite high, they have not yet been observed in the laboratory, as their energy is below thresholds of most detection methods, and due to extremely low neutrino interaction cross-sections at sub-eV energies. In contrast, boron-8 solar neutrinos—which are emitted with a higher energy—have been detected definitively despite having a space density that is lower than that of relic neutrinos by some 6 orders of magnitude.

3.1.4   Detection

Main article: Neutrino detector

Neutrinos cannot be detected directly, because they do not ionize the materials they are passing through (they do not carry electric charge and other proposed effects, like the MSW effect, do not produce traceable radiation). A unique reaction to identify antineutrinos, sometimes referred to as inverse beta decay, as applied by Reines and Cowan (see below), requires a very large detector in order to detect a significant number of neutrinos. All detection methods require the neutrinos to carry a minimum threshold energy. So far, there is no detection method for low-energy neutrinos, in the sense that potential neutrino interactions (for example by the MSW effect) cannot be uniquely distinguished from other causes. Neutrino detectors are often built underground in order to isolate the detector from cosmic rays and other background radiation.

Antineutrinos were first detected in the 1950s near a nuclear reactor. Reines and Cowan used two targets containing a solution of cadmium chloride in water. Two scintillation detectors were placed next to the cadmium targets. Antineutrinos with an energy above the threshold of 1.8 MeV caused charged current interactions with the protons in the water, producing positrons and neutrons. This is very much like $\beta+$ decay, where energy is used to convert a proton into a neutron, a positron ($e+$) and an electron neutrino (ν e) is emitted:

From known $\beta+$ decay:

$$\text{Energy} + p \rightarrow n + e+ + \nu e$$

In the Cowan and Reines experiment, instead of an outgoing neutrino, you have an incoming antineutrino (ν e) from a nuclear reactor:

$$\text{Energy} (>1.8 \text{ MeV}) + p + \nu e \rightarrow n + e+$$

The resulting positron annihilation with electrons in the detector material created photons with an energy of about 0.5 MeV. Pairs of photons in coincidence could be detected by the two scintillation detectors above and below the target.

The neutrons were captured by cadmium nuclei resulting in gamma rays of about 8 MeV that were detected a few microseconds after the photons from a positron annihilation event.

Since then, various detection methods have been used. Super Kamiokande is a large volume of water surrounded by photomultiplier tubes that watch for the Cherenkov radiation emitted when an incoming neutrino creates an electron or muon in the water. The Sudbury Neutrino Observatory is similar, but uses heavy water as the detecting medium, which uses the same effects, but also allows the additional reaction any-flavor neutrino photo-dissociation of deuterium, resulting in a free neutron which is then detected from gamma radiation after chlorine-capture. Other detectors have consisted of large volumes of chlorine or gallium which are periodically checked for excesses of argon or germanium, respectively, which are created by electron-neutrinos interacting with the original substance. MINOS uses a solid plastic scintillator coupled to photomultiplier tubes, while Borexino uses a liquid pseudocumene scintillator also watched by photomultiplier tubes and the proposed NOvA detector will use liquid scintillator watched by avalanche photodiodes. The IceCube Neutrino Observatory uses 1 km^3 of the Antarctic ice sheet near the south pole with photomultiplier tubes distributed throughout the volume.

3.1.5   Motivation for scientific interest

Neutrinos' low mass and neutral charge mean they interact exceedingly weakly with other particles and fields. This feature of weak interaction interests scientists because it means neutrinos can be used to probe environments that other radiation (such as light or radio waves) cannot penetrate.

Using neutrinos as a probe was first proposed in the mid-20th century as a way to detect conditions at the core of the Sun. The solar core cannot be imaged directly because electromagnetic radiation (such as light) is diffused by the great amount and density of matter surrounding the core. On the other hand, neutrinos pass through the Sun with few interactions. Whereas photons emitted from the solar core may require 40,000 years to diffuse to the outer layers of the Sun, neutrinos generated in stellar fusion reactions at the core cross this distance practically unimpeded at nearly the speed of light.[*][65][*][66]

Neutrinos are also useful for probing astrophysical sources beyond the Solar System because they are the only known particles that are not significantly attenuated by their travel through the interstellar medium. Optical photons can be obscured or diffused by dust, gas, and background radiation. High-energy cosmic rays, in the form of swift protons and atomic nuclei, are unable to travel more than about 100 megaparsecs due to the Greisen–Zatsepin–Kuzmin limit (GZK cutoff). Neutrinos, in contrast, can travel even greater distances barely attenuated.

The galactic core of the Milky Way is fully obscured by dense gas and numerous bright objects. Neutrinos produced in the galactic core might be measurable by Earth-based neutrino telescopes.

Another important use of the neutrino is in the observation of supernovae, the explosions that end the lives of highly massive stars. The core collapse phase of a supernova is an extremely dense and energetic event. It is so dense that no known particles are able to escape the advancing core front except for neutrinos. Consequently, supernovae are known to release approximately 99% of their radiant energy in a short (10-second) burst of neutrinos.*[67] These neutrinos are a very useful probe for core collapse studies.

The rest mass of the neutrino (see above) is an important test of cosmological and astrophysical theories (see *Dark matter*). The neutrino's significance in probing cosmological phenomena is as great as any other method, and is thus a major focus of study in astrophysical communities.*[68]

The study of neutrinos is important in particle physics because neutrinos typically have the lowest mass, and hence are examples of the lowest-energy particles theorized in extensions of the Standard Model of particle physics.

In November 2012 American scientists used a particle accelerator to send a coherent neutrino message through 780 feet of rock. This marks the first use of neutrinos for communication, and future research may permit binary neutrino messages to be sent immense distances through even the densest materials, such as the Earth's core.*[69]

3.1.6 See also

- List of neutrino experiments

3.1.7 Notes

[1] More specifically, the electron neutrino.

[2] Niels Bohr was notably opposed to this interpretation of beta decay and was ready to accept that energy, momentum and angular momentum were not conserved quantities.

[3] These events necessitated renaming Pauli's less massive, momentum-conserving particle.

[4] In this context, "light neutrino" means neutrinos with less than half the mass of the Z boson.

[5] Typically about one third of the heat which is deposited in a reactor core is available to be converted to electricity, and a 4000 MW reactor would produce only 2700 MW of actual

heat, with the rest being converted to its 1300 MW of electric power production.

3.1.8 References

[1] "Astronomers Accurately Measure the Mass of Neutrinos for the First Time" . *scitechdaily.com*. Image credit:NASA, ESA, and J. Lotz, M. Mountain, A. Koekemoer, and the HFF Team (STScI). February 10, 2014. Archived from the original on May 7, 2014. Retrieved May 7, 2014.

[2] Foley, James A. (February 10, 2014). "Mass of Neutrinos Accurately Calculated for First Time, Physicists Report" . *natureworldnews.com*. Image credit: . via Wikimedia Commons. Archived from the original on May 7, 2014. Retrieved May 7, 2014.

[3] Battye, Richard A.; Moss, Adam (2014). "Evidence for Massive Neutrinos from Cosmic Microwave Background and Lensing Observations" . *Physical Review Letters* **112** (5): 051303. arXiv:1308.5870v2. Bibcode:2014PhRvL.112e1303B. doi:10.1103/PhysRevLett.112.051303. PMID 24580586.

[4] "Neutrino" . *Glossary for the Research Perspectives of the Max Planck Society*. Max Planck Gesellschaft. Retrieved 2012-03-27.

[5] Dodelson, Scott; Widrow, Lawrence M. (1994). "Sterile neutrinos as dark matter" **72** (17).

[6] Bahcall, John N.; Serenelli, Aldo M.; Basu, Sarbani (2005). "New Solar Opacities, Abundances, Helioseismology, and Neutrino Fluxes" . *The Astrophysical Journal* **621** (1): L85–8. arXiv:astro-ph/0412440. Bibcode:2005ApJ...621L..85B. doi:10.1086/428929.

[7] Olive, K. A. "Sum of Neutrino Masses" (PDF). *Chinese Physics C*.

[8] Brown, Laurie M. (1978). "The idea of the neutrino" . *Physics Today* **31** (9): 23–8. Bibcode:1978PhT....31i..23B. doi:10.1063/1.2995181.

[9] E. Amaldi (1984). "From the discovery of the neutron to the discovery of nuclear fission" . *Phys. Rep.* **111** (1–4): 306.

[10] F. Close (2010). *Neutrino.* Oxford University Press. ISBN 978-0-19-957459-9.

[11] E. Fermi (1934). "Versuch einer Theorie der β-Strahlen. I" . *Zeitschrift für Physik A* **88** (3–4): 161. Bibcode:1934ZPhy...88..161F. doi:10.1007/BF01351864. Translated in F. L. Wilson (1968). "Fermi's Theory of Beta Decay" (PDF). *American Journal of Physics* **36** (12): 1150. Bibcode:1968AmJPh..36.1150W. doi:10.1119/1.1974382.

[12] K.-C. Wang (1942). "A Suggestion on the Detection of the Neutrino" . *Physical Review* **61** (1–2): 97. Bibcode:1942PhRv...61...97W. doi:10.1103/PhysRev.61.97.

[13] C. L. Cowan Jr.; F. Reines; F. B. Harrison; H. W. Kruse et al. (1956). "Detection of the Free Neutrino: a Confirmation". *Science* **124** (3212): 103–4. Bibcode:1956Sci...124..103C. doi:10.1126/science.124.3212.103. PMID 17796274.

[14] K. Winter (2000). *Neutrino physics*. Cambridge University Press. p. 38ff. ISBN 978-0-521-65003-8. This source reproduces the 1956 paper.

[15] "The Nobel Prize in Physics 1995". The Nobel Foundation. Retrieved 29 June 2010.

[16] I. V. Anicin (2005). "The Neutrino – Its Past, Present and Future". arXiv:physics/0503172.

[17] M. Maltoni; T. Schwetz; M. Tórtola; J. W. F. Valle (2004). "Status of global fits to neutrino oscillations". *New Journal of Physics* **6** (1): 122. arXiv:hep-ph/0405172. Bibcode:2004NJPh....6..122M. doi:10.1088/1367-2630/6/1/122.

[18] Particle Data Group; Eidelman, S.; Hayes, K. G.; Olive, K. A.; Aguilar-Benitez, M.; Amsler, C.; Asner, D.; Babu, K. S.; Barnett, R. M.; Beringer, J.; Burchat, P. R.; Carone, C. D.; Caso, S.; Conforto, G.; Dahl, O.; d'Ambrosio, G.; Doser, M.; Feng, J. L.; Gherghetta, T.; Gibbons, L.; Goodman, M.; Grab, C.; Groom, D. E.; Gurtu, A.; Hagiwara, K.; Hernández-Rey, J. J.; Hikasa, K.; Honscheid, K.; Jawahery, H. et al. (2004). "Review of Particle Physics". *Physics Letters B* **592**: 1–5. arXiv:astro-ph/0406663. Bibcode:2004PhLB..592....1P. doi:10.1016/j.physletb.2004.06.001.

[19] S.M. Caroll (25 March 2009). "Ada Lovelace Day: Chien-Shiung Wu". *Discover Magazine*. Retrieved 2011-09-23.

[20] Kolbe, E.; Langanke, K.; Fuller, G. M. (2004). "Neutrino-Induced Fission of Neutron-Rich Nuclei". *Physical Review Letters* **92** (11): 111101. arXiv:astro-ph/0308350. Bibcode:2004PhRvL..92k1101K. doi:10.1103/PhysRevLett.92.111101. PMID 15089120.

[21] Kelić, A.; Zinner, N.; Kolbe, E.; Langanke, K.; Schmidt, K.-H. (2005). "Cross sections and fragment distributions from neutrino-induced fission on r-process nuclei". *Physics Letters B* **616** (1–2): 48–58. arXiv:hep-ex/0312045. Bibcode:2005PhLB..616...48K. doi:10.1016/j.physletb.2005.04.074.

[22] Karagiorgi, G.; Aguilar-Arevalo, A.; Conrad, J. M.; Shaevitz, M. H.; Whisnant, K.; Sorel, M.; Barger, V. (2007). "LeptonicCPviolation studies at MiniBooNE in the (3+2) sterile neutrino oscillation hypothesis". *Physical Review D* **75**: 013011. arXiv:hep-ph/0609177. Bibcode:2007PhRvD..75a3011K. doi:10.1103/PhysRevD.75.013011.

[23] M. Alpert (2007). "Dimensional Shortcuts". *Scientific American*. Retrieved 2009-10-31.

[24] Mueller, Th. A.; Lhuillier, D.; Fallot, M.; Letourneau, A.; Cormon, S.; Fechner, M.; Giot, L.; Lasserre, T.; Martino, J.; Mention, G.; Porta, A.; Yermia, F. (2011). "Improved predictions of reactor antineutrino spectra". *Physical Review C* **83** (5): 054615. arXiv:1101.2663. Bibcode:2011PhRvC..83e4615M. doi:10.1103/PhysRevC.83.054615.

[25] Mention, G.; Fechner, M.; Lasserre, Th.; Mueller, Th. A.; Lhuillier, D.; Cribier, M.; Letourneau, A. (2011). "Reactor antineutrino anomaly". *Physical Review D* **83** (7): 073006. arXiv:1101.2755. Bibcode:2011PhRvD..83g3006M. doi:10.1103/PhysRevD.83.073006.

[26] R. Cowen (2 February 2010). "Ancient Dawn's Early Light Refines the Age of the Universe". *Science News*. Retrieved 2010-02-03.

[27] neutrinos.llnl.gov "LLNL/SNL Applied Antineutrino Physics Project. LLNL-WEB-204112". 2006.

[28] apc.univ-paris7.fr "Applied Antineutrino Physics 2007 workshop". 2007.

[29] "New Tool To Monitor Nuclear Reactors Developed". ScienceDaily. 13 March 2008. Retrieved 2008-03-16.

[30] Alan Kostelecký, V.; Mewes, Matthew (2004). "Lorentz andCPTviolation in neutrinos". *Physical Review D* **69**: 016005. arXiv:hep-ph/0309025. Bibcode:2004PhRvD..69a6005A. doi:10.1103/PhysRevD.69.016005.

[31] C. Giunti; C.W. Kim (2007). *Fundamentals of neutrino physics and astrophysics*. Oxford University Press. p. 255. ISBN 0-19-850871-9.

[32] "Neutrino shape-shift points to new physics" *Physics News*, 19 July 2013.

[33] "Neutrino 'flavour' flip confirmed" *BBC News*, 19 July 2013.

[34] Adamson, P.; Andreopoulos, C.; Arms, K. E.; Armstrong, R.; Auty, D. J.; Avvakumov, S.; Ayres, D. S.; Baller, B.; Barish, B.; Barnes, P. D.; Barr, G.; Barrett, W. L.; Beall, E.; Becker, B. R.; Belias, A.; Bergfeld, T.; Bernstein, R. H.; Bhattacharya, D.; Bishai, M.; Blake, A.; Bock, B.; Bock, G. J.; Boehm, J.; Boehnlein, D. J.; Bogert, D.; Border, P. M.; Bower, C.; Buckley-Geer, E.; Cabrera, A. et al. (2007). "Measurement of neutrino velocity with the MINOS detectors and NuMI neutrino beam". *Physical Review D* **76** (7): 072005. arXiv:0706.0437. Bibcode:2007PhRvD..76g2005A. doi:10.1103/PhysRevD.76.072005.

[35] D. Overbye (22 September 2011). "Tiny neutrinos may have broken cosmic speed limit". *New York Times*. That group found, although with less precision, that the neutrino speeds were consistent with the speed of light.

[36] Hesla, Leah (June 8, 2012). "MINOS reports new measurement of neutrino velocity". Fermilab today. Retrieved April 2, 2015.

[37] Antonello, M.; Aprili, P.; Baiboussinov, B.; Baldo Ceolin, M.; Benetti, P.; Calligarich, E.; Canci, N.; Centro, S.; Cesana, A.; Cieślik, K.; Cline, D.B.; Cocco, A.G.; Dabrowska, A.; Dequal, D.; Dermenev, A.; Dolfini, R.; Farnese, C.; Fava, A.; Ferrari, A.; Fiorillo, G.; Gibin, D.; Gigli Berzolari, A.; Gninenko, S.; Guglielmi, A.; Haranczyk, M.; Holeczek, J.; Ivashkin, A.; Kisiel, J.; Kochanek, I. et al. (2012). "Measurement of the neutrino velocity with the ICARUS detector at the CNGS beam". *Physics Letters B* **713**: 17–22. arXiv:1203.3433. Bibcode:2012PhLB..713...17I. doi:10.1016/j.physletb.2012.05.033.

[38] "Neutrinos sent from CERN to Gran Sasso respect the cosmic speed limit, experiments confirm" (Press release). CERN. June 8, 2012. Retrieved April 2, 2015.

[39] Hut, P.; Olive, K.A. (1979). "A cosmological upper limit on the mass of heavy neutrinos". *Physics Letters B* **87** (1–2): 144–6. Bibcode:1979PhLB...87..144H. doi:10.1016/0370-2693(79)90039-X.

[40] Goobar, Ariel; Hannestad, Steen; Mörtsell, Edvard; Tu, Huitzu (2006). "The neutrino mass bound from WMAP 3 year data, the baryon acoustic peak, the SNLS supernovae and the Lyman-α forest". *Journal of Cosmology and Astroparticle Physics* **2006** (6): 019. arXiv:astro-ph/0602155. Bibcode:2006JCAP...06..019G. doi:10.1088/1475-7516/2006/06/019.

[41] Fukuda, Y.; Hayakawa, T.; Ichihara, E.; Inoue, K.; Ishihara, K.; Ishino, H.; Itow, Y.; Kajita, T.; Kameda, J.; Kasuga, S.; Kobayashi, K.; Kobayashi, Y.; Koshio, Y.; Martens, K.; Miura, M.; Nakahata, M.; Nakayama, S.; Okada, A.; Oketa, M.; Okumura, K.; Ota, M.; Sakurai, N.; Shiozawa, M.; Suzuki, Y.; Takeuchi, Y.; Totsuka, Y.; Yamada, S.; Earl, M.; Habig, A. et al. (1998). "Measurements of the Solar Neutrino Flux from Super-Kamiokande's First 300 Days". *Physical Review Letters* **81** (6): 1158. arXiv:hep-ex/9805021. Bibcode:1998PhRvL..81.1158F. doi:10.1103/PhysRevLett.81.1158.

[42] Mohapatra, R N; Antusch, S; Babu, K S; Barenboim, G; Chen, M-C; De Gouvêa, A; De Holanda, P; Dutta, B; Grossman, Y; Joshipura, A; Kayser, B; Kersten, J; Keum, Y Y; King, S F; Langacker, P; Lindner, M; Loinaz, W; Masina, I; Mocioiu, I; Mohanty, S; Murayama, H; Pascoli, S; Petcov, S T; Pilaftsis, A; Ramond, P; Ratz, M; Rodejohann, W; Shrock, R; Takeuchi, T et al. (2007). "Theory of neutrinos: A white paper". *Reports on Progress in Physics* **70** (11): 1757. arXiv:hep-ph/0510213. Bibcode:2007RPPh...70.1757M. doi:10.1088/0034-4885/70/11/R02.

[43] Araki, T.; Eguchi, K.; Enomoto, S.; Furuno, K.; Ichimura, K.; Ikeda, H.; Inoue, K.; Ishihara, K.; Iwamoto, T.; Kawashima, T.; Kishimoto, Y.; Koga, M.; Koseki, Y.; Maeda, T.; Mitsui, T.; Motoki, M.; Nakajima, K.; Ogawa, H.; Owada, K.; Ricol, J.-S.; Shimizu, I.; Shirai, J.; Suekane, F.; Suzuki, A.; Tada, K.; Tajima, O.; Tamae, K.; Tsuda, Y.; Watanabe, H. et al. (2005). "Measurement of Neutrino Oscillation with KamLAND: Evidence of Spectral

[Distortion]". *Physical Review Letters* **94** (8): 081801. arXiv:hep-ex/0406035. Bibcode:2005PhRvL..94h1801A. doi:10.1103/PhysRevLett.94.081801. PMID 15783875.

[44] "MINOS experiment sheds light on mystery of neutrino disappearance" (Press release). Fermilab. 30 March 2006. Retrieved 2007-11-25.

[45] Amsler, C.; Doser, M.; Antonelli, M.; Asner, D.M.; Babu, K.S.; Baer, H.; Band, H.R.; Barnett, R.M.; Bergren, E.; Beringer, J.; Bernardi, G.; Bertl, W.; Bichsel, H.; Biebel, O.; Bloch, P.; Blucher, E.; Blusk, S.; Cahn, R.N.; Carena, M.; Caso, C.; Ceccucci, A.; Chakraborty, D.; Chen, M.-C.; Chivukula, R.S.; Cowan, G.; Dahl, O.; d'Ambrosio, G.; Damour, T.; De Gouvêa, A. et al. (2008). "Review of Particle Physics". *Physics Letters B* **667**: 1. Bibcode:2008PhLB..667....1P. doi:10.1016/j.physletb.2008.07.018.

[46] Nieuwenhuizen, Th. M. (2009). "Do non-relativistic neutrinos constitute the dark matter?". *EPL* **86** (5): 59001. arXiv:0812.4552. Bibcode:2009EL.....8659001N. doi:10.1209/0295-5075/86/59001.

[47] "The most sensitive analysis on the neutrino mass [...] is compatible with a neutrino mass of zero. Considering its uncertainties this value corresponds to an upper limit on the electron neutrino mass of $m < 2.2$ eV/c^2 (95% Confidence Level)" The Mainz Neutrino Mass Experiment

[48] Agafonova, N.; Aleksandrov, A.; Altinok, O.; Ambrosio, M.; Anokhina, A.; Aoki, S.; Ariga, A.; Ariga, T.; Autiero, D.; Badertscher, A.; Bagulya, A.; Bendhabi, A.; Bertolin, A.; Besnier, M.; Bick, D.; Boyarkin, V.; Bozza, C.; Brugière, T.; Brugnera, R.; Brunet, F.; Brunetti, G.; Buontempo, S.; Cazes, A.; Chaussard, L.; Chernyavsky, M.; Chiarella, V.; Chon-Sen, N.; Chukanov, A.; Ciesielski, R. et al. (2010). "Observation of a first $\nu\tau$ candidate event in the OPERA experiment in the CNGS beam". *Physics Letters B* **691** (3): 138–45. arXiv:1006.1623. Bibcode:2010PhLB..691..138A. doi:10.1016/j.physletb.2010.06.022.

[49] Thomas, Shaun A.; Abdalla, Filipe B.; Lahav, Ofer (2010). "Upper Bound of 0.28 eV on Neutrino Masses from the Largest Photometric Redshift Survey". *Physical Review Letters* **105** (3): 031301. arXiv:0911.5291. Bibcode:2010PhRvL.105c1301T. doi:10.1103/PhysRevLett.105.031301. PMID 20867754.

[50] Planck Collaboration, P. A. R.; Ade, P. A. R.; Aghanim, N.; Armitage-Caplan, C.; Arnaud, M.; Ashdown, M.; Atrio-Barandela, F.; Aumont, J.; Baccigalupi, C.; Banday, A. J.; Barreiro, R. B.; Bartlett, J. G.; Battaner, E.; Benabed, K.; Benoît, A.; Benoit-Lévy, A.; Bernard, J.-P.; Bersanelli, M.; Bielewicz, P.; Bobin, J.; Bock, J. J.; Bonaldi, A.; Bond, J. R.; Borrill, J.; Bouchet, F. R.; Bridges, M.; Bucher, M.; Burigana, C.; Butler, R. C. et al. (2013). "Planck 2013 results. XVI. Cosmological parameters". *Astronomy & Astrophysics* **1303**: 5076. arXiv:1303.5076. Bibcode:2013arXiv1303.5076P. doi:10.1051/0004-6361/201321591.

[51] Battye, Richard A.; Moss, Adam (2014). "Evidence for Massive Neutrinos from Cosmic Microwave Background and Lensing Observations". *Physical Review Letters* **112** (5): 051303. arXiv:1308.5870. Bibcode:2014PhRvL.112e1303B. doi:10.1103/PhysRevLett.112.051303. PMID 24580586.

[52] A. Gando et al. (KamLAND-Zen Collaboration) (Feb 7, 2013). "Limit on Neutrinoless ββ Decay of Xe136 from the First Phase of KamLAND-Zen and Comparison with the Positive Claim in Ge76". *Phys. Rev. Lett.* *110, 062502*. Bibcode:2013PhRvL.110f2502G. doi:10.1103/PhysRevLett.110.062502.

[53] Lucio, J. L.; Rosado, A.; Zepeda, A. (1985). "Characteristic size for the neutrino". *Physical Review D* **31** (5): 1091–1096. Bibcode:1985PhRvD..31.1091L. doi:10.1103/PhysRevD.31.1091. PMID 9955801.

[54] Choi, Charles Q. (2 June 2009). "Particles Larger Than Galaxies Fill the Universe?". *National Geographic News.*

[55] B. Kayser (2005). "Neutrino mass, mixing, and flavor change" (PDF). Particle Data Group. Retrieved 2007-11-25.

[56] S.M. Bilenky; C. Giunti (2001). "Lepton Numbers in the framework of Neutrino Mixing". *International Journal of Modern Physics A* **16** (24): 3931–3949. arXiv:hep-ph/0102320. Bibcode:2001IJMPA..16.3931B. doi:10.1142/S0217751X01004967.

[57] "Nuclear Fission and Fusion, and Nuclear Interactions". NLP National Physical Laboratory. 2008. Retrieved 2009-06-25.

[58] A. Bernstein; Wang, Y.; Gratta, G.; West, T. (2002). "Nuclear reactor safeguards and monitoring with antineutrino detectors". *Journal of Applied Physics* **91** (7): 4672. arXiv:nucl-ex/0108001. Bibcode:2002JAP....91.4672B. doi:10.1063/1.1452775.

[59] A. Bandyopadhyay et al. (ISS Physics Working Group) et al. (2007). "Physics at a future Neutrino Factory and superbeam facility". *Reports on Progress in Physics* **72** (10): 6201. arXiv:0710.4947. Bibcode:2009RPPh...72j6201B. doi:10.1088/0034-4885/72/10/106201.

[60] F. Reines; C. Cowan, Jr. (1997). "The Reines-Cowan Experiments: Detecting the Poltergeist" (PDF). *Los Alamos Science* **25**: 3.

[61] S. A. Colgate & R. H. White (1966). "The Hydrodynamic Behavior of Supernova Explosions". *The Astrophysical Journal* **143**: 626. Bibcode:1966ApJ...143..626C. doi:10.1086/148549.

[62] A.K. Mann (1997). *Shadow of a star: The neutrino story of Supernova 1987A*. W. H. Freeman. p. 122. ISBN 0-7167-3097-9.

[63] Products of the 1987A supernova

[64] Diameter of neutrino-generating core, and total neutrino power of SN 1987A

[65] J.N. Bahcall (1989). *Neutrino Astrophysics*. Cambridge University Press. ISBN 0-521-37975-X.

[66] D.R. David Jr. (2003). "Nobel Lecture: A half-century with solar neutrinos". *Reviews of Modern Physics* **75** (3): 10. Bibcode:2003RvMP...75..985D. doi:10.1103/RevModPhys.75.985.

[67] "Physics – Supernova Starting Gun: Neutrinos". Focus.aps.org. 2009-07-17. Retrieved 2012-04-05.

[68] G.B. Gelmini; A. Kusenko; T.J. Weiler (May 2010). "Through Neutrino Eyes". *Scientific American* **302** (5): 38–45. Bibcode:2010SciAm.302e..38G. doi:10.1038/scientificamerican0510-38.

[69] Stancil, D. D.; Adamson, P.; Alania, M.; Aliaga, L. et al. (2012). "Demonstration of Communication Using Neutrinos" (PDF). *Modern Physics Letters A* **27** (12): 1250077. arXiv:1203.2847. Bibcode:2012MPLA...2750077S. doi:10.1142/S0217732312500770. Lay summary – *Popular Science* (March 15, 2012).

3.1.9 Bibliography

- Adam, T.; *et al.* (OPERA collaboration) (2011). "Measurement of the neutrino velocity with the OPERA detector in the CNGS beam". arXiv:1109.4897 [hep-ex].

- Alberico, W. M.; Bilenky, S. M. (2004). "Neutrino Oscillations, Masses And Mixing". *Physics of Particles and Nuclei* **35**: 297–323. arXiv:hep-ph/0306239. Bibcode:2003hep.ph....6239A.

- Bahcall, J. N. (1989). *Neutrino Astrophysics*. Cambridge University Press. ISBN 0-521-35113-8.

- Bumfiel, G. (1 October 2001). "The Milky Way's Hidden Black Hole". *Scientific American*. Retrieved 2010-04-23.

- Close, F. (2010). *Neutrino*. Oxford University Press. ISBN 978-0-19-957459-9.

- Griffiths, D. J. (1987). *Introduction to Elementary Particles*. John Wiley & Sons. ISBN 0-471-60386-4.

- Perkins, D. H. (1999). *Introduction to High Energy Physics*. Cambridge University Press. ISBN 0-521-62196-8.

- Povh, B. (1995). *Particles and Nuclei: An Introduction to the Physical Concepts*. Springer-Verlag. ISBN 0-387-59439-6.

- Riazuddin (2005). "Neutrinos" (PDF). National Center for Physics.

- Schopper, H. F. (1966). *Weak interactions and nuclear beta decay*. North-Holland.

- Tammann, G. A.; Thielemann, F. K.; Trautmann, D. (2003). "Opening new windows in observing the Universe". Europhysics News. Retrieved 2006-06-08.

- Tipler, P.; Llewellyn, R. (2002). *Modern Physics* (4th ed.). W. H. Freeman. ISBN 0-7167-4345-0.

- Tomonaga, S.-I. (1997). *The Story of Spin*. University of Chicago Press.

- Zuber, K. (2003). *Neutrino Physics*. IOP Publishing. ISBN 978-0-7503-0750-5.

3.1.10 External links

- "What's a Neutrino?", Dave Casper (University of California, Irvine)

- Neutrino unbound: On-line review and e-archive on Neutrino Physics and Astrophysics

- Nova: The Ghost Particle: Documentary on US public television from WGBH

- Measuring the density of the earth's core with neutrinos

- Universe submerged in a sea of chilled neutrinos, *New Scientist*, 5 March 2008

- What's a neutrino?

- Search for neutrinoless double beta decay with enriched 76Ge in Gran Sasso 1990–2003

- Neutrino caught in the act of changing from muon-type to tau-type, CERN press release

- Cosmic Weight Gain: A Wispy Particle Bulks Up by George Johnson

- Neutrino 'ghost particle' sized up by astronomers BBC News 22 June 2010

- Pillar of physics challenged

- Merrifield, Michael; Copeland, Ed; Bowley, Roger (2010). "Neutrinos". *Sixty Symbols*. Brady Haran for the University of Nottingham.

- The Neutrino with Dr. Clyde L. Cowan (Lecture on Project Poltergeist by Clyde Cowan)

- Nuclear Reactor as the Source of Antineutrinos

3.2 Electron neutrino

The **electron neutrino** (ν_e) is a subatomic lepton elementary particle which has no net electric charge. Together with the electron it forms the first generation of leptons, hence its name *electron neutrino*. It was first hypothesized by Wolfgang Pauli in 1930, to account for missing momentum and missing energy in beta decay, and was discovered in 1956 by a team led by Clyde Cowan and Frederick Reines (see Cowan–Reines neutrino experiment).[1]

3.2.1 Proposal

In the early 1900s, theories predicted that the electrons resulting from beta decay should have been emitted at a specific energy. However, in 1914, James Chadwick showed that electrons were instead emitted in a continuous spectrum.[1]

$$n0 \rightarrow p+ + e-$$

The early understanding of beta decay

In 1930, Wolfgang Pauli theorized that an undetected particle was carrying away the observed difference between the energy, momentum, and angular momentum of the initial and final particles.[nb 1][2]

$$n0 \rightarrow p+ + e- + \nu 0e$$

Pauli's version of beta decay

Pauli's letter

On 4 December 1930, Pauli wrote a letter to the Physical Institute of the Federal Institute of Technology, Zürich, in which he proposed the electron neutrino as a potential solution to solve the problem of the continuous beta decay spectrum. An excerpt of the letter reads:[1]

> Dear radioactive ladies and gentlemen,
> As the bearer of these lines [...] will explain more exactly, considering the 'false' statistics of N-14 and Li-6 nuclei, as well as the continuous β-spectrum, I have hit upon a desperate remedy to save the "exchange theorem" of statistics and the energy theorem. Namely [there is] the possibility that there could exist in the nuclei electrically neutral particles that I wish to call neutrons,[nb 2] which have spin 1/2 and obey the

exclusion principle, and additionally differ from light quanta in that they do not travel with the velocity of light: The mass of the neutron must be of the same order of magnitude as the electron mass and, in any case, not larger than 0.01 proton mass. The continuous β-spectrum would then become understandable by the assumption that in β decay a neutron is emitted together with the electron, in such a way that the sum of the energies of neutron and electron is constant.

[...]

But I don't feel secure enough to publish anything about this idea, so I first turn confidently to you, dear radioactives, with a question as to the situation concerning experimental proof of such a neutron, if it has something like about 10 times the penetrating capacity of a γ ray.

I admit that my remedy may appear to have a small *a priori* probability because neutrons, if they exist, would probably have long ago been seen. However, only those who wager can win, and the seriousness of the situation of the continuous β-spectrum can be made clear by the saying of my honored predecessor in office, Mr. Debye, [...] "One does best not to think about that at all, like the new taxes." [...] So, dear radioactives, put it to test and set it right. [...]

With many greetings to you, also to Mr. Back, your devoted servant,

W. Pauli

A translated reprint of the full letter can be found in the September 1978 issue of *Physics Today*.*[3]

3.2.2 Discovery

Main article: Cowan–Reines neutrino experiment

The electron neutrino was discovered by Clyde Cowan and Frederick Reines in 1956.*[1]*[4]

3.2.3 Name

Pauli originally named his proposed light particle a *neutron*. When James Chadwick discovered a much more massive nuclear particle in 1932 and also named it a neutron, this left the two particles with the same name. Enrico Fermi, who developed the theory of beta decay, coined the term *neutrino* in 1934 to resolve the confusion. It was a pun on *neutrone*, the Italian equivalent of *neutron*: the *-one* ending can be an augmentative in Italian, so *neutrone* could be read

as the "large neutral thing"; *-ino* replaces the augmentative suffix with a diminutive one. *[5]

Upon the prediction and discovery of a second neutrino, it became important to distinguish between different types of neutrinos. Pauli's neutrino is now identified as the *electron neutrino*, while the second neutrino is identified as the *muon neutrino*.

3.2.4 Electron antineutrino

Like all fermions, the electron neutrino has a corresponding antiparticle, the electron antineutrino (ν e), which differs only in that some of its properties have equal magnitude but opposite sign. The process of beta decay produces both beta particles and electron antineutrinos. Wolfgang Pauli proposed the existence of these particles, in 1930, to ensure that beta decay conserved energy (the electrons in beta decay have a continuum of energies and momentum (the momentum of the electron and recoil nucleus – in beta decay – do not add up to zero).

3.2.5 Notes

[1] Niels Bohr was notably opposed to this interpretation of beta decay and was ready to accept that energy, momentum and angular momentum were not conserved quantities.

[2] See *Name*.

3.2.6 See also

- Muon neutrino
- PMNS matrix
- Tau neutrino

3.2.7 References

[1] "The Reines-Cowan Experiments: Detecting the Poltergeist" (PDF). *Los Alamos Science* **25**: 3. 1997. Retrieved 2010-02-10.

[2] K. Riesselmann (2007). "Logbook: Neutrino Invention". *Symmetry Magazine* **4** (2).

[3] L.M. Brown (1978). "The idea of the neutrino". *Physics Today* **31** (9): 23. Bibcode:1978PhT....31i..23B. doi:10.1063/1.2995181.

[4] F. Reines, C.L. Cowan, Jr. (1956). "The Neutrino". *Nature* **178** (4531): 446. Bibcode:1956Natur.178..446R. doi:10.1038/178446a0.

[5] M.F. L'Annunziata (2007). *Radioactivity*. Elsevier. p. 100. ISBN 978-0-444-52715-8.

3.2.8 Further reading

- F. Reines, C.L. Cowan, Jr. (1956). "The Neutrino". *Nature* **178** (4531): 446. Bibcode:1956Natur.178..446R. doi:10.1038/178446a0.

- C.L. Cowan, Jr., F. Reines, F.B. Harrison, H.W. Kruse, A.D. McGuire (1956). "Detection of the Free Neutrino: A Confirmation". *Science* **124** (3212): 103–4. Bibcode:1956Sci...124..103C. doi:10.1126/science.124.3212.103. PMID 17796274.

3.3 Muon neutrino

The **muon neutrino** is a subatomic lepton elementary particle which has the symbol ν
μ and no net electric charge. Together with the muon it forms the second generation of leptons, hence its name *muon neutrino*. It was first hypothesized in the early 1940s by several people, and was discovered in 1962 by Leon Lederman, Melvin Schwartz and Jack Steinberger. The discovery was rewarded with the 1988 Nobel Prize in Physics.

3.3.1 Discovery

In 1962 Leon M. Lederman, Melvin Schwartz and Jack Steinberger established by performing an experiment at the Brookhaven National Laboratory[1] that more than one type of neutrino exists by first detecting interactions of the muon neutrino (already hypothesised with the name *neutretto*[2]), which earned them the 1988 Nobel Prize.[3]

3.3.2 Speed

Main article: Faster-than-light neutrino anomaly

In September 2011, OPERA researchers reported that muon neutrinos were apparently traveling at faster than light speed. This result was confirmed again in a second experiment in November 2011. These results have been viewed skeptically by the scientific community at large, and more experiments have/are investigating the phenomenon. In March 2012, the ICARUS team published results directly contradicting the results of OPERA.[4]

Later in July 2012 the apparent anomalous super-luminous propagation of neutrinos was traced to a faulty element of the fibre optic timing system in Gran-Sasso. After it was corrected the neutrinos appeared to travel with the speed of light within the errors of the experiment.[5]

3.3.3 See also

- Electron neutrino
- Neutrino oscillation
- PMNS matrix
- Tau neutrino

3.3.4 References

[1] G. Danby, J.-M. Gaillard, K. Goulianos, L. M. Lederman, N. B. Mistry, M. Schwartz, J. Steinberger (1962). "Observation of high-energy neutrino reactions and the existence of two kinds of neutrinos". *Physical Review Letters* **9**: 36. Bibcode:1962PhRvL...9...36D. doi:10.1103/PhysRevLett.9.36.

[2] I.V. Anicin (2005). "The Neutrino – Its Past, Present and Future". *SFIN (Institute of Physics, Belgrade) year XV, Series A: Conferences, No. A* **2**: 3–59. arXiv:physics/0503172. Bibcode:2005physics...3172A.

[3] "The Nobel Prize in Physics 1988". The Nobel Foundation. Retrieved 2010-02-11.

[4] M. Antonello et at. (2012). "Measurement of the neutrino velocity with the ICARUS detector at the CNGS beam". http://arxiv.org/abs/1203.3433v3

[5] "OPERA experiment reports anomaly in flight time of neutrinos from CERN to Gran Sasso (UPDATE 8 June 2012)". *CERN press office*. 8 June 2012. Retrieved 19 April 2013.

3.3.5 Further reading

- Leon M. Lederman (1988). "Observations in Particle Physics from Two Neutrinos to the Standard Model" (PDF). *Nobel Lectures*. The Nobel Foundation. Retrieved 2010-02-11.

- Melvin Schwartz (1988). "The First High Energy Neutrino Experiment" (PDF). *Nobel Lectures*. The Nobel Foundation. Retrieved 2010-02-11.

- Jack Steinberger (1988). "Experiments with High-Energy Neutrino Beams" (PDF). *Nobel Lectures*. The Nobel Foundation. Retrieved 2010-02-11.

3.4 Tauon neutrino

The **tau neutrino** or **tauon neutrino** is a subatomic elementary particle which has the symbol ν
τ and no net electric charge. Together with the tau, it forms the third generation of leptons, hence its name *tau neutrino*.

Its existence was immediately implied after the tau particle was detected in a series of experiments between 1974 and 1977 by Martin Lewis Perl with his colleagues at the SLAC–LBL group.[*][1] The discovery of the tau neutrino was announced in July 2000 by the DONUT collaboration.[*][2][*][3]

3.4.1 Discovery

Main article: DONUT

The tau neutrino is last of the leptons, and is the second most recent particle of the Standard Model to be discovered. The DONUT experiment (which stands for *Direct Observation of the Nu Tau*) from Fermilab was built during the 1990s to specifically detect the tau neutrino. These efforts came to fruition in July 2000, when the DONUT collaboration reported its detection.[*][2][*][3]

3.4.2 See also

- Electron neutrino

- Muon neutrino

- PMNS matrix

3.4.3 References

[1] M. L. Perl; Abrams, G.; Boyarski, A.; Breidenbach, M.; Briggs, D.; Bulos, F.; Chinowsky, W.; Dakin, J.; Feldman, G.; Friedberg, C.; Fryberger, D.; Goldhaber, G.; Hanson, G.; Heile, F.; Jean-Marie, B.; Kadyk, J.; Larsen, R.; Litke, A.; Lüke, D.; Lulu, B.; Lüth, V.; Lyon, D.; Morehouse, C.; Paterson, J.; Pierre, F.; Pun, T.; Rapidis, P.; Richter, B.; Sadoulet, B. et al. (1975). "Evidence for Anomalous Lepton Production in e+e− Annihilation". *Physical Review Letters* **35** (22): 1489. Bibcode:1975PhRvL..35.1489P. doi:10.1103/PhysRevLett.35.1489.

[2] "Physicists Find First Direct Evidence for Tau Neutrino at Fermilab" (Press release). Fermilab. 20 July 2000.

[3] K. Kodama *et al.* (DONUT Collaboration; Kodama; Ushida; Andreopoulos; Saoulidou; Tzanakos; Yager; Baller; Boehnlein; Freeman; Lundberg; Morfin; Rameika; Yun; Song; Yoon; Chung; Berghaus; Kubantsev; Reay; Sidwell; Stanton; Yoshida; Aoki; Hara; Rhee; Ciampa; Erickson; Graham et al. (2001). "Observation of tau neutrino interactions". *Physics Letters B* **504** (3): 218. arXiv:hep-ex/0012035. Bibcode:2001PhLB..504..218D. doi:10.1016/S0370-2693(01)00307-0.

3.5 Neutrino detector

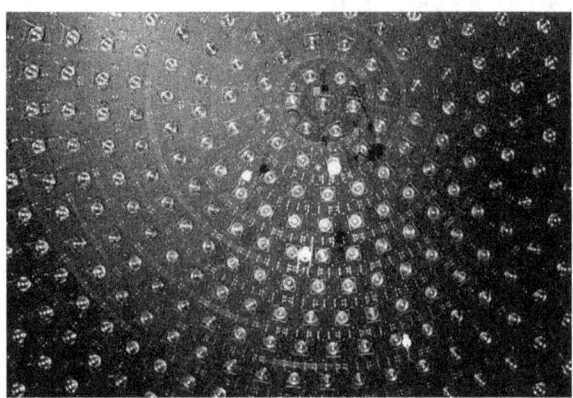

The inside of the MiniBooNE neutrino detector

A **neutrino detector** is a physics apparatus designed to study neutrinos. Because neutrinos only weakly interact with other particles of matter, neutrino detectors must be very large in order to detect a significant number of neutrinos. Neutrino detectors are often built underground, to isolate the detector from cosmic rays and other background radiation.[*][1] The field of neutrino astronomy is still very much in its infancy – the only confirmed extraterrestrial sources so far are the Sun and supernova SN1987A. Neutrino observatories will "give astronomers fresh eyes with which to study the universe." [*][2]

Various detection methods have been used. Super Kamiokande is a large volume of water surrounded by phototubes that watch for the Cherenkov radiation emitted when an incoming neutrino creates an electron or muon in the water. The Sudbury Neutrino Observatory is similar, but uses heavy water as the detecting medium. Other detectors have consisted of large volumes of chlorine or gallium which are periodically checked for excesses of argon or germanium, respectively, which are created by neutrinos interacting with the original substance. MINOS uses a solid plastic scintillator watched by phototubes, Borexino uses a liquid pseudocumene scintillator also watched by phototubes while the proposed NOvA detector will use a liquid scintillator watched by avalanche photodiodes.

The proposed acoustic detection of neutrinos via the thermoacoustic effect is the subject of dedicated studies done by the ANTARES, IceCube and KM3NeT collaborations.

3.5.1 Theory

Neutrinos are omnipresent in nature such that in just one second, tens of billions of them "pass through every square centimetre of our bodies without us ever noticing." [*][3] Despite this, they are extremely "difficult to detect" and may originate from events in the universe such as "colliding black holes, gamma ray bursts from exploding stars, and/or violent events at the cores of distant galaxies," according to some speculation by scientists.*[4] There are three types of neutrinos or what scientists term "flavors": electron, muon and tau neutrinos, which are named after the type of particle that arises after neutrino collisions; as neutrinos propagate through space, the neutrinos "oscillate between the three available flavours." *[3] Neutrinos only have a "smidgen of weight" according to the laws of physics, perhaps less than a "millionth as much as an electron." *[1] Neutrinos can interact via the neutral current (involving the exchange of a Z boson) or charged current (involving the exchange of a W boson) weak interactions.

- In a neutral current interaction, the neutrino leaves the detector after having transferred some of its energy and momentum to a target particle. If the target particle is charged and sufficiently light (e.g. an electron), it may be accelerated to a relativistic speed and consequently emit Cherenkov radiation, which can be observed directly. All three neutrino flavors can participate regardless of the neutrino energy. However, no neutrino flavor information is left behind.

- In a charged current interaction, the neutrino transforms into its partner lepton (electron, muon, or tau).*[5] However, if the neutrino does not have sufficient energy to create its heavier partner's mass, the charged current interaction is unavailable to it. Solar and reactor neutrinos have enough energy to create electrons. Most accelerator-based neutrino beams can also create muons, and a few can create taus. A detector which can distinguish among these leptons can reveal the flavor of the incident neutrino in a charged current interaction. Because the interaction involves the exchange of a charged boson, the target particle also changes character (e.g., neutron → proton).

3.5.2 Detection techniques

Scintillators

Antineutrinos were first detected near the Savannah River nuclear reactor in 1956. Frederick Reines and Clyde Cowan used two targets containing a solution of cadmium chloride in water. Two scintillation detectors were placed next to the cadmium targets. Antineutrinos with an energy above the threshold of 1.8 MeV caused charged current "inverse beta-decay" interactions with the protons in the water, producing positrons and neutrons. The resulting positron annihilations with electrons created pairs of coincident photons with an energy of about 0.5 MeV each, which could be detected by

the two scintillation detectors above and below the target. The neutrons were captured by cadmium nuclei resulting in delayed gamma rays of about 8 MeV that were detected a few microseconds after the photons from a positron annihilation event.

This experiment was designed by Cowan and Reines to give a unique signature for antineutrinos, to prove the existence of these particles. It was not the experimental goal to measure the total antineutrino flux. The detected antineutrinos thus all carried an energy greater 1.8 MeV, which is the threshold for the reaction channel used (1.8 MeV is the energy needed to create a positron and a neutron from a proton). Only about 3% of the antineutrinos from a nuclear reactor carry enough energy for the reaction to occur.

A more recently built and much larger KamLAND detector used similar techniques to study antineutrino oscillations from 53 Japanese nuclear power plants. A smaller, but more pure Borexino detector was able to measure the Beryllium neutrinos from the Sun.

Radiochemical methods

Chlorine detectors, based on the method suggested by Bruno Pontecorvo, consist of a tank filled with a chlorine containing fluid such as tetrachloroethylene. A neutrino converts a chlorine−37 atom into one of argon−37 via the charged current interaction. The threshold neutrino energy for this reaction is 0.814 MeV. The fluid is periodically purged with helium gas which would remove the argon. The helium is then cooled to separate out the argon, and the argon atoms are counted based on their electron capture radioactive decays. A chlorine detector in the former Homestake Mine near Lead, South Dakota, containing 520 short tons (470 metric tons) of fluid, was the first to detect the solar neutrinos, and made the first measurement of the deficit of electron neutrinos from the sun (see Solar neutrino problem).

A similar detector design, with a much lower detection threshold of 0.233 MeV, uses a gallium → germanium transformation which is sensitive to lower energy neutrinos. A neutrino is able to react with an atom of gallium−71, converting it into an atom of the unstable isotope germanium−71. The germanium was then chemically extracted and concentrated. Neutrinos were thus detected by measuring the radioactive decay of germanium. This latter method is nicknamed the "Alsace-Lorraine" technique because of the reaction sequence (gallium-germanium-gallium) involved. These radiochemical detection methods are useful only for counting neutrinos; no neutrino direction or energy information is available. The SAGE experiment in Russia used about 50 tons, and the GALLEX/GNO experiments in Italy about 30 tons, of gallium as reaction mass. This experiment is difficult to scale up due to the prohibitive cost of gallium. Larger experiments have therefore turned to a cheaper reaction mass.

Cherenkov detectors

"Ring-imaging" Cherenkov detectors take advantage of a phenomenon called Cherenkov light. Cherenkov radiation is produced whenever charged particles such as electrons or muons are moving through a given detector medium somewhat faster than the speed of light in that medium. In a Cherenkov detector, a large volume of clear material such as water or ice is surrounded by light-sensitive photomultiplier tubes. A charged lepton produced with sufficient energy and moving through such a detector does travel somewhat faster than the speed of light in the detector medium (although somewhat slower than the speed of light in a vacuum). The charged lepton generates a visible "optical shockwave" of Cherenkov radiation. This radiation is detected by the photomultiplier tubes and shows up as a characteristic ring-like pattern of activity in the array of photomultiplier tubes. As neutrinos can interact with atomic nuclei to produce charged leptons which emit Cherenkov radiation, this pattern can be used to infer direction, energy, and (sometimes) flavor information about incident neutrinos.

Two water-filled detectors of this type (Kamiokande and IMB) recorded a neutrino burst from supernova 1987A.[6] Scientists detected 19 neutrinos from an explosion of a star inside the Large Magellanic Cloud—only 19 out of the billion trillion trillion trillion trillion neutrinos emitted by the supernova.[1] The Kamiokande detector was able to detect the burst of neutrinos associated with this supernova, and in 1988 it was used to directly confirm the production of solar neutrinos. The largest such detector is the water-filled Super-Kamiokande. This detector uses 50,000 tons of pure water surrounded by 11,000 photomultiplier tubes buried 1 km underground.

The Sudbury Neutrino Observatory (SNO) uses 1,000 tonnes of ultrapure heavy water contained in a 12-metre-diameter vessel made of acrylic plastic surrounded by a cylinder of ultrapure ordinary water 22 metres in diameter and 34 metres high.[5] In addition to the neutrino interactions visible in a regular water detector, the deuterium in heavy water can be broken up by a neutrino. The resulting free neutron is subsequently captured, releasing a burst of gamma rays that can be detected. All three neutrino flavors participate equally in this dissociation reaction.

The MiniBooNE detector employs pure mineral oil as its detection medium. Mineral oil is a natural scintillator, so charged particles without sufficient energy to produce Cherenkov light still produce scintillation light. Low energy

muons and protons, invisible in water, can be detected.

An illustration of the Antares neutrino detector deployed under water.

Located at a depth of about 2.5 km in the Mediterranean Sea, the ANTARES (*Astronomy with a Neutrino Telescope and Abyss environmental RESearch*) is fully operational since May 30, 2008. Consisting of an array of twelve separate 350-meter-long vertical detector strings 70 meters apart, each with 75 photomultiplier optical modules, this detector uses the surrounding sea water as the detector medium. The next generation deep sea neutrino telescope KM3NeT will have a total instrumented volume of about 5 km^3. The detector will be distributed over three installation sites in the Mediterranean. Implementation of the first phase of the telescope as started in 2013.

The Antarctic Muon And Neutrino Detector Array (AMANDA) operated from 1996 to 2004. This detector used photomultiplier tubes mounted in strings buried deep (1.5–2 km) inside Antarctic glacial ice near the South Pole. The ice itself is used as the detector medium. The direction of incident neutrinos is determined by recording the arrival time of individual photons using a three-dimensional array of detector modules each containing one photomultiplier tube. This method allows detection of neutrinos above 50 GeV with a spatial resolution of approximately 2 degrees. AMANDA was used to generate neutrino maps of the northern sky in order to search for extraterrestrial neutrino sources and to search for dark matter. AMANDA is currently being upgraded to the IceCube observatory, eventually increasing the volume of the detector array to one cubic kilometer.[7]

Radio detectors

The Radio Ice Cerenkov Experiment uses antennas to detect Cerenkov radiation from high-energy neutrinos in Antarctica. The Antarctic Impulse Transient Antenna

(ANITA) is a balloon-born device flying over Antarctica and detecting Askaryan radiation produced by ultra-high energy neutrinos interacting with the ice below.

Tracking calorimeters

Tracking calorimeters such as the MINOS detectors use alternating planes of absorber material and detector material. The absorber planes provide detector mass while the detector planes provide the tracking information. Steel is a popular absorber choice, being relatively dense and inexpensive and having the advantage that it can be magnetised. The NOvA proposal suggests eliminating the absorber planes in favor of using a very large active detector volume. The active detector is often liquid or plastic scintillator, read out with photomultiplier tubes, although various kinds of ionisation chambers have also been used.

Tracking calorimeters are only useful for high energy (GeV range) neutrinos. At these energies, neutral current interactions appear as a shower of hadronic debris and charged current interactions are identified by the presence of the charged lepton's track (possibly alongside some form of hadronic debris.) A muon produced in a charged current interaction leaves a long penetrating track and is easy to spot. The length of this muon track and its curvature in the magnetic field provide energy and charge ($\mu-$ versus $\mu+$) information. An electron in the detector produces an electromagnetic shower which can be distinguished from hadronic showers if the granularity of the active detector is small compared to the physical extent of the shower. Tau leptons decay essentially immediately to either pions or another charged lepton and cannot be observed directly in this kind of detector. (To directly observe taus, one typically looks for a kink in tracks in photographic emulsion.)

3.5.3 Background suppression

Most neutrino experiments must address the flux of cosmic rays that bombard the Earth's surface.

The higher energy (>50 MeV or so) neutrino experiments often cover or surround the primary detector with a "veto" detector which reveals when a cosmic ray passes into the primary detector, allowing the corresponding activity in the primary detector to be ignored ("vetoed").

For lower energy experiments, the cosmic rays are not directly the problem. Instead, the spallation neutrons and radioisotopes produced by the cosmic rays may mimic the desired physics signals. For these experiments, the solution is to locate the detector deep underground so that the earth above can reduce the cosmic ray rate to tolerable levels.

3.5.4 Telescopes

Neutrino detectors can be aimed at astrophysics observations, many astrophysics events being believed to emit neutrinos.

Underwater neutrino telescopes:

- DUMAND (1976–1995; cancelled)

- Baikal (1993 on)

- ANTARES (2006 on)

- KM3NeT (future telescope; under construction since 2013)

- NESTOR Project (under development since 1998)

Under-ice neutrino telescopes :

- AMANDA (1996–2009, superseded by IceCube)

- IceCube (2004 on)*[2]

- DeepCore and PINGU, an existing extension and a proposed extension of IceCube.

Underground neutrino telescopes:

- Soudan lab, in Soudan, Minnesota*[8]

Miscellaneous :

- GALLEX (1991–1997; ended)

- Tauwer experiment (construction date to be determined)

3.5.5 See also

- List of neutrino experiments

3.5.6 References

[1] KENNETH CHANG (April 26, 2005). "Tiny, Plentiful and Really Hard to Catch". The New York Times. Retrieved 2011-06-16. In 1987, astronomers counted 19 neutrinos from an explosion of a star in the nearby Large Magellanic Cloud, 19 out of the billion trillion trillion trillion trillion neutrinos that flew from the supernova.

[2] Ian Sample (23 January 2011). "The hunt for neutrinos in the Antarctic". The Guardian. Retrieved 2011-06-16. The $272m (£170m) IceCube instrument is not your typical telescope. Instead of collecting light from the stars, planets or other celestial objects, IceCube looks for ghostly particles called neutrinos that hurtle across space with high-energy cosmic rays. If all goes to plan, the observatory will reveal where these mysterious rays come from, and how they get to be so energetic. But that is just the start. Neutrino observatories such as IceCube will ultimately give astronomers fresh eyes with which to study the universe.

[3] Pierre Le Hir (22 March 2011). "Tracking down the crafty neutrino". Guardian Weekly. Retrieved 2011-06-16. But they are nevertheless almost undetectable: in just one second several tens of billions of neutrinos pass through every square centimetre of our bodies without us ever noticing. ... No magnetic field diverts them from their course, shooting straight ahead at almost the speed of light. ... Almost nothing stops them. ... Neutrinos are remarkably tricky customers. There are three types or flavours: electron, muon and tau neutrinos, named after three other particles to which they give rise when they collide with an atom.

[4] Dr David Whitehouse, BBC News Online science editor (15 July 2003). "Icebound telescope probes the Universe". BBC News. Retrieved 2011-06-16. Sensors in the ice have detected the rare and fleeting flashes of light caused when neutrinos interact with the ice. ... Amanda 2 (Antarctic Muon and Neutrino Detector Array - 2) is designed to look not up, but down, through the Earth to the sky of the Northern Hemisphere.

[5] Dr David Whitehouse, BBC News Online science editor (22 April 2002). "Experiment confirms Sun theories". BBC News. Retrieved 2011-06-16. New evidence confirms last year's indication that one type of neutrino emerging from the Sun's core does switch to another type en route to the Earth. ... The data were obtained from the underground Sudbury Neutrino Observatory (SNO) in Canada. ... Neutrinos are ghostly particles with no electric charge and very little mass. They are known to exist in three types related to three different charged particles - the electron and its lesser-known relatives, the muon and the tau. ...

[6] MALCOLM W. BROWNE (February 28, 1995). "Four Telescopes in Neutrino Hunt". The New York Times. Retrieved 2011-06-16. NEUTRINO astronomy was given a strong push in 1987 when a supernova in a galaxy only one-quarter of a million light-years away from Earth flared into view—the closest supernova in 400 years.

[7] J.P. (Dec 1, 2010). "Hang on, that's not a neutrino". The Economist. Retrieved 2011-06-16. The largest, IceCube, sits deep underneath the South Pole in a cubic kilometre of perfectly clear, bubble-free ancient ice and is set to start working in earnest early next year. All rely on detecting the flickers of light emitted on the exceedingly rare occasions when a neutrino does interact with an atom of ice or water.

[8] "Minnesota neutrino project to get under way this month". USA Today. Feb 11, 2005. Retrieved 2011-06-16. Later this month, Fermi National Accelerator Laboratory near Chicago will begin shooting trillions of subatomic "neutrino" particles through 450 miles of solid earth, their target a detector at the Soudan Underground Laboratory beneath this Iron Range town.

3.6   Neutrino oscillation

Neutrino oscillation is a quantum mechanical phenomenon whereby a neutrino created with a specific lepton flavor (electron, muon or tau) can later be measured to have a different flavor. The probability of measuring a particular flavor for a neutrino varies periodically as it propagates through space.[1]

First predicted by Bruno Pontecorvo in 1957,[2] neutrino oscillation has since been observed by a multitude of experiments in several different contexts. Also, it turned out to be the resolution to the long-standing solar neutrino problem.

Neutrino oscillation is of great theoretical and experimental interest, since observation of the phenomenon implies that the neutrino has a non-zero mass, which was not included as part of the original Standard Model of particle physics.[1]

3.6.1   Observations

A great deal of evidence for neutrino oscillation has been collected from many sources, over a wide range of neutrino energies and with many different detector technologies.[3]

Solar neutrino oscillation

The first experiment that detected the effects of neutrino oscillation was Ray Davis's Homestake Experiment in the late 1960s, in which he observed a deficit in the flux of solar neutrinos with respect to the prediction of the Standard Solar Model, using a chlorine-based detector.[4] This gave rise to the Solar neutrino problem. Many subsequent radiochemical and water Cherenkov detectors confirmed the deficit, but neutrino oscillation was not conclusively identified as the source of the deficit until the Sudbury Neutrino Observatory provided clear evidence of neutrino flavor change in 2001.[5]

Solar neutrinos have energies below 20 MeV and travel approximately 1 A.U. between the source in the Sun and detector on the Earth. At energies above 5 MeV, solar neutrino oscillation actually takes place in the Sun through a resonance known as the MSW effect, a different process from the vacuum oscillation described later in this article.[1]

Atmospheric neutrino oscillation

Large detectors such as IMB, MACRO, and Kamiokande II observed a deficit in the ratio of the flux of muon to electron flavor atmospheric neutrinos (see *muon decay*). The Super Kamiokande experiment provided a very precise measurement of neutrino oscillation in an energy range of hundreds of MeV to a few TeV, and with a baseline of the diameter of the Earth; the first experimental evidence for atmospheric neutrino oscillations was announced in 1998.[6]

Reactor neutrino oscillation

Many experiments have searched for oscillation of electron anti-neutrinos produced at nuclear reactors. Such oscillations give the value of the parameter θ_{13}. The KamLAND experiment, started in 2002, has made a high precision observation of reactor neutrino oscillation. Neutrinos produced in nuclear reactors have energies similar to solar neutrinos, of around a few MeV. The baselines of these experiments have ranged from tens of meters to over 100 km.

In 2012, the Daya Bay team announced a discovery that $\theta_{13} \neq 0$ at 5.2σ significance.[7] RENO soon confirmed the result.[8]

Beam neutrino oscillation

Neutrino beams produced at a particle accelerator offer the greatest control over the neutrinos being studied. Many experiments have taken place which study the same neutrino oscillations which take place in atmospheric neutrino oscillation, using neutrinos with a few GeV of energy and several hundred km baselines. The MINOS, K2K, and Super-K experiments have all independently observed muon neutrino disappearance over such long baselines.[1]

Data from the LSND experiment appear to be in conflict with the oscillation parameters measured in other experiments. Results from the MiniBooNE appeared in Spring 2007 and contradicted the results from LSND, although they could support the existence of a fourth neutrino type, the sterile neutrino.[1]

In 2010, the INFN and CERN announced the observation of a tau particle in a muon neutrino beam in the OPERA detector located at Gran Sasso, 730 km away from the source in Geneva.[9]

The currently-running T2K experiment uses a neutrino beam directed through 295 km of earth, and will measure

the parameter θ_{13}. The experiment uses the Super-K detector. NOvA is a similar effort. This detector will use the same beam as MINOS and will have a baseline of 810 km.

3.6.2 Theory

Neutrino oscillation arises from a mixture between the flavor and mass eigenstates of neutrinos. That is, the three neutrino states that interact with the charged leptons in weak interactions are each a different superposition of the three neutrino states of definite mass. Neutrinos are created in weak processes in their flavor eigenstates[nb 1]. As a neutrino propagates through space, the quantum mechanical phases of the three mass states advance at slightly different rates due to the slight differences in the neutrino masses. This results in a changing mixture of mass states as the neutrino travels, but a different mixture of mass states corresponds to a different mixture of flavor states. So a neutrino born as, say, an electron neutrino will be some mixture of electron, mu, and tau neutrino after traveling some distance. Since the quantum mechanical phase advances in a periodic fashion, after some distance the state will nearly return to the original mixture, and the neutrino will be again mostly electron neutrino. The electron flavor content of the neutrino will then continue to oscillate as long as the quantum mechanical state maintains coherence. Since mass differences between neutrino flavors are small in comparison with long coherence length for neutrino oscillations this microscopic quantum effect becomes observable over macroscopic distances.

On July 19, 2013 the results from the T2K experiment presented at the European Physical Society Conference on High Energy Physics in Stockholm, Sweden, confirmed the theory.[11][12]

Pontecorvo–Maki–Nakagawa–Sakata matrix

Main article: Pontecorvo–Maki–Nakagawa–Sakata matrix

The idea of neutrino oscillation was first put forward in 1957 by Bruno Pontecorvo, who proposed that neutrino-antineutrino transitions may occur in analogy with neutral kaon mixing.[2] Although such matter-antimatter oscillation has not been observed, this idea formed the conceptual foundation for the quantitative theory of neutrino flavor oscillation, which was first developed by Maki, Nakagawa, and Sakata in 1962[13] and further elaborated by Pontecorvo in 1967.[14] One year later the solar neutrino deficit was first observed,[15] and that was followed by the famous paper of Gribov and Pontecorvo published in 1969 titled "Neutrino astronomy and lepton charge".[16]

The concept of neutrino mixing is a natural outcome of gauge theories with massive neutrinos and its structure can be characterized in general.*[17] In its simplest form it is expressed as a unitary transformation relating the flavor and mass eigenbasis can be written

$$|\nu_\alpha\rangle = \sum_i U^*_{\alpha i} |\nu_i\rangle$$

$$|\nu_i\rangle = \sum_\alpha U_{\alpha i} |\nu_\alpha\rangle$$

where

- $|\nu_\alpha\rangle$ is a neutrino with definite flavor. α = e (electron), μ (muon) or τ (tauon).

- $|\nu_i\rangle$ is a neutrino with definite mass m_i, $i = 1, 2, 3$.

- The asterisk (*) represents a complex conjugate. For antineutrinos, the complex conjugate should be dropped from the first equation, and added to the second.

$U_{\alpha i}$ represents the *Pontecorvo–Maki–Nakagawa–Sakata matrix* (also called the *PMNS matrix*, *lepton mixing matrix*, or sometimes simply the *MNS matrix*). It is the analogue of the CKM matrix describing the analogous mixing of quarks. If this matrix were the identity matrix, then the flavor eigenstates would be the same as the mass eigenstates. However, experiment shows that it is not.

When the standard three neutrino theory is considered, the matrix is 3×3. If only two neutrinos are considered, a 2×2 matrix is used. If one or more sterile neutrinos are added (see later) it is 4×4 or larger. In the 3×3 form, it is given by:*[18]

$$U = \begin{bmatrix} U_{e1} & U_{e2} & U_{e3} \\ U_{\mu 1} & U_{\mu 2} & U_{\mu 3} \\ U_{\tau 1} & U_{\tau 2} & U_{\tau 3} \end{bmatrix}$$

$$= \begin{bmatrix} 1 & 0 & 0 \\ 0 & c_{23} & s_{23} \\ 0 & -s_{23} & c_{23} \end{bmatrix} \begin{bmatrix} c_{13} & 0 & s_{13}e^{-i\delta} \\ 0 & 1 & 0 \\ -s_{13}e^{i\delta} & 0 & c_{13} \end{bmatrix} \begin{bmatrix} c_{12} & s_{12} & 0 \\ -s_{12} & c_{12} & 0 \\ 0 & 0 & 1 \end{bmatrix} \begin{bmatrix} 1 & 0 & 0 \\ 0 & e^{i\alpha_1/2} & 0 \\ 0 & 0 & e^{i\alpha_2/2} \end{bmatrix}$$

$$= \begin{bmatrix} c_{12}c_{13} & s_{12}c_{13} & s_{13}e^{-i\delta} \\ -s_{12}c_{23} - c_{12}s_{23}s_{13}e^{i\delta} & c_{12}c_{23} - s_{12}s_{23}s_{13}e^{i\delta} & s_{23}c_{13} \\ s_{12}s_{23} - c_{12}c_{23}s_{13}e^{i\delta} & -c_{12}s_{23} - s_{12}c_{23}s_{13}e^{i\delta} & c_{23}c_{13} \end{bmatrix} \begin{bmatrix} 1 & 0 & 0 \\ 0 & e^{i\alpha_1/2} & 0 \\ 0 & 0 & e^{i\alpha_2/2} \end{bmatrix}$$

where $c_{ij} = \cos\theta_{ij}$ and $s_{ij} = \sin\theta_{ij}$. The phase factors α_1 and α_2 are physically meaningful only if neutrinos are Majorana particles —i.e. if the neutrino is identical to its antineutrino (whether or not they are is unknown) —and do not enter into oscillation phenomena regardless. If neutrinoless double beta decay occurs, these factors influence its rate. The

phase factor δ is non-zero only if neutrino oscillation violates CP symmetry. This is expected, but not yet observed experimentally. If experiment shows this 3×3 matrix to be not unitary, a sterile neutrino or some other new physics is required.

Propagation and interference

Since $|\nu_i\rangle$ are mass eigenstates, their propagation can be described by plane wave solutions of the form

$$|\nu_i(t)\rangle = e^{-i(E_i t - \vec{p}_i \cdot \vec{x})}|\nu_i(0)\rangle,$$

where

- quantities are expressed in natural units ($c = 1$, $\hbar = 1$)

- E_i is the energy of the mass-eigenstate i,

- t is the time from the start of the propagation,

- $\vec{p}_i$ is the three-dimensional momentum,

- $\vec{x}$ is the current position of the particle relative to its starting position

In the ultrarelativistic limit, $|\vec{p}_i| = p_i \gg m_i$, we can approximate the energy as

$$E_i = \sqrt{p_i^2 + m_i^2} \simeq p_i + \frac{m_i^2}{2p_i} \approx E + \frac{m_i^2}{2E},$$

where E is the total energy of the particle.

This limit applies to all practical (currently observed) neutrinos, since their masses are less than 1 eV and their energies are at least 1 MeV, so the Lorentz factor γ is greater than 10^6 in all cases. Using also $t \approx L$, where L is the distance traveled and also dropping the phase factors, the wavefunction becomes:

$$|\nu_i(L)\rangle = e^{-im_i^2 L/2E}|\nu_i(0)\rangle.$$

Eigenstates with different masses propagate at different speeds. The heavier ones lag behind while the lighter ones pull ahead. Since the mass eigenstates are combinations of flavor eigenstates, this difference in speed causes interference between the corresponding flavor components of each mass eigenstate. Constructive interference causes it to be possible to observe a neutrino created with a given flavor to change its flavor during its propagation. The probability that a neutrino originally of flavor α will later be observed as having flavor β is

$$P_{\alpha \to \beta} = |\langle \nu_\beta | \nu_\alpha(t) \rangle|^2 = \left| \sum_i U_{\alpha i}^* U_{\beta i} e^{-im_i^2 L/2E} \right|^2 .$$

This is more conveniently written as

$$P_{\alpha \to \beta} = \delta_{\alpha\beta} \quad - \quad 4\sum_{i>j} \mathrm{Re}(U_{\alpha i}^* U_{\beta i} U_{\alpha j} U_{\beta j}^*) \sin^2(\tfrac{\Delta m_{ij}^2 L}{4E})$$
$$+ \quad 2\sum_{i>j} \mathrm{Im}(U_{\alpha i}^* U_{\beta i} U_{\alpha j} U_{\beta j}^*) \sin(\tfrac{\Delta m_{ij}^2 L}{2E}),$$

where $\Delta m_{ij}^2 \equiv m_i^2 - m_j^2$. The phase that is responsible for oscillation is often written as (with c and $\hbar$ restored)

$$\frac{\Delta m^2 c^3 L}{4\hbar E} = \frac{\mathrm{GeV\,fm}}{4\hbar c} \times \frac{\Delta m^2}{\mathrm{eV}^2} \frac{L}{\mathrm{km}} \frac{\mathrm{GeV}}{E} \approx 1.27 \times \frac{\Delta m^2}{\mathrm{eV}^2} \frac{L}{\mathrm{km}} \frac{\mathrm{GeV}}{E},$$

where 1.27 is unitless. In this form, it is convenient to plug in the oscillation parameters since:

- The mass differences, Δm^2, are known to be on the order of $1 \times 10^{*}-4$ eV^2

- Oscillation distances, L, in modern experiments are on the order of kilometers

- Neutrino energies, E, in modern experiments are typically on order of MeV or GeV.

If there is no CP-violation (δ is zero), then the second sum is zero. Otherwise, the CP asymmetry can be given as

$$A_{\mathrm{CP}}^{(\alpha\beta)} = P(\nu_\alpha \to \nu_\beta) - P(\bar\nu_\alpha \to \bar\nu_\beta) = 4\sum_{i>j} \mathrm{Im}\left(U_{\alpha i}^* U_{\beta i} U_{\alpha j} U_{\beta j}^*\right) \sin\left(\frac{\Delta m_{ij}^2 L}{2E}\right)$$

In terms of Jarlskog invariant

$$\mathrm{Im}\left(U_{\alpha i} U_{\beta i}^* U_{\alpha j}^* U_{\beta j}\right) = J \sum_{\gamma,k} \varepsilon_{\alpha\beta\gamma} \varepsilon_{ijk}$$

the CP asymmetry is expressed as

$$A_{\mathrm{CP}}^{(\alpha\beta)} = 16 J \sum_\gamma \varepsilon_{\alpha\beta\gamma} \sin\left(\frac{\Delta m_{21}^2 L}{4E}\right) \sin\left(\frac{\Delta m_{32}^2 L}{4E}\right) \sin\left(\frac{\Delta m_{13}^2 L}{4E}\right)$$

Two neutrino case

The above formula is correct for any number of neutrino generations. Writing it explicitly in terms of mixing angles is extremely cumbersome if there are more than two neutrinos that participate in mixing. Fortunately, there are several cases in which only two neutrinos participate significantly. In this case, it is sufficient to consider the mixing matrix

$$U = \begin{pmatrix} \cos\theta & \sin\theta \\ -\sin\theta & \cos\theta \end{pmatrix}.$$

Then the probability of a neutrino changing its flavor is

$$P_{\alpha \to \beta, \alpha \neq \beta} = \sin^2(2\theta) \sin^2\left(\frac{\Delta m^2 L}{4E}\right) \text{ (natural units)}.$$

Or, using SI units and the convention introduced above

$$P_{\alpha \to \beta, \alpha \neq \beta} = \sin^2(2\theta) \sin^2\left(1.27 \frac{\Delta m^2 L}{E} \frac{[\mathrm{eV}^2]}{[\mathrm{GeV}]} [\mathrm{km}]\right).$$

This formula is often appropriate for discussing the transition $\nu_\mu \leftrightarrow \nu_\tau$ in atmospheric mixing, since the electron neutrino plays almost no role in this case. It is also appropriate for the solar case of $\nu_e \leftrightarrow \nu_x$, where ν_x is a superposition of ν_μ and ν_τ. These approximations are possible because the mixing angle θ_{13} is very small and because two of the mass states are very close in mass compared to the third.

Classical analogue of neutrino oscillation

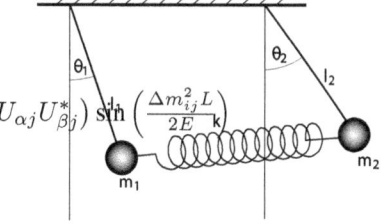

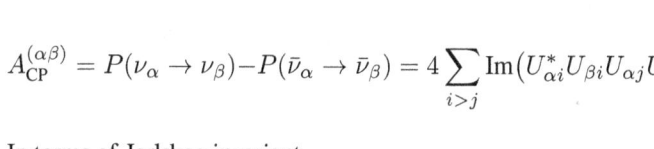

Spring-coupled pendulums

Time evolution of the pendulums

Lower frequency normal mode

Higher frequency normal mode

The basic physics behind neutrino oscillation can be found in any system of coupled harmonic oscillators. A simple example is a system of two pendulums connected by a weak spring (a spring with a small spring constant). The first pendulum is set in motion by the experimenter while the second begins at rest. Over time, the second pendulum begins to swing under the influence of the spring, while the first pendulum's amplitude decreases as it loses energy to the second. Eventually all of the system's energy is transferred to the second pendulum and the first is at rest. The process then reverses. The energy oscillates between the two pendulums repeatedly until it is lost to friction.

The behavior of this system can be understood by looking at its normal modes of oscillation. If the two pendulums are identical then one normal mode consists of both pendulums swinging in the same direction with a constant distance between them, while the other consists of the pendulums swinging in opposite (mirror image) directions. These normal modes have (slightly) different frequencies because the second involves the (weak) spring while the first does not. The initial state of the two-pendulum system is a combination of both normal modes. Over time, these normal modes drift out of phase, and this is seen as a transfer of motion from the first pendulum to the second.

The description of the system in terms of the two pendulums is analogous to the flavor basis of neutrinos. These are the parameters that are most easily produced and detected (in the case of neutrinos, by weak interactions involving the W boson). The description in terms of normal modes is analogous to the mass basis of neutrinos. These modes do not interact with each other when the system is free of outside influence.

When the pendulums are not identical the analysis is slightly more complicated. In the small-angle approximation, the potential energy of a single pendulum system is $\frac{1}{2}\frac{mg}{L}x^2$, where g is the standard gravity, L is the length of the pendulum, m is the mass of the pendulum, and x is the horizontal displacement of the pendulum. As an isolated system the pendulum is a harmonic oscillator with a frequency of $\sqrt{g/L}$. The potential energy of a spring is $\frac{1}{2}kx^2$ where k is the spring constant and x is the displacement. With a mass attached it oscillates with a period of $\sqrt{k/m}$. With two pendulums (labeled a and b) of equal mass but possibly unequal lengths and connected by a spring, the total potential energy is

$$V = \frac{m}{2}\left(\frac{g}{L_a}x_a^2 + \frac{g}{L_b}x_b^2 + \frac{k}{m}(x_b - x_a)^2 \right).$$

This is a quadratic form in x_a and x_b, which can also be written as a matrix product:

$$V = \frac{m}{2}\begin{pmatrix} x_a & x_b \end{pmatrix}\begin{pmatrix} \frac{g}{L_a} + \frac{k}{m} & -\frac{k}{m} \\ -\frac{k}{m} & \frac{g}{L_b} + \frac{k}{m} \end{pmatrix}\begin{pmatrix} x_a \\ x_b \end{pmatrix}.$$

The 2×2 matrix is real symmetric and so (by the spectral theorem) it is "orthogonally diagonalizable". That is, there is an angle θ such that if we define

$$\begin{pmatrix} x_a \\ x_b \end{pmatrix} = \begin{pmatrix} \cos\theta & \sin\theta \\ -\sin\theta & \cos\theta \end{pmatrix}\begin{pmatrix} x_1 \\ x_2 \end{pmatrix}$$

then

$$V = \frac{m}{2}\begin{pmatrix} x_1 & x_2 \end{pmatrix}\begin{pmatrix} \lambda_1 & 0 \\ 0 & \lambda_2 \end{pmatrix}\begin{pmatrix} x_1 \\ x_2 \end{pmatrix}$$

where λ_1 and λ_2 are the eigenvalues of the matrix. The variables x_1 and x_2 describe normal modes which oscillate with frequencies of $\sqrt{\lambda_1}$ and $\sqrt{\lambda_2}$. When the two pendulums are identical ($L_a = L_b$), θ is $45°$.

The angle θ is analogous to the Cabibbo angle (though that angle applies to quarks rather than neutrinos).

When the number of oscillators (particles) is increased to three, the orthogonal matrix can no longer be described by a single angle; instead, three are required (Euler angles). Furthermore, in the quantum case, the matrices may be complex. This requires the introduction of complex phases in addition to the rotation angles, which are associated with CP violation but do not influence the observable effects of neutrino oscillation.

3.6.3   Theory, graphically

Two neutrino probabilities in vacuum

In the approximation where only two neutrinos participate in the oscillation, the probability of oscillation follows a simple pattern:

The blue curve shows the probability of the original neutrino retaining its identity. The red curve shows the probability of conversion to the other neutrino. The maximum probability of conversion is equal to $\sin^2 2\theta$. The frequency of the oscillation is controlled by Δm^2.

Three neutrino probabilities

If three neutrinos are considered, the probability for each neutrino to appear is somewhat complex. Here are shown the probabilities for each initial flavor, with one plot showing a long range to display the slow "solar" oscillation and

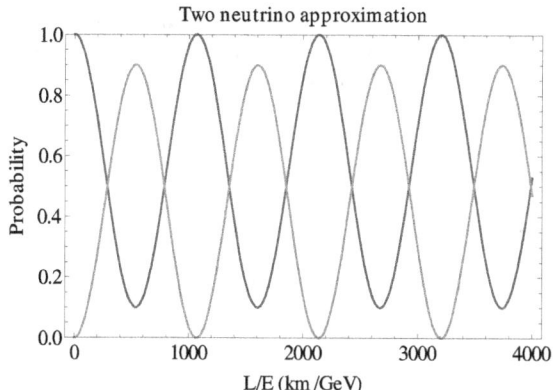

the other zoomed in to display the fast "atmospheric" oscillation. The oscillation parameters used here are consistent with current measurements, but since some parameters are still quite uncertain, these graphs are only qualitatively correct in some aspects. These values were used:

- $\sin^2 2\theta_{13} = 0.10$ (Controls the size of the small wiggles.)

- $\sin^2 2\theta_{23} = 0.97$.

- $\sin^2 2\theta_{12} = 0.861$.

- $\delta = 0$ (If it is actually large, these probabilities will be somewhat distorted and different for neutrinos and antineutrinos.)

- $\Delta m^2_{12} = 7.59 \times 10^{-5}$ eV^2.

- $\Delta m^2_{32} \approx \Delta m^2_{13} = 2.32 \times 10^{-3}$ eV^2.

- Normal mass hierarchy.

3.6.4 Observed values of oscillation parameters

- $\sin^2(2\theta_{13}) = 0.093 \pm 0.008$.[19] PDG combination of Daya Bay, RENO, and Chooz results.

- $\sin^2(2\theta_{12}) = 0.846 + 0.021 - 0.021$.[19] This corresponds to θ_{sol} (solar), obtained from KamLand, solar, reactor and accelator data.

- $\sin^2(2\theta_{23}) > 0.92$ at 90% confidence level, corresponding to $\theta_{23} \equiv \theta_{atm} = 45 \pm 7.1°$ (atmospheric)[20]

- $\Delta m^2_{21} \equiv \Delta m^2_{sol} = 7.53 + 0.18 - 0.18 \times 10^{-5}$ eV^2[19]

- $|\Delta m^2_{31}| \approx |\Delta m^2_{32}| \equiv \Delta m^2_{atm} = 2.44 + 0.06 - 0.06 \times 10^{-3}$ eV^2 (normal mass hierarchy)[19]

- δ, α_1, α_2, and the sign of Δm^2_{32} are currently unknown

Solar neutrino experiments combined with KamLAND have measured the so-called solar parameters Δm^2_{sol} and $\sin^2 \theta_{sol}$. Atmospheric neutrino experiments such as Super-Kamiokande together with the K2K and MINOS long baseline accelerator neutrino experiment have determined the so-called atmospheric parameters Δm^2_{atm} and $\sin^2 \theta_{atm}$. The last mixing angle, θ_{13}, has been measured by the experiments Daya Bay, Double Chooz and RENO as $\sin^2 2\theta_{13}$.

For atmospheric neutrinos (where the relevant difference of masses is about $\Delta m^2 = 2.4 \times 10^{-3}$ eV^2 and the typical energies are ~1 GeV), oscillations become visible for neutrinos traveling several hundred km, which means neutrinos that reach the detector from below the horizon.

The mixing parameter θ_{13} is measured using electron anti-neutrinos from nuclear reactors. The rate of anti-neutrino interactions is measured in detectors sited near the reactors to determine the flux prior to any significant oscillations and then it is measured in far detectors (sited km from the reactors). The oscillation is observed as an apparent disappearance of electron anti-neutrinos in the far detectors (*i.e.* the interaction rate at the far site is lower than predicted from the observed rate at the near site).

From atmospheric and solar neutrino oscillation experiments, it is known that two mixing angles of the MNS matrix are large and the third is smaller. This is in sharp contrast to the CKM matrix in which all three angles are small and hierarchically decreasing. Nothing is known about the CP-violating phase of the MNS matrix.

If the neutrino mass proves to be of Majorana type (making the neutrino its own antiparticle), it is possible that the MNS matrix has more than one phase.

Since experiments observing neutrino oscillation measure the squared mass difference and not absolute mass, one can claim that the lightest neutrino mass is exactly zero, without contradicting observations. This is however regarded as unlikely by theorists.

3.6.5 Origins of neutrino mass

The question of how neutrino masses arise has not been answered conclusively. In the Standard Model of particle physics, fermions only have mass because of interactions with the Higgs field (see *Higgs boson*). These interactions involve both left- and right-handed versions of the fermion (see *chirality*). However, only left-handed neutrinos have been observed so far.

Neutrinos may have another source of mass through the Majorana mass term. This type of mass applies for electrically-neutral particles since otherwise it would allow particles to turn into anti-particles, which would violate conservation of electric charge.

The smallest modification to the Standard Model, which only has left-handed neutrinos, is to allow these left-handed neutrinos to have Majorana masses. The problem with this is that the neutrino masses are surprisingly smaller than the rest of the known particles (at least 500,000 times smaller than the mass of an electron), which, while it does not invalidate the theory, is widely regarded as unsatisfactory as this construction offers no insight into the origin of the neutrino mass scale.

The next simplest addition would be to add into the Standard Model right-handed neutrinos that interact with the left-handed neutrinos and the Higgs field in an analogous way to the rest of the fermions. These new neutrinos would interact with the other fermions solely in this way, so are not phenomenologically excluded. The problem of the disparity of the mass scales remains.

Seesaw mechanism

Main article: Seesaw mechanism

The most popular conjectured solution currently is the *seesaw mechanism*, where right-handed neutrinos with very large Majorana masses are added. If the right-handed neutrinos are very heavy, they induce a very small mass for the left-handed neutrinos, which is proportional to the inverse of the heavy mass.

If it is assumed that the neutrinos interact with the Higgs field with approximately the same strengths as the charged fermions do, the heavy mass should be close to the GUT scale. Note that, in the Standard Model there is just one fundamental mass scale (which can be taken as the scale of $SU(2)_L \times U(1)_Y$ breaking) and all masses (such as the electron or the mass of the Z boson) have to originate from this one.

There are other varieties of seesaw[21] and there is currently great interest in the so-called low-scale seesaw schemes, such as the inverse seesaw mechanism.[22]

The addition of right-handed neutrinos has the effect of adding new mass scales, unrelated to the mass scale of the Standard Model, hence the observation of heavy right-handed neutrinos would reveal physics beyond the Standard Model. Right-handed neutrinos would help to explain the origin of matter through a mechanism known as leptogenesis.

Other sources

There are alternative ways to modify the standard model that are similar to the addition of heavy right-handed neutrinos (e.g., the addition of new scalars or fermions in triplet states) and other modifications that are less similar (e.g., neutrino masses from loop effects and/or from suppressed couplings). One example of the last type of models is provided by certain versions supersymmetric extensions of the standard model of fundamental interactions, where R parity is not a symmetry. There, the exchange of supersymmetric particles such as squarks and sleptons can break the lepton number and lead to neutrino masses. These interactions are normally excluded from theories as they come from a class of interactions that lead to unacceptably rapid proton decay if they are all included. These models have little predictive power and are not able to provide a cold dark matter candidate.

3.6.6 Oscillations in the Early Universe

During the early universe when particle concentrations were high and temperatures hot, neutrino oscillations can behave differently.[23] Depending on neutrino mixing-angle parameters and masses, a broad spectrum of behavior may arise including vacuum-like neutrino oscillations, smooth evolution, or self-maintained coherence. The physics for this system is non-trivial and involves neutrino oscillations in a dense neutrino gas.

3.6.7 See also

- MSW effect

- Majoron

- Neutral kaon mixing

- Neutral particle oscillation

3.6.8 Notes

[1] More formally, the neutrinos are emitted in an entangled state with the other bodies in the decay or reaction, and the mixed state is properly described by a density matrix. However, for all practical situations, the other particles in the decay may be well localized in time and space (e.g. to

within a nuclear distance), leaving their momentum with a large spread. When these partner states are projected out, the neutrino is left in a state that for all intents and purposes behaves as the simple superposition of mass states described here. See *[10] for more information.

3.6.9 References

[1] Barger, Vernon; Marfatia, Danny; Whisnant, Kerry Lewis (2012). *The Physics of Neutrinos.* Princeton University Press. ISBN 0-691-12853-7.

[2] B. Pontecorvo (1957). "Mesonium and anti-mesonium". *Zh. Eksp. Teor. Fiz.* **33**: 549–551. reproduced and translated in *Sov. Phys. JETP* **6**: 429. 1957. Missing or empty |title= (help) and B. Pontecorvo (1967). "Neutrino Experiments and the Problem of Conservation of Leptonic Charge". *Zh. Eksp. Teor. Fiz.* **53**: 1717. reproduced and translated in Pontecorvo, B. (1968). "Neutrino Experiments and the Problem of Conservation of Leptonic Charge". *Sov. Phys. JETP* **26**: 984. Bibcode:1968JETP...26..984P.

[3] M. C. Gonzalez-Garcia and Michele Maltoni (2008). "Phenomenology with Massive Neutrinos". *Physics Reports* **460**: 1–129. arXiv:0704.1800. Bibcode:2008PhR...460....1G. doi:10.1016/j.physrep.2007.12.004.

[4] Davis, Raymond; Harmer, Don S.; Hoffman, Kenneth C. (1968). "Search for Neutrinos from the Sun". *Physical Review Letters* **20** (21): 1205–1209. Bibcode:1968PhRvL..20.1205D. doi:10.1103/PhysRevLett.20.1205. ISSN 0031-9007.

[5] Q. Ahmad (SNO Collaboration), Q. et al. (2001). "Measurement of the Rate of $\nu_e + d \rightarrow p + p + e$<sup– Interactions Produced by ^{8}B Solar Neutrinos at the Sudbury Neutrino Observatory". *Physical Review Letters* **87** (7). arXiv:nucl-ex/0106015. Bibcode:2001PhRvL..87g1301A. doi:10.1103/PhysRevLett.87.071301. ISSN 0031-9007.

[6] Y. Fukudae (Super-Kamiokande Collaboration) et al. (1998). "Evidence for Oscillation of Atmospheric Neutrinos". *Physical Review Letters* **81** (8): 1562–1567. arXiv:hep-ex/9807003. Bibcode:1998PhRvL..81.1562F. doi:10.1103/PhysRevLett.81.1562. ISSN 0031-9007.

[7] F. P. An (Daya Bay Collaboration) et al. (2012). "Observation of Electron-Antineutrino Disappearance at Daya Bay". *Physical Review Letters* **108** (17). arXiv:1203.1669. Bibcode:2012PhRvL.108q1803A. doi:10.1103/PhysRevLett.108.171803. ISSN 0031-9007.

[8] Kim, Soo-Bong; for RENO collaboration (2012). "Observation of Reactor Electron Antineutrino Disappearance in the RENO Experiment". arXiv:1204.0626v2 [hep-ex].

[9] N. Agafonova (OPERA Collaboration) et al. (2010). "Observation of a first ντ candidate event in the OPERA experiment in the CNGS beam". *Physics Letters B* **691** (3): 138–145. arXiv:1006.1623. Bibcode:2010PhLB..691..138A. doi:10.1016/j.physletb.2010.06.022. ISSN 0370-2693.

[10] Andrew G. Cohen, Sheldon L. Glashow, and Zoltan Ligeti (2009). "Disentangling neutrino oscillations". *Physics Letters B* **678** (2): 191. arXiv:0810.4602. Bibcode:2009PhLB..678..191C. doi:10.1016/j.physletb.2009.06.020.

[11] "Neutrino shape-shift points to new physics" *Physics News*, 19 July 2013.

[12] "Neutrino 'flavour' flip confirmed" *BBC News*, 19 July 2013.

[13] Z. Maki, M. Nakagawa, and S. Sakata (1962). "Remarks on the Unified Model of Elementary Particles". *Progress of Theoretical Physics* **28** (5): 870. Bibcode:1962PThPh..28..870M. doi:10.1143/PTP.28.870.

[14] B. Pontecorvo (1967). "Neutrino Experiments and the Problem of Conservation of Leptonic Charge". *Zh. Eksp. Teor. Fiz.* **53**: 1717. reproduced and translated in Pontecorvo, B. (1968). "Neutrino Experiments and the Problem of Conservation of Leptonic Charge". *Sov. Phys. JETP* **26**: 984. Bibcode:1968JETP...26..984P.

[15] Raymond Davis Jr., Don S. Harmer, and Kenneth C. Hoffman (1968). "Search for Neutrinos from the Sun". *Physical Review Letters* **20** (21): 1205. Bibcode:1968PhRvL..20.1205D. doi:10.1103/PhysRevLett.20.1205.

[16] V. Gribov and B. Pontecorvo (1969). "Neutrino astronomy and lepton charge". *Physics Letters B* **28** (7): 493. Bibcode:1969PhLB...28..493G. doi:10.1016/0370-2693(69)90525-5.

[17] . J. Schechter, J.W.F. Valle; Valle (1980). "Neutrino Masses in SU(2) x U(1) Theories". *Physical Review D* **22** (9): 2227. Bibcode:1980PhRvD..22.2227S. doi:10.1103/PhysRevD.22.2227.

[18] S. Eidelman et al. (2004). "Particle Data Group - The Review of Particle Physics". *Physics Letters B* **592** (1): 1. arXiv:astro-ph/0406663. Bibcode:2004PhLB..592....1P. doi:10.1016/j.physletb.2004.06.001. Chapter 15: *Neutrino mass, mixing, and flavor change*. Revised September 2005.

[19] K.A. Olive (Particle Data Group) et al. (2014). "2014 Review of Particle Physics".

[20] K. Nakamura et al. (2010). "Review of Particle Physics". *Journal of Physics G* **37** (7A): 1. Bibcode:2010JPhG...37g5021N. doi:10.1088/0954-3899/37/7a/075021.

[21] J. W. F. Valle (2006). "Neutrino physics overview". *Journal of Physics: Conference Series* **53** (1): 473. arXiv:hep-ph/0608101. Bibcode:2006JPhCS..53..473V. doi:10.1088/1742-6596/53/1/031.

[22] R.N. Mohapatra and J. W. F. Valle (1986). "Neutrino Mass and Baryon Number Nonconservation in Superstring Models". *Physical Review D* **34** (5): 1642. Bibcode:1986PhRvD..34.1642M. doi:10.1103/PhysRevD.34.1642.

[23] Kostelecký, Alan; Samuel, Stuart (1994). "Nonlinear neutrino oscillations in the expanding universe". *Phys. Rev. D* **49** (4): 1740–1757. Bibcode:1994PhRvD..49.1740K. doi:10.1103/PhysRevD.49.1740.

3.6.10   Further reading

- Gonzalez-Garcia; Nir (2002). "Neutrino Masses and Mixing: Evidence and Implications". *Reviews of Modern Physics* **75** (2): 345–402. arXiv:hep-ph/0202058. Bibcode:2003RvMP...75..345G. doi:10.1103/RevModPhys.75.345.

- Maltoni; Schwetz; Tortola; Valle (2004). "Status of global fits to neutrino oscillations". *New Journal of Physics* **6**: 122. arXiv:hep-ph/0405172. Bibcode:2004NJPh....6..122M. doi:10.1088/1367-2630/6/1/122.

- Fogli; Lisi; Marrone; Montanino; Palazzo; Rotunno (2012). "Global analysis of neutrino masses, mixings, and phases: Entering the era of leptonic CP violation searches". *Physical Review D* **86** (1): 013012. arXiv:1205.5254. doi:10.1103/PhysRevD.86.013012.

- Forero; Tortola; Valle (2012). "Global status of neutrino oscillation parameters after Neutrino-2012". *Physical Review D* **86** (7): 073012. arXiv:1205.4018. Bibcode:2012PhRvD..86g3012F. doi:10.1103/PhysRevD.86.073012.

3.6.11   External links

- Maury Goodman, "The Neutrino Oscillation Industry" (2006). *(Provides links to many other neutrino oscillation websites.)*

- Review Articles on arxiv.org

3.7   PMNS matrix

In particle physics, the **Pontecorvo–Maki–Nakagawa–Sakata matrix** (**PMNS matrix**), **Maki–Nakagawa–Sakata matrix** (**MNS matrix**), **lepton mixing matrix**, or **neutrino mixing matrix**, is a unitary matrix[note 1] which contains information on the mismatch of quantum states of neutrinos when they propagate freely and when they take part in the weak interactions. It is important in the understanding of neutrino oscillation. This matrix was introduced in 1962 by Ziro Maki, Masami Nakagawa and Shoichi Sakata,[1] to explain the neutrino oscillations predicted by Bruno Pontecorvo.[2]

3.7.1   The PMNS matrix

The Standard Model of particle physics contains three generations or "flavors" of neutrinos, ν_e, ν_μ, and ν_τ labeled according to the charged leptons with which they partner in the charged-current weak interaction. These three eigenstates of the weak interaction form a complete, orthonormal basis for the Standard Model neutrino. Similarly, one can construct an eigenbasis out of three neutrino states of definite mass, ν_1, ν_2, and ν_3, which diagonalize the neutrino's free-particle Hamiltonian. Observations of neutrino oscillation have experimentally determined that for neutrinos, like the quarks, these two eigenbases are not the same - they are "rotated" relative to each other. Each flavor state can thus be written as a superposition of mass eigenstates, and vice-versa. The PMNS matrix, with components U_{ai} corresponding to the amplitude of mass eigenstate i in flavor a, parameterizes the unitary transformation between the two bases:

$$\begin{bmatrix} \nu_e \\ \nu_\mu \\ \nu_\tau \end{bmatrix} = \begin{bmatrix} U_{e1} & U_{e2} & U_{e3} \\ U_{\mu1} & U_{\mu2} & U_{\mu3} \\ U_{\tau1} & U_{\tau2} & U_{\tau3} \end{bmatrix} \begin{bmatrix} \nu_1 \\ \nu_2 \\ \nu_3 \end{bmatrix}.$$

The vector on the left represents a generic neutrino state expressed in the flavor basis, and on the right is the PMNS matrix multiplied by a vector representing the same neutrino state in the mass basis. A neutrino of a given flavor α is thus a "mixed" state of neutrinos with different mass: if one could measure directly that neutrino's mass, it would be found to have mass m_i with probability $|U_{ai}|^2$.

The PMNS matrix for antineutrinos is identical to the matrix for neutrinos under CPT symmetry.

Due to the difficulties of detecting neutrinos, it is much more difficult to determine the individual coefficients than in the equivalent matrix for the quarks (the CKM matrix).

Assumptions

As noted above, PMNS matrix is unitary (i.e. the sum of the square of the values in each row and in each column, which represent the probabilities of different possible events given the same starting point, add up to 100%) in the simplest Standard Model case in which there are three generations of neutrinos with Dirac mass that oscillate between three neutrino mass eigenvalues, an assumption that is made when best fit values for its parameters are calculated.

The PMNS matrix is not necessarily unitary and additional parameters are necessary to describe all possible neutrino mixing parameters, in other models of neutrino oscillation and mass generation, such as the see-saw model, and in general, in the case of neutrinos that have Majorana mass rather than Dirac mass.

There are also additional mass parameters and mixing angles in a simple extension of the PMNS matrix in which there are more than three flavors of neutrinos, regardless of the character of neutrino mass. As of July 2014, scientists studying neutrino oscillation are actively considering fits of the experimental neutrino oscillation data to an extended PMNS matrix with a fourth, light "sterile" neutrino and four mass eigenvalues, although the current experimental data tends to disfavor that possibility.[3][4][5]

Parameterization

In general, there are nine degrees of freedom in any three by three matrix, and in the PMNS matrix, because it is a matrix whose directly physically observable values (the square of the respective entries) are real numbers between zero and 1 form a unitary matrix, the matrix can thus be fully described by four free parameters from which all physically observable properties of the matrix can be discerned.[6] The PMNS matrix is most commonly parameterized by three mixing angles (θ_{12}, θ_{23} and θ_{13}) and a single phase called δ_{CP} related to charge-parity violations (i.e. differences in the rates of oscillation between two states with opposite starting points which makes the order in time in which events take place necessary to predict their oscillation rates), in which case the matrix can be written as:

$$\begin{bmatrix} 1 & 0 & 0 \\ 0 & c_{23} & s_{23} \\ 0 & -s_{23} & c_{23} \end{bmatrix} \begin{bmatrix} c_{13} & 0 & s_{13}e^{-i\delta_{CP}} \\ 0 & 1 & 0 \\ -s_{13}e^{i\delta_{CP}} & 0 & c_{13} \end{bmatrix} \begin{bmatrix} c_{12} & s_{12} & 0 \\ -s_{12} & c_{12} & 0 \\ 0 & 0 & 1 \end{bmatrix}$$

$$= \begin{bmatrix} c_{12}c_{13} & s_{12}c_{13} & s_{13}e^{-i\delta_{CP}} \\ -s_{12}c_{23} - c_{12}s_{23}s_{13}e^{i\delta_{CP}} & c_{12}c_{23} - s_{12}s_{23}s_{13}e^{i\delta_{CP}} & s_{23}c_{13} \\ s_{12}s_{23} - c_{12}c_{23}s_{13}e^{i\delta_{CP}} & -c_{12}s_{23} - s_{12}c_{23}s_{13}e^{i\delta_{CP}} & c_{23}c_{13} \end{bmatrix}$$

where s_{ij} and c_{ij} are used to denote $\sin\theta_{ij}$ and $\cos\theta_{ij}$ respectively. In the case of Majorana neutrinos, two extra complex phases are needed, as the phase of Majorana fields cannot be freely redefined due to the condition $\nu = \nu^c$. An infinite number of possible parameterizations exist; one other common example being the Wolfenstein parameterization.

The mixing angles have been measured by a variety of experiments (see neutrino mixing for a description). The CP-violating phase δ_{CP} has not been measured directly, but estimates can be obtained by fits using the other measurements.

Experimentally measured parameter values

As of July 2014, the current best directly measured values are:[7][8]

$$\sin^2 2\theta_{12} = 0.857 \pm 0.024$$
$$\sin^2 2\theta_{23} > 0.95$$
$$\sin^2 2\theta_{13} = 0.095 \pm 0.010$$

while the current best-fit values, using direct and indirect measurements, from NuFit are:[9][10]

$$\theta_{12}[°] = 33.36^{+0.81}_{-0.78}$$
$$\theta_{23}[°] = 40.0^{+2.1}_{-1.5} \text{ or } 50.4^{+1.3}_{-1.3}$$
$$\theta_{13}[°] = 8.66^{+0.44}_{-0.46}$$
$$\delta_{CP}[°] = 300^{+66}_{-138}$$

Notes regarding the best fit parameter values

- These best fit values imply that there is much more neutrino mixing than there is mixing between the quark flavors in the CKM matrix (in the CKM matrix, the corresponding mixing angles are $\theta_{12} = 13.04°\pm0.05°$, $\theta_{23} = 2.38°\pm0.06°$, $\theta_{13} = 0.201°\pm0.011°$).

- These values are inconsistent with tribimaximal neutrino mixing (i.e. $\theta_{12} = \theta_{23} = 45°$, $\theta_{13} = 0°$) at a statistical significance of more than five standard deviations. Tribimaximal neutrino mixing was a common assumption in theoretical physics papers analyzing neutrino oscillation before more precise measurements were available.

- A value of θ_{23} equal to exactly 45 degrees, which would imply maximal mixing between the second and third neutrino mass eigenstates, is ruled out with a statistical significance in excess of 2 standard deviations.[10]

- The alternative choices for θ_{23} are referred to as "first quadrant" and "second quadrant" values. The data favor the first quadrant value over the second quadrant value with a statistical significance of 1.5 standard deviations in a "normal mass hierarchy" context (i.e. where the second neutrino mass eigenstate is lighter than the third neutrino mass eigenstate), but there is not a statistically significant preference between the two values in the case of an "inverted mass hierarchy" (i.e. where the second neutrino mass eigenstate is heavier than the third neutrino mass eigenstate).[10] This is the only PMNS matrix parameter which is strongly sensitive to the mass hierarchy of the neutrino masses given the currently available experimental data.[10]

- The extent to which the best fit value for δ_{CP} is meaningful should not be overstated. The best fit value for δ_{CP} is consistent with zero at the 0.9 standard deviation level, since in circular coordinates 0 degrees and 360 degrees are equivalent. Generally speaking, in particle physics, experimental results that are within 2 standard deviations of each other are called "consistent" with each other. Currently, all possible values for δ_{CP} are with 1.8 standard deviations of the best fit values, so all possible values of δ_{CP} are "consistent" with the experimental data, even though those values closer to the best fit value are somewhat more likely to be correct.

3.7.2 See also

- Neutrino oscillations

- Koide formula

- Cabibbo–Kobayashi–Maskawa matrix

3.7.3 Notes

[1] The PMNS matrix is not unitary in the seesaw model.

3.7.4 References

[1] Maki, Z; Nakagawa, M.; Sakata, S. (1962). "Remarks on the Unified Model of Elementary Particles". *Progress of Theoretical Physics* **28**: 870. Bibcode:1962PThPh..28..870M. doi:10.1143/PTP.28.870.

[2] Pontecorvo, B. (1957). "Inverse beta processes and nonconservation of lepton charge". *Zhurnal Éksperimental' noĭ i Teoreticheskoĭ Fiziki* **34**: 247. reproduced and translated in *Soviet Physics JETP* **7**: 172. 1958.

[3] Kayser, Boris (February 13, 2014). "Are There Sterile Neutrinos?". arXiv:1402.3028 [hep-ph].

[4] Esmaili, Arman; Kemp, Ernesto; Peres, O. L. G.; Tabrizi, Zahra (30 Oct 2013). "Probing light sterile neutrinos in medium baseline reactor experiments". arXiv:1308.6218 [hep-ph].

[5] F.P. An, *et al.*(Daya Bay collaboration) (July 27, 2014). "Search for a Light Sterile Neutrino at Daya Bay". arXiv:1407.7259 [hep-ex].

[6] Valle, J. W. F. (2006). "Neutrino physics overview". *Journal of Physics: Conference Series* **53**: 473. arXiv:hep-ph/0608101. Bibcode:2006JPhCS..53..473V. doi:10.1088/1742-6596/53/1/031.

[7] J. Beringer *et al.* (Particle Data Group) (2012 and 2013 partial update for the 2014 edition). "PDGLive: Neutrino Mixing". Particle Data Group. Retrieved 2014-08-21. Check date values in: |date= (help)

[8] J. Beringer *et al.* (Particle Data Group) (2012). "Review of Particle Physics". *Physical Review D* **86**: 010001. Bibcode:2012PhRvD..86a0001B. doi:10.1103/PhysRevD.86.010001.

[9] Gonzalez-Garcia, M. C.; Maltoni, M.; Salvado, J.; Schwetz, T. (June 2014). "NuFit 1.3". Retrieved 2014-07-09.

[10] Gonzalez-Garcia, M. C.; Maltoni, Michele; Salvado, Jordi; Schwetz, Thomas (21 December 2012). "Global fit to three neutrino mixing: Critical look at present precision". *Journal of High Energy Physics* **2012** (12): 123. arXiv:1209.3023. Bibcode:2012JHEP...12..123G. doi:10.1007/JHEP12(2012)123.

3.8 Solar neutrino problem

The **solar neutrino problem** was a major discrepancy between measurements of the numbers of neutrinos flowing through the Earth and theoretical models of the solar interior, lasting from the mid-1960s to about 2002. The discrepancy has since been resolved by new understanding of neutrino physics, requiring a modification of the Standard Model of particle physics – specifically, neutrino oscillation. Essentially, as neutrinos have mass, they can change from the type that had been expected to be produced in the Sun's interior into two types that would not be caught by the detectors in use at the time.

3.8.1 Introduction

The Sun is a natural nuclear fusion reactor, powered by a proton–proton chain reaction which converts four hydrogen nuclei (protons) into alpha particles, neutrinos, positrons and energy. The excess energy is released as gamma rays and as kinetic energy of the particles and as neutrinos — which travel from the Sun's core to Earth without any appreciable absorption by the Sun's outer layers.

As neutrino detectors became sensitive enough to measure the flow of neutrinos from the Sun, it became clear that the number detected was lower than that predicted by models of the solar interior. In various experiments, the number of detected neutrinos was between one third and one half of the predicted number. This came to be known as the *solar neutrino problem*.

3.8.2 Measurements

In the late 1960s, Ray Davis's and John N. Bahcall's Homestake Experiment was the first to measure the flux of neutrinos from the Sun and detect a deficit. The experiment used a chlorine-based detector. Many subsequent radiochemical and water Cherenkov detectors confirmed the deficit, including the Sudbury Neutrino Observatory.

The expected number of solar neutrinos had been computed based on the standard solar model which Bahcall had helped to establish and which gives a detailed account of the Sun's internal operation.

In 2002 Ray Davis and Masatoshi Koshiba won part of the Nobel Prize in Physics for experimental work that found the number of solar neutrinos was around a third of the number predicted by the standard solar model.*[1]

3.8.3 Proposed solutions

Changes to the solar model

Early attempts to explain the discrepancy proposed that the models of the Sun were wrong, i.e. the temperature and pressure in the interior of the Sun were substantially different from what was believed. For example, since neutrinos measure the amount of current nuclear fusion, it was suggested that the nuclear processes in the core of the Sun might have temporarily shut down. Since it takes thousands of years for heat energy to move from the core to the surface of the Sun, this would not immediately be apparent.

However, these solutions were rendered untenable by advances in both helioseismology, the study of how waves propagate through the Sun, and improved neutrino measurements.

Helioseismology observations made it possible to measure the interior temperatures of the Sun; these agreed with the standard solar models. (There are unresolved problems of the structure of what was found with helioseismology. Instead of the old "pot-on-the-stove" model of vertical convection, horizontal jet streams were found in the top layer of the convective zone. Small ones were found around each pole and larger ones extended to the equator. As might be expected, these had different speeds.)

Detailed observations of the neutrino spectrum from the more advanced neutrino observatories also produced results which no adjustment of the solar model could accommodate. In effect, overall lower neutrino flux (which the Homestake experiment results found) required a reduction in the solar core temperature. However, details in the energy spectrum of the neutrinos required a higher core temperature. This happens because different energy neutrinos are produced by different nuclear reactions, whose rates have different dependence upon the temperature; in order to match parts of the neutrino spectrum a higher temperature is needed. An exhaustive analysis of alternatives found that no combination of adjustments of the solar model was capable of producing the observed neutrino energy spec-

trum, and all adjustments that could be made to the model worsened some aspect of the discrepancies.*[2]

3.8.4 Resolution

Main article: Neutrino oscillation

The solar neutrino problem was resolved with an improved understanding of the properties of neutrinos. According to the Standard Model of particle physics, there are three different kinds of neutrinos:

- *electron neutrinos* (which are the ones produced in the Sun and the ones detected by the above-mentioned experiments, in particular the chlorine-detector Homestake Mine experiment),

- *muon neutrinos*, and

- *tau neutrinos.*

Through the 1970s, it was widely believed that neutrinos were massless and their types were invariant. However, in 1968 Pontecorvo proposed that if neutrinos had mass, then they could change from one type to another.*[3] Thus, the "missing" solar neutrinos could be electron neutrinos which changed into other types along the way to Earth and therefore were not seen by the detectors in the Homestake Mine and contemporary neutrino observatories.

The supernova 1987A produced an indication that neutrinos might have mass, because of the difference in time of arrival of the neutrinos detected at Kamiokande and IMB.*[4] However, because very few neutrino events were detected it was difficult to draw any conclusions with certainty. In addition, whether neutrinos have mass or not could have been more definitively established had Kamiokande and IMB both had high precision timers which would have recorded how long it took the neutrino burst to travel through the Earth. If neutrinos were massless, they would travel at the speed of light; if they had mass, they would travel at velocities slightly less than that of light. Because the detectors were not intended for supernova neutrino detection, however, this was not done.

The first strong evidence for neutrino oscillation came in 1998 from the Super-Kamiokande collaboration in Japan.*[5] It produced observations consistent with muon-neutrinos (produced in the upper atmosphere by cosmic rays) changing into tau-neutrinos. What was proved was that fewer neutrinos were detected coming through the Earth than could be detected coming directly above the detector. Not only that, their observations only concerned muon neutrinos coming from the interaction of cosmic rays with the Earth's atmosphere. No tau neutrinos were observed at Super-Kamiokande.

The convincing evidence for solar neutrino oscillation came in 2001 from the Sudbury Neutrino Observatory (SNO) in Canada. It detected all types of neutrinos coming from the Sun,[*][6] and was able to distinguish between electron-neutrinos and the other two flavors (but could not distinguish the muon and tau flavours), by uniquely using heavy water as the detection medium. After extensive statistical analysis, it was found that about 35% of the arriving solar neutrinos are electron-neutrinos, with the others being muon- or tau-neutrinos.[*][7] The total number of detected neutrinos agrees quite well with the earlier predictions from nuclear physics, based on the fusion reactions inside the Sun.

3.8.5   See also

- Neutral particle oscillation

3.8.6   References

[1] "The Nobel Prize in Physics 2002". Retrieved 2006-07-18.

[2] Haxton, W.C. Annual Reviews of Astronomy and Astrophysics, vol 33, pp. 459–504, 1995.

[3] Gribov, V. (1969).   "Neutrino astronomy and lepton charge".   *Physics Letters B* **28** (7): 493–496. Bibcode:1969PhLB...28..493G.   doi:10.1016/0370-2693(69)90525-5.

[4] W. David Arnett & Jonathan L. Rosner (1987).   "Neutrino mass limits from SN1987A".   *Physical Review Letters* **58** (18): 1906. Bibcode:1987PhRvL..58.1906A. doi:10.1103/PhysRevLett.58.1906.

[5] Detecting Massive Neutrinos; August 1999; *Scientific American*; by Kearns, Kajita, Totsuka.

[6] Q.R. Ahmad, et al., "Measurement of the rate of interactions produced by 8B solar neutrinos at the Sudbury Neutrino Observatory," *Physical Review Letters* 87, 071301 (2001)

[7] Arthur B. McDonald, Joshua R. Klein and David L. Wark, 'Solving the Solar Neutrino Problem', *Scientific American*, vol. 288, no. 4 (April 2003), pp. 40–49

3.8.7   External links

- Solar neutrino data
- Solving the Mystery of the Missing Neutrinos
- Raymond Davis Jr.'s logbook
- Nova – The Ghost Particle
- The Solar Neutrino Problem by John N. Bahcall
- The Solar Neutrino Problem, by L. Stockman
- A set of photos of different Neutrino detectors
- John Bahcall's web site

3.9   Sterile neutrino

Sterile neutrinos (or **inert neutrinos**) are hypothetical particles (neutral leptons – neutrinos) that interact only via gravity and do not interact via any of the fundamental interactions of the Standard Model. The term *sterile neutrino* is used to distinguish them from the known *active neutrinos* in the Standard Model, which are charged under the weak interaction.

This term usually refers to neutrinos with right-handed chirality (see right-handed neutrino), which may be added to the Standard Model. Occasionally it is used in a more general sense for any neutral fermion.

The existence of right-handed neutrinos is theoretically well-motivated, as all other known fermions have been observed with left and right chirality, and they can explain the observed active neutrino masses in a natural way. The mass of the right-handed neutrinos themselves is unknown and could have any value between 10^{15} GeV and less than one eV.[*][1]

The number of sterile neutrino types is unknown. This is in contrast to the number of active neutrino types, which has to equal that of charged leptons and quark generations to ensure the anomaly freedom of the electroweak interaction.

The search for sterile neutrinos is an active area of particle physics. If they exist and their mass is smaller than the energies of particles in the experiment, they can be produced in the laboratory, either by mixing between active and sterile neutrinos or in high energy particle collisions. If they are heavier, the only directly observable consequence of their existence would be the observed active neutrino masses. They may, however, be responsible for a number of unexplained phenomena in physical cosmology and astrophysics, including dark matter, baryogenesis or dark radiation.[*][1]

Sterile neutrinos may be Neutral Heavy Leptons (NHLs, or Heavy Neutral Leptons, HNLs).

3.9.1   Motivation

See also: Neutrino: Chirality and Neutrino oscillation

Experimental results show that all produced and observed neutrinos have left-handed helicities (spins antiparallel to momenta), and all antineutrinos have right-handed helicities, within the margin of error. In the massless limit, it means that only one of two possible chiralities is observed for either particle. These are the only helicities (and chiralities) included in the Standard Model of particle interactions.

Recent experiments such as neutrino oscillation, however, have shown that neutrinos have a non-zero mass, which is not predicted by the Standard Model and suggests new, unknown physics. This unexpected mass explains neutrinos with right-handed helicity and antineutrinos with left-handed helicity: since they do not move at the speed of light, their helicity is not relativistic invariant (it is possible to move faster than them and observe the opposite helicity). Yet all neutrinos have been observed with left-handed *chirality*, and all antineutrinos right-handed. Chirality is a fundamental property of particles and *is* relativistic invariant: it is the same regardless of the particle's speed and mass in every reference frame. The question, thus, remains: can neutrinos and antineutrinos be differentiated only by chirality? Or do right-handed neutrinos and left-handed antineutrinos exist as separate particles?

3.9.2 Properties

Such particles would belong to a singlet representation with respect to the strong interaction and the weak interaction, having zero electric charge, zero weak hypercharge, zero weak isospin, and, as with the other leptons, no color, although they do have a B-L of −1. If the standard model is embedded in a hypothetical SO(10) grand unified theory, they can be assigned an X charge of −5. The left-handed anti-neutrino has a B-L of 1 and an X charge of 5.

Due to the lack of charge, sterile neutrinos would not interact electromagnetically, weakly, or strongly, making them extremely difficult to detect. They have Yukawa interactions with ordinary leptons and Higgs bosons, which via the Higgs mechanism lead to mixing with ordinary neutrinos. In experiments involving energies larger than their mass they would participate in all processes in which ordinary neutrinos take part, but with a quantum mechanical probability that is suppressed by the small mixing angle. That makes it possible to produce them in experiments if they are light enough. They would also interact gravitationally due to their mass, however, and if they are heavy enough, they could explain cold dark matter or warm dark matter. In some grand unification theories, such as SO(10), they also interact via gauge interactions which are extremely suppressed at ordinary energies because their gauge boson is extremely massive. They do not appear at all in some other GUTs, such as the Georgi–Glashow model (i.e. all its SU(5)

charges or quantum numbers are zero).

Mass

All particles are initially massless under the Standard Model, since there are no Dirac mass terms in the Standard Model's Lagrangian. The only mass terms are generated by the Higgs mechanism, which produces non-zero Yukawa couplings between the left-handed components of fermions, the Higgs field, and their right-handed components. This occurs when the **SU**(2) doublet Higgs field ϕ acquires its non-zero vacuum expectation value, ν , spontaneously breaking its $SU(2)_L \times U(1)$ symmetry, and thus yielding non-zero Yukawa couplings:

$$\mathcal{L}(\psi) = \bar{\psi}(i\not{\partial})\psi - G\bar{\psi}_L\phi\psi_R$$

Such is the case for charged leptons, like the electron; but within the standard model, the right-handed neutrino does not exist, so even with a Yukawa coupling neutrinos remain massless. In other words, there are no mass terms for neutrinos under the Standard Model: the model only contains a left-handed neutrino and its antiparticle, a right-handed antineutrino, for each generation, produced in weak eigenstates during weak interactions. See neutrino masses in the Standard Model for a detailed explanation.

In the seesaw mechanism, one eigenvector of the neutrino mass matrix, which includes sterile neutrinos, is predicted to be significantly heavier than the other.

A sterile neutrino would have the same weak hypercharge, weak isospin, and mass as its antiparticle. For any charged particle, for example the electron, this is not the case: its antiparticle, the positron, has opposite electric charge, among other opposite charges. Similarly, an up quark has a charge of $+\frac{2}{3}$ and (for example) a color charge of red, while its antiparticle has an electric charge of $-\frac{2}{3}$ and a color charge of anti-red.

Dirac and Majorana terms Sterile neutrinos allow the introduction of a **Dirac mass** term as usual. This can yield the observed neutrino mass, but it requires that the strength of the Yukawa coupling be much weaker for the electron neutrino than the electron, without explanation. Similar problems (although less severe) are observed in the quark sector, where the top and bottom masses differ by a factor 40.

Unlike for the left-handed neutrino, a **Majorana mass** term can be added for a sterile neutrino without violating local symmetries (weak isospin and weak hypercharge) since it has no weak charge. However, this would still violate total lepton number.

It is possible to include **both** Dirac and Majorana terms: this is done in the seesaw mechanism (below). In addition to satisfying the Majorana equation, if the neutrino were also its own antiparticle, then it would be the first Majorana fermion. In that case, it could annihilate with another neutrino, allowing neutrinoless double beta decay. The other case is that it is a Dirac fermion, which is not its own antiparticle.

To put this in mathematical terms, we have to make use of the transformation properties of particles. For free fields, a Majorana field is defined as an eigenstate of charge conjugation. However, neutrinos interact only via the weak interactions, which are not invariant under charge conjugation (C), so an interacting Majorana neutrino cannot be an eigenstate of C. The generalized definition is: "a Majorana neutrino field is an eigenstate of the CP transformation" . Consequently, Majorana and Dirac neutrinos would behave differently under CP transformations (actually Lorentz and CPT transformations). Also, a massive Dirac neutrino would have nonzero magnetic and electric dipole moments, whereas a Majorana neutrino would not. However, the Majorana and Dirac neutrinos are different only if their rest mass is not zero. For Dirac neutrinos, the dipole moments are proportional to mass and would vanish for a massless particle. Both Majorana and Dirac mass terms however can appear in the mass Lagrangian.

Seesaw mechanism

Main article: Seesaw mechanism

In addition to the left-handed neutrino, which couples to its family charged lepton in weak charged currents, if there is also a right-handed sterile neutrino partner, a weak isosinglet with no charge, then it is possible to add a Majorana mass term without violating electroweak symmetry. Both neutrinos have mass and handedness is no longer preserved (thus "left or right-handed neutrino" means that the state is mostly left or right-handed). To get the neutrino mass eigenstates, we have to diagonalize the general mass matrix M:

$$m_\nu = \begin{pmatrix} 0 & m_D \\ m_D & M_{NHL} \end{pmatrix}$$

where M_{NHL} is big and m_D is of intermediate size terms.

Apart from empirical evidence, there is also a theoretical justification for the seesaw mechanism in various extensions to the Standard Model. Both Grand Unification Theories (GUTs) and left-right symmetrical models predict the following relation:

$$m_\nu \ll m_D \ll M_{NHL}$$

According to GUTs and left-right models, the right-handed neutrino is extremely heavy: $M_{NHL} \approx 10^5$—10^{12} GeV, while the smaller eigenvalue is approximately equal to

$$m_\nu \approx \frac{m_D^2}{M_{NHL}}$$

This is the seesaw mechanism: as the sterile right-handed neutrino gets heavier, the normal left-handed neutrino gets lighter. The left-handed neutrino is a mixture of two Majorana neutrinos, and this mixing process is how sterile neutrino mass is generated.

3.9.3   Detection attempts

The production and decay of sterile neutrinos could happen through the mixing with virtual ("off mass shell") neutrinos. There were several experiments set up to discover or observe NHLs, for example the NuTeV (E815) experiment at Fermilab or LEP-l3 at CERN. They all lead to establishing limits to observation, rather than actual observation of those particles. If they are indeed a constituent of dark matter, sensitive X-ray detectors would be needed to observe the radiation emitted by their decays.[2]

Sterile neutrinos may mix with ordinary neutrinos via a Dirac mass after electroweak symmetry breaking, in analogy to quarks and charged leptons. Sterile neutrinos and (in more-complicated models) ordinary neutrinos may also have Majorana masses. In type 1 seesaw mechanism both Dirac and Majorana masses are used to drive ordinary neutrino masses down and make the sterile neutrinos much heavier than the Standard Model's interacting neutrinos. In some models the heavy neutrinos can be as heavy as the GUT scale ($\approx 10^{15}$ GeV). In other models they could be lighter than the weak gauge bosons W and Z as in the so-called νMSM model where their masses are between GeV and keV. A light (with the mass ≈ 1 eV) sterile neutrino was suggested as a possible explanation of the results of the Liquid Scintillator Neutrino Detector experiment. On April 11, 2007, researchers at the MiniBooNE experiment at Fermilab announced that they had not found any evidence supporting the existence of such a sterile neutrino.[3] More-recent results and analysis have provided some support for the existence of the sterile neutrino.[4][5] Two separate detectors near a nuclear reactor in France found 3% of anti-neutrinos missing. They suggested the existence of a 4th neutrino with a mass of 0.7 keV.[6] Sterile neutrinos are also candidates for dark radiation. Daya Bay has also searched for a light sterile neutrino and excluded some mass regions.[*][7]

The number of neutrinos and the masses of the particles can have large-scale effects that shape the appearance of the cosmic microwave background. The total number of neutrino species, for instance, affects the rate at which the cosmos expanded in its earliest epochs: more neutrinos means a faster expansion. The Planck Satellite 2013 data release found no evidence of additional neutrino-like particles.[*][8]

3.9.4 See also

- MiniBooNE at Fermilab

3.9.5 References

Notes

References

[1] Marco Drewes (2013). "The Phenomenology of Right Handed Neutrinos". *International Journal of Modern Physics E* **22** (8): 1330019. arXiv:1303.6912. Bibcode:2013IJMPE..2230019D. doi:10.1142/S0218301313300191.

[2] Battison, Leila (2011-09-16). "Dwarf galaxies suggest dark matter theory may be wrong". BBC News. Retrieved 2011-09-18.

[3] First_Results (PDF)

[4] Scientific American: "Dimensional Shortcuts", August 2007

[5] Bulbul, E.; Markevitch, M.; Foster, A.; Smith, R.K.; Loewenstein, M.; Randall, S.W. (2014). "Detection of an Unidentified Emission Line in the Stacked X-ray Spectrum of Galaxy Clusters". *The Astrophysical Journal* **789** (1): 13. arXiv:1402.2301v2. Bibcode:2014ApJ...789...13B. doi:10.1088/0004-637X/789/1/13.

[6] The Reactor Antineutrino Anomaly

[7] "Search for a Light Sterile Neutrino at Daya Bay". *Phys. Rev. Lett. 113, 141802.* 1 October 2014. arXiv:1407.7259. Bibcode:2014PhRvL.113n1802A. doi:10.1103/PhysRevLett.113.141802.

[8] Ade, P.A.R.; et al. (Planck Collaboration) (2013). "Planck 2013 results. XVI. Cosmological parameters". arXiv:1303.5076 [astro-ph.CO].

Bibliography

- M. Drewes (2013). "The Phenomenology of Right Handed Neutrinos". *International Journal of Modern Physics E.* arXiv:1303.6912. Bibcode:2013IJMPE..2230019D. doi:10.1142/S0218301313300191.

- A. Merle (2013). "keV Neutrino Model Building". *International Journal of Modern Physics D* **22** (10): 1330020. arXiv:1302.2625. Bibcode:2013IJMPD..2230020M. doi:10.1142/S0218271813300206.

- A. G. Vaitaitis et al. (1999). "Search for Neutral Heavy Leptons in a High-Energy Neutrino Beam". *Physical Review Letters* **83** (24): 4943–4946. arXiv:hep-ex/9908011. Bibcode:1999PhRvL..83.4943V. doi:10.1103/PhysRevLett.83.4943.

- J. A. Formaggio; J. Conrad; M. Shaevitz; A. Vaitaitis (1998). "Helicity effects in neutral heavy lepton decays". *Physical Review D* **57** (11): 7037–7040. Bibcode:1998PhRvD..57.7037F. doi:10.1103/PhysRevD.57.7037.

- K. Nakamura; Particle Data Group (2010). "Review of Particle Physics". *Journal of Physics G* **37** (75021): 075021. Bibcode:2010JPhG...37g5021N. doi:10.1088/0954-3899/37/7A/075021.

3.9.6 External links

- The NuTeV experiment at Fermilab

- The L3 Experiment at CERN

- Experiment Nixes Fourth Neutrino (April 2007 Scientific American)

Chapter 4

Properties

4.1 Lepton number

In particle physics, the **lepton number** is the number of leptons minus the number of antileptons.

In equation form,

$$L = n_\ell - n_{\bar\ell}$$

so all leptons have assigned a value of +1, antileptons −1, and non-leptonic particles 0. Lepton number (sometimes also called lepton charge) is an additive quantum number, which means that its sum is preserved in interactions (as opposed to multiplicative quantum numbers such as parity, where the product is preserved instead).

Beside the leptonic number, **leptonic family numbers** are also defined:

- L_e , the **electronic number** for the electron and the electron neutrino;

- L_μ , the **muonic number** for the muon and the muon neutrino;

- L_τ , the **tauonic number** for the tau and the tau neutrino;

with the same assigning scheme as the leptonic number: +1 for particles of the corresponding family, −1 for the antiparticles, and 0 for leptons of other families or non-leptonic particles.

An example is the muon decay. Like many lepton interactions, muon decay is a Weak Interaction. This is cited as a test for special relativity testing the time dilation effect

4.1.1 Violations of the lepton number conservation laws

In the Standard Model, leptonic family numbers (LF numbers) would be preserved if neutrinos were massless. Since neutrino oscillations have been observed, neutrinos do have a tiny nonzero mass and conservation laws for LF numbers are therefore only approximate. This means the conservation laws are violated, although because of the smallness of the neutrino mass they still hold to a very large degree for interactions containing charged leptons. However, the (total) lepton number conservation law must still hold (under the Standard Model). Thus, it is possible to see rare muon decays such as $\mu \rightarrow e\gamma$ or $\mu N \rightarrow eN$.[1]

Because the lepton number conservation law in fact is violated by chiral anomalies, there are problems applying this symmetry universally over all energy scales. However, the quantum number $B - L$ is much more likely to work and is seen in different models such as the Pati–Salam model.

Experiments such as MEGA and SINDRUM have searched for lepton number violation in muon decays to electrons; MEG set the current branching limit of order $10^{*}-13$ and plans to lower to limit to $10^{*}-14$ after 2016. Some BSM theories such as SUSY predict branching ratios of order $10^{*}-12$ to $10^{*}-14$.[1] The Mu2e experiment in construction has a planned sensitivity of order $10^{*}-17$.

4.1.2 References

[1] "New Limit on the Lepton-Flavor-Violating Decay mu to e+gamma" . *PRL*. 21 Oct 2011. arXiv:1107.5547. Bibcode:2011PhRvL.107q1801A. doi:10.1103/PhysRevLett.107.171801.

- Griffiths, David J. (1987). *Introduction to Elementary Particles*. Wiley, John & Sons, Inc. ISBN 0-471-60386-4.

- Tipler, Paul; Llewellyn, Ralph (2002). *Modern Physics (4th ed.)*. W. H. Freeman. ISBN 0-7167-4345-0.

- M. Raidal et al. (2008). *Eur. Phys. J. C 57, 13*. Missing or empty |title= (help)

4.2 Chirality

A **chiral** phenomenon is one that is not identical to its mirror image (see the article on mathematical chirality). The spin of a particle may be used to define a **handedness**, or helicity, for that particle which, in the case of a massless particle, is the same as chirality. A symmetry transformation between the two is called parity. Invariance under parity by a Dirac fermion is called **chiral symmetry**.

An experiment on the weak decay of cobalt−60 nuclei carried out by Chien-Shiung Wu and collaborators in 1957 demonstrated that parity is not a symmetry of the universe.

4.2.1 Chirality and helicity

See also: Helicity (particle physics)

The helicity of a particle is right-handed if the direction of its spin is the same as the direction of its motion. It is left-handed if the directions of spin and motion are opposite. By convention for rotation, a standard clock, with its spin vector defined by the rotation of its hands, tossed with its face directed forwards, has left-handed helicity. Mathematically, helicity is the sign of the projection of the spin vector onto the momentum vector: left is negative, right is positive.

The chirality of a particle is more abstract. It is determined by whether the particle transforms in a right- or left-handed representation of the Poincaré group. (However, some representations, such as Dirac spinors, have both right- and left-handed components. In cases like this, we can define projection operators that project out either the right or left hand components and discuss the right- and left-handed portions of the representation.)

For massless particles —such as the photon, the gluon, and the (hypothetical) graviton—chirality is the same as helicity; a given massless particle appears to spin in the same direction along its axis of motion regardless of point of view of the observer.

For massive particles —such as electrons, quarks, and neutrinos—chirality and helicity must be distinguished. In the case of these particles, it is possible for an observer to change to a reference frame that overtakes the spinning particle, in which case the particle will then appear to move backwards, and its helicity (which may be thought of as 'apparent chirality') will be reversed.

A *massless* particle moves with the speed of light, so a real observer (who must always travel at less than the speed of light) cannot be in any reference frame where the particle appears to reverse its relative direction, meaning that all real observers see the same chirality. Because of this, the direction of spin of massless particles is not affected by a Lorentz boost (change of viewpoint) in the direction of motion of the particle, and the sign of the projection (helicity) is fixed for all reference frames: the helicity of massless particles is a relativistic invariant (i.e. a quantity whose value is the same in all inertial reference frames).

With the discovery of neutrino oscillation, which implies that neutrinos have mass, the only observed massless particle is the photon. The gluon is also expected to be massless, although the assumption that it is has not been conclusively tested. Hence, these are the only two particles now known for which helicity could be identical to chirality, and only one of them has been confirmed by measurement. All other observed particles have mass and thus may have different helicities in different reference frames. It is still possible that as-yet unobserved particles, like the graviton, might be massless, and hence have invariant helicity like the photon.

4.2.2 Chiral theories

Only left-handed fermions interact with the weak interaction. In most circumstances, two left-handed fermions interact more strongly than right-handed or opposite-handed fermions, implying that the universe has a preference for left-handed chirality, which violates a symmetry of the other forces of nature.

Chirality for a Dirac fermion ψ is defined through the operator γ^5, which has eigenvalues ±1. Any Dirac field can thus be projected into its left- or right-handed component by acting with the projection operators $(1-\gamma^5)/2$ or $(1+\gamma^5)/2$ on ψ.

The coupling of the charged weak interaction to fermions is proportional to the first projection operator, which is responsible for this interaction's parity symmetry violation.

A common source of confusion is due to conflating this operator with the helicity operator. Since the helicity of massive particles is frame-dependent, it might seem that the same particle would interact with the weak force according to one frame of reference, but not another. The resolution to this false paradox is that *the chirality operator is equivalent to helicity for massless fields only*, for which helicity is not frame-dependent. By contrast, for massive particles, *chirality is not the same as helicity*, so there is no frame dependence of the weak interaction: a particle that couples the weak force in one frame, does so in every frame.

A theory that is asymmetric with respect to chiralities is called a *chiral theory*, while a non-chiral (i.e., parity-symmetric) theory is sometimes called a *vector theory*. Many pieces of the Standard Model of physics are non-chiral, which is traceable to anomaly cancellation in chiral theories. Quantum chromodynamics is an example of a *vector theory*, since both chiralities of all quarks appear in the theory, and couple to gluons in the same way.

The electroweak theory, developed in the mid 20th century, is an example of a *chiral theory*. Originally, it assumed that neutrinos were massless, and only assumed the existence of left-handed neutrinos (along with their complementary right-handed antineutrinos). After the observation of neutrino oscillations, which imply that neutrinos are massive like all other fermions, the revised theories of the electroweak interaction now include both right- and left-handed neutrinos. However, it is still a chiral theory, as it does not respect parity symmetry.

The exact nature of the neutrino is still unsettled and so the electroweak theories that have been proposed are somewhat different, but most accommodate the chirality of neutrinos in the same way as was already done for all other fermions.

4.2.3 Chiral symmetry

Vector gauge theories with massless Dirac fermion fields ψ exhibit chiral symmetry, i.e., rotating the left-handed and the right-handed components independently makes no difference to the theory. We can write this as the action of rotation on the fields:

$$\psi_L \to e^{i\theta_L}\psi_L \text{ and } \psi_R \to \psi_R$$

or

$$\psi_L \to \psi_L \text{ and } \psi_R \to e^{i\theta_R}\psi_R.$$

With N flavors, we have unitary rotations instead: $U(N)_L \times U(N)_R$.

More generally, we write the right-handed and left-handed states as a projection operator acting on a spinor. The right-handed and left-handed projection operators are

$$P_R = \frac{1 + \gamma^5}{2}$$

and

$$P_L = \frac{1 - \gamma^5}{2}$$

Massive fermions do not exhibit chiral symmetry, as the mass term in the Lagrangian, $m\,\psi$, breaks chiral symmetry explicitly.

Spontaneous chiral symmetry breaking may also occur in some theories, as it most notably does in quantum chromodynamics.

The chiral symmetry transformation can be divided into a component that treats the left-handed and the right-handed parts equally, known as **vector symmetry**, and a component that actually treats them differently, known as **axial symmetry**.[*][1] A scalar field model encoding chiral symmetry and its breaking is the sigma model.

The most common application is expressed as equal treatment of clockwise and counter-clockwise rotations from a fixed frame of reference.

The general principle is often referred to by the name **chiral symmetry**. The rule is absolutely valid in the classical mechanics of Newton and Einstein, but results from quantum mechanical experiments show a difference in the behavior of left-chiral versus right-chiral subatomic particles.

Example: *u* and *d* quarks in QCD

Consider quantum chromodynamics (QCD) with two *massless* quarks *u* and *d* (massive fermions do not exhibit chiral symmetry). The Lagrangian reads

$$\mathcal{L} = \overline{u}\,i\slashed{D}\,u + \overline{d}\,i\slashed{D}\,d + \mathcal{L}_{\text{gluons}}\,.$$

In terms of left-handed and right-handed spinors, it reads

$$\mathcal{L} = \overline{u}_L\,i\slashed{D}\,u_L + \overline{u}_R\,i\slashed{D}\,u_R + \overline{d}_L\,i\slashed{D}\,d_L + \overline{d}_R\,i\slashed{D}\,d_R + \mathcal{L}_{\text{gluons}}\,.$$

(Here, *i* is the imaginary unit and $\slashed{D}$ the Dirac operator.)

Defining

$$q = \begin{bmatrix} u \\ d \end{bmatrix},$$

it can be written as

$$\mathcal{L} = \overline{q}_L\,i\slashed{D}\,q_L + \overline{q}_R\,i\slashed{D}\,q_R + \mathcal{L}_{\text{gluons}}\,.$$

The Lagrangian is unchanged under a rotation of q_L by any 2 x 2 unitary matrix L, and q_R by any 2 x 2 unitary matrix R.

This symmetry of the Lagrangian is called *flavor chiral symmetry*, and denoted as $U(2)_L \times U(2)_R$. It decomposes into

$$SU(2)_L \times SU(2)_R \times U(1)_V \times U(1)_A .$$

The singlet vector symmetry, $U(1)_V$, acts as

$$q_L \to e^{i\theta} q_L \qquad q_R \to e^{i\theta} q_R ,$$

and corresponds to baryon number conservation.

The singlet axial group $U(1)_A$ acts as

$$q_L \to e^{i\theta} q_L \qquad q_R \to e^{-i\theta} q_R ,$$

and it does not correspond to a conserved quantity, because it is explicitly violated due to a quantum anomaly.

The remaining chiral symmetry $SU(2)_L \times SU(2)_R$ turns out to be spontaneously broken by a quark condensate $\langle \bar{q}_R^a q_L^b \rangle = v\delta^{ab}$ formed through nonperturbative action of QCD gluons, into the diagonal vector subgroup $SU(2)_V$ known as isospin. The Goldstone bosons corresponding to the three broken generators are the three pions. As a consequence, the effective theory of QCD bound states like the baryons, must now include mass terms for them, ostensibly disallowed by unbroken chiral symmetry. Thus, this chiral symmetry breaking induces the bulk of hadron masses, such as those for the nucleons−−in effect, the bulk of the mass of all visible matter.

In the real world, because of the nonvanishing and differing masses of the quarks, $SU(2)_L \times SU(2)_R$ is only an approximate symmetry*[2] to begin with, and therefore the pions are not massless, but have small masses: they are pseudo-Goldstone bosons.*[3]

More Flavors

For more "light" quark species, N flavors in general, the corresponding chiral symmetries are $U(N)_L \times U(N)_R$, decomposing into

$$SU(N)_L \times SU(N)_R \times U(1)_V \times U(1)_A ,$$

and exhibiting a very analogous chiral symmetry breaking pattern.

Most usually, N=3 is taken, the u, d, and s quarks taken to be light (the Eightfold way (physics)), so then approximately massless for the symmetry to be meaningful to a lowest order, while the other three quarks are sufficiently heavy to barely have a residual chiral symmetry be visible for practical purposes.

Particle Physics

In theoretical physics, the electroweak model breaks parity maximally. All its fermions are chiral Weyl fermions, which means that the charged weak gauge bosons only couple to left-handed quarks and leptons. (Note that the neutral electroweak Z boson already couples to left *and* right-handed fermions.) Some theorists found this objectionable, and so proposed a GUT extension of the weak force which has new, high energy W' and Z' bosons which couple with right handed quarks and leptons.

$$\frac{[SU(2)_W \times U(1)_Y]}{\mathbb{Z}_2}$$

to

$$\frac{SU(2)_L \times SU(2)_R \times U(1)_{B-L}}{\mathbb{Z}_2}.$$

Here, $SU(2)_L$ (pronounced SU(2) left) is none other than $SU(2)_W$ and B−L is the baryon number minus the lepton number. An advantage of this model over the Standard Model is that the electric charge formula in this model is given by

$$Q = I_{3L} + I_{3R} + \frac{B-L}{2}$$

where $I_{3L,R}$ are the weak isospin values of the fields in the theory.

There is also the chromodynamic $SU(3)_C$. The idea was to restore parity by introducing a **left-right symmetry**. This is a group extension of $\mathbf{Z}_2$ (the left-right symmetry) by

$$\frac{SU(3)_C \times SU(2)_L \times SU(2)_R \times U(1)_{B-L}}{\mathbb{Z}_6}$$

to the semidirect product

$$\frac{SU(3)_C \times SU(2)_L \times SU(2)_R \times U(1)_{B-L}}{\mathbb{Z}_6} \rtimes \mathbb{Z}_2.$$

This has two connected components where $\mathbf{Z}_2$ acts as an automorphism, which is the composition of an involutive outer automorphism of $SU(3)_C$ with the interchange of the left and right copies of $SU(2)$ with the reversal of $U(1)_{B-L}$. It was shown by Rabindra N. Mohapatra and Goran Senjanovic in 1975 that left-right symmetry can be spontaneously broken to give a chiral low energy theory, which is the Standard Model of Glashow, Weinberg and Salam and it also connects the small observed neutrino masses to the breaking of left-right symmetry via the seesaw mechanism.

In this setting the chiral quarks

$$(3,2,1)_{1/3}$$

and

$$(\bar{3},1,2)_{-\frac{1}{3}}$$

are unified into an irrep

$$(3,2,1)_{\frac{1}{3}} \oplus (\bar{3},1,2)_{-\frac{1}{3}}.$$

The leptons are also unified into an irrep

$$(1,2,1)_{-1} \oplus (1,1,2)_1.$$

The Higgs bosons needed to implement the breaking of left-right symmetry down to the Standard Model are

$$(1,3,1)_2 \oplus (1,1,3)_2.$$

This then predicts three sterile neutrinos, which is perfectly consistent with current neutrino oscillation data. Within the seesaw mechanism, the sterile neutrinos become super-heavy without affecting physics at low energies.

Because the left-right symmetry is spontaneously broken, left-right models predict domain walls.

This left-right symmetry idea first appeared in the Pati–Salam model (1974), Mohapatra–Pati models (1975) and later in trinification (1984).

4.2.4   See also

- Electroweak theory
- Chirality (chemistry)
- Chirality (mathematics)
- Chiral symmetry breaking
- Handedness
- Spinors and Dirac fields

4.2.5   References

[1] Ta-Pei Cheng and Ling-Fong Li, *Gauge Theory of Elementary Particle Physics*, (Oxford 1984) ISBN 978-0198519614

[2] Gell-Mann, M.; Renner, B. (1968).   "Behavior of Current Divergences under SU_{3}×SU_{3}". *Physical Review* **175** (5): 2195.   Bibcode:1968PhRv..175.2195G. doi:10.1103/PhysRev.175.2195.

[3] Peskin, Michael; Schroeder, Daniel (1995). *An Introduction to Quantum Field Theory*. Westview Press. p. 670. ISBN 0-201-50397-2.

- Walter Greiner and Berndt Müller (2000). *Gauge Theory of Weak Interactions*. Springer. ISBN 3-540-67672-4.

- Gordon L. Kane (1987). *Modern Elementary Particle Physics*. Perseus Books. ISBN 0-201-11749-5.

- Kondepudi, Dilip K.; Hegstrom, Roger A. (January 1990). "The Handedness of the Universe". *Scientific American* **262** (1): 108–115.

- Winters, Jeffrey (November 1995).   "Looking for the Right Hand". *Discover*. Retrieved 12 September 2015.

4.2.6   External links

- To see a summary of the differences and similarities between chirality and helicity (those covered here and more) in chart form, one may go to Pedagogic Aids to Quantum Field Theory and click on the link near the bottom of the page entitled "Chirality and Helicity Summary". To see an in depth discussion of the two with examples, which also shows how chirality and helicity approach the same thing as speed approaches that of light, click the link entitled "Chirality and Helicity in Depth" on the same page.

- History of science: parity violation

- Helicity, Chirality, Mass, and the Higgs (Quantum Diaries blog)

- Chirality vs helicity chart (Robert D. Klauber)

4.3   Helicity

This article is about helicity in particle physics. For other uses, see Helicity (disambiguation).
See also: Chirality (physics)

In particle physics, **helicity** is the projection of the angular momentum onto the direction of momentum. The angular momentum $J\rightarrow$ is the sum of an orbital momentum $L\rightarrow$ and a spin $S\rightarrow$, and by definition its relation to linear momentum $p\rightarrow$ is

$$\vec{L} = \vec{r} \times \vec{p},$$

so its component in the direction of $p\rightarrow$ is zero. Thus, helicity is also the projection of the spin onto the direction of momentum. This quantity is conserved.[*][1]

Because the eigenvalues of spin with respect to an axis have discrete values, the eigenvalues of helicity are also discrete. For a particle of spin S, the eigenvalues of helicity are S, $S − 1$, ..., $−S$. The measured helicity of a spin S particle will range from $−S$ to $+S$.[*][2][*]:12

For massless spin-$\frac{1}{2}$ particles, helicity is equivalent to the chirality operator multiplied by $\hbar/2$. By contrast, for massive particles, distinct chirality states (e.g., as occur in the weak interaction charges) have both positive and negative helicity components, in ratios proportional to the mass of the particle.

4.3.1 Little group

In 3 + 1 dimensions, the little group for a massless particle is the double cover of SE(2). This has unitary representations which are invariant under the SE(2) "translations" and transform as $e^{*}ih\theta$ under a SE(2) rotation by θ. This is the helicity h representation. There is also another unitary representation which transforms non-trivially under the SE(2) translations. This is the *continuous spin* representation.

In $d + 1$ dimensions, the little group is the double cover of SE($d − 1$) (the case where $d \leq 2$ is more complicated because of anyons, etc.). As before, there are unitary representations which don't transform under the SE($d − 1$) "translations" (the "standard" representations) and "continuous spin" representations.

4.3.2 See also

- Wigner's classification

- Pauli–Lubanski pseudovector

4.3.3 References

[1] Landau, L D; Lifshitz, E M (2013). *Quantum mechanics. A shorter course of theoretical physics* **2**. Elsevier. pp. 273–274. ISBN 9781483187228.

[2] Troshin, S. M.; Tyurin, N. E. (1994). *Spin phenomena in particle interactions*. Singapore: World Scientific. ISBN 9789810216924.

- Povh, Bogdan; Lavelle, Martin; Rith, Klaus; Scholz, Christoph; Zetsche, Frank (2008). *Particles and nuclei an introduction to the physical concepts* (6th ed.). Berlin: Springer. ISBN 9783540793687.

- Schwartz, Matthew D. (2014). "Chirality, helicity and spin". *Quantum field theory and the standard model*. Cambridge: Cambridge University Press. pp. 185–187. ISBN 9781107034730.

- Taylor, John (1992). "Gauge theories in particle physics". In Davies, Paul. *The new physics* (1st pbk. ed.). Cambridge, [England]: Cambridge University Press. pp. 458–480. ISBN 9780521438315.

4.4 Michel parameters

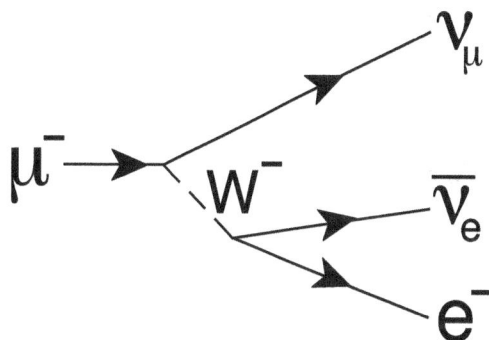

Feynman diagram of the muon decay

The **Michel parameters**, usually denoted by ρ, η, ξ and δ, are four parameters used in describing the phase space distribution of leptonic decays of charged leptons, $l_i^- \rightarrow l_j^- \nu_i \bar{\nu}_j$. They are named after the physicist Louis Michel. Sometimes instead of δ, the product $\xi\delta$ is quoted. Within the Standard Model of electroweak interactions, these parameters are expected to be

$$\rho = \frac{3}{4}, \quad \eta = 0, \quad \xi = 1, \quad \xi\delta = \frac{3}{4}.$$

Precise measurements of energy and angular distributions of the daughter leptons in decays of polarized muons and tau leptons are so far in good agreement with these predictions of the Standard Model.

4.4.1 Muon decay

See also: Muon § Muon decay

Let us consider the decay of the positive muon:

$$\mu^+ \to e^+ + \nu_e + \bar{\nu}_\mu.$$

In the muon rest frame, energy and angular distributions of the positrons emitted in the decay of a polarised muon expressed in terms of Michel parameters are the following, neglecting electron and neutrino masses and the radiative corrections:

$$\frac{d^2\Gamma}{x^2 dx d\cos\theta} \sim (3-3x)+\frac{2}{3}\rho(4x-3)+P_\mu\xi\cos\theta[(1-x)+\frac{2}{3}\delta(4x-3)]$$

where P_μ is muon polarisation, $x = E_e/E_e^{max}$, and θ is the angle between muon spin direction and positron momentum direction.[*][1] For the decay of the negative muon, the sign of the term containing $cos\theta$ should be inverted.

For the decay of the positive muon, the expected decay distribution for the Standard Model values of Michel parameters is

$$\frac{d^2\Gamma}{dx d\cos\theta} \sim x^2[(3 - 2x) - P_\mu\cos\theta(1 - 2x)].$$

Integration of this expression over electron energy gives the angular distribution of the daughter positrons:

$$\frac{d\Gamma}{d\cos\theta} \sim 1 + \frac{1}{3}P_\mu\cos\theta.$$

The positron energy distribution integrated over the polar angle is

$$\frac{d\Gamma}{dx} \sim (3x^2 - 2x^3).$$

4.4.2 References

[1] R. Bayes *et al.* (TWIST collaboration) (2011). "Experimental Constraints on Left-Right Symmetric Models from Muon Decay". *Physical Review Letters* **106** (4): 041804. Bibcode:2011PhRvL.106d1804B. doi:10.1103/PhysRevLett.106.041804.

• Lecture on Lepton Universality by Michel Davier at the 1997 SLAC Summer Institute.

• Electroweak Couplings, Lepton Universality, and the Origin of Mass: An Experimental Perspective, article by John Swain, from the Proceedings of the Third Latin American Symposium on High Energy Physics.

4.5 Weak isospin

In particle physics, **weak isospin** is a quantum number relating to the weak interaction, and parallels the idea of isospin under the strong interaction. Weak isospin is usually given the symbol T or I with the third component written as T_z, T_3, I_z or I_3.[*][1] Weak isospin is a complement of the weak hypercharge, which unifies weak interactions with electromagnetic interactions. It can be understood as the eigenvalue of a charge operator.

The weak isospin conservation law relates the conservation of T_3; all weak interactions must preserve T_3. It is also conserved by the other interactions and is therefore a conserved quantity in general. For this reason T_3 is more important than T and often the term "weak isospin" refers to the "3rd component of weak isospin".

4.5.1 Relation with chirality

Fermions with negative chirality (also called left-handed fermions) have $T = \frac{1}{2}$ and can be grouped into doublets with $T_3 = \pm\frac{1}{2}$ that behave the same way under the weak interaction. For example, up-type quarks (u, c, t) have $T_3 = +\frac{1}{2}$ and always transform into down-type quarks (d, s, b), which have $T_3 = -\frac{1}{2}$, and vice versa. On the other hand, a quark never decays weakly into a quark of the same T_3. Something similar happens with left-handed leptons, which exist as doublets containing a charged lepton (e−, μ−, τ−) with $T_3 = -\frac{1}{2}$ and a neutrino (ν

e, ν

μ, ν

τ) with $T_3 = \frac{1}{2}$.

Fermions with positive chirality (also called right-handed fermions) have $T = 0$ and form singlets that do not undergo weak interactions.

Electric charge, Q, is related to weak isospin, T_3, and weak hypercharge, Y_W, by

$$Q = T_3 + \frac{Y_W}{2}.$$

4.5.2 Weak isospin and the W bosons

The symmetry associated with spin is SU(2). This requires gauge bosons to transform between weak isospin charges:

bosons W+, W− and W0. This implies that W bosons have a $T = 1$, with three different values of T_3.

- W+ boson ($T_3 = +1$) is emitted in transitions {($T_3 = +\frac{1}{2}$) → ($T_3 = -\frac{1}{2}$)},

- W− boson ($T_3 = −1$) is emitted in transitions {($T_3 = -\frac{1}{2}$) → ($T_3 = +\frac{1}{2}$)}.

- W0 boson ($T_3 = 0$) would be emitted in reactions where T_3 does not change. However, under electroweak unification, the W0 boson mixes with the weak hypercharge gauge boson B, resulting in the observed Z0 boson and the photon of Quantum Electrodynamics.

4.5.3 See also

- Field theoretical formulation of standard model

- Weak hypercharge

4.5.4 References

[1] Ambiguities: I is also used as sign for the 'normal' isospin, same for the third component I_3 aka I_z. T is also used as the sign for Topness. This article uses T and T_3.

4.6 Weak hypercharge

The **weak hypercharge** in particle physics is a quantum number relating the electric charge and the third component of weak isospin. It is conserved (only terms that are overall weak-hypercharge neutral are allowed in the Lagrangian) and is similar to the Gell-Mann–Nishijima formula for the hypercharge of strong interactions (which is not conserved in weak interactions). It is frequently denoted Y_W and corresponds to the gauge symmetry U(1).*[1]

4.6.1 Definition

Weak hypercharge is the generator of the U(1) component of the electroweak gauge group, SU(2)×U(1) and its associated quantum field B mixes with the W^3 electroweak quantum field to produce the observed Z gauge boson and the photon of quantum electrodynamics.

Weak hypercharge, usually written as Y_W, satisfies the equality:

$$Q = T_3 + \frac{Y_W}{2}$$

where Q is the electrical charge (in elementary charge units) and T_3 is the third component of weak isospin. Rearranging, the weak hypercharge can be explicitly defined as:

$$Y_W = 2(Q - T_3)$$

Note: sometimes weak hypercharge is scaled so that

$$Y_W = Q - T_3$$

although this is a minority usage.*[2]

Hypercharge assignments in the Standard Model are determined up to a twofold ambiguity by demanding cancellation of all anomalies.

4.6.2 Baryon and lepton number

Weak hypercharge is related to baryon number minus lepton number via:

$$X + 2Y_W = 5(B - L)$$

where X is a GUT-associated conserved quantum number. Since weak hypercharge is always conserved this implies that baryon number minus lepton number is also always conserved, within the Standard Model and most extensions.

Neutron decay

n → p + e− + ν
e

Hence neutron decay conserves baryon number B and lepton number L separately, so also the difference $B - L$ is conserved.

Proton decay

Proton decay is a prediction of many grand unification theories.

p+ → e+ + π0 → e+ + 2γ

Hence proton decay conserves $B - L$, even though it violates both lepton number and baryon number conservation.

4.6.3 See also

- Standard Model (mathematical formulation)

4.6.4 Notes

[1] J. F. Donoghue, E. Golowich, B. R. Holstein (1994). *Dynamics of the standard model*. Cambridge University Press. p. 52. ISBN 0-521-47652-6.

[2] M. R. Anderson (2003). *The mathematical theory of cosmic strings*. CRC Press. p. 12. ISBN 0-7503-0160-0.

4.7 "B" − "L"

"B-L" redirects here. For other uses, see BL (disambiguation).

In high energy physics, $B - L$ (pronounced "bee minus ell") is the difference between the baryon number (B) and the lepton number (L).

4.7.1 Details

This quantum number is the charge of a global/gauge U(1) symmetry in some Grand Unified Theory models, called $U(1)_{B-L}$. Unlike baryon number alone or lepton number alone, this hypothetical symmetry would not be broken by chiral anomalies or gravitational anomalies, as long as this symmetry is global, which is why this symmetry is often invoked. If $B - L$ exists as a symmetry, it has to be spontaneously broken to give the neutrinos a nonzero mass if we assume the seesaw mechanism. The gauge bosons associated to this symmetry are commonly called X and Y bosons.

The anomalies that would break baryon number conservation and lepton number conservation individually cancel in such a way that $B - L$ is always conserved. One hypothetical example is proton decay where a proton ($B = 1$; $L = 0$) would decay into a pion ($B = 0$, $L = 0$) and positron ($B = 0$; $L = -1$).

Weak hypercharge Y
W is related to $B - L$ via:

$$X + 2Y$$
$$W = 5(B - L)$$

where X is the U(1) symmetry Grand Unified Theory-associated conserved quantum number.

4.7.2 See also

- Baryogenesis

- Leptogenesis

- Majoron

- Proton decay

- X and Y bosons

- X (charge)

- Leptoquark

4.8 "X"

In particle physics, the **X-charge** (or simply X) is a conserved quantum number associated with the SO(10) grand unification theory.

X is related to the difference between the baryon number B and the lepton number L (that is $B - L$), and the weak hypercharge Y_W via the relation:

$$X = 5(B - L) - 2Y_W$$

4.8.1 See also

- Standard Model (mathematical formulation)

- Noether's theorem

- X and Y bosons

4.8.2 Notes

4.9 Weak interaction

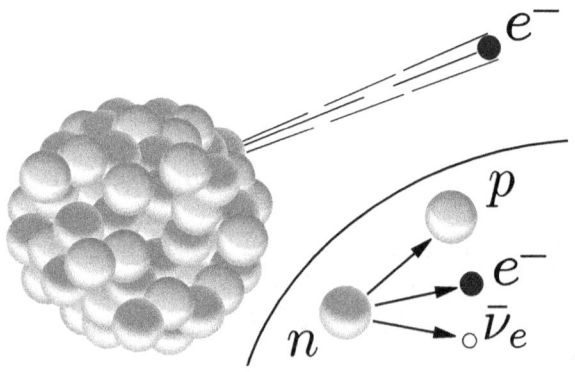

The radioactive beta decay is possible due to the weak interaction, which transforms a neutron into: a proton, an electron, and an electron antineutrino.

In particle physics, the **weak interaction** is the mechanism responsible for the **weak force** or **weak nuclear force**, one of the four known fundamental interactions of nature, alongside the strong interaction, electromagnetism, and gravitation. The weak interaction is responsible for the radioactive decay of subatomic particles, and it plays an essential role in nuclear fission. The theory of the weak interaction is sometimes called **quantum flavordynamics (QFD)**, in analogy with the terms QCD and QED, but the term is rarely used because the weak force is best understood in terms of electro-weak theory (EWT).[1]

In the Standard Model of particle physics, the weak interaction is caused by the emission or absorption of W and Z bosons. All known fermions interact through the weak interaction. Fermions are particles that have half-integer spin (one of the fundamental properties of particles). A fermion can be an elementary particle, such as the electron, or it can be a composite particle, such as the proton. The masses of W^+, W^-, and Z bosons are each far greater than that of protons or neutrons, consistent with the short range of the weak force. The force is termed *weak* because its field strength over a given distance is typically several orders of magnitude less than that of the strong nuclear force and electromagnetic force.

During the quark epoch, the electroweak force split into the electromagnetic and weak forces. Important examples of weak interaction include beta decay, and the production, from hydrogen, of deuterium needed to power the sun's thermonuclear process. Most fermions will decay by a weak interaction over time. Such decay also makes radiocarbon dating possible, as carbon-14 decays through the weak interaction to nitrogen-14. It can also create radioluminescence, commonly used in tritium illumination, and in the related field of betavoltaics.[2]

Quarks, which make up composite particles like neutrons and protons, come in six "flavours" – up, down, strange, charm, top and bottom – which give those composite particles their properties. The weak interaction is unique in that it allows for quarks to swap their flavour for another. For example, during beta minus decay, a down quark decays into an up quark, converting a neutron to a proton. Also the weak interaction is the only fundamental interaction that breaks parity-symmetry, and similarly, the only one to break CP-symmetry.

4.9.1 History

In 1933, Enrico Fermi proposed the first theory of the weak interaction, known as Fermi's interaction. He suggested that beta decay could be explained by a four-fermion interaction, involving a contact force with no range.[3][4]

However, it is better described as a non-contact force field having a finite range, albeit very short. In 1968, Sheldon Glashow, Abdus Salam and Steven Weinberg unified the electromagnetic force and the weak interaction by showing them to be two aspects of a single force, now termed the electro-weak force.

The existence of the W and Z bosons was not directly confirmed until 1983.

4.9.2 Properties

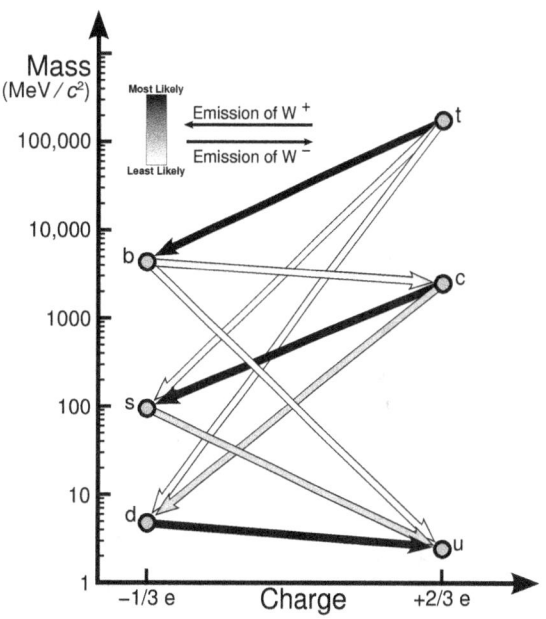

A diagram depicting the various decay routes due to the weak interaction and some indication of their likelihood. The intensity of the lines are given by the CKM parameters.

The weak interaction is unique in a number of respects:

1. It is the only interaction capable of changing the flavor of quarks (i.e., of changing one type of quark into another).

2. It is the only interaction that violates **P** or parity-symmetry. It is also the only one that violates **CP** symmetry.

3. It is propagated by carrier particles (known as gauge bosons) that have significant masses, an unusual feature which is explained in the Standard Model by the Higgs mechanism.

Due to their large mass (approximately 90 GeV/c^2[5]) these carrier particles, termed the W and Z bosons, are short-lived: they have a lifetime of under $1\times10^{*}-24$ seconds.[6] The weak interaction has a coupling constant (an indicator of interaction strength) of between $10^{*}-7$ and $10^{*}-6$, compared to the strong interaction's coupling constant of about 1 and the electromagnetic coupling constant of about $10^{*}-2$;[7] consequently the weak interaction is weak in terms of strength.[8] The weak interaction has a very short range (around $10^{*}-17-10^{*}-16$ m[8]).[7] At distances around $10^{*}-18$ meters, the weak interaction has a strength of a similar magnitude to the electromagnetic force, but this starts to decrease exponentially with increasing distance. At distances of around $3\times10^{*}-17$ m, the weak interaction is 10,000 times weaker than the electromagnetic.[9]

The weak interaction affects all the fermions of the Standard Model, as well as the Higgs boson; neutrinos interact through gravity and the weak interaction only, and neutrinos were the original reason for the name *weak force*.[8] The weak interaction does not produce bound states (nor does it involve binding energy) – something that gravity does on an astronomical scale, that the electromagnetic force does at the atomic level, and that the strong nuclear force does inside nuclei.[10]

Its most noticeable effect is due to its first unique feature: flavor changing. A neutron, for example, is heavier than a proton (its sister nucleon), but it cannot decay into a proton without changing the flavor (type) of one of its two *down* quarks to *up*. Neither the strong interaction nor electromagnetism permit flavour changing, so this must proceed by **weak decay**; without weak decay, quark properties such as strangeness and charm (associated with the quarks of the same name) would also be conserved across all interactions. All mesons are unstable because of weak decay.[11] In the process known as beta decay, a *down* quark in the neutron can change into an *up* quark by emitting a virtual W− boson which is then converted into an electron and an electron antineutrino.[12] Another example is the electron capture, a common variant of radioactive decay, where a proton (up quark) and an electron within an atom interact, and are changed to a neutron (down quark) and an electron neutrino.

Due to the large mass of a boson, weak decay is much more unlikely than strong or electromagnetic decay, and hence occurs less rapidly. For example, a neutral pion (which decays electromagnetically) has a life of about $10^{*}-16$ seconds, while a charged pion (which decays through the weak interaction) lives about $10^{*}-8$ seconds, a hundred million times longer.[13] In contrast, a free neutron (which also decays through the weak interaction) lives about 15 minutes.[12]

Weak isospin and weak hypercharge

Main article: Weak isospin

All particles have a property called weak isospin (T_3), which serves as a quantum number and governs how that particle interacts in the weak interaction. Weak isospin therefore plays the same role in the weak interaction as electric charge does in electromagnetism, and color charge in the strong interaction. All fermions have a weak isospin value of either $+\frac{1}{2}$ or $-\frac{1}{2}$. For example, the up quark has a T_3 of $+\frac{1}{2}$ and the down quark $-\frac{1}{2}$. A quark never decays through the weak interaction into a quark of the same T_3: quarks with a T_3 of $+\frac{1}{2}$ decay into quarks with a T_3 of $-\frac{1}{2}$ and vice versa.

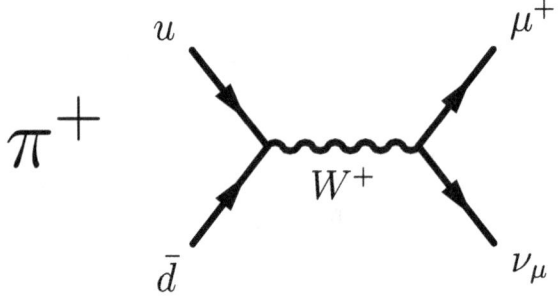

π+ decay through the weak interaction

In any given interaction, weak isospin is conserved: the sum of the weak isospin numbers of the particles entering the interaction equals the sum of the weak isospin numbers of the particles exiting that interaction. For example, a (left-handed) π+, with a weak isospin of 1 normally decays into a ν μ (+1/2) and a μ+ (as a right-handed antiparticle, +1/2).[13]

Following the development of the electroweak theory, another property, weak hypercharge, was developed. It is dependent on a particle's electrical charge and weak isospin, and is defined as:

$$Y_W = 2(Q - T_3)$$

where Y_W is the weak hypercharge of a given type of particle, Q is its electrical charge (in elementary charge units) and T_3 is its weak isospin. Whereas some particles have a weak isospin of zero, all particles, except gluons, have non-zero weak hypercharge. Weak hypercharge is the generator of the U(1) component of the electroweak gauge group.

4.9.3 Interaction types

There are two types of weak interaction (called *vertices*). The first type is called the "charged-current interaction" because it is mediated by particles that carry an electric charge (the W+ or W− bosons), and is responsible for the beta decay phenomenon. The second type is called the "neutral-current interaction" because it is mediated by a neutral particle, the Z boson.

Charged-current interaction

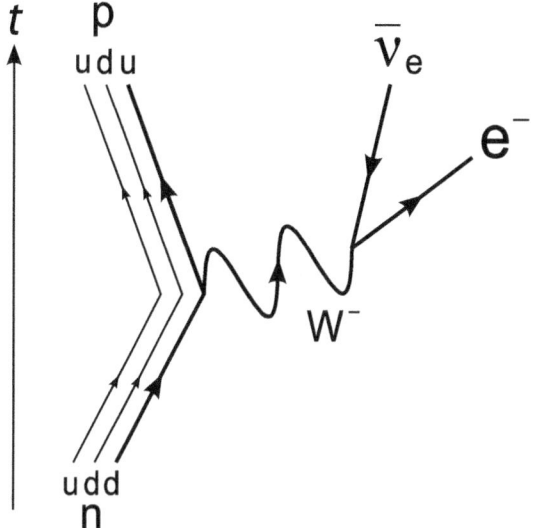

The Feynman diagram for beta-minus decay of a neutron into a proton, electron and electron anti-neutrino, via an intermediate heavy W− boson

In one type of charged current interaction, a charged lepton (such as an electron or a muon, having a charge of −1) can absorb a W+ boson (a particle with a charge of +1) and be thereby converted into a corresponding neutrino (with a charge of 0), where the type ("family") of neutrino (electron, muon or tau) is the same as the type of lepton in the interaction, for example:

$$\mu^- + W^+ \to \nu_\mu$$

Similarly, a down-type quark (d with a charge of $-1/3$) can be converted into an up-type quark (u, with a charge of $+2/3$), by emitting a W− boson or by absorbing a W+ boson. More precisely, the down-type quark becomes a quantum superposition of up-type quarks: that is to say, it has a possibility of becoming any one of the three up-type quarks, with the probabilities given in the CKM matrix tables. Conversely, an up-type quark can emit a W+ boson – or absorb a W− boson – and thereby be converted into a down-type quark, for example:

$$d \to u + W^-$$
$$d + W^+ \to u$$
$$c \to s + W^+$$
$$c + W^- \to s$$

The W boson is unstable so will rapidly decay, with a very short lifetime. For example:

$$W^- \to e^- + \bar{\nu}_e$$
$$W^+ \to e^+ + \nu_e$$

Decay of the W boson to other products can happen, with varying probabilities.[*][15]

In the so-called beta decay of a neutron (see picture, above), a down quark within the neutron emits a virtual W− boson and is thereby converted into an up quark, converting the neutron into a proton. Because of the energy involved in the process (i.e., the mass difference between the down quark and the up quark), the W− boson can only be converted into an electron and an electron-antineutrino.[*][16] At the quark level, the process can be represented as:

$$d \to u + e^- + \bar{\nu}_e$$

Neutral-current interaction

In neutral current interactions, a quark or a lepton (e.g., an electron or a muon) emits or absorbs a neutral Z boson. For example:

$$e^- \to e^- + Z^0$$

Like the W boson, the Z boson also decays rapidly,[*][15] for example:

$$Z^0 \to b + \bar{b}$$

4.9.4 Electroweak theory

Main article: Electroweak interaction

The Standard Model of particle physics describes the electromagnetic interaction and the weak interaction as two different aspects of a single electroweak interaction, the theory of which was developed around 1968 by Sheldon Glashow, Abdus Salam and Steven Weinberg. They were awarded the 1979 Nobel Prize in Physics for their work.[17] The Higgs mechanism provides an explanation for the presence of three massive gauge bosons (the three carriers of the weak interaction) and the massless photon of the electromagnetic interaction.[18]

According to the electroweak theory, at very high energies, the universe has four massless gauge boson fields similar to the photon and a complex scalar Higgs field doublet. However, at low energies, gauge symmetry is spontaneously broken down to the $U(1)$ symmetry of electromagnetism (one of the Higgs fields acquires a vacuum expectation value). This symmetry breaking would produce three massless bosons, but they become integrated by three photon-like fields (through the Higgs mechanism) giving them mass. These three fields become the W+, W− and Z bosons of the weak interaction, while the fourth gauge field, which remains massless, is the photon of electromagnetism.[18]

This theory has made a number of predictions, including a prediction of the masses of the Z and W bosons before their discovery. On 4 July 2012, the CMS and the ATLAS experimental teams at the Large Hadron Collider independently announced that they had confirmed the formal discovery of a previously unknown boson of mass between 125–127 GeV/c^2, whose behaviour so far was "consistent with" a Higgs boson, while adding a cautious note that further data and analysis were needed before positively identifying the new boson as being a Higgs boson of some type. By 14 March 2013, the Higgs boson was tentatively confirmed to exist.[19]

4.9.5 Violation of symmetry

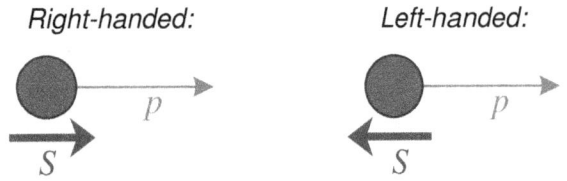

Right-handed: Left-handed:

Left- and right-handed particles: p is the particle's momentum and S is its spin. Note the lack of reflective symmetry between the states.

The laws of nature were long thought to remain the same under mirror reflection, the reversal of one spatial axis. The results of an experiment viewed via a mirror were expected to be identical to the results of a mirror-reflected copy of the experimental apparatus. This so-called law of parity conservation was known to be respected by classical gravitation, electromagnetism and the strong interaction; it

was assumed to be a universal law.[20] However, in the mid-1950s Chen Ning Yang and Tsung-Dao Lee suggested that the weak interaction might violate this law. Chien Shiung Wu and collaborators in 1957 discovered that the weak interaction violates parity, earning Yang and Lee the 1957 Nobel Prize in Physics.[21]

Although the weak interaction used to be described by Fermi's theory, the discovery of parity violation and renormalization theory suggested that a new approach was needed. In 1957, Robert Marshak and George Sudarshan and, somewhat later, Richard Feynman and Murray Gell-Mann proposed a **V−A** (vector minus axial vector or left-handed) Lagrangian for weak interactions. In this theory, the weak interaction acts only on left-handed particles (and right-handed antiparticles). Since the mirror reflection of a left-handed particle is right-handed, this explains the maximal violation of parity. Interestingly, the **V−A** theory was developed before the discovery of the Z boson, so it did not include the right-handed fields that enter in the neutral current interaction.

However, this theory allowed a compound symmetry **CP** to be conserved. **CP** combines parity **P** (switching left to right) with charge conjugation **C** (switching particles with antiparticles). Physicists were again surprised when in 1964, James Cronin and Val Fitch provided clear evidence in kaon decays that CP symmetry could be broken too, winning them the 1980 Nobel Prize in Physics.[22] In 1973, Makoto Kobayashi and Toshihide Maskawa showed that CP violation in the weak interaction required more than two generations of particles,[23] effectively predicting the existence of a then unknown third generation. This discovery earned them half of the 2008 Nobel Prize in Physics.[24] Unlike parity violation, CP violation occurs in only a small number of instances, but remains widely held as an answer to the difference between the amount of matter and antimatter in the universe; it thus forms one of Andrei Sakharov's three conditions for baryogenesis.[25]

4.9.6 See also

- Weakless Universe – the postulate that weak interactions are not anthropically necessary

- Gravity

- Nuclear force

- Electromagnetism

4.9.7 References

Citations

[1] Griffiths, David (2009). *Introduction to Elementary Particles*. pp. 59–60. ISBN 978-3-527-40601-2.

[2] "The Nobel Prize in Physics 1979: Press Release". *NobelPrize.org*. Nobel Media. Retrieved 22 March 2011.

[3] Fermi, Enrico (1934). "Versuch einer Theorie der β-Strahlen. I". *Zeitschrift für Physik A* **88** (3–4): 161–177. Bibcode:1934ZPhy...88..161F. doi:10.1007/BF01351864.

[4] Wilson, Fred L. (December 1968). "Fermi's Theory of Beta Decay". *American Journal of Physics* **36** (12): 1150–1160. Bibcode:1968AmJPh..36.1150W. doi:10.1119/1.1974382.

[5] W.-M. Yao *et al.* (Particle Data Group) (2006). "Review of Particle Physics: Quarks" (PDF). *Journal of Physics G* **33**: 1–1232. arXiv:astro-ph/0601168. Bibcode:2006JPhG...33....1Y. doi:10.1088/0954-3899/33/1/001.

[6] Peter Watkins (1986). *Story of the W and Z*. Cambridge: Cambridge University Press. p. 70. ISBN 978-0-521-31875-4.

[7] "Coupling Constants for the Fundamental Forces". *HyperPhysics*. Georgia State University. Retrieved 2 March 2011.

[8] J. Christman (2001). "The Weak Interaction" (PDF). *Physnet*. Michigan State University.

[9] "Electroweak". *The Particle Adventure*. Particle Data Group. Retrieved 3 March 2011.

[10] Walter Greiner; Berndt Müller (2009). *Gauge Theory of Weak Interactions*. Springer. p. 2. ISBN 978-3-540-87842-1.

[11] Cottingham & Greenwood (1986, 2001), p.29

[12] Cottingham & Greenwood (1986, 2001), p.28

[13] Cottingham & Greenwood (1986, 2001), p.30

[14] Baez, John C.; Huerta, John (2009). "The Algebra of Grand Unified Theories". *Bull.Am.Math.Soc.* **0904**: 483–552. arXiv:0904.1556. Bibcode:2009arXiv0904.1556B. doi:10.1090/s0273-0979-10-01294-2. Retrieved 15 October 2013.

[15] K. Nakamura *et al.* (Particle Data Group) (2010). "Gauge and Higgs Bosons" (PDF). *Journal of Physics G* **37**. doi:10.1088/0954-3899/37/7a/075021.

[16] K. Nakamura *et al.* (Particle Data Group) (2010). "n" (PDF). *Journal of Physics G* **37**: 7. doi:10.1088/0954-3899/37/7a/075021.

[17] "The Nobel Prize in Physics 1979". *NobelPrize.org*. Nobel Media. Retrieved 26 February 2011.

[18] C. Amsler *et al.* (Particle Data Group) (2008). "Review of Particle Physics – Higgs Bosons: Theory and Searches" (PDF). *Physics Letters B* **667**: 1–6. Bibcode:2008PhLB..667....1P. doi:10.1016/j.physletb.2008.07.018.

[19] "New results indicate that new particle is a Higgs boson | CERN". Home.web.cern.ch. Retrieved 20 September 2013.

[20] Charles W. Carey (2006). "Lee, Tsung-Dao". *American scientists*. Facts on File Inc. p. 225. ISBN 9781438108070.

[21] "The Nobel Prize in Physics 1957". *NobelPrize.org*. Nobel Media. Retrieved 26 February 2011.

[22] "The Nobel Prize in Physics 1980". *NobelPrize.org*. Nobel Media. Retrieved 26 February 2011.

[23] M. Kobayashi, T. Maskawa (1973). "CP-Violation in the Renormalizable Theory of Weak Interaction". *Progress of Theoretical Physics* **49** (2): 652–657. Bibcode:1973PThPh..49..652K. doi:10.1143/PTP.49.652.

[24] "The Nobel Prize in Physics 1980". *NobelPrize.org*. Nobel Media. Retrieved 17 March 2011.

[25] Paul Langacker (2001) [1989]. "Cp Violation and Cosmology". In Cecilia Jarlskog. *CP violation*. London, River Edge: World Scientific Publishing Co. p. 552. ISBN 9789971505615.

General readers

- R. Oerter (2006). *The Theory of Almost Everything: The Standard Model, the Unsung Triumph of Modern Physics*. Plume. ISBN 978-0-13-236678-6.

- B.A. Schumm (2004). *Deep Down Things: The Breathtaking Beauty of Particle Physics*. Johns Hopkins University Press. ISBN 0-8018-7971-X.

Texts

- D.A. Bromley (2000). *Gauge Theory of Weak Interactions*. Springer. ISBN 3-540-67672-4.

- G.D. Coughlan, J.E. Dodd, B.M. Gripaios (2006). *The Ideas of Particle Physics: An Introduction for Scientists* (3rd ed.). Cambridge University Press. ISBN 978-0-521-67775-2.

- W. N. Cottingham; D. A. Greenwood (2001) [1986]. *An introduction to nuclear physics* (2nd ed.). Cambridge University Press. p. 30. ISBN 978-0-521-65733-4.

- D.J. Griffiths (1987). *Introduction to Elementary Particles*. John Wiley & Sons. ISBN 0-471-60386-4.

- G.L. Kane (1987). *Modern Elementary Particle Physics*. Perseus Books. ISBN 0-201-11749-5.

- D.H. Perkins (2000). *Introduction to High Energy Physics*. Cambridge University Press. ISBN 0-521-62196-8.

Chapter 5

Appendix A – Related topics

5.1 Antiparticle

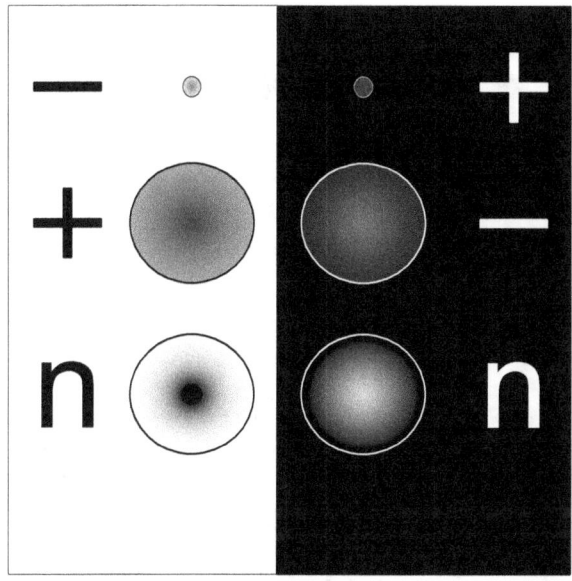

Illustration of electric charge of particles (left) and antiparticles (right). From top to bottom; electron/positron, proton/antiproton, neutron/antineutron.

Corresponding to most kinds of particles, there is an associated antimatter **antiparticle** with the same mass and opposite charge (including electric charge). For example, the antiparticle of the electron is the positively charged positron, which is produced naturally in certain types of radioactive decay.

The laws of nature are very nearly symmetrical with respect to particles and antiparticles. For example, an antiproton and a positron can form an antihydrogen atom, which is believed to have the same properties as a hydrogen atom. This leads to the question of why the formation of matter after the Big Bang resulted in a universe consisting almost entirely of matter, rather than being a half-and-half mixture of matter and antimatter. The discovery of Charge Parity violation helped to shed light on this problem by showing that this symmetry, originally thought to be perfect, was only approximate.

Particle-antiparticle pairs can annihilate each other, producing photons; since the charges of the particle and antiparticle are opposite, total charge is conserved. For example, the positrons produced in natural radioactive decay quickly annihilate themselves with electrons, producing pairs of gamma rays, a process exploited in positron emission tomography.

Antiparticles are produced naturally in beta decay, and in the interaction of cosmic rays in the Earth's atmosphere. Because charge is conserved, it is not possible to create an antiparticle without either destroying a particle of the same charge (as in beta decay) or creating a particle of the opposite charge. The latter is seen in many processes in which both a particle and its antiparticle are created simultaneously, as in particle accelerators. This is the inverse of the particle-antiparticle annihilation process.

Although particles and their antiparticles have opposite charges, electrically neutral particles need not be identical to their antiparticles. The neutron, for example, is made out of quarks, the antineutron from antiquarks, and they are distinguishable from one another because neutrons and antineutrons annihilate each other upon contact. However, other neutral particles are their own antiparticles, such as photons, hypothetical gravitons, and some WIMPs.

5.1.1 History

Experiment

In 1932, soon after the prediction of positrons by Paul Dirac, Carl D. Anderson found that cosmic-ray collisions produced these particles in a cloud chamber—a particle detector in which moving electrons (or positrons) leave behind trails as they move through the gas. The electric charge-to-mass ratio of a particle can be measured by observing the radius of curling of its cloud-chamber track in a magnetic field. Positrons, because of the direction that their paths curled, were at first mistaken for electrons travelling in the opposite direction. Positron paths in a cloud-chamber trace the same helical path as an electron but rotate in the opposite direction with respect to the magnetic field direction due to their having the same magnitude of charge-to-mass ratio but with opposite charge and, therefore, opposite signed charge-to-mass ratios.

The antiproton and antineutron were found by Emilio Segrè and Owen Chamberlain in 1955 at the University of California, Berkeley. Since then, the antiparticles of many other subatomic particles have been created in particle accelerator experiments. In recent years, complete atoms of antimatter have been assembled out of antiprotons and positrons, collected in electromagnetic traps.*[1]

Dirac's Hole theory

... the development of quantum field theory made the interpretation of antiparticles as holes unnecessary, even though it lingers on in many textbooks.

Steven Weinberg*[2]

Solutions of the Dirac equation contained negative energy quantum states. As a result, an electron could always radiate energy and fall into a negative energy state. Even worse, it could keep radiating infinite amounts of energy because there were infinitely many negative energy states available. To prevent this unphysical situation from happening, Dirac proposed that a "sea" of negative-energy electrons fills the universe, already occupying all of the lower-energy states so that, due to the Pauli exclusion principle, no other electron could fall into them. Sometimes, however, one of these negative-energy particles could be lifted out of this Dirac sea to become a positive-energy particle. But, when lifted out, it would leave behind a *hole* in the sea that would act exactly like a positive-energy electron with a reversed charge. These he interpreted as "negative-energy electrons" and attempted to identify them with protons in his 1930 paper *A Theory of Electrons and Protons**[3] However, these "negative-energy electrons" turned out to be positrons, and not protons.

This picture implied an infinite negative charge for the universe--a problem of which Dirac was aware. Dirac tried to argue that we would perceive this as the normal state of zero charge. Another difficulty was the difference in masses of the electron and the proton. Dirac tried to argue that this was due to the electromagnetic interactions with the sea, until Hermann Weyl proved that hole theory was completely symmetric between negative and positive charges. Dirac also predicted a reaction $e- + p+ \rightarrow \gamma + \gamma$, where an electron and a proton annihilate to give two photons. Robert Oppenheimer and Igor Tamm proved that this

would cause ordinary matter to disappear too fast. A year later, in 1931, Dirac modified his theory and postulated the positron, a new particle of the same mass as the electron. The discovery of this particle the next year removed the last two objections to his theory.

However, the problem of infinite charge of the universe remains. Also, as we now know, bosons also have antiparticles, but since bosons do not obey the Pauli exclusion principle (only fermions do), hole theory does not work for them. A unified interpretation of antiparticles is now available in quantum field theory, which solves both these problems.

5.1.2   Particle-antiparticle annihilation

Main article: Annihilation

If a particle and antiparticle are in the appropriate quan-

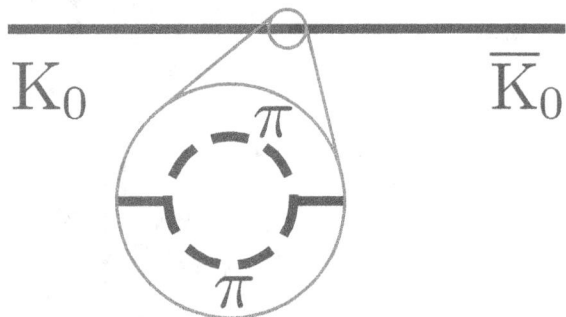

An example of a virtual pion pair that influences the propagation of a kaon, causing a neutral kaon to mix with the antikaon. This is an example of renormalization in quantum field theory—the field theory being necessary because of the change in particle number.

tum states, then they can annihilate each other and produce other particles. Reactions such as $e- + e+ \rightarrow \gamma + \gamma$ (the two-photon annihilation of an electron-positron pair) are an example. The single-photon annihilation of an electron-positron pair, $e- + e+ \rightarrow \gamma$, cannot occur in free space because it is impossible to conserve energy and momentum together in this process. However, in the Coulomb field of a nucleus the translational invariance is broken and single-photon annihilation may occur.*[4] The reverse reaction (in free space, without an atomic nucleus) is also impossible for this reason. In quantum field theory, this process is allowed only as an intermediate quantum state for times short enough that the violation of energy conservation can be accommodated by the uncertainty principle. This opens the way for virtual pair production or annihilation in which a one particle quantum state may *fluctuate* into a two particle state and back. These processes are important in the vacuum state and renormalization of a quantum field theory. It also opens the way for neutral particle mixing through processes such as the one pictured here, which is a complicated example of mass renormalization.

5.1.3 Properties of antiparticles

Quantum states of a particle and an antiparticle can be interchanged by applying the charge conjugation (**C**), parity (**P**), and time reversal (**T**) operators. If $|p, \sigma, n\rangle$ denotes the quantum state of a particle (**n**) with momentum **p**, spin **J** whose component in the z-direction is σ, then one has

$$CPT |p, \sigma, n\rangle = (-1)^{J-\sigma} |p, -\sigma, n^c\rangle,$$

where **n*c** denotes the charge conjugate state, *i.e.*, the antiparticle. This behaviour under **CPT** is the same as the statement that the particle and its antiparticle lie in the same irreducible representation of the Poincaré group. Properties of antiparticles can be related to those of particles through this. If **T** is a good symmetry of the dynamics, then

$$T |p, \sigma, n\rangle \propto |-p, -\sigma, n\rangle,$$

$$CP |p, \sigma, n\rangle \propto |-p, \sigma, n^c\rangle,$$

$$C |p, \sigma, n\rangle \propto |p, \sigma, n^c\rangle,$$

where the proportionality sign indicates that there might be a phase on the right hand side. In other words, particle and antiparticle must have

- the same mass **m**

- the same spin state **J**

- opposite electric charges **q** and **-q**.

5.1.4 Quantum field theory

This section draws upon the ideas, language and notation of canonical quantization of a quantum field theory.

One may try to quantize an electron field without mixing the annihilation and creation operators by writing

$$\psi(x) = \sum_k u_k(x) a_k e^{-iE(k)t},$$

where we use the symbol k to denote the quantum numbers p and σ of the previous section and the sign of the energy, $E(k)$, and a_k denotes the corresponding annihilation operators. Of course, since we are dealing with fermions, we have to have the operators satisfy canonical anti-commutation relations. However, if one now writes down the Hamiltonian

$$H = \sum_k E(k) a_k^\dagger a_k,$$

then one sees immediately that the expectation value of H need not be positive. This is because $E(k)$ can have any sign whatsoever, and the combination of creation and annihilation operators has expectation value 1 or 0.

So one has to introduce the charge conjugate *antiparticle* field, with its own creation and annihilation operators satisfying the relations

$$b_{k'} = a_k^\dagger \text{ and } b_{k'}^\dagger = a_k,$$

where k has the same p, and opposite σ and sign of the energy. Then one can rewrite the field in the form

$$\psi(x) = \sum_{k_+} u_k(x) a_k e^{-iE(k)t} + \sum_{k_-} u_k(x) b_k^\dagger e^{-iE(k)t},$$

where the first sum is over positive energy states and the second over those of negative energy. The energy becomes

$$H = \sum_{k_+} E_k a_k^\dagger a_k + \sum_{k_-} |E(k)| b_k^\dagger b_k + E_0,$$

where E_0 is an infinite negative constant. The vacuum state is defined as the state with no particle or antiparticle, *i.e.*, $a_k |0\rangle = 0$ and $b_k |0\rangle = 0$. Then the energy of the vacuum is exactly E_0. Since all energies are measured relative to the vacuum, **H** is positive definite. Analysis of the properties of a_k and b_k shows that one is the annihilation operator for particles and the other for antiparticles. This is the case of a fermion.

This approach is due to Vladimir Fock, Wendell Furry and Robert Oppenheimer. If one quantizes a real scalar field, then one finds that there is only one kind of annihilation operator; therefore, real scalar fields describe neutral bosons. Since complex scalar fields admit two different kinds of annihilation operators, which are related by conjugation, such fields describe charged bosons.

Feynman–Stueckelberg interpretation

By considering the propagation of the negative energy modes of the electron field backward in time, Ernst Stueckelberg reached a pictorial understanding of the fact that the particle and antiparticle have equal mass **m** and spin **J** but opposite charges **q**. This allowed him to rewrite perturbation theory precisely in the form of diagrams. Richard Feynman later gave an independent systematic derivation of these diagrams from a particle formalism, and they are now called Feynman diagrams. Each line of a diagram represents a particle propagating either backward or forward in time. This technique is the most widespread method of computing amplitudes in quantum field theory today.

Since this picture was first developed by Ernst Stueckelberg, and acquired its modern form in Feynman's work, it is called the *Feynman-Stueckelberg interpretation* of antiparticles to honor both scientists.

As a consequence of this interpretation, Villata argued that the assumption of antimatter as CPT-transformed matter would imply that the gravitational interaction between matter and antimatter is repulsive.*[5]

5.1.5   See also

- Gravitational interaction of antimatter

- Parity, charge conjugation and time reversal symmetry.

- CP violations and the baryon asymmetry of the universe.

- Quantum field theory and the list of particles

- Baryogenesis

5.1.6   References

[1] http://news.nationalgeographic.com/news/2010/11/101118-antimatter-trapped-engines-bombs-nature-science-comb

[2] Weinberg, Steve. *The quantum theory of fields, Volume 1 : Foundations*. p. 14. ISBN 0-521-55001-7.

[3] Dirac, Paul (1930). "A Theory of Electrons and Protons". *Proceedings of the Royal Society A* **126** (801): 360–365. Bibcode:1930RSPSA.126..360D. doi:10.1098/rspa.1930.0013.

[4] Sodickson, L.; W. Bowman; J. Stephenson (1961). "Single-Quantum Annihilation of Positrons". *Physical Review* **124** (6): 1851–1861. Bibcode:1961PhRv..124.1851S. doi:10.1103/PhysRev.124.1851.

[5] M. Villata, CPT symmetry and antimatter gravity in general relativity, 2011, EPL (Europhysics Letters) 94, 20001

- Feynman, R. P. (1987). "The reason for antiparticles". In R. P. Feynman and S. Weinberg. *The 1986 Dirac memorial lectures*. Cambridge University Press. ISBN 0-521-34000-4.

- Weinberg, S. (1995). *The Quantum Theory of Fields, Volume 1: Foundations*. Cambridge University Press. ISBN 0-521-55001-7.

5.2   Beta decay

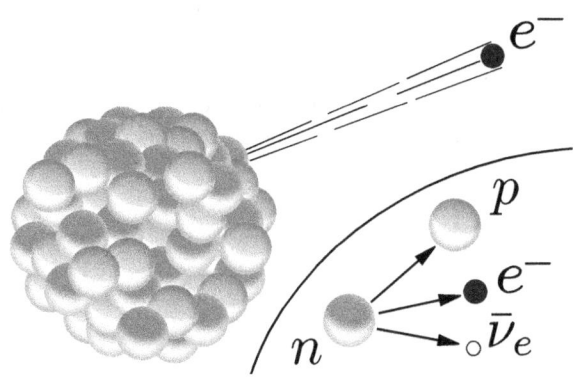

β− decay in an atomic nucleus (the accompanying antineutrino is omitted). The inset shows beta decay of a free neutron. In both processes, the intermediate emission of a virtual W− boson (which then decays to electron and antineutrino) is not shown.

In nuclear physics, **beta decay** (β-decay) is a type of radioactive decay in which a proton is transformed into a neutron, or vice versa, inside an atomic nucleus. This process allows the atom to move closer to the optimal ratio of protons and neutrons. As a result of this transformation, the nucleus emits a detectable beta particle, which is an electron or positron.*[1]

Beta decay is mediated by the weak force. There are two types of beta decay, known as *beta minus* and *beta plus*. Beta minus (β*−) decay produces an electron and electron antineutrino, while beta plus (β*+) decay produces a positron and electron neutrino; β*+ decay is thus also known as positron emission.*[2]

An example of electron emission (β*− decay) is the decay of carbon-14 into nitrogen-14:

$$^{14}_{6}C \rightarrow\ ^{14}_{7}N + e{-} + \nu_e$$

In this form of decay, the original element becomes a new chemical element in a process known as nuclear transmutation. This new element has an unchanged mass number A, but an atomic number Z that is increased by one. As in all nuclear decays, the decaying element (in this case 14

6C) is known as the *parent nuclide* while the resulting element (in this case 14

7N) is known as the *daughter nuclide*. The emitted electron or positron is known as a beta particle.

An example of positron emission (β^*+ decay) is the decay of magnesium-23 into sodium-23:

23

12Mg → 23

11Na + e+ + ν

e

In contrast to β^*- decay, β^*+ decay is accompanied by the emission of an electron neutrino and a positron. β^*+ decay also results in nuclear transmutation, with the resulting element having an atomic number that is decreased by one.

Electron capture is sometimes included as a type of beta decay, because the basic nuclear process, mediated by the weak force, is the same. In electron capture, an inner atomic electron is captured by a proton in the nucleus, transforming it into a neutron, and an electron neutrino is released. An example of electron capture is the decay of krypton-81 into bromine-81:

81

36Kr + e− → 81

35Br + ν

e

Electron capture is a competing (simultaneous) decay process for all nuclei that can undergo β^*+ decay. The converse, however, is not true: electron capture is the *only* type of decay that is allowed in proton-rich nuclides that do not have sufficient energy to emit a positron and neutrino.[3]

5.2.1 β^*- decay

In $\beta-$ decay, the weak interaction converts an atomic nucleus into a nucleus with atomic number increased by one, while emitting an electron (e−) and an electron antineutrino (ν

e).

The generic equation is:

A

ZX → A

Z+1X' + e− + ν

e[1]

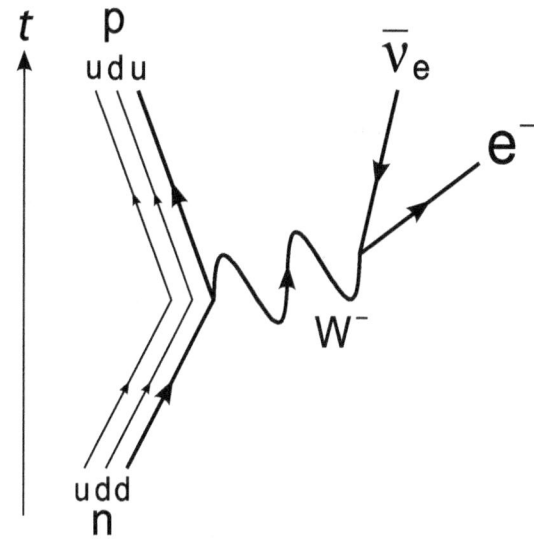

The Feynman diagram for $\beta-$ decay of a neutron into a proton, electron, and electron antineutrino via an intermediate $W-$ boson.

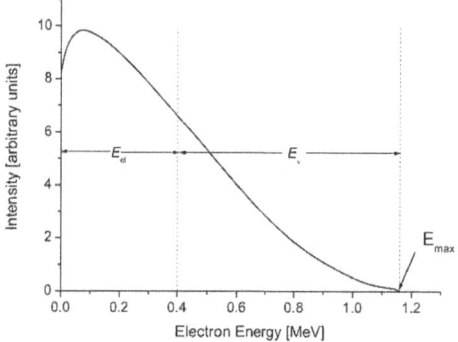

A beta spectrum, showing a typical division of energy between electron and antineutrino

where A and Z are the mass number and atomic number of the decaying nucleus, and X and X' are the initial and final elements, respectively.

Another example is when the free neutron (1

0n) decays by $\beta-$ decay into a proton (p):

n → p + e− + ν

e.

At the fundamental level (as depicted in the Feynman diagram on the left), this is caused by the conversion of the negatively charged ($-\frac{1}{3}$ e) down quark to the positively charged ($+\frac{2}{3}$ e) up quark by emission of a W− boson; the W− boson subsequently decays into an electron and an electron antineutrino:

d → u + e− + ν
e.

The beta spectrum is a continuous spectrum: the total decay energy is divided between the electron and the antineutrino. In the figure to the right, this is shown, by way of example, for an electron of 0.4 MeV energy. In this example, the antineutrino then gets the remainder: 0.76 MeV, since the total decay energy is assumed to be 1.16 MeV.

β− decay generally occurs in neutron-rich nuclei.*[4]

5.2.2   β*+ decay

Main article: Positron emission

In β+ decay, or "positron emission", the weak interaction converts an atomic nucleus into a nucleus with atomic number decreased by one, while emitting a positron (e+) and an electron neutrino (ν
e). The generic equation is:

A
ZX → A
Z−1X' + e+ + ν
e*[1]

This may be considered as the decay of a proton inside the nucleus to a neutron

p → n + e+ + ν
e*[1]

However, β+ decay cannot occur in an isolated proton because it requires energy due to the mass of the neutron being greater than the mass of the proton. β+ decay can only happen inside nuclei when the daughter nucleus has a greater binding energy (and therefore a lower total energy) than the mother nucleus. The difference between these energies goes into the reaction of converting a proton into a neutron, a positron and a neutrino and into the kinetic energy of these particles. In an opposite process to negative beta decay, the weak interaction converts a proton into a neutron by converting an up quark into a down quark by having it emit a W+ or absorb a W−.

5.2.3   Electron capture (K-capture)

Main article: Electron capture

In all cases where β+ decay of a nucleus is allowed energetically, so is electron capture, the process in which the same nucleus captures an atomic electron with the emission of a neutrino:

A
ZX + e− → A
Z−1X' + ν
e

The emitted neutrino is mono-energetic. In proton-rich nuclei where the energy difference between initial and final states is less than $2m_ec^2$, β+ decay is not energetically possible, and electron capture is the sole decay mode.*[3]

If the captured electron comes from the innermost shell of the atom, the K-shell, which has the highest probability to interact with the nucleus, the process is called K-capture.*[5] If it comes from the L-shell, the process is called L-capture, etc.

5.2.4   Competition of beta decay types

Three types of beta decay in competition are illustrated by the single isotope copper-64 (29 protons, 35 neutrons), which has a half-life of about 12.7 hours. This isotope has one unpaired proton and one unpaired neutron, so either the proton or the neutron can decay. This particular nuclide (though not all nuclides in this situation) is almost equally likely to decay through proton decay by positron emission (18%) or electron capture (43%), as through neutron decay by electron emission (39%).

5.2.5   Helicity (polarization) of neutrinos, electrons and positrons emitted in beta decay

After the discovery of parity non-conservation (see history below), it was found that, in beta decay, electrons are emitted mostly with negative helicity, i.e., they move, naively speaking, like left-handed screws driven into a material (they have negative longitudinal polarization).*[6] Conversely, positrons have mostly positive helicity, i.e., they move like right-handed screws. Neutrinos (emitted in positron decay) have positive helicity, while antineutrinos (emitted in electron decay) have negative helicity.*[7]

The higher the energy of the particles, the higher their polarization.

5.2.6   Energy release

The Q value is defined as the total energy released in a given nuclear decay. In beta decay, Q is therefore also the sum of the kinetic energies of the emitted beta particle, neutrino, and recoiling nucleus. (Because of the large mass of the nucleus compared to that of the beta particle and neutrino, the kinetic energy of the recoiling nucleus can generally be neglected.) Beta particles can therefore be emitted with any kinetic energy ranging from 0 to Q.[1] A typical Q is around 1 MeV, but can range from a few keV to a few tens of MeV.

Since the rest mass of the electron is 511 keV, the most energetic beta particles are ultrarelativistic, with speeds very close to the speed of light.

β^- decay

Consider the generic equation for beta decay

$$^A_Z X \rightarrow\, ^A_{Z+1}X' + e- + \nu_e.$$

The Q value for this decay is

$$Q = \left[m_N\left(^A_Z X\right) - m_N\left(^A_{Z+1}X'\right) - m_e - m_{\overline{\nu}_e}\right]c^2$$

where $m_N\left(^A_Z X\right)$ is the mass of the nucleus of the $^A_Z X$ atom, m_e is the mass of the electron, and $m_{\overline{\nu}_e}$ is the mass of the electron antineutrino. In other words, the total energy released is the mass energy of the initial nucleus, minus the mass energy of the final nucleus, electron, and antineutrino. The mass of the nucleus m_N is related to the standard atomic mass m by

$$m\left(^A_Z X\right)c^2 = m_N\left(^A_Z X\right)c^2 + Zm_ec^2 - \sum_{i=1}^{Z}B_i$$

That is, the total atomic mass is the mass of the nucleus, plus the mass of the electrons, minus the binding energy B_i of each electron. Substituting this into our original equation, while neglecting the nearly-zero antineutrino mass and difference in electron binding energy, which is very small for high-Z atoms, we have

$$Q = \left[m\left(^A_Z X\right) - m\left(^A_{Z+1}X'\right)\right]c^2$$

This energy is carried away as kinetic energy by the electron and neutrino.

Because the reaction will proceed only when the Q-value is positive, β^- decay can occur when the mass of atom $^A_Z X$ is greater than the mass of atom $^A_{Z+1}X'$.[8]

β^+ decay

The equations for β^+ decay are similar, with the generic equation

$$^A_Z X \rightarrow\, ^A_{Z-1}X' + e+ + \nu_e$$

giving

$$Q = \left[m_N\left(^A_Z X\right) - m_N\left(^A_{Z-1}X'\right) - m_e - m_{\nu_e}\right]c^2$$

However, in this equation, the electron masses do not cancel, and we are left with

$$Q = \left[m\left(^A_Z X\right) - m\left(^A_{Z-1}X'\right) - 2m_e\right]c^2$$

Because the reaction will proceed only when the Q-value is positive, β^+ decay can occur when the mass of atom $^A_Z X$ exceeds that of $^A_{Z-1}X'$ by at least twice the mass of the electron.[8]

Electron capture

The analogous calculation for electron capture must take into account the binding energy of the electrons. This is because the atom will be left in an excited state after capturing the electron, and the binding energy of the captured innermost electron is significant. Using the generic equation for electron capture

$$^A_Z X + e- \rightarrow\, ^A_{Z-1}X' + \nu_e$$

we have

$$Q = \left[m_N\left(^A_Z X\right) + m_e - m_N\left(^A_{Z-1}X'\right) - m_{\nu_e}\right]c^2$$

which simplifies to

$$Q = \left[m\left(^A_Z X\right) - m\left(^A_{Z-1}X'\right)\right]c^2 - B_n$$

where B_n is the binding energy of the captured electron.

Because the binding energy of the electron is much less than the mass of the electron, nuclei that can undergo β^+ decay can always also undergo electron capture, but the reverse is not true.[8]

5.2.7   Nuclear transmutation

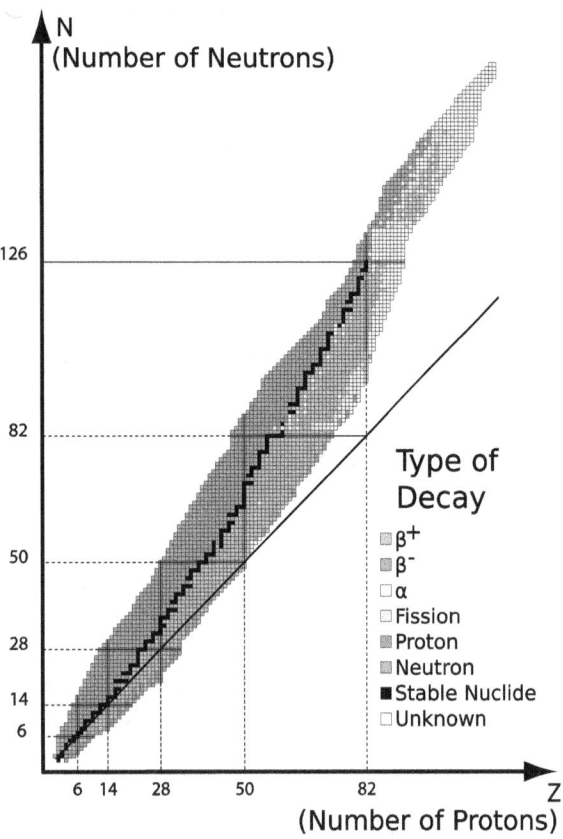

If the proton and neutron are part of an atomic nucleus, these decay processes transmute one chemical element into another. For example:

Beta decay does not change the number A of nucleons in the nucleus, but changes only its charge Z. Thus the set of all nuclides with the same A can be introduced; these *isobaric* nuclides may turn into each other via beta decay. Among them, several nuclides (at least one for any given mass number A) are beta stable, because they present local minima of the mass excess: if such a nucleus has (A, Z) numbers, the neighbour nuclei (A, Z−1) and (A, Z+1) have higher mass excess and can beta decay into (A, Z), but not vice versa. For all odd mass numbers A, there is only one known beta-stable isobar. For even A, there are up to three different beta-stable isobars experimentally known; for example, 96 40Zr, 96 42Mo, and 96 44Ru are all beta-stable. There are about 355 known beta-decay stable nuclides total.[9]

Usually, unstable nuclides are clearly either "neutron rich" or "proton rich", with the former undergoing beta decay and the latter undergoing electron capture (or more rarely, due to the higher energy requirements, positron decay). However, in a few cases of odd-proton, odd-neutron radionuclides, it may be energetically favorable for the radionuclide to decay to an even-proton, even-neutron isobar either by undergoing beta-positive or beta-negative decay. An often-cited example is 64 29Cu, which decays by positron emission/electron capture 61% of the time to 64 28Ni, and 39% of the time by (negative) beta decay to 64 30Zn.[10]

Most naturally occurring isotopes on Earth are beta stable. Those that are not have half-lives ranging from under a second to periods of time significantly greater than the age of the universe. One common example of a long-lived isotope is the odd-proton odd-neutron nuclide 40 19K, which undergoes all three types of beta decay ($\beta-$, $\beta+$ and electron capture) with a half-life of 1.277×10^9 years.[11]

5.2.8   Double beta decay

Main article: Double beta decay

Some nuclei can undergo double beta decay ($\beta\beta$ decay) where the charge of the nucleus changes by two units. Double beta decay is difficult to study, as the process has an extremely long half-life. In nuclei for which both β decay and $\beta\beta$ decay are possible, the rarer $\beta\beta$ decay process is effectively impossible to observe. However, in nuclei where β decay is forbidden but $\beta\beta$ decay is allowed, the process can be seen and a half-life measured.[12] Thus, $\beta\beta$ decay is usually studied only for beta stable nuclei. Like single beta decay, double beta decay does not change A; thus, at least one of the nuclides with some given A has to be stable with regard to both single and double beta decay.

"Ordinary" double beta decay results in the emission of two electrons and two antineutrinos. If neutrinos are Majorana particles (i.e., they are their own antiparticles), then a decay known as neutrinoless double beta decay will occur. Most neutrino physicists believe that neutrinoless double beta decay has never been observed.[12]

5.2.9   Bound-state $\beta^* -$ decay

A very small minority of free neutron decays (about four per million) are so-called "two-body decays", in which the proton, electron and antineutrino are produced, but the electron fails to gain the 13.6 eV energy necessary to escape the proton, and therefore simply remains bound to it, as a neutral hydrogen atom.[13] In this type of beta decay, in essence all of the neutron decay energy is carried off by the antineutrino.

For fully ionized atoms (bare nuclei), it is possible in likewise manner for electrons to fail to escape the atom, and to be emitted from the nucleus into low-lying atomic bound states (orbitals). This can not occur for neutral atoms whose low-lying bound states are already filled by electrons.

The phenomenon in fully ionized atoms was first observed for ^{163}Dy*66+ in 1992 by Jung et al. of the Darmstadt Heavy-Ion Research group. Although neutral ^{163}Dy is a stable isotope, the fully ionized ^{163}Dy*66+ undergoes β decay into the K and L shells with a half-life of 47 days.[14]

Another possibility is that a fully ionized atom undergoes greatly accelerated β decay, as observed for ^{187}Re by Bosch et al., also at Darmstadt. Neutral ^{187}Re does undergo β decay with a half-life of 42×10^9 years, but for fully ionized ^{187}Re*75+ this is shortened by a factor of 10^9 to only 32.9 years.[15] For comparison the variation of decay rates of other nuclear processes due to chemical environment is less than 1%.

5.2.10   Forbidden transitions

Beta decays can be classified according to the L-value of the emitted radiation. When $L > 0$, the decay is referred to as "forbidden". Nuclear selection rules require high L-values to be accompanied by changes in nuclear spin (J) and parity (π). The selection rules for the Lth forbidden transitions are:

$$\Delta J = L - 1, L, L + 1; \Delta \pi = (-1)^L,$$

where $\Delta \pi = 1$ or -1 corresponds to no parity change or parity change, respectively. The special case of a transition between isobaric analogue states, where the structure of the final state is very similar to the structure of the initial state, is referred to as "superallowed" for beta decay, and proceeds very quickly. The following table lists the ΔJ and $\Delta \pi$ values for the first few values of L:

5.2.11   Fermi transitions

A **Fermi transition** is a beta decay in which the spins of the emitted electron (positron) and anti-neutrino (neutrino) couple to total spin $S = 0$, leading to an angular momentum change $\Delta J = 0$ between the initial and final states of the nucleus (assuming an allowed transition $\Delta L = 0$). In the non-relativistic limit, the nuclear part of the operator for a Fermi transition is given by

$$\mathcal{O}_F = G_V \sum_a \hat{\tau}_{a\pm}$$

with G_V the weak vector coupling constant, $\tau_\pm$ the isospin raising and lowering operators, and a running over all protons and neutrons in the nucleus.

5.2.12   Gamow-Teller transitions

A **Gamow-Teller transition** is a beta decay in which the spins of the emitted electron (positron) and anti-neutrino (neutrino) couple to total spin $S = 1$, leading to an angular momentum change $\Delta J = 0, \pm 1$ between the initial and final states of the nucleus (assuming an allowed transition). In this case, the nuclear part of the operator is given by

$$\mathcal{O}_{GT} = G_A \sum_a \hat{\sigma}_a \hat{\tau}_{a\pm}$$

with G_A the weak axial-vector coupling constant, and σ the spin Pauli matrices, which can produce a spin-flip in the decaying nucleon.

5.2.13   Beta emission spectrum

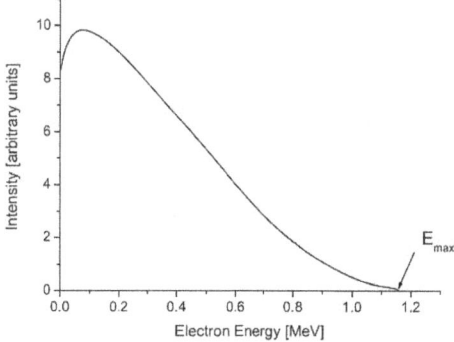

Beta spectrum of ^{210}Bi. E_{max} = Q = 1.16 MeV is the maximum energy

Beta decay can be considered as a perturbation as described in quantum mechanics, and thus Fermi's Golden Rule can be applied. This leads to an expression for the kinetic energy spectrum $N(T)$ of emitted betas as follows:[16]

$$N(T) = C_L(T) F(Z, T) p E (Q - T)^2$$

where T is the kinetic energy, C_L is a shape function that depends on the forbiddenness of the decay (it is constant for allowed decays), $F(Z, T)$ is the Fermi Function (see below) with Z the charge of the final-state nucleus, $E = T + mc^2$ is the total energy, $p = \sqrt{(E/c)^2 - (mc)^2}$ is the momentum, and Q is the Q value of the decay. The kinetic energy of the emitted neutrino is given approximately by Q minus the kinetic energy of the beta.

As an example, the beta decay spectrum of ^{210}Bi (originally called RaE) is shown to the right.

Fermi function

The Fermi function that appears in the beta spectrum formula accounts for the Coulomb attraction / repulsion between the emitted beta and the final state nucleus. Approximating the associated wavefunctions to be spherically symmetric, the Fermi function can be analytically calculated to be:[17]

$$F(Z,T) = \frac{2(1+S)}{\Gamma(1+2S)^2}(2p\rho)^{2S-2}e^{\pi\eta}|\Gamma(S+i\eta)|^2,$$

where $S = \sqrt{1 - \alpha^2 Z^2}$ (α is the fine-structure constant), $\eta = \pm \alpha ZE/pc$ (+ for electrons, − for positrons), $\varrho = r_N/\hbar$ (r_N is the radius of the final state nucleus), and Γ is the Gamma function.

For non-relativistic betas ($Q \ll m_e c^2$), this expression can be approximated by:[18]

$$F(Z,T) \approx \frac{2\pi\eta}{1 - e^{-2\pi\eta}}.$$

Other approximations can be found in the literature.[19][20]

Kurie plot

A **Kurie plot** (also known as a **Fermi–Kurie plot**) is a graph used in studying beta decay developed by Franz N. D. Kurie, in which the square root of the number of beta particles whose momenta (or energy) lie within a certain narrow range, divided by the Fermi function, is plotted against beta-particle energy.[21][22] It is a straight line for allowed transitions and some forbidden transitions, in accord with the Fermi beta-decay theory. The energy-axis (x-axis) intercept of a Kurie plot corresponds to the maximum energy imparted to the electron/positron (the decay's Q-value). With a Kurie plot one can find the limit on the effective mass of a neutrino.[23]

5.2.14   History

Discovery and characterization of β* − decay

Radioactivity was discovered in 1896 by Henri Becquerel in uranium, and subsequently observed by Marie and Pierre Curie in thorium and in the new elements polonium and radium. In 1899, Ernest Rutherford separated radioactive emissions into two types: alpha and beta (now beta minus), based on penetration of objects and ability to cause ionization. Alpha rays could be stopped by thin sheets of paper or aluminium, whereas beta rays could penetrate several millimetres of aluminium. (In 1900, Paul Villard identified a still more penetrating type of radiation, which Rutherford identified as a fundamentally new type in 1903, and termed gamma rays).

In 1900, Becquerel measured the mass-to-charge ratio (m/e) for beta particles by the method of J.J. Thomson used to study cathode rays and identify the electron. He found that m/e for a beta particle is the same as for Thomson's electron, and therefore suggested that the beta particle is in fact an electron.

In 1901, Rutherford and Frederick Soddy showed that alpha and beta radioactivity involves the transmutation of atoms into atoms of other chemical elements. In 1913, after the products of more radioactive decays were known, Soddy and Kazimierz Fajans independently proposed their radioactive displacement law, which states that beta (i.e., β−) emission from one element produces another element one place to the right in the periodic table, while alpha emission produces an element two places to the left.

Neutrinos in beta decay

Historically, the study of beta decay provided the first physical evidence of the neutrino. Measurements of the beta particle (electron) kinetic energy spectrum in 1911 by Lise Meitner and Otto Hahn and in 1913 by Jean Danysz showed multiple lines on a diffuse background, offering the first hint of a continuous spectrum.[24] In 1914, James Chadwick used a magnetic spectrometer with one of Hans Geiger's new counters to make a more accurate measurement and showed that the spectrum was continuous.[24][25] This was in apparent contradiction to the law of conservation of energy, since if beta decay were simply electron emission as assumed at the time, then the energy of the emitted electron should equal the energy difference between the initial and final nuclear states and lead to a narrow energy distribution, as observed for both alpha and gamma decay.[26] For beta decay, however, the observed broad continuous spectrum suggested that energy is lost in the beta decay process.

In 1920–1927, Charles Drummond Ellis (along with James Chadwick and colleagues) further established that the beta decay spectrum is continuous, ending all controversies. It also had an effective upper bound in energy, which was a severe blow to Bohr's suggestion that conservation of energy might be true only in a statistical sense, and might be violated in any given decay. Now the problem of how to account for the variability of energy in known beta decay products, as well as for conservation of momentum and angular momentum in the process, became acute.

A second problem related to the conservation of angular momentum. Molecular band spectra showed that the nuclear spin of nitrogen-14 is 1 (i.e. equal to the reduced Planck constant), and more generally that the spin is integral for nuclei of even mass number and half-integral for nuclei of odd mass number, as later explained by the proton-neutron model of the nucleus.[26] Beta decay leaves the mass number unchanged, so that the change of nuclear spin must be an integer. However the electron spin is 1/2, so that angular momentum would not be conserved if beta decay were simply electron emission.

In a famous letter written in 1930, Wolfgang Pauli suggested that, in addition to electrons and protons, atomic nuclei also contained an extremely light neutral particle, which he called the neutron. He suggested that this "neutron" was also emitted during beta decay (thus accounting for the known missing energy, momentum, and angular momentum) and had simply not yet been observed. In 1931, Enrico Fermi renamed Pauli's "neutron" to neutrino and, in 1934, he published a very successful model of beta decay in which neutrinos were produced. The neutrino interaction with matter was so weak that detecting it proved a severe experimental challenge, which was finally met in 1956 in the Cowan–Reines neutrino experiment.[27] However, the properties of neutrinos were (with a few minor modifications) as predicted by Pauli and Fermi.

Discovery of other types of beta decay

In 1934, Frédéric and Irène Joliot-Curie bombarded aluminium with alpha particles to effect the nuclear reaction $^4_2$He + $^{27}_{13}$Al → $^{30}_{15}$P + $^1_0$n, and observed that the product isotope $^{30}_{15}$P emits a positron identical to those found in cosmic rays by Carl David Anderson in 1932. This was the first example of β+ decay (positron emission), which they termed artificial radioactivity since $^{30}_{15}$P is a short-lived nuclide which does not exist in nature.

The theory of electron capture was first discussed by Gian-Carlo Wick in a 1934 paper, and then developed by Hideki Yukawa and others. K-electron capture was first observed in 1937 by Luis Alvarez, in the nuclide ^{48}V.[28][29][30] Alvarez went on to study electron capture in ^{67}Ga and other nuclides.[28][31][32]

Non-conservation of parity

In 1956, Chien-Shiung Wu and coworkers proved in the Wu experiment that parity is not conserved in beta decay.[33][34] This surprising fact had been postulated shortly before in an article by Tsung-Dao Lee and Chen Ning Yang.[35]

5.2.15 See also

- Double beta decay
- Electron capture
- Neutrino
- Alpha decay
- Betavoltaics
- Particle radiation
- Radionuclide
- Tritium illumination, a form of fluorescent lighting powered by beta decay
- Pandemonium effect
- Total absorption spectroscopy

5.2.16 References

- Tuli, J. K. (2011). *Nuclear Wallet Cards* (PDF) (8th ed.). Brookhaven National Laboratory.

[1] Konya, J.; Nagy, N. M. (2012). *Nuclear and Radiochemistry*. Elsevier. pp. 74–75. ISBN 978-0-12-391487-3.

[2] Basdevant, Jean-Louis; Rich, James; Spiro, Michael (2005). *Fundamentals in Nuclear Physics: From Nuclear Structure to Cosmology*. Springer. ISBN 978-0387016726.

[3] Zuber, Kai (2011). *Neutrino Physics* (2 ed.). CRC Press. p. 466. ISBN 9781420064711.

[4] Loveland, Walter D. (2005). *Modern Nuclear Chemistry*. Wiley. p. 232. ISBN 0471115320.

[5] Tatjana Jevremovic (21 April 2009). *Nuclear Principles in Engineering*. Springer Science & Business Media. p. 201. ISBN 978-0-387-85608-7.

[6] H. Frauenfelder, R. Bobone, E. Von Goeler, N. Levine, H. R. Lewis, R. N. Peacock, A. Rossi and G. DePasquali, Physical Review 106 (1957) 386

[7] E. J. Konopinski and M. E. Rose, *The Theory of nuclear Beta Decay*, in: *Alpha-, Beta- and Gamma-Ray Spectroscopy*, ed. by Kai Siegbahn, Vol. 2, North-Holland Publishing Company, Amsterdam, 1966

[8] Kenneth S. Krane (5 November 1987). *Introductory Nuclear Physics*. Wiley. ISBN 978-0-471-80553-3.

[9] "Interactive Chart of Nuclides". National Nuclear Data Center, Brookhaven National Laboratory. Retrieved 2014-09-18.

[10] "WWW Table of Radioactive Isotopes, Copper 64". *LBNL Isotopes Project*. Lawrence Berkeley National Laboratory. Retrieved 2014-09-18.

[11] "WWW Table of Radioactive Isotopes, Potassium 40". *LBNL Isotopes Project*. Lawrence Berkeley National Laboratory. Retrieved 2014-09-18.

[12] S.M. Bilenky (October 5, 2010). "Neutrinoless double beta-decay". *Physics of Particles and Nuclei* **41** (5). arXiv:1001.1946. Bibcode:2010PPN....41..690B. doi:10.1134/S1063779610050035.

[13] An Overview Of Neutron Decay J. Byrne in Quark-Mixing, CKM Unitarity (H.Abele and D.Mund, 2002), see p.XV

[14] Jung, M. et al. (1992). "First observation of bound-state $\beta^* -$ decay". *Physical Review Letters* **69** (15): 2164–2167. Bibcode:1992PhRvL..69.2164J. doi:10.1103/PhysRevLett.69.2164. PMID 10046415.

[15] Bosch, F. et al. (1996). "Observation of bound-state beta minus decay of fully ionized ^{187}Re: ^{187}Re–^{187}Os Cosmochronometry". *Physical Review Letters* **77** (26): 5190–5193. Bibcode:1996PhRvL..77.5190B. doi:10.1103/PhysRevLett.77.5190. PMID 10062738.

[16] Nave, C. R. "Energy and Momentum Spectra for Beta Decay". *HyperPhysics*. Retrieved 2013-03-09.

[17] Fermi, E. (1934). "Versuch einer Theorie der β-Strahlen. I". *Zeitschrift für Physik* **88** (3–4): 161–177. Bibcode:1934ZPhy...88..161F. doi:10.1007/BF01351864.

[18] Mott, N. F.; Massey, H. S. W. (1933). *The Theory of Atomic Collisions*. Clarendon Press. LCCN 34001940.

[19] Venkataramaiah, P.; Gopala, K.; Basavaraju, A.; Suryanarayana, S. S.; Sanjeeviah, H. (1985). "A simple relation for the Fermi function". *Journal of Physics G* **11** (3): 359–364. Bibcode:1985JPhG...11..359V. doi:10.1088/0305-4616/11/3/014.

[20] Schenter, G. K.; Vogel, P. (1983). "A simple approximation of the fermi function in nuclear beta decay". *Nuclear Science and Engineering* **83** (3): 393–396. OSTI 5307377.

[21] Kurie, F. N. D.; Richardson, J. R.; Paxton, H. C. (1936). "The Radiations Emitted from Artificially Produced Radioactive Substances. I. The Upper Limits and Shapes of the β-Ray Spectra from Several Elements". *Physical Review* **49** (5): 368–381. Bibcode:1936PhRv...49..368K. doi:10.1103/PhysRev.49.368.

[22] Kurie, F. N. D. (1948). "On the Use of the Kurie Plot". *Physical Review* **73** (10): 1207. Bibcode:1948PhRv...73.1207K. doi:10.1103/PhysRev.73.1207.

[23] Rodejohann, Werner (2012). "Neutrinoless double beta decay and neutrino physics". arXiv:1206.2560v2.

[24] Jensen, Carsten (2000). *Controversy and Consensus: Nuclear Beta Decay 1911-1934*. Birkhäuser Verlag. ISBN 3-7643-5313-9.

[25] Chadwick, James (1914). "Intensitätsverteilung im magnetischen Spektren der β-Strahlen von Radium B + C". *Verhandlungen der Deutschen Physikalischen Gesellschaft* (in German) (Deutsche Physikalische Gesellschaft) **16**: 383–391.

[26] Brown, Laurie M. (1978). "The idea of the neutrino". *Physics Today* **31** (9): 23–8. Bibcode:1978PhT....31i..23B. doi:10.1063/1.2995181.

[27] C. L Cowan Jr., F. Reines, F. B. Harrison, H. W. Kruse, A. D McGuire (July 20, 1956). "Detection of the Free Neutrino: a Confirmation". *Science* **124** (3212): 103–4. Bibcode:1956Sci...124..103C. doi:10.1126/science.124.3212.103. PMID 17796274.

[28] Segré, E. (1987). "K-Electron Capture by Nuclei". In Trower, P. W. *Discovering Alvarez: Selected Works of Luis W. Alvarez*. University of Chicago Press. pp. 11–12. ISBN 978-0-226-81304-2.

[29] "The Nobel Prize in Physics 1968: Luis Alvarez". The Nobel Foundation. Retrieved 2009-10-07.

[30] Alvarez, L. W. (1937). "Nuclear K Electron Capture". *Physical Review* **52** (2): 134–135. Bibcode:1937PhRv...52..134A. doi:10.1103/PhysRev.52.134.

[31] Alvarez, L. W. (1938). "Electron Capture and Internal Conversion in Gallium 67". *Physical Review* **53** (7): 606. Bibcode:1938PhRv...53..606A. doi:10.1103/PhysRev.53.606.

[32] Alvarez, L. W. (1938). "The Capture of Orbital Electrons by Nuclei". *Physical Review* **54** (7): 486–497. Bibcode:1938PhRv...54..486A. doi:10.1103/PhysRev.54.486.

[33] C. S. Wu; E. Ambler; R. W. Hayward; D. D. Hoppes; R. P. Hudson (1957). "Experimental Test of Parity Conservation in Beta Decay". *Physical Review* **105**: 1413–1415. Bibcode:1957PhRv..105.1413W. doi:10.1103/PhysRev.105.1413.

[34] http://blogs.scientificamerican.
com/guest-blog/2013/10/15/
channeling-ada-lovelace-chien-shiung-wu-courage

[35] T. D. Lee, C. N. Yang (1956). "Question of Parity Conservation in Weak Interactions". *Physical Review* **104**: 254–258. Bibcode:1956PhRv..104..254L. doi:10.1103/PhysRev.104.254.

5.2.17 Bibliography

- Sin-Itiro Tomonaga (1997). *The Story of Spin*. University of Chicago Press.

5.2.18 External links

- **The Live Chart of Nuclides - IAEA** with filter on decay type

- Definition of Beta Disintegration (Decay) at Science Dictionary

5.3 Double beta decay

Double beta decay is a radioactive decay process where a nucleus releases two beta rays as a single process.

5.3.1 History

The idea of double beta decay was first proposed by Maria Goeppert-Mayer in 1935.[*][1] In 1937 Ettore Majorana theoretically demonstrated that all results of beta decay theory remain unchanged if the neutrino is its own anti-particle, i.e. a Majorana particle. In 1939 Wendell H. Furry proposed if neutrinos are a Majorana particle, double beta decay can proceed without emission of any neutrino, via the process now called the neutrinoless beta decay.[*][2]

In 1930–40s parity violation in weak interactions was not known and consequently calculations showed that neutrinoless double beta decay should be much more likely to occur than ordinary double beta decay (if neutrinos are Majorana particles). The predicted half-lives were on the order of $10^{*}15–16$ years. Efforts to observe the process in laboratory date back to at least 1948 when Edward L. Fireman made the first attempt to measure the half-life of the 124Sn isotope. Radiometric experiments through about 1960 produced negative results or false positives, not confirmed by later experiments. In 1950 for the first time the half-life of the 130Te was measured by geochemical methods, to be $1.4 \times 10^{*}21$ years, reasonably close to the modern value.[*][2]

In 1956 after the V-A nature of weak interactions was established it became clear the half-life of neutrinoless double beta decay would significantly exceed that of ordinary double beta decay. Despite significant progress in experimental techniques in 1960–70s, double beta decay was not observed in a laboratory until the 1980s. Experiments had only been able to establish the lower bound for the half-life —about 10^{21} years. On the other hand, geochemical experiments detected double beta decay of 82Se and 128Te.[*][2]

Double beta decay was first observed in a laboratory in 1987 by the group of Michael Moe at UC Irvine on 82Se. Since then many experiments have observed ordinary double beta decay in other isotopes. None of those experiments has produced positive results for the neutrinoless process, raising the half-life lower bound to 10^{25} years. Geochemical experiments continued through the 1990s, producing positive results for a few more isotopes.[*][2] Double beta decay is the rarest known kind of radioactive decay; as of 2012 it has been observed for only 12 isotopes (including double electron capture in 130Ba observed in 2001), and all have a mean lifetime over 10^{18} yr (table below).[*][2]

5.3.2 Ordinary double beta decay

In double beta decay, two neutrons in the nucleus are converted to protons, and two electrons and two electron antineutrinos are emitted. The process can be thought as a sum of two beta minus decays. In order for (double) beta decay to be possible, the final nucleus must have a larger binding energy than the original nucleus. For some nuclei, such as germanium-76, the nucleus one atomic number higher has a smaller binding energy, preventing single beta decay. However, the nucleus with atomic number two higher, selenium-76, has a larger binding energy, so double beta decay is allowed.

For some nuclei, the process occurs as conversion of two protons to neutrons, emitting two electron neutrinos and absorbing two orbital electrons (double electron capture). If the mass difference between the parent and daughter atoms is more than 1.022 MeV/c² (two electron masses), another decay is accessible, capture of one orbital electron and emission of one positron. When the mass difference is more than 2.044 MeV/c² (four electron masses), emission of two positrons is possible. These theoretical decay branches have not been observed.

Known double beta decay isotopes

There are 35 naturally occurring isotopes capable of double beta decay. The decay can be observed in practice if the single beta decay is forbidden by energy conservation. This happens for even-Z, even-N isotopes, which are more stable due to spin-coupling, seen by the pairing term in the semi-empirical mass formula.

Many isotopes are theoretically expected to double beta decay. In most cases, the double beta decay is so rare its nearly impossible to observe against the background. However, the double beta decay of 238U (also an alpha emitter) has been measured radiochemically.. Two of the nuclides (48Ca and 96Zr) from the table below can also single beta decay but this is extremely suppressed and never been observed.

Eleven isotopes have been experimentally observed undergoing two-neutrino double beta decay.[*][3] The table below contains nuclides with the latest experimentally measured half-lives, as of December 2012.[*][3]

Note:In the table above where two errors are specified the first one is statistical error and the second is systematic.[*][3]

5.3.3   Neutrinoless double beta decay

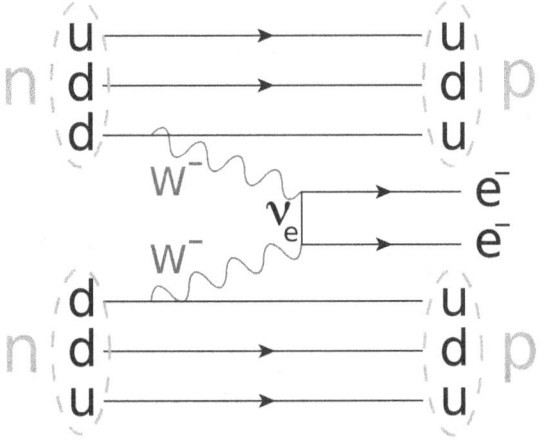

Feynman diagram of neutrinoless double beta decay, with two neutrons decaying to two protons. The only emitted products in this process are two electrons, which can occur if the neutrino and antineutrino are the same particle (i.e. Majorana neutrinos) so the same neutrino can be emitted and absorbed within the nucleus. In conventional double beta decay, two antineutrinos —one arising from each W vertex —are emitted from the nucleus, in addition to the two electrons. The detection of neutrinoless double beta decay is thus a sensitive test of whether neutrinos are Majorana particles.

The processes described in the previous section are also known as two-neutrino double beta decay, as two neutrinos (or antineutrinos) are emitted. If the neutrino is a Majorana particle (meaning that the antineutrino and the neutrino are actually the same particle), and at least one type of neutrino has non-zero mass (which has been established by the neutrino oscillation experiments), then it is possible for neutrinoless double beta decay to occur. In the simplest theo-

retical treatment, light neutrino exchange, the two neutrinos annihilate each other, or equivalently, a nucleon absorbs the neutrino emitted by another nucleon.

The neutrinos in the above diagram are virtual particles. With only two electrons in the final state, the electrons total kinetic energy would be approximately the binding energy difference of the initial and final nuclei (with the nucleus recoil accounting for the rest). To a very good approximation, the electrons are emitted back-to-back. The decay rate for this process is approximately given by

$$\Gamma = \quad G|M|^2|m_{\beta\beta}|^2,$$

where G is the two-body phase-space factor, M is the nuclear matrix element, and $m_{\beta\beta}$ is the effective Majorana neutrino mass given by

$$m_{\beta\beta} = \sum_{i=1}^{3} m_i U_{ei}^2.$$

In this expression, m_i is the neutrino masses (of the i^*th mass eigenstate), and the U_{ei} are elements of the lepton mixing Pontecorvo–Maki–Nakagawa–Sakata (PMNS) matrix. Therefore, observing neutrinoless double beta decay, in addition to confirming the Majorana neutrino nature, would give information on the absolute neutrino mass scale, potentially the neutrino mass hierarchy, and Majorana phases in the PMNS matrix.[*][5][*][6]

The deep significance of the process stems from the "black-box theorem" which that observing neutrinoless double beta decay implies at least one neutrino is a Majorana particle, irrespective of whether the process is engendered by neutrino exchange.[*][7]

Experiments

Numerous experiments have searched for neutrinoless double beta decay. Recent and proposed experiments include:

- Completed experiments:
  - Gotthard TPC
  - Heidelberg-Moscow
  - IGEX
  - NEMO

- Experiments currently taking data:
  - COBRA, ^{116}Cd in room temperature CdZnTe crystals

- CUORE(CUORE-0), ^{130}Te in TeO$_2$ crystals.
- DCBA, testing a magnetic tracking detector at KEK
- EXO, a ^{136}Xe search
- GERDA, a ^{76}Ge detector
- KamLAND-Zen, a ^{136}Xe search
- Majorana, using high purity ^{76}Ge p-type point-contact detectors
- XMASS using liquid Xe

- Proposed/future experiments:

  - CANDLES, ^{48}Ca in CaF$_2$, at Kamioka Observatory
  - MOON, developing ^{100}Mo detectors
  - AMoRE, ^{100}Mo enriched CaMoO4 crystals at YangYang underground laboratory[*][8]
  - LUMINEU, exploring ^{100}Mo enriched ZnMoO$_4$ crystals at LSM, France.
  - NEXT, a Xenon TPC. NEXT-DEMO ran and NEXT-100 will run in 2016.
  - SNO+, a liquid scintillator, will study ^{130}Te
  - SuperNEMO, a NEMO upgrade, will study ^{82}Se
  - TIN.TIN, a ^{124}Sn detector at INO

Status

Early experiments did claim neutrinoless decay, but modern searches have set limits disfavoring those results. Recent published lower bounds for Ge and Xe indicate no sign of neutrinoless decay.

Heidelberg-Moscow Controversy Heidelberg-Moscow collaboration initially released limits on neutrinoless beta decay in germanium-76.[*][1] Then some members claimed detection in 2001.[*][9] This claim was criticized by outside physicists[*][1][*][10][*][11] as well as other members of the collaboration.[*][12] In 2006 a refined estimate by the same authors stated the half-life was 2.3×10[*]25 years.[*][13] More sensitive experiments are expected to resolve the controversy.[*][1][*][14]

Current Results As of 2014, GERDA has reached much lower background, obtaining a half-life limit of 2.1×10[*]25 years with 21.6 kg*yr exposure.[*][15] IGEX and HDM data increase the limit to 3×10[*]25 yr and rule out detection at high confidence. Searches with ^{136}Xe, Kamland-Zen and EXO-200, yielded a limit of 2.6×10[*]25 yr. Using the latest nuclear matrix elements, the ^{136}Xe results also disfavor the HM claim.

5.3.4 See also

- Double electron capture
- Beta decay
- Neutrino
- Particle radiation
- Radioactive isotope

5.3.5 References

[1] Giuliani, A.; Poves, A. (2012). "Neutrinoless Double-Beta Decay". *Advances in High Energy Physics* **2012**: 1. doi:10.1155/2012/857016.

[2] Barabash, A. S. (2011). "Experiment double beta decay: Historical review of 75 years of research". *Physics of Atomic Nuclei* **74** (4): 603–613. arXiv:1104.2714. Bibcode:2011PAN....74..603B. doi:10.1134/S1063778811030070.

[3] Beringer, J.; Arguin, J.; Barnett, R.; Copic, K.; Dahl, O.; Groom, D.; Lin, C.; Lys, J.; Murayama, H.; Wohl, C. G.; Yao, W. -M.; Zyla, P. A.; Amsler, C.; Antonelli, M.; Asner, D. M.; Baer, H.; Band, H. R.; Basaglia, T.; Bauer, C. W.; Beatty, J. J.; Belousov, V. I.; Bergren, E.; Bernardi, G.; Bertl, W.; Bethke, S.; Bichsel, H.; Biebel, O.; Blucher, E.; Blusk, S.; Brooijmans, G. (2012). "Review of Particle Physics". *Physical Review D* **86**. Bibcode:2012PhRvD..86a0001B. doi:10.1103/PhysRevD.86.010001.

[4] Agostini, M.; Allardt, M.; Andreotti, E.; Bakalyarov, A. M.; Balata, M.; Barabanov, I.; Heider, M. B.; Barros, N.; Baudis, L.; Bauer, C.; Becerici-Schmidt, N.; Bellotti, E.; Belogurov, S.; Belyaev, S. T.; Benato, G.; Bettini, A.; Bezrukov, L.; Bode, T.; Brudanin, V.; Brugnera, R.; Budjáš, D.; Caldwell, A.; Cattadori, C.; Chernogorov, A.; Cossavella, F.; Demidova, E. V.; Denisov, A.; Domula, A.; Egorov, V. et al. (2013). "Measurement of the half-life of the two-neutrino double beta decay of76Ge with the GERDA experiment". *Journal of Physics G: Nuclear and Particle Physics* **40** (3): 035110. doi:10.1088/0954-3899/40/3/035110.

[5] K. Grotz and H.V. Klapdor, „The Weak Interaction in Nuclear, Particle and Astrophysics ", Adam Hilger, Bristol, 1990, 461 ps.

[6] H.V. Klapdor, A. Staudt „Non-accelerator Particle Physics", 2.edition, Institute of Physics Publishing, Bristol, Philadelphia, 1998, 535 ps.

[7] Schechter, J.; J. W. F. Valle (1982-06-01). "Neutrinoless Double beta Decay in SU(2) ⊗ U(1) theories". *Physical Review D* 25 : 2951. Bibcode:1982PhRvD..25.2951S.doi: 10.1103/PhysRevD.25.2951

[8] N. D. Khanbekov (September 2013). "AMoRE: Collaboration for searches for the neutrinoless double-beta decay of the isotope of 100Mo with the aid of 40Ca100MoO4 as a cryogenic scintillation detector". *Physics of Atomic Nuclei, Volume 76, Issue 9.* Bibcode:2013PAN....76.1086K. doi:10.1134/S1063778813090093.

[9] Klapdor-Kleingrothaus, H. V.; Dietz, A.; Harney, H. L.; Krivosheina, I. V. (2001). "Evidence for Neutrinoless Double Beta Decay". *Modern Physics Letters A* **16** (37): 2409. doi:10.1142/S0217732301005825.

[10] Aalseth, C. E.; Avignone, F. T.; Barabash, A.; Boehm, F.; Brodzinski, R. L.; Collar, J. I.; Doe, P. J.; Ejiri, H.; Elliott, S. R.; Fiorini, E.; Gaitskell, R. J.; Gratta, G.; Hazama, R.; Kazkaz, K.; King, G. S.; Kouzes, R. T.; Miley, H. S.; Moe, M. K.; Morales, A.; Morales, J.; Piepke, A.; Robertson, R. G. H.; Tornow, W.; Vogel, P.; Warner, R. A.; Wilkerson, J. F. (2002). "Comment on "evidence for Neutrinoless Double Beta Decay"". *Modern Physics Letters A* **17** (22): 1475. doi:10.1142/S0217732302007715.

[11] Zdesenko, Y. G.; Danevich, F. A.; Tretyak, V. I. (2002). "Has neutrinoless double β decay of 76Ge been really observed?". *Physics Letters B* **546** (3–4): 206. doi:10.1016/S0370-2693(02)02705-3.

[12] A. M. Bakalyarov, A. Y. Balysh, S. T. Belyaev, V. I. Lebedev, and S. V. Zhukov, (2003). "Results of the experiment on investigation of Germanium-76 double beta decay". *Proceedings of the NANP, Dubna, Russia.*

[13] Klapdor-Kleingrothaus, H. V.; Krivosheina, I. V. (2006). "THE EVIDENCE FOR THE OBSERVATION OF 0νββ DECAY: THE IDENTIFICATION OF 0νββ EVENTS FROM THE FULL SPECTRA". *Modern Physics Letters A* **21** (20): 1547. doi:10.1142/S0217732306020937.

[14] Schwingenheuer, B. (2013). "Status and prospects of searches for neutrinoless double beta decay". *Annalen der Physik* **525** (4): 269. doi:10.1002/andp.201200222.

[15] Agostini, M.; Allardt, M.; Andreotti, E.; Bakalyarov, A. M.; Balata, M.; Barabanov, I.; Barnabé Heider, M.; Barros, N.; Baudis, L.; Bauer, C.; Becerici-Schmidt, N.; Bellotti, E.; Belogurov, S.; Belyaev, S. T.; Benato, G.; Bettini, A.; Bezrukov, L.; Bode, T.; Brudanin, V.; Brugnera, R.; Budjáš, D.; Caldwell, A.; Cattadori, C.; Chernogorov, A.; Cossavella, F.; Demidova, E. V.; Domula, A.; Egorov, V.; Falkenstein, R. et al. (2013). "Results on Neutrinoless Double-β Decay of ^{76}Ge from Phase I of the GERDA Experiment". *Physical Review Letters* **111** (12). doi:10.1103/PhysRevLett.111.122503.

5.3.6   External links

• Double beta decay on arxiv.org

Enrico Fermi

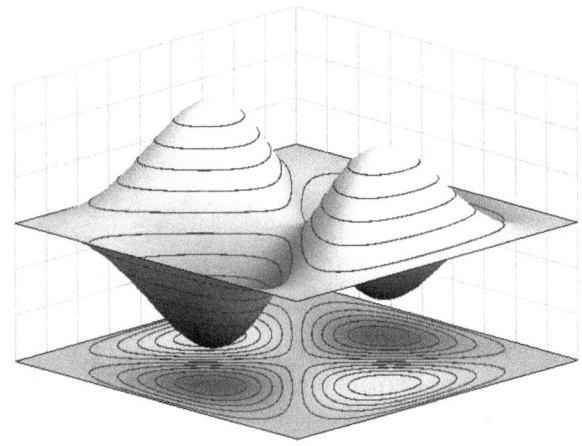

Antisymmetric wavefunction for a (fermionic) 2-particle state in an infinite square well potential.

5.4   Fermion

In particle physics, a **fermion** (a name coined by Paul Dirac*[1] from the surname of Enrico Fermi) is any particle characterized by Fermi–Dirac statistics. These particles obey the Pauli exclusion principle. Fermions include all quarks and leptons, as well as any composite particle made of an odd number of these, such as all baryons and many atoms and nuclei. Fermions differ from bosons, which obey Bose–Einstein statistics.

A fermion can be an elementary particle, such as the electron, or it can be a composite particle, such as the proton. According to the spin-statistics theorem in any reasonable relativistic quantum field theory, particles with integer spin are bosons, while particles with half-integer spin are neutrons fermions.

Besides this spin characteristic, fermions have another specific property: they possess conserved baryon or lepton quantum numbers. Therefore what is usually referred as the spin statistics relation is in fact a spin statistics-quantum number relation.*[2]

As a consequence of the Pauli exclusion principle, only one fermion can occupy a particular quantum state at any given time. If multiple fermions have the same spatial probability distribution, then at least one property of each fermion, such as its spin, must be different. Fermions are usually associated with matter, whereas bosons are generally force carrier particles, although in the current state of particle physics the distinction between the two concepts is unclear. At low temperature fermions show superfluidity for uncharged particles and superconductivity for charged particles. Composite fermions, such as protons and neutrons, are the key building blocks of everyday matter. Weakly interacting fermions can also display bosonic behavior under extreme conditions, such as superconductivity.

5.4.1 Elementary fermions

The Standard Model recognizes two types of elementary fermions, quarks and leptons. In all, the model distinguishes 24 different fermions. There are six quarks (up, down, strange, charm, bottom and top quarks), and six leptons (electron, electron neutrino, muon, muon neutrino, tau particle and tau neutrino), along with the corresponding antiparticle of each of these.

Mathematically, fermions come in three types - Weyl fermions (massless), Dirac fermions (massive), and Majorana fermions (each its own antiparticle). Most Standard Model fermions are believed to be Dirac fermions, although it is unknown at this time whether the neutrinos are Dirac or Majorana fermions. Dirac fermions can be treated as a combination of two Weyl fermions.*[3]*:106 So far there is no known example of Weyl fermion in particle physics. In July 2015, Weyl fermions have been experimentally realized in Weyl semimetals.

5.4.2 Composite fermions

See also: List of particles § Composite particles

Composite particles (such as hadrons, nuclei, and atoms) can be bosons or fermions depending on their constituents. More precisely, because of the relation between spin and statistics, a particle containing an odd number of fermions is itself a fermion. It will have half-integer spin.

Examples include the following:

- A baryon, such as the proton or neutron, contains three fermionic quarks and thus it is a fermion.

- The nucleus of a carbon-13 atom contains six protons and seven neutrons and is therefore a fermion.

- The atom helium-3 (^{3}He) is made of two protons, one neutron, and two electrons, and therefore it is a fermion.

The number of bosons within a composite particle made up of simple particles bound with a potential has no effect on whether it is a boson or a fermion.

Fermionic or bosonic behavior of a composite particle (or system) is only seen at large (compared to size of the system) distances. At proximity, where spatial structure begins to be important, a composite particle (or system) behaves according to its constituent makeup.

Fermions can exhibit bosonic behavior when they become loosely bound in pairs. This is the origin of superconductivity and the superfluidity of helium-3: in superconducting materials, electrons interact through the exchange of phonons, forming Cooper pairs, while in helium-3, Cooper pairs are formed via spin fluctuations.

The quasiparticles of the fractional quantum Hall effect are also known as composite fermions, which are electrons with an even number of quantized vortices attached to them.

Skyrmions

Main article: Skyrmion

In a quantum field theory, there can be field configurations of bosons which are topologically twisted. These are coherent states (or solitons) which behave like a particle, and they can be fermionic even if all the constituent particles are bosons. This was discovered by Tony Skyrme in the early 1960s, so fermions made of bosons are named skyrmions after him.

Skyrme's original example involved fields which take values on a three-dimensional sphere, the original nonlinear sigma model which describes the large distance behavior of pions. In Skyrme's model, reproduced in the large N or string approximation to quantum chromodynamics (QCD), the proton and neutron are fermionic topological solitons of the pion field.

Whereas Skyrme's example involved pion physics, there is a much more familiar example in quantum electrodynamics with a magnetic monopole. A bosonic monopole with the smallest possible magnetic charge and a bosonic version of the electron will form a fermionic dyon.

The analogy between the Skyrme field and the Higgs field of the electroweak sector has been used*[4] to postulate that all fermions are skyrmions. This could explain why all known fermions have baryon or lepton quantum numbers and provide a physical mechanism for the Pauli exclusion principle.

5.4.3 See also

5.4.4 Notes

[1] Notes on Dirac's lecture *Developments in Atomic Theory* at Le Palais de la Découverte, 6 December 1945, UK-NATARCHI Dirac Papers BW83/2/257889. See note 64 on page 331 in "The Strangest Man: The Hidden Life of Paul Dirac, Mystic of the Atom" by Graham Farmelo

[2] Physical Review D volume 87, page 0550003, year 2013, author Weiner, Richard M., title "Spin-statistics-quantum number connection and supersymmetry" arxiv:1302.0969

[3] T. Morii; C. S. Lim; S. N. Mukherjee (1 January 2004). *The Physics of the Standard Model and Beyond*. World Scientific. ISBN 978-981-279-560-1.

[4] Weiner, Richard M. (2010). "The Mysteries of Fermions". *International Journal of Theoretical Physics* **49** (5): 1174–1180. arXiv:0901.3816. Bibcode:2010IJTP...49.1174W. doi:10.1007/s10773-010-0292-7.

5.5 Flavour

In particle physics, **flavour** or **flavor** refers to a species of an elementary particle. The Standard Model counts six flavours of quarks and six flavours of leptons. They are conventionally parameterized with *flavour quantum numbers* that are assigned to all subatomic particles, including composite ones. For hadrons, these quantum numbers depend on the numbers of constituent quarks of each particular flavour.

5.5.1 Intuitive description

Elementary particles are not eternal and indestructible. Unlike in classical mechanics, where forces only change a particle's momentum, the weak force can alter the essence of a particle, even an elementary particle. This means that it can convert one quark to another quark with different mass and electric charge, and the same for leptons. From the point of view of quantum mechanics, changing the flavour of a particle by the weak force is no different in principle from changing its spin by electromagnetic interaction, and should be described with quantum numbers as well. In particular, flavour states may undergo quantum superposition.

In atomic physics the principal quantum number of an electron specifies the electron shell in which it resides, which determines the energy level of the whole atom. In an analogous way, the five flavour quantum numbers of a quark specify which of six flavours (u, d, s, c, b, t) it has, and when these quarks are combined this results in different types of baryons and mesons with different masses, electric charges, and decay modes.

5.5.2 Flavour symmetry

If there are two or more particles which have identical interactions, then they may be interchanged without affecting the physics. Any (complex) linear combination of these two particles give the same physics, as long as they are orthogonal or perpendicular to each other. In other words, the theory possesses symmetry transformations such as $M \begin{pmatrix} u \\ d \end{pmatrix}$, where u and d are the two fields, and M is any 2×2 unitary matrix with a unit determinant. Such matrices form a Lie group called SU(2) (see special unitary group). This is an example of flavour symmetry.

In quantum chromodynamics, flavour is a global symmetry. In the electroweak theory, on the other hand, this symmetry is broken, and flavour changing processes exist, such as quark decay or neutrino oscillations.

5.5.3 Flavour quantum numbers

Leptons

All leptons carry a lepton number $L = 1$. In addition, leptons carry weak isospin, T_3, which is $-1/2$ for the three charged leptons (i.e. electron, muon and tau) and $+1/2$ for the three associated neutrinos. Each doublet of a charged lepton and a neutrino consisting of opposite T_3 are said to constitute one generation of leptons. In addition, one defines a quantum number called weak hypercharge, Y_W, which is -1 for all left-handed leptons.*[1] Weak isospin and weak hypercharge are gauged in the Standard Model.

Leptons may be assigned the six flavour quantum numbers: electron number, muon number, tau number, and corresponding numbers for the neutrinos. These are conserved in strong and electromagnetic interactions, but violated by weak interactions. Therefore, such flavour quantum numbers are not of great use. A separate quantum number for each generation is more useful: electronic lepton number (+1 for electrons and electron neutrinos), muonic lepton number (+1 for muons and muon neutrinos), and tauonic lepton number (+1 for tau leptons and tau neutrinos). However, even these numbers are not absolutely conserved, as neutrinos of different generations can mix; that is, a neutrino of one flavour can transform into another flavour. The strength of such mixings is specified by a matrix called the Pontecorvo–Maki–Nakagawa–Sakata matrix (PMNS matrix).

Quarks

All quarks carry a baryon number $B = 1/3$. They also all carry weak isospin, $T_3 = \pm 1/2$. The positive-T_3 quarks (up, charm, and top quarks) are called *up-type quarks* and negative-T_3 quarks (down, strange, and bottom quarks) are called *down-type quarks*. Each doublet of up and down type quarks constitutes one generation of quarks.

For all the quark flavour quantum numbers (strangeness, charm, topness and bottomness) the convention is that the flavour charge and the electric charge of a quark have the same sign. Thus any flavour carried by a charged meson has the same sign as its charge. Quarks have the following flavour quantum numbers:

- Isospin, less ambiguously known as "isobaric spin", which has value $I_3 = 1/2$ for the up quark and $I_3 = -1/2$ for the down quark.

- Strangeness (S): Defined as $S = -(n_s - n_{\bar{s}})$, where n_s represents the number of strange quarks (s) and $n_{\bar{s}}$ represents the number of strange antiquarks (s). This quantum number was introduced by Murray Gell-Mann. This definition gives the strange quark a strangeness of -1 for the above-mentioned reason.

- Charm (C): Defined as $C = (n_c - n_{\bar{c}})$, where n_c represents the number of charm quarks (c) and $n_{\bar{c}}$ represents the number of charm antiquarks. Is +1 for the charm quark.

- Bottomness (B′): Also called 'beauty'. Defined as $B' = -(n_b - n_{\bar{b}})$, where n_b represents the number of bottom quarks (b) and $n_{\bar{b}}$ represents the number of bottom antiquarks.

- Topness (T): Also called 'truth'. Defined as $T = (n_t - n_{\bar{t}})$, where n_t represents the number of top quarks (t) and $n_{\bar{t}}$ represents the number of top antiquarks. However, because of the extremely short half-life of the top quark, by the time it can interact strongly it has already decayed to another flavour of quark (usually to a bottom quark). For that reason the top quark doesn't hadronize, that is it never forms any meson or baryon.

These five quantum numbers, together with baryon number (which is not a flavour quantum number) completely specify numbers of all 6 quark flavours separately (as $n_q - n_{\bar{q}}$, i.e. an antiquark is counted with the minus sign). They are conserved by both the electromagnetic and strong interactions (but not the weak interaction). From them can be built the derived quantum numbers:

- Hypercharge (Y): $Y = B + S + C + B' + T$

- Electric charge: $Q = I_3 + 1/2 Y$ (see Gell-Mann–Nishijima formula)

The terms "strange" and "strangeness" predate the discovery of the quark, but continued to be used after its discovery for the sake of continuity (i.e. the strangeness of each type of hadron remained the same); strangeness of anti-particles being referred to as +1, and particles as −1 as per the original definition. Strangeness was introduced to explain the rate of decay of newly discovered particles, such as the kaon, and was used in the Eightfold Way classification of hadrons and in subsequent quark models. These quantum numbers are preserved under strong and electromagnetic interactions, but not under weak interactions.

For first-order weak decays, that is processes involving only one quark decay, these quantum numbers (e.g. charm) can only vary by 1 ($|C| = \pm 1$); $\Delta B' = \pm 1$. Since first-order processes are more common than second-order processes (involving two quark decays), this can be used as an approximate "selection rule" for weak decays.

A quark of a given flavour is an eigenstate of the weak interaction part of the Hamiltonian: it will interact in a definite way with the W and Z bosons. On the other hand, a fermion of a fixed mass (an eigenstate of the kinetic and strong interaction parts of the Hamiltonian) is normally a superposition of various flavours. As a result, the flavour content of a quantum state may change as it propagates freely. The transformation from flavour to mass basis for quarks is given by the Cabibbo–Kobayashi–Maskawa matrix (CKM matrix). This matrix is analogous to the PMNS matrix for neutrinos, and defines the strength of flavour changes under weak interactions of quarks.

The CKM matrix allows for CP violation if there are at least three generations.

Antiparticles and hadrons

Flavour quantum numbers are additive. Hence antiparticles have flavour equal in magnitude to the particle but opposite in sign. Hadrons inherit their flavour quantum number from their valence quarks: this is the basis of the classification in the quark model. The relations between the hypercharge, electric charge and other flavour quantum numbers hold for hadrons as well as quarks.

5.5.4   Quantum chromodynamics

Flavour symmetry is closely related to chiral symmetry. This part of the article is best read along with the one on chirality.

Quantum chromodynamics (QCD) contains six flavours of quarks. However, their masses differ and as a result they are not strictly interchangeable with each other. The up and down flavours are close to having equal masses, and the theory of these two quarks possesses an approximate SU(2) symmetry (isospin symmetry).

Under some circumstances, the masses of the quarks can be neglected entirely. One can then make flavour transformations independently on the left- and right-handed parts of each quark field. The flavour group is then a chiral group $SU_L(N_f) \times SU_R(N_f)$.

If all quarks had non-zero but equal masses, then this chiral symmetry is broken to the *vector symmetry* of the "diagonal flavour group" $SU(N_f)$, which applies the same transformation to both helicities of the quarks. Such a reduction of the symmetry is called *explicit symmetry breaking*. The amount of explicit symmetry breaking is controlled by the current quark masses in QCD.

Even if quarks are massless, chiral flavour symmetry can be spontaneously broken if the vacuum of the theory contains a chiral condensate (as it does in low-energy QCD). This gives rise to an effective mass for the quarks, often identified with the valence quark mass in QCD.

Symmetries of QCD

Analysis of experiments indicate that the current quark masses of the lighter flavours of quarks are much smaller than the QCD scale, Λ_{QCD}, hence chiral flavour symmetry is a good approximation to QCD for the up, down and strange quarks. The success of chiral perturbation theory and the even more naive chiral models spring from this fact. The valence quark masses extracted from the quark model are much larger than the current quark mass. This indicates that QCD has spontaneous chiral symmetry breaking with the formation of a chiral condensate. Other phases of QCD may break the chiral flavour symmetries in other ways.

5.5.5   Conservation laws

All of the various charges discussed above are conserved by the fact that the charge operator is best understood as the generator of a symmetry that commutes with the Hamiltonian. Thus, the eigenvalues of the various charge operators are conserved.

Absolutely conserved flavour quantum numbers are: (including the baryon number for completeness)

- electric charge (Q)

- weak isospin (I_3)

- baryon number (B)

- lepton number (L)

In some theories, the individual baryon and lepton number conservation can be violated, if the difference between them ($B - L$) is conserved (see chiral anomaly). All other flavour quantum numbers are violated by the electroweak interactions. Strong interactions conserve all flavours.

5.5.6   History

Some of the historical events that lead to the development of flavour symmetry are discussed in the article on isospin.

5.5.7   See also

- Standard Model (mathematical formulation)

- Cabibbo–Kobayashi–Maskawa matrix

- Strong CP problem and chirality (physics)

- Chiral symmetry breaking and quark matter

- Quark flavour tagging, such as B-tagging, is an example of particle identification in experimental particle physics.

5.5.8   References

[1] See table in S. Raby, R. Slanky (1997). "Neutrino Masses: How to add them to the Standard Model" (PDF). *Los Alamos Science* (25): 64.

5.5.9   Further reading

- Lessons in Particle Physics Luis Anchordoqui and Francis Halzen, University of Wisconsin, 18th Dec. 2009

5.5.10   External links

- The particle data group.

5.6 Generation

In particle physics, a **generation** (or **family**) is a division of the elementary particles. Between generations, particles differ by their (flavour) quantum number and mass, but their interactions are identical.

There are three generations according to the Standard Model of particle physics. Each generation is divided into two types of leptons and two types of quarks. The two leptons may be classified into one with electric charge −1 (electron-like) and one neutral (neutrino); the two quarks may be classified into one with charge −$\frac{1}{3}$ (down-type) and one with charge +$\frac{2}{3}$ (up-type).

5.6.1 Overview

Each member of a higher generation has greater mass than the corresponding particle of the previous generation, with the possible exception of the neutrinos (whose small but non-zero masses have not been accurately determined). For example, the first-generation electron has a mass of only 0.511 MeV/c^2, the second-generation muon has a mass of 106 MeV/c^2, and the third-generation tau has a mass of 1777 MeV/c^2 (almost twice as heavy as a proton). This mass hierarchy causes particles of higher generations to decay to the first generation, which explains why everyday matter (atoms) is made of particles from the first generation. Electrons surround a nucleus made of protons and neutrons, which contain up and down quarks. The second and third generations of charged particles do not occur in normal matter and are only seen in extremely high-energy environments such as cosmic rays or particle accelerators. The term *generation* was first introduced by Haim Harari in Les Houches Summer School, 1976.[*][1] [*][2]

Neutrinos of all generations stream throughout the universe but rarely interact with normal matter.[*][3] It is hoped that a comprehensive understanding of the relationship between the generations of the leptons may eventually explain the ratio of masses of the fundamental particles, and shed further light on the nature of mass generally, from a quantum perspective.[*][4]

5.6.2 Fourth generation

Fourth and further generations are considered to be unlikely. Some of the arguments against the possibility of a fourth generation are based on the subtle modifications of precision electroweak observables that extra generations would induce; such modifications are strongly disfavored by measurements. Furthermore, a fourth generation with a "light" neutrino (one with a mass less than about 45

GeV/c^2) has been ruled out by measurements of the widths of the Z boson at CERN's Large Electron–Positron Collider (LEP).[*][5] Nonetheless, searches at high-energy colliders for particles from a fourth generation continue, but as yet no evidence has been observed.[*][6] In such searches, fourth-generation particles are denoted by the same symbols as third-generation ones with an added prime (e.g. b' and t').

According to the results of the statistical analysis by researchers from CERN, and Humboldt University of Berlin, the existence of further fermions can be excluded with a probability of 99.99999% (5.3 sigma). The researchers combined latest data collected by the particle accelerators LHC and Tevatron with many known measurements results relating to particles, such as the Z-boson or the top-quark. The most important data used for this analysis come from the discovery of the Higgs particle. In the Standard Model, the Higgs particle gives all other particles their mass. As additional fermions were not detected directly in accelerator experiments, they have to be heavier than the fermions known so far. Hence, these fermions would also interact with the Higgs particle more strongly. This interaction would have modified the properties of the Higgs particle such that this particle would not have been detected.[*][7]

5.6.3 See also

- Metric expansion of space
- Spacetime
- Supersymmetry
- World line

5.6.4 References

[1] Harari, H. (1977). "Beyond charm". In Balian, R.; Llewellyn-Smith, C.H. *Weak and Electromagnetic Interactions at High Energy, Les Houches, France, Jul 5- Aug 14, 1976.* Les Houches Summer School Proceedings **29**. North-Holland. p. 613.

[2] Harari H. (1977). "Three generations of quarks and leptons" (PDF). In E. van Goeler, Weinstein R. (eds.). *Proceedings of the XII Rencontre de Moriond.* p. 170. SLAC-PUB-1974.

[3] "Experiment confirms famous physics model" (Press release). MIT News Office. 18 April 2007.

[4] M.H. Mac Gregor (2006). "A 'Muon Mass Tree' with α-quantized Lepton, Quark, and Hadron Masses". arXiv:hep-ph/0607233 [hep-ph].

[5] D. Decamp *et al.* (ALEPH collaboration) (1989). "Determination of the number of light neutrino species". *Physics*

Letters B **231** (4): 519. Bibcode:1989PhLB..231..519D. doi:10.1016/0370-2693(89)90704-1.

[6] C. Amsler *et al.* (Particle Data Group) (2008). "Review of Particle Physics: b′ (4th Generation) Quarks, Searches for" (PDF). *Physics Letters B* **667** (1): 1–1340. Bibcode:2008PhLB..667....1P. doi:10.1016/j.physletb.2008.07.018.

[7] *12 matter particles suffice in nature* Dec 13, 2012 Phys.Org

5.7 Leptoquark

Leptoquarks are hypothetical particles that carry information between quarks and leptons of a given generation that allow quarks and leptons to interact. They are color-triplet bosons that carry both lepton and baryon numbers. They are encountered in various extensions of the Standard Model, such as technicolor theories or GUTs based on Pati–Salam model, SU(5) or E_6, etc. Their quantum numbers like spin, (fractional) electric charge and weak isospin vary among theories.

Leptoquarks, predicted to be nearly as heavy as an atom of lead, could only be created at high energies, and would decay rapidly. A third generation leptoquark, for example, might decay into a bottom quark and a tau lepton. Some theorists propose that the 'leptoquark' observed by HERA and DESY could be a new force that bonds positrons and quarks or be examples of preons found at high energies.*[1] Leptoquarks could explain the reason for the three generations of matter. Furthermore, leptoquarks could explain why the same number of quarks and leptons exist and many other similarities between the quark and the lepton sectors. At high energies, when leptons that do not feel the strong force and quarks that cannot be separately observed because of the strong force become one, it could form a more fundamental particle and describe a higher symmetry. There would be three kinds of leptoquarks made of the leptons and quarks of each generation.

The LHeC project to add an electron ring to collide bunches with the existing LHC proton ring is proposed as a project to look for higher-generation leptoquarks.*[2]

5.7.1 Existence

In 1997, an excess of events at the HERA accelerator created a stir in the particle physics community, because one possible explanation of the excess was the involvement of leptoquarks. However, more recent studies performed both at HERA and at the Tevatron with larger samples of data ruled out this possibility for masses of the leptoquark up to 275-325 GeV.*[3] Second generation leptoquarks were

also looked for and not found.*[4] For leptoquarks to be proven to exist, the missing energy in particle collisions attributed to neutrinos would have to be excessively energetic. It is likely that the creation of leptoquarks would mimic the creation of massive quarks.*[5]

5.7.2 See also

- Quark–lepton complementarity

- X and Y bosons

5.7.3 References

[1] Scientific American

[2] Birmingham LHeC project page

[3] H1 Collaboration; Andreev, V.; Anthonis, T.; Aplin, S.; Asmone, A.; Astvatsatourov, A.; Babaev, A.; Backovic, S.; Bähr, J.; Baghdasaryan, A.; Baranov, P.; Barrelet, E.; Bartel, W.; Baudrand, S.; Baumgartner, S.; Becker, J.; Beckingham, M.; Behnke, O.; Behrendt, O.; Belousov, A.; Berger, Ch.; Berger, N.; Bizot, J.C.; Boenig, M.-O.; Boudry, V.; Bracinik, J.; Brandt, G.; Brisson, V.; Brown, D.P. et al. (2005). "Search for Leptoquark Bosons in ep Collisions at HERA". *Physics Letters B* **629**: 9–19. arXiv:hep-ex/0506044. Bibcode:2005PhLB..629....9H. doi:10.1016/j.physletb.2005.09.048.

[4] The Search for Leptoquarks.

[5] Search for Third Generation Leptoquarks

5.8 Koide formula

The **Koide formula** is an unexplained empirical equation discovered by Yoshio Koide in 1981. It relates the masses of the three charged leptons so well that it predicted the mass of the tau.

5.8.1 Formula

The Koide formula is:

$$Q = \frac{m_e + m_\mu + m_\tau}{(\sqrt{m_e} + \sqrt{m_\mu} + \sqrt{m_\tau})^2} \approx \frac{2}{3}.$$

It is clear that $1/3 < Q < 1$. The superior bound follows if we assume that the square roots can not be negative. By Cauchy-Schwarz $1/3Q$ can be interpreted as the squared cosine of the angle between the vector

$(\sqrt{m_e}, \sqrt{m_\mu}, \sqrt{m_\tau})$

and the vector

$(1, 1, 1)$.

The mystery is in the physical value. The masses of the electron, muon, and tau are measured respectively as $m_e = 0.510998910(13)$ MeV/c^2, $m_\mu = 105.658367(4)$ MeV/c^2, and $m_\tau = 1776.84(17)$ MeV/c^2, where the digits in parentheses are the uncertainties in the last figures.[1] This gives $Q = 0.666659(10)$.[2] Not only is this result odd in that three apparently random numbers should give a simple fraction, but also that Q is exactly halfway between the two extremes of $1/3$ (should the three masses be equal) and 1 (should one mass dominate).

While the original formula appeared in the context of preon models, other ways have been found to produce it (both by Sumino and by Koide, see references below). As a whole, however, understanding remains incomplete.

Similar matches have been found for quarks depending on running masses, and for triplets of quarks not of the same flavour.[3][4][5] With alternating quarks, chaining Koide equations for consecutive triplets, it is possible to reach a result of 173.263947(6) GeV for the mass of the top quark.[6]

5.8.2 Similar Formulae

There are similar empirical formulae which relate other masses. For example the bottom and top quark masses of approximately 5 GeV and 174 GeV satisfy:

$$Q = \frac{m_B + m_T}{(\sqrt{m_B} + \sqrt{m_T})^2} \approx \frac{3}{4}.$$

Where again the fraction 3/4 is exactly in the middle of $1/2 < Q < 1$, although the masses of these quarks are known less accurately.

Quark masses depend on the energy scale used to measure them, which makes an analysis more complicated.

Taking the heaviest three quarks, charm (1290 MeV), bottom (4370 MeV) and top (174100 MeV), gives a much closer match:

$$Q = \frac{m_C + m_B + m_T}{(\sqrt{m_C} + \sqrt{m_B} + \sqrt{m_T})^2} \approx \frac{2}{3}.$$

It is possible to get exactly 2/3 within the experimental uncertainties of the masses (as of 2015). This was noticed by R&Z in the first version of their paper[7] but the observation was removed in the published version,[8] so the first published mention is from F. G. Cao.[9]

The masses of the lightest quarks (up, down, strange) are not known well enough to test the formula but if up is taken equal to zero; in such case the formula -including the masslessness of up quark- reduces to one of Harari-Haut-Weyers,[10] known well before Koide.

5.8.3 Running of particle masses

In quantum field theory, quantities like coupling constant and mass "run" with the energy scale. That is, their value depends on the energy scale at which the observation occurs, in a way described by a renormalization group equation (RGE). One usually expects relationships between such quantities to be simple at high energies (where some symmetry is unbroken) but not at low energies, where the RG flow will have produced complicated deviations from the high energy relation. The Koide relation is exact (within experimental error) for the pole masses, which are low-energy quantities defined at different energy scales. For this reason, many physicists regard the relation as "numerology" (e.g.[11]). However, the Japanese physicist Yukinari Sumino has constructed an effective field theory in which a new gauge symmetry causes the pole masses to exactly satisfy the relation.[12] Goffinet's doctoral thesis gives a discussion on pole masses and how the Koide formula can be reformulated without taking the square roots of masses.[13]

5.8.4 See also

- CKM matrix

- Clifford algebra

- Generation

- Higgs mechanism

- Higgsless model

- PMNS matrix

- Quark–lepton complementarity

- Seesaw mechanism

- Technicolor

5.8.5   References

[1] Amsler, C.; et al.   (Particle Data Group) (2008, and 2009 partial update).   "Review of Particle Physics – Leptons" (PDF). *Physics Letters B* **667** (1-5): 1.   Bibcode:2008PhLB..667....1P. doi:10.1016/j.physletb.2008.07.018.   Check date values in: |date= (help)

[2] Since the uncertainties in m_e and m_μ are much smaller than that in m_τ, the uncertainty in Q was calculated as $\Delta Q = \frac{\partial Q}{\partial m_\tau} \Delta m_\tau$.

[3] Rodejohann, W.; Zhang, H. (2011).   "Extension of an empirical charged lepton mass relation to the neutrino sector" .   *Physics Letters B* **698** (2): 152–156.   arXiv:1101.5525.   Bibcode:2011PhLB..698..152R. doi:10.1016/j.physletb.2011.03.007.

[4] Rosen, G. (2007).   "Heuristic development of a Dirac-Goldhaber model for lepton and quark structure" (PDF).   *Modern Physics Letters A* **22** (4):   283–288.   Bibcode:2007MPLA...22..283R. doi:10.1142/S0217732307022621.

[5] Kartavtsev, A. (2011).   "A remark on the Koide relation for quarks" .   arXiv:1111.0480 [hep-ph].

[6] Rivero, A. (2011).   "A new Koide tuple: Strange-charm-bottom" .   arXiv:1111.7232 [hep-ph].

[7] Rodejohann, W.; Zhang, H. (2011).   "Extension of an empirical charged lepton mass relation to the neutrino sector" .   arXiv:1101.5525 [hep-ph].

[8] Rodejohann, W.; Zhang, H. (2011).   "Extension of an empirical charged lepton mass relation to the neutrino sector" .   *Physics Letters B* **698** (2): 152–156.   arXiv:1101.5525.   Bibcode:2011PhLB..698..152R. doi:10.1016/j.physletb.2011.03.007.

[9] Cao, F. G. (2012).   "Neutrino masses from lepton and quark mass relations and neutrino oscillations" .   *Physical Review D* **85** (11):   113003. arXiv:1205.4068.   Bibcode:2012PhRvD..85k3003C. doi:10.1103/PhysRevD.85.113003.

[10] Harari, H.; Haut, H.; Weyers, J. (1978).   "Quark Masses And Cabibbo Angles" .   *Physics Letters B* **78**: 459.   Bibcode:1978PhLB...78..459H. doi:10.1016/0370-2693(78)90485-9.

[11] Motl, L. (16 January 2012).   "Could the Koide formula be real?". *The Reference Frame*. Retrieved 2014-07-10.

[12] Sumino, Y. (2009).   "Family Gauge Symmetry as an Origin of Koide's Mass Formula and Charged Lepton Spectrum" .   *Journal of High Energy Physics* **2009** (5): 75.   arXiv:0812.2103.   Bibcode:2009JHEP...05..075S. doi:10.1088/1126-6708/2009/05/075.

[13] Goffinet, F. (2008). *A bottom-up approach to fermion masses* (PDF) (PhD Thesis). Université catholique de Louvain.

5.8.6   Further reading

- Koide, Y. (1983).   "New view of quark and lepton mass hierarchy" .   *Physical Review D* **28** (1):   252–254.   Bibcode:1983PhRvD..28..252K. doi:10.1103/PhysRevD.28.252.

  - Koide, Y. (1984).   "Erratum: New view of quark and lepton mass hierarchy" .   *Physical Review D* **29** (7): 1544.   Bibcode:1984PhRvD..29Q1544K. doi:10.1103/PhysRevD.29.1544.

- Koide, Y. (1983).   "A fermion-boson composite model of quarks and leptons" . *Physics Letters B* **120** (1–3):   161–165.   Bibcode:1983PhLB..120..161K. doi:10.1016/0370-2693(83)90644-5.

- Oneda, S.; Koide, Y. (1991). *Asymptotic symmetry and its implication in elementary particle physics.* World Scientific. ISBN 981-02-0498-1.

- Foot, R. (1994).   "A note on Koide's lepton mass relation" .   arXiv:hep-ph/9402242 [hep-ph].

- Koide, Y. (2000).   "Quark and lepton mass matrices with a cyclic permutation invariant form" .   arXiv:hep-ph/0005137 [hep-ph].

- Rivero, A.; Gsponer, A. (2005). "The strange formula of Dr. Koide" .   arXiv:hep-ph/0505220 [hep-ph].

- Koide, Y. (2005).   "Challenge to the mystery of the charged lepton mass" .   arXiv:hep-ph/0506247 [hep-ph].

- Li, N.; Ma, B.-Q. (2006).   "Energy scale independence for quark and lepton masses" .   arXiv:hep-ph/0601031 [hep-ph].

- Brannen, C. (2010).   "Spin Path Integrals and Generations" (PDF). *Foundations of Physics* **40**: 1681. arXiv:1006.3114.   Bibcode:2010FoPh...40.1681B. doi:10.1007/s10701-010-9465-8.   (See the article's *references* links to "The lepton masses" and "Recent results from the MINOS experiment" .)

- Kocik, J. (2012). "The Koide lepton mass formula and geometry of circles" .   arXiv:1201.2067 [physics.gen-ph].

5.8.7   External links

- Wolfram Alpha, link solves for the predicted tau mass from the Koide formula

5.9 Majorana fermion

Not to be confused with Majoron.

A **Majorana fermion** (/maɪəˈrɒnə ˈfɛərmiːɒn/[1]), also

Ettore Majorana hypothesised the existence of Majorana fermions in 1937

referred to as a **Majorana particle**, is a fermion that is its own antiparticle. They were hypothesized by Ettore Majorana in 1937. The term is sometimes used in opposition to a Dirac fermion, which describes fermions that are not their own antiparticles.

All of the Standard Model fermions except the neutrino behave as Dirac fermions at low energy (after electroweak symmetry breaking), but the (massive) nature of the neutrino is not settled and it may be either Dirac or Majorana. In condensed matter physics, Majorana fermions exist as quasiparticle excitations in superconductors and can be used to form Majorana bound states governed by non-abelian statistics.

5.9.1 Theory

The concept goes back to Majorana's suggestion in 1937[2] that neutral spin$-1/2$ particles can be described by a real wave equation (the Majorana equation), and would therefore be identical to their antiparticle (because the wave functions of particle and antiparticle are related by complex conjugation).

The difference between Majorana fermions and Dirac fermions can be expressed mathematically in terms of the creation and annihilation operators of second quantization. The creation operator $\gamma_j^\dagger$ creates a fermion in quantum state j (described by a *real* wave function), whereas the annihilation operator γ_j annihilates it (or, equivalently, creates the corresponding antiparticle). For a Dirac fermion the operators $\gamma_j^\dagger$ and γ_j are distinct, whereas for a Majorana fermion they are identical. Majorana Fermions are the first exhibit of supersymmetry as their real and imaginary wave functions are the same. They could theoretically solve the Higgs boson mass problem.

5.9.2 Elementary particle

Because particles and antiparticles have opposite conserved charges, in order to be a Majorana fermion, namely, it is its own antiparticle, it is necessarily uncharged. All of the elementary fermions of the Standard Model have gauge charges, so they cannot have fundamental Majorana masses. However, the right-handed sterile neutrinos introduced to explain neutrino oscillation could have Majorana masses. If they do, then at low energy (after electroweak symmetry breaking), by the seesaw mechanism, the neutrino fields would naturally behave as six Majorana fields, with three expected to have very high masses (comparable to the GUT scale) and the other three expected to have very low masses (comparable to 1 eV). If right-handed neutrinos exist but do not have a Majorana mass, the neutrinos would instead behave as three Dirac fermions and their antiparticles with masses coming directly from the Higgs interaction, like the other Standard Model fermions.

The seesaw mechanism is appealing because it would naturally explain why the observed neutrino masses are so small. However, if the neutrinos are Majorana then they violate the conservation of lepton number and even B − L.

Neutrinoless double beta decay, which can be viewed as two beta decay events with the produced antineutrinos immediately annihilating with one another, is only possible if neutrinos are their own antiparticles.[3] Experiments are underway to search for this type of decay.[4]

The high-energy analog of the neutrinoless double beta decay process is the production of same sign charged lepton pairs at hadron colliders;[5] it is being searched for by both the ATLAS and CMS experiments at the Large Hadron Collider. In theories based on left–right symmetry, there is a deep connection between these processes.[6] In the most accepted explanation of the smallness of neutrino mass, the seesaw mechanism, the neutrino is naturally a Majorana fermion.

Majorana fermions cannot possess intrinsic electric or magnetic moments, only toroidal moments.[*][7][*][8][*][9] Such minimal interaction with electromagnetic fields makes them potential candidates for cold dark matter.[*][10][*][11] The hypothetical neutralino of supersymmetric models is a Majorana fermion.

5.9.3   Majorana bound states

In superconducting materials, Majorana fermions can emerge as (non-fundamental) quasiparticles (which are more commonly referred as Bogoliubov quasiparticles in condensed matter.).   This becomes possible because a quasiparticle in a superconductor is its own antiparticle. Majorana fermions (i.e. the Bogoliubov quasiparticles) in superconductors were observed by many experiments many years ago.

Mathematically, the superconductor imposes electron hole "symmetry" on the quasiparticle excitations, relating the creation operator $\gamma(E)$ at energy E to the annihilation operator $\gamma^\dagger(-E)$ at energy $-E$. Majorana fermions can be bound to a defect at zero energy, and then the combined objects are called Majorana bound states or Majorana zero modes.[*][12] This name is more appropriate than Majorana fermion (although the distinction is not always made in the literature), because the statistics of these objects is no longer fermionic. Instead, the Majorana bound states are an example of non-abelian anyons: interchanging them changes the state of the system in a way that depends only on the order in which the exchange was performed. The non-abelian statistics that Majorana bound states possess allows them to be used as a building block for a topological quantum computer.[*][13]

A quantum vortex in certain superconductors or superfluids can trap midgap states, so this is one source of Majorana bound states.[*][14][*][15][*][16] Shockley states at the end points of superconducting wires or line defects are an alternative, purely electrical, source.[*][17] An altogether different source uses the fractional quantum Hall effect as a substitute for the superconductor.[*][18]

Experiments in superconductivity

In 2008, Fu and Kane provided a groundbreaking development by theoretically predicting that Majorana bound states can appear at the interface between topological insulators and superconductors.[*][19][*][20] Many proposals of a similar spirit soon followed, where it was shown that Majorana bound states can appear even without any topological insulator. An intense search to provide experimental evidence

of Majorana bound states in superconductors[*][21][*][22] first produced some positive results in 2012.[*][23][*][24] A team from the Kavli Institute of Nanoscience at Delft University of Technology in the Netherlands reported an experiment involving indium antimonide nanowires connected to a circuit with a gold contact at one end and a slice of superconductor at the other.   When exposed to a moderately strong magnetic field the apparatus showed a peak electrical conductance at zero voltage that is consistent with the formation of a pair of Majorana bound states, one at either end of the region of the nanowire in contact with the superconductor.[*][25] This type of bounded state with zero energy was soon detected by several other groups in similar hybrid devices.[*][26][*][27][*][28][*][29]

This experiment from Delft marks a possible verification of independent 2010 theoretical proposals from two groups[*][30][*][31] predicting the solid state manifestation of Majorana bound states in semiconducting wires.   However, it was also pointed out that some other trivial non-topological bounded states[*][32] could highly mimic the zero voltage conductance peak of Majorana bound state.

In 2014, evidence of Majorana bound states was observed using a low-temperature scanning tunneling microscope, by scientists at Princeton University.[*][33][*][34] It was suggested that Majorana bound states appeared at the edges of a chain of iron atoms formed on the surface of superconducting lead. Physicist Jason Alicea of California Institute of Technology, not involved in the research, said the study offered "compelling evidence" for Majorana fermions but that "we should keep in mind possible alternative explanations—even if there are no immediately obvious candidates".[*][35]

5.9.4   References

[1] "Quantum Computation possible with Majorana Fermions" on YouTube, uploaded 19 April 2013, retrieved 5 October 2014; and also based on the physicist's name's pronunciation.

[2] Majorana, Ettore; Maiani, Luciano (2006).   "A symmetric theory of electrons and positrons".   In Bassani, Giuseppe Franco. *Ettore Majorana Scientific Papers*. pp. 201–33.   doi:10.1007/978-3-540-48095-2_10. ISBN 978-3-540-48091-4.   Translated from: Majorana, Ettore (1937).   "Teoria simmetrica dell'elettrone e del positrone".   *Il Nuovo Cimento* (in Italian) **14** (4): 171–84. doi:10.1007/bf02961314.

[3] Schechter, J.; Valle, J.W.F. (1982).   "Neutrinoless Double beta Decay in SU(2) x U(1) Theories".   *Physical Review D* **25** (11): 2951.   Bibcode:1982PhRvD..25.2951S. doi:10.1103/PhysRevD.25.2951.   (subscription required (help)).

[4] Rodejohann, Werner (2011). "Neutrino-less Double Beta Decay and Particle Physics". *International Journal of Modern Physics* **E20** (9): 1833. arXiv:1106.1334. Bibcode:2011IJMPE..20.1833R. doi:10.1142/S0218301311020186. (registration required (help)).

[5] Keung, Wai-Yee; Senjanović, Goran (1983). "Majorana Neutrinos and the Production of the Right-Handed Charged Gauge Boson". *Physical Review Letters* **50** (19): 1427. Bibcode:1983PhRvL..50.1427K. doi:10.1103/PhysRevLett.50.1427. (subscription required (help)).

[6] Tello, Vladimir; Nemevšek, Miha; Nesti, Fabrizio; Senjanović, Goran; Vissani, Francesco (2011). "Left-Right Symmetry: from LHC to Neutrinoless Double Beta Decay". *Physical Review Letters* **106** (15): 151801. arXiv:1011.3522. Bibcode:2011PhRvL.106o1801T. doi:10.1103/PhysRevLett.106.151801. (subscription required (help)).

[7] Kayser, Boris; Goldhaber, Alfred S. (1983). "CPT and CP properties of Majorana particles, and the consequences". *Physical Review D* **28** (9): 2341–2344. Bibcode:1983PhRvD..28.2341K. doi:10.1103/PhysRevD.28.2341. (subscription required (help)).

[8] Radescu, E. E. (1985). "On the electromagnetic properties of Majorana fermions". *Physical Review D* **32** (5): 1266–1268. Bibcode:1985PhRvD..32.1266R. doi:10.1103/PhysRevD.32.1266. (subscription required (help)).

[9] Boudjema, F.; Hamzaoui, C.; Rahal, V.; Ren, H. C. (1989). "Electromagnetic Properties of Generalized Majorana Particles". *Physical Review Letters* **62** (8): 852–854. Bibcode:1989PhRvL..62..852B. doi:10.1103/PhysRevLett.62.852. (subscription required (help)).

[10] Pospelov, Maxim; ter Veldhuis, Tonnis (2000). "Direct and indirect limits on the electro-magnetic form factors of WIMPs". *Physics Letters B* **480**: 181–186. arXiv:hep-ph/0003010. Bibcode:2000PhLB..480..181P. doi:10.1016/S0370-2693(00)00358-0.

[11] Ho, Chiu Man; Scherrer, Robert J. (2013). "Anapole Dark Matter". *Physics Letters B* **722** (8): 341–346. arXiv:1211.0503. Bibcode:2013PhLB..722..341H. doi:10.1016/j.physletb.2013.04.039.

[12] Wilczek, Frank (2009). "Majorana returns" (PDF). *Nature Physics* **5** (9): 614–618. Bibcode:2009NatPh...5..614W. doi:10.1038/nphys1380.

[13] Nayak, Chetan; Simon, Steven H.; Stern, Ady; Freedman, Michael; Das Sarma, Sankar (2008). "Non-Abelian anyons and topological quantum computation". *Reviews of Modern Physics* **80** (3): 1083.

arXiv:0707.1889. Bibcode:2008RvMP...80.1083N. doi:10.1103/RevModPhys.80.1083.

[14] N.B. Kopnin; M.M. Salomaa (1991). "Mutual friction in superfluid ^{3}He: Effects of bound states in the vortex core". *Physical Review B* **44** (17): 9667. Bibcode:1991PhRvB..44.9667K. doi:10.1103/PhysRevB.44.9667.

[15] Volovik, G. E. (1999). "Fermion zero modes on vortices in chiral superconductors". *JETP Letters* **70** (9): 609–614. arXiv:cond-mat/9909426. Bibcode:1999JETPL..70..609V. doi:10.1134/1.568223.

[16] Read, N.; Green, Dmitry (2000). "Paired states of fermions in two dimensions with breaking of parity and time-reversal symmetries and the fractional quantum Hall effect". *Physical Review B* **61** (15): 10267. arXiv:cond-mat/9906453. Bibcode:2000PhRvB..6110267R. doi:10.1103/PhysRevB.61.10267.

[17] Kitaev, A. Yu (2001). "Unpaired Majorana fermions in quantum wires". *Physics-Uspekhi (supplement)* **44** (131): 131. arXiv:cond-mat/0010440. Bibcode:2001PhyU...44..131K. doi:10.1070/1063-7869/44/10S/S29.

[18] Moore, Gregory; Read, Nicholas (August 1991). "Nonabelions in the fractional quantum Hall effect". *Nuclear Physics B* **360** (2–3): 362. Bibcode:1991NuPhB.360..362M. doi:10.1016/0550-3213(91)90407-O.

[19] Fu, Liang; Kane, Charles L. (2008). "Superconducting Proximity Effect and Majorana Fermions at the Surface of a Topological Insulator". *Physical Review Letters* **10** (9): 096407. arXiv:0707.1692. Bibcode:2008PhRvL.100i6407F. doi:10.1103/PhysRevLett.100.096407.

[20] Fu, Liang; Kane, Charles L. (2009). "Josephson current and noise at a superconductor/quantum-spin-Hall-insulator/superconductor junction". *Physical Review B* **79** (16): 161408. arXiv:0804.4469. Bibcode:2009PhRvB..79p1408F. doi:10.1103/PhysRevB.79.161408. (subscription required (help)).

[21] Alicea, Jason (2012). "New directions in the pursuit of Majorana fermions in solid state systems". *Reports on Progress in Physics* **75** (7): 076501. arXiv:1202.1293. Bibcode:2012RPPh...75g6501A. doi:10.1088/0034-4885/75/7/076501. PMID 22790778. (subscription required (help)).

[22] Beenakker, C. W. J. (April 2013). "Search for Majorana fermions in superconductors". *Annual Review of Condensed Matter Physics* **4** (113): 113–136. arXiv:1112.1950. Bibcode:2013ARCMP...4..113B. doi:10.1146/annurev-conmatphys-030212-184337. (subscription required (help)).

[23] Reich, Eugenie Samuel (28 February 2012). "Quest for quirky quantum particles may have struck gold". *Nature News*. doi:10.1038/nature.2012.10124.

[24] Amos, Jonathan (13 April 2012). "Majorana particle glimpsed in lab". *BBC News*. Retrieved 15 April 2012.

[25] Mourik, V.; Zuo, K.; Frolov, S. M.; Plissard, S. R.; Bakkers, E. P. A. M.; Kouwenhoven, L. P. (12 April 2012). "Signatures of Majorana fermions in hybrid superconductor-semiconductor nanowire devices". *Science* **336** (6084): 1003–1007. arXiv:1204.2792. Bibcode:2012Sci...336.1003M. doi:10.1126/science.1222360.

[26] Deng, M.T.; Yu, C.L.; Huang, G.Y.; Larsson, M.; Caroff, P.; Xu, H.Q. (28 November 2012). "Anomalous zero-bias conductance peak in a Nb-InSb nanowire-Nb hybrid device". *Nano Letters* **12** (12): 6414–6419. Bibcode:2012NanoL..12.6414D. doi:10.1021/nl303758w.

[27] Das, A.; Ronen, Y.; Most, Y.; Oreg, Y.; Heiblum, M.; Shtrikman, H. (11 November 2012). "Zero-bias peaks and splitting in an Al-InAs nanowire topological superconductor as a signature of Majorana fermions." . *Nature Physics* **8** (12): 887–895. arXiv:1205.7073. Bibcode:2012NatPh...8..887D. doi:10.1038/nphys2479.

[28] Churchill, H. O. H.; Fatemi, V.; Grove-Rasmussen, K.; Deng, M.T.; Caroff, P.; Xu, H.Q.; Marcus, C.M. (6 June 2013). "Superconductor-nanowire devices from tunneling to the multichannel regime: Zero-bias oscillations and magnetoconductance crossover". *PHYSICAL REVIEW B* **87** (24): 241401(R). arXiv:1303.2407. Bibcode:2013PhRvB..87x1401C. doi:10.1103/PhysRevB.87.241401.

[29] Deng, M.T.; Yu, C.L.; Huang, G.Y.; Larsson, Marcus; Caroff, P.; Xu, H.Q. (11 November 2014). "Parity independence of the zero-bias conductance peak in a nanowire based topological superconductor-quantum dot hybrid device". *Scientific Reports* **4**: 7261. arXiv:1406.4435. Bibcode:2014NatSR...4E7261D. doi:10.1038/srep07261.

[30] Lutchyn, Roman M.; Sau, Jay D.; Das Sarma, S. (August 2010). "Majorana Fermions and a Topological Phase Transition in Semiconductor-Superconductor Heterostructures". *Physical Review Letters* **105** (7): 077001. arXiv:1002.4033. Bibcode:2010PhRvL.105g7001L. doi:10.1103/PhysRevLett.105.077001.

[31] Oreg, Yuval; Refael, Gil; von Oppen, Felix (October 2010). "Helical Liquids and Majorana Bound States in Quantum Wires". *Physical Review Letters* **105** (17): 177002. arXiv:1003.1145. Bibcode:2010PhRvL.105q7002O. doi:10.1103/PhysRevLett.105.177002.

[32] Lee, E. J. H.; Jiang, X.; Houzet, M.; Aguado, R.; Lieber, C.M.; Franceschi, S.D. (15 December 2013). "Spin-resolved Andreev levels and parity crossings in hybrid superconductor–semiconductor

nanostructures". *Nature Nanotechnology* **9**: 79–84. arXiv:1302.2611. Bibcode:2014NatNa...9...79L. doi:10.1038/nnano.2013.267.

[33] Nadj-Perge, Stevan; Drozdov, Ilya K.; Li, Jian; Chen, Hua; Jeon, Sangjun; Seo, Jungpil; MacDonald, Allan H.; Bernevig, B. Andrei; Yazdani, Ali (2 October 2014). "Observation of Majorana fermions in ferromagnetic atomic chains on a superconductor". *Science*. arXiv:1410.3453. Bibcode:2014Sci...346..602N. doi:10.1126/science.1259327. (subscription required (help)).

[34] "Majorana fermion: Physicists observe elusive particle that is its own antiparticle". Phys.org. October 2, 2014. Retrieved 3 October 2014.

[35] "New Particle Is Both Matter and Antimatter". *Scientific American*. October 2, 2014. Retrieved 3 October 2014.

5.9.5   Further reading

- Pal, Palash B. (2011) [12 October 2010]. "Dirac, Majorana and Weyl fermions". *American Journal of Physics* **79** (5): 485. arXiv:1006.1718. Bibcode:2011AmJPh..79..485P. doi:10.1119/1.3549729. (subscription required (help)).

5.10   MSW effect

The **Mikheyev–Smirnov–Wolfenstein effect** (often referred to as *matter effect*) is a particle physics process which can act to modify neutrino oscillations in matter. Work in 1978, by American physicist Lincoln Wolfenstein, and 1986, by Soviet physicists Stanislav Mikheyev and Alexei Smirnov, led to an understanding of this effect. Later in 1986, Stephen Parke of Fermilab provided the first full analytic treatment of this effect.

5.10.1   Explanation

The presence of electrons in matter changes the energy levels of the propagation eigenstates (mass eigenstates) of neutrinos due to charged current coherent forward scattering of the electron neutrinos (i.e., weak interactions). The coherent forward scattering is analogous to the electromagnetic process leading to the refractive index of light in a medium. This means that neutrinos in matter have a different effective mass than neutrinos in vacuum, and since neutrino oscillations depend upon the squared mass difference of the neutrinos, neutrino oscillations may be different in matter than they are in vacuum. With antineutrinos, the conceptual point is the same but the effective charge that the weak interaction couples to (called *weak isospin*) has an opposite sign.

The effect is important at the very large electron densities of the Sun where electron neutrinos are produced. The high-energy neutrinos seen, for example, in SNO (Sudbury Neutrino Observatory) and in Super-Kamiokande, are produced mainly as the higher mass eigenstate in matter ν_{2m}, and remain as such as the density of solar material changes. (When neutrinos go through the *MSW resonance* the neutrinos have the maximal probability to change their nature, but it happens that this probability is negligibly small—this is sometimes called propagation in the adiabatic regime). Thus, the neutrinos of high energy leaving the sun are in a vacuum propagation eigenstate, ν_2, that has a reduced overlap with the electron neutrino $\nu_e = \nu_1 \cos\theta + \nu_2 \sin\theta$ seen by charged current reactions in the detectors.

5.10.2 Experimental evidence

For high-energy solar neutrinos the MSW effect is important, and leads to the expectation that $P_{ee} = \sin^2\theta$, where θ is the solar mixing angle. This was dramatically confirmed in the Sudbury Neutrino Observatory (SNO), which has resolved the solar neutrino problem. SNO measured the flux of Solar electron neutrinos to be ~34% of the total neutrino flux (the electron neutrino flux measured via the charged current reaction, and the total flux via the neutral current reaction). The SNO results agree well with the expectations. Earlier, Kamiokande and Super-Kamiokande measured a mixture of charged current and neutral current reactions, that also support the occurrence of the MSW effect with a similar suppression, but with less confidence.

For the low-energy solar neutrinos, on the other hand, the matter effect is negligible, and the formalism of oscillations in vacuum is valid. The size of the source (i.e. the Solar core) is significantly larger than the oscillation length, therefore, averaging over the oscillation factor, one obtains $P_{ee} = 1 - \sin^2 2\theta / 2$. For $\theta = 34°$ this corresponds to a survival probability of $P_{ee} \approx 60\%$. This is consistent with the experimental observations of low energy Solar neutrinos by the Homestake experiment (the first experiment to reveal the solar neutrino problem), followed by GALLEX, GNO, and SAGE (collectively, gallium radiochemical experiments), and, more recently, the Borexino experiment. These experiments provided further evidence of the MSW effect.

These results are further supported by the reactor experiment KamLAND, that alone is able to provide also a measurement of the parameters of oscillation that is consistent with all other measurements.

The transition between the low energy regime (the MSW effect is negligible) and the high energy regime (the oscillation probability is determined by matter effects) lies in the region of about 2 MeV for the Solar neutrinos.

The MSW effect can also modify neutrino oscillations in the Earth, and future search for new oscillations and/or leptonic CP violation may make use of this property.

5.10.3 See also

• Neutrino oscillations

5.10.4 References

• G. Brooijmans (28 July 1998). "Neutrino Oscillations in Matter: the MSW Effect" . *A New Limit on $\nu_\mu \rightarrow \nu_\tau$ Oscillations*. Université catholique de Louvain. p. 40. Retrieved 2010-04-24.

• P. Langacker (27 November 1995). "Mikheyev–Smirnov–Wolfenstein (MSW)". *Solar Neutrinos*. University of Pennsylvania. Retrieved 2010-04-24.

• B. Schwarzschild (2003). "Antineutrinos From Distant Reactors Simulate the Disappearance of Solar Neutrinos" . *Physics Today* **56**: 14. Bibcode:2003PhT....56c..14S. doi:10.1063/1.1570758.

• L. Wolfenstein (1978). "Neutrino oscillations in matter" . *Physical Review D* **17** (9): 2369. Bibcode:1978PhRvD..17.2369W. doi:10.1103/PhysRevD.17.2369.

5.11 Quark

This article is about the particle. For other uses, see Quark (disambiguation).

A **quark** (/ˈkwɔrk/ or /ˈkwɑrk/) is an elementary particle and a fundamental constituent of matter. Quarks combine to form composite particles called hadrons, the most stable of which are protons and neutrons, the components of atomic nuclei.[1] Due to a phenomenon known as *color confinement*, quarks are never directly observed or found in isolation; they can be found only within hadrons, such as baryons (of which protons and neutrons are examples), and mesons.[2][3] For this reason, much of what is known about quarks has been drawn from observations of the hadrons themselves.

Quarks have various intrinsic properties, including electric charge, mass, color charge and spin. Quarks are the only elementary particles in the Standard Model of particle physics to experience all four fundamental interactions, also known as *fundamental forces* (electromagnetism, gravitation, strong interaction, and weak interaction), as well as the only known particles whose electric charges are not integer multiples of the elementary charge.

There are six types of quarks, known as *flavors*: up, down, strange, charm, top, and bottom.[*][4] Up and down quarks have the lowest masses of all quarks. The heavier quarks rapidly change into up and down quarks through a process of particle decay: the transformation from a higher mass state to a lower mass state. Because of this, up and down quarks are generally stable and the most common in the universe, whereas strange, charm, bottom, and top quarks can only be produced in high energy collisions (such as those involving cosmic rays and in particle accelerators). For every quark flavor there is a corresponding type of antiparticle, known as an *antiquark*, that differs from the quark only in that some of its properties have equal magnitude but opposite sign.

The quark model was independently proposed by physicists Murray Gell-Mann and George Zweig in 1964.[*][5] Quarks were introduced as parts of an ordering scheme for hadrons, and there was little evidence for their physical existence until deep inelastic scattering experiments at the Stanford Linear Accelerator Center in 1968.[*][6][*][7] Accelerator experiments have provided evidence for all six flavors. The top quark was the last to be discovered at Fermilab in 1995.[*][5]

5.11.1 Classification

See also: Standard Model

The Standard Model is the theoretical framework describing all the currently known elementary particles. This

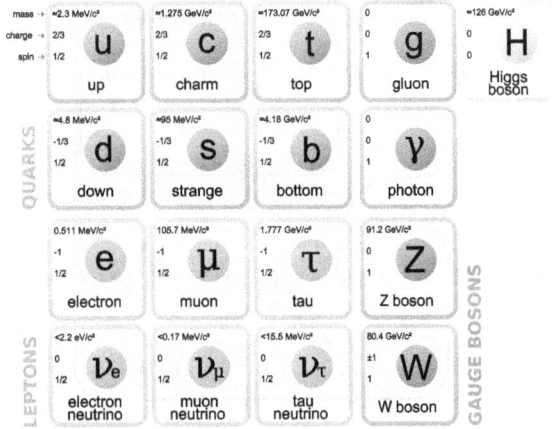

Six of the particles in the Standard Model are quarks (shown in purple). Each of the first three columns forms a generation *of matter.*

model contains six flavors of quarks (q), named up (u), down (d), strange (s), charm (c), bottom (b), and top (t).[*][4] Antiparticles of quarks are called *antiquarks*, and are denoted by a bar over the symbol for the corresponding quark, such as u for an up antiquark. As with antimatter in general, antiquarks have the same mass, mean lifetime, and spin as their respective quarks, but the electric charge and other charges have the opposite sign.[*][8]

Quarks are spin-$\frac{1}{2}$ particles, implying that they are fermions according to the spin-statistics theorem. They are subject to the Pauli exclusion principle, which states that no two identical fermions can simultaneously occupy the same quantum state. This is in contrast to bosons (particles with integer spin), any number of which can be in the same state.[*][9] Unlike leptons, quarks possess color charge, which causes them to engage in the strong interaction. The resulting attraction between different quarks causes the formation of composite particles known as *hadrons* (see "Strong interaction and color charge" below).

The quarks which determine the quantum numbers of hadrons are called *valence quarks*; apart from these, any hadron may contain an indefinite number of virtual (or *sea*) quarks, antiquarks, and gluons which do not influence its quantum numbers.[*][10] There are two families of hadrons: baryons, with three valence quarks, and mesons, with a valence quark and an antiquark.[*][11] The most common baryons are the proton and the neutron, the building blocks of the atomic nucleus.[*][12] A great number of hadrons are known (see list of baryons and list of mesons), most of them differentiated by their quark content and the properties these constituent quarks confer. The existence of "exotic" hadrons with more valence quarks, such as tetraquarks (qqqq) and pentaquarks (qqqqq), has been conjectured[*][13] but not proven.[*][nb 1][*][13][*][14] However, on 13 July 2015, the LHCb collaboration at CERN reported results consistent with pentaquark states.[*][15]

Elementary fermions are grouped into three generations, each comprising two leptons and two quarks. The first generation includes up and down quarks, the second strange and charm quarks, and the third bottom and top quarks. All searches for a fourth generation of quarks and other elementary fermions have failed,[*][16] and there is strong indirect evidence that no more than three generations exist.[*][nb 2][*][17] Particles in higher generations generally have greater mass and less stability, causing them to decay into lower-generation particles by means of weak interactions. Only first-generation (up and down) quarks occur commonly in nature. Heavier quarks can only be created in high-energy collisions (such as in those involving cosmic rays), and decay quickly; however, they are thought to have been present during the first fractions of a second after the Big Bang, when the universe was in an extremely hot and dense phase (the quark epoch). Studies of heavier quarks are conducted in artificially created conditions, such as in particle accelerators.[*][18]

Having electric charge, mass, color charge, and flavor, quarks are the only known elementary particles that engage in all four fundamental interactions of contemporary physics: electromagnetism, gravitation, strong interaction, and weak interaction.[*][12] Gravitation is too weak to be relevant to individual particle interactions except at extremes of energy (Planck energy) and distance scales (Planck distance). However, since no successful quantum theory of gravity exists, gravitation is not described by the Standard Model.

See the table of properties below for a more complete overview of the six quark flavors' properties.

5.11.2   History

Murray Gell-Mann at TED in 2007. Gell-Mann and George Zweig proposed the quark model in 1964.

The quark model was independently proposed by physicists Murray Gell-Mann[*][19] (pictured) and George Zweig[*][20][*][21] in 1964.[*][5] The proposal came shortly after Gell-Mann's 1961 formulation of a particle classification system known as the *Eightfold Way*—or, in more technical terms, SU(3) flavor symmetry.[*][22] Physicist Yuval Ne'eman had independently developed a scheme similar to the Eightfold Way in the same year.[*][23][*][24]

At the time of the quark theory's inception, the "particle zoo" included, amongst other particles, a multitude of

hadrons. Gell-Mann and Zweig posited that they were not elementary particles, but were instead composed of combinations of quarks and antiquarks. Their model involved three flavors of quarks, up, down, and strange, to which they ascribed properties such as spin and electric charge.[*][19][*][20][*][21] The initial reaction of the physics community to the proposal was mixed. There was particular contention about whether the quark was a physical entity or a mere abstraction used to explain concepts that were not fully understood at the time.[*][25]

In less than a year, extensions to the Gell-Mann–Zweig model were proposed. Sheldon Lee Glashow and James Bjorken predicted the existence of a fourth flavor of quark, which they called *charm*. The addition was proposed because it allowed for a better description of the weak interaction (the mechanism that allows quarks to decay), equalized the number of known quarks with the number of known leptons, and implied a mass formula that correctly reproduced the masses of the known mesons.[*][26]

In 1968, deep inelastic scattering experiments at the Stanford Linear Accelerator Center (SLAC) showed that the proton contained much smaller, point-like objects and was therefore not an elementary particle.[*][6][*][7][*][27] Physicists were reluctant to firmly identify these objects with quarks at the time, instead calling them "partons"—a term coined by Richard Feynman.[*][28][*][29][*][30] The objects that were observed at SLAC would later be identified as up and down quarks as the other flavors were discovered.[*][31] Nevertheless, "parton" remains in use as a collective term for the constituents of hadrons (quarks, antiquarks, and gluons).

The strange quark's existence was indirectly validated by SLAC's scattering experiments: not only was it a necessary component of Gell-Mann and Zweig's three-quark model, but it provided an explanation for the kaon (K) and pion (π) hadrons discovered in cosmic rays in 1947.[*][32]

In a 1970 paper, Glashow, John Iliopoulos and Luciano Maiani presented further reasoning for the existence of the as-yet undiscovered charm quark.[*][33][*][34] The number of supposed quark flavors grew to the current six in 1973, when Makoto Kobayashi and Toshihide Maskawa noted that the experimental observation of CP violation[*][nb 3][*][35] could be explained if there were another pair of quarks.

Charm quarks were produced almost simultaneously by two teams in November 1974 (see November Revolution)—one at SLAC under Burton Richter, and one at Brookhaven National Laboratory under Samuel Ting. The charm quarks were observed bound with charm antiquarks in mesons. The two parties had assigned the discovered meson two different symbols, J and ψ; thus, it became formally known as the J/ψ meson. The discovery finally convinced the physics

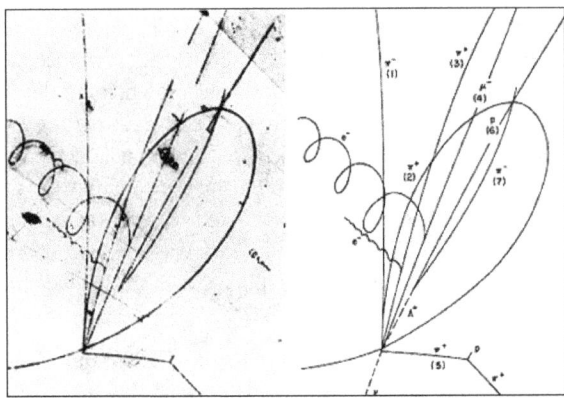

*Photograph of the event that led to the discovery of the Σ++
c baryon, at the Brookhaven National Laboratory in 1974*

community of the quark model's validity.[30]

In the following years a number of suggestions appeared for extending the quark model to six quarks. Of these, the 1975 paper by Haim Harari[36] was the first to coin the terms *top* and *bottom* for the additional quarks.[37]

In 1977, the bottom quark was observed by a team at Fermilab led by Leon Lederman.[38][39] This was a strong indicator of the top quark's existence: without the top quark, the bottom quark would have been without a partner. However, it was not until 1995 that the top quark was finally observed, also by the CDF[40] and DØ[41] teams at Fermilab.[5] It had a mass much larger than had been previously expected,[42] almost as large as that of a gold atom.[43]

5.11.3   Etymology

For some time, Gell-Mann was undecided on an actual spelling for the term he intended to coin, until he found the word *quark* in James Joyce's book *Finnegans Wake*:

> Three quarks for Muster Mark!
> Sure he has not got much of a bark
> And sure any he has it's all beside the mark.
> —James Joyce, *Finnegans Wake*[44]

Gell-Mann went into further detail regarding the name of the quark in his book *The Quark and the Jaguar*:[45]

> In 1963, when I assigned the name "quark" to the fundamental constituents of the nucleon, I had the sound first, without the spelling, which could have been "kwork". Then, in one of my occasional perusals of *Finnegans Wake*, by James Joyce, I came across the word "quark" in the phrase "Three quarks for Muster Mark". Since "quark" (meaning, for one thing, the cry of the gull) was clearly intended to rhyme with "Mark", as well as "bark" and other such words, I had to find an excuse to pronounce it as "kwork". But the book represents the dream of a publican named Humphrey Chimpden Earwicker. Words in the text are typically drawn from several sources at once, like the "portmanteau" words in "Through the Looking-Glass". From time to time, phrases occur in the book that are partially determined by calls for drinks at the bar. I argued, therefore, that perhaps one of the multiple sources of the cry "Three quarks for Muster Mark" might be "Three quarts for Mister Mark", in which case the pronunciation "kwork" would not be totally unjustified. In any case, the number three fitted perfectly the way quarks occur in nature.

Zweig preferred the name *ace* for the particle he had theorized, but Gell-Mann's terminology came to prominence once the quark model had been commonly accepted.[46]

The quark flavors were given their names for several reasons. The up and down quarks are named after the up and down components of isospin, which they carry.[47] Strange quarks were given their name because they were discovered to be components of the strange particles discovered in cosmic rays years before the quark model was proposed; these particles were deemed "strange" because they had unusually long lifetimes.[48] Glashow, who co-proposed charm quark with Bjorken, is quoted as saying, "We called our construct the 'charmed quark', for we were fascinated and pleased by the symmetry it brought to the subnuclear world." [49] The names "bottom" and "top", coined by Harari, were chosen because they are "logical partners for up and down quarks".[36][37][48] In the past, bottom and top quarks were sometimes referred to as "beauty" and "truth" respectively, but these names have somewhat fallen out of use.[50] While "truth" never did catch on, accelerator complexes devoted to massive production of bottom quarks are sometimes called "beauty factories".[51]

5.11.4   Properties

Electric charge

See also: Electric charge

Quarks have fractional electric charge values – either $1/3$

or $^2/_3$ times the elementary charge (e), depending on flavor. Up, charm, and top quarks (collectively referred to as *up-type quarks*) have a charge of $+^2/_3$ e, while down, strange, and bottom quarks (*down-type quarks*) have $-^1/_3$ e. Antiquarks have the opposite charge to their corresponding quarks; up-type antiquarks have charges of $-^2/_3$ e and down-type antiquarks have charges of $+^1/_3$ e. Since the electric charge of a hadron is the sum of the charges of the constituent quarks, all hadrons have integer charges: the combination of three quarks (baryons), three antiquarks (antibaryons), or a quark and an antiquark (mesons) always results in integer charges.[*][52] For example, the hadron constituents of atomic nuclei, neutrons and protons, have charges of 0 e and +1 e respectively; the neutron is composed of two down quarks and one up quark, and the proton of two up quarks and one down quark.[*][12]

Spin

See also: Spin (physics)

Spin is an intrinsic property of elementary particles, and its direction is an important degree of freedom. It is sometimes visualized as the rotation of an object around its own axis (hence the name "spin"), though this notion is somewhat misguided at subatomic scales because elementary particles are believed to be point-like.[*][53]

Spin can be represented by a vector whose length is measured in units of the reduced Planck constant $\hbar$ (pronounced "h bar"). For quarks, a measurement of the spin vector component along any axis can only yield the values $+\hbar/2$ or $-\hbar/2$; for this reason quarks are classified as spin-$\frac{1}{2}$ particles.[*][54] The component of spin along a given axis – by convention the z axis – is often denoted by an up arrow ↑ for the value $+\frac{1}{2}$ and down arrow ↓ for the value $-\frac{1}{2}$, placed after the symbol for flavor. For example, an up quark with a spin of $+\frac{1}{2}$ along the z axis is denoted by u↑.[*][55]

Weak interaction

Main article: Weak interaction
 A quark of one flavor can transform into a quark of another flavor only through the weak interaction, one of the four fundamental interactions in particle physics. By absorbing or emitting a W boson, any up-type quark (up, charm, and top quarks) can change into any down-type quark (down, strange, and bottom quarks) and vice versa. This flavor transformation mechanism causes the radioactive process of beta decay, in which a neutron (n) "splits" into a proton (p), an electron (e−) and an electron antineutrino (ν e) (see picture). This occurs when one of the down quarks in the neutron (udd) decays into an up quark by emitting a

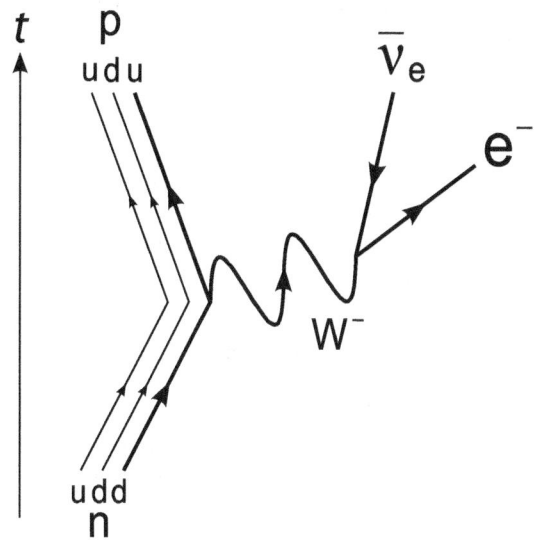

Feynman diagram of beta decay with time flowing upwards. The CKM matrix (discussed below) encodes the probability of this and other quark decays.

virtual W− boson, transforming the neutron into a proton (uud). The W− boson then decays into an electron and an electron antineutrino.[*][56]

Both beta decay and the inverse process of *inverse beta decay* are routinely used in medical applications such as positron emission tomography (PET) and in experiments involving neutrino detection.

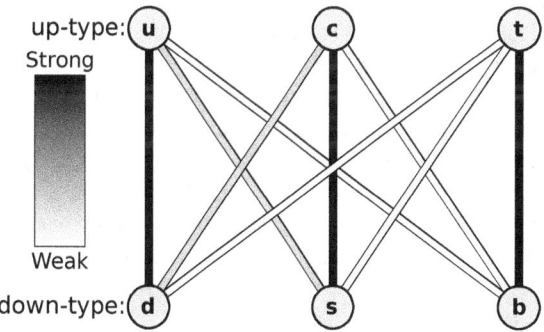

The strengths of the weak interactions between the six quarks. The "intensities" of the lines are determined by the elements of the CKM matrix.

While the process of flavor transformation is the same for all quarks, each quark has a preference to transform into the quark of its own generation. The relative tendencies of all flavor transformations are described by a mathematical table, called the Cabibbo–Kobayashi–Maskawa matrix (CKM matrix). Enforcing unitarity, the approximate magnitudes of the entries of the CKM matrix are:[57]

$$\begin{bmatrix} |V_{ud}| & |V_{us}| & |V_{ub}| \\ |V_{cd}| & |V_{cs}| & |V_{cb}| \\ |V_{td}| & |V_{ts}| & |V_{tb}| \end{bmatrix} \approx \begin{bmatrix} 0.974 & 0.225 & 0.003 \\ 0.225 & 0.973 & 0.041 \\ 0.009 & 0.040 & 0.999 \end{bmatrix},$$

where V_{ij} represents the tendency of a quark of flavor i to change into a quark of flavor j (or vice versa).[nb 4]

There exists an equivalent weak interaction matrix for leptons (right side of the W boson on the above beta decay diagram), called the Pontecorvo–Maki–Nakagawa–Sakata matrix (PMNS matrix).[58] Together, the CKM and PMNS matrices describe all flavor transformations, but the links between the two are not yet clear.[59]

Strong interaction and color charge

See also: Color charge and Strong interaction

According to quantum chromodynamics (QCD), quarks possess a property called *color charge*. There are three types of color charge, arbitrarily labeled *blue*, *green*, and *red*.[nb 5] Each of them is complemented by an anticolor – *antiblue*, *antigreen*, and *antired*. Every quark carries a color, while every antiquark carries an anticolor.[60]

The system of attraction and repulsion between quarks charged with different combinations of the three colors is called strong interaction, which is mediated by force carrying particles known as *gluons*; this is discussed at length below. The theory that describes strong interactions is called quantum chromodynamics (QCD). A quark, which will have a single color value, can form a bound system with an antiquark carrying the corresponding anticolor. The result of two attracting quarks will be color neutrality: a quark with color charge ξ plus an antiquark with color charge $-\xi$ will result in a color charge of 0 (or "white" color) and the formation of a meson. This is analogous to the additive color model in basic optics. Similarly, the combination of three quarks, each with different color charges, or three antiquarks, each with anticolor charges, will result in the same "white" color charge and the formation of a baryon or antibaryon.[61]

In modern particle physics, gauge symmetries – a kind of symmetry group – relate interactions between particles (see gauge theories). Color SU(3) (commonly abbreviated to SU(3)$_c$) is the gauge symmetry that relates the color charge in quarks and is the defining symmetry for quantum chromodynamics.[62] Just as the laws of physics are independent of which directions in space are designated x, y, and z, and remain unchanged if the coordinate axes are rotated to a new orientation, the physics of quantum chromodynamics is independent of which directions in three-dimensional color space are identified as blue, red, and green. SU(3)$_c$ color transformations correspond to "rotations" in color space (which, mathematically speaking, is a complex space). Every quark flavor f, each with subtypes f_B, f_G, f_R corresponding to the quark colors,[63] forms a triplet: a three-component quantum field which transforms under the fundamental representation of SU(3)$_c$.[64] The requirement that SU(3)$_c$ should be local – that is, that its transformations be allowed to vary with space and time – determines the properties of the strong interaction, in particular the existence of eight gluon types to act as its force carriers.[62][65]

Mass

See also: Invariant mass

Two terms are used in referring to a quark's mass: *current quark mass* refers to the mass of a quark by itself, while *constituent quark mass* refers to the current quark mass plus the mass of the gluon particle field surrounding the quark.[66] These masses typically have very different values. Most of a hadron's mass comes from the gluons that bind the constituent quarks together, rather than from the quarks themselves. While gluons are inherently massless, they possess energy – more specifically, quantum chromodynamics binding energy (QCBE) – and it is this that contributes so greatly to the overall mass of the hadron (see mass in special relativity). For example, a proton has a mass of approximately 938 MeV/c^2, of which the rest mass of its three valence quarks only contributes about 11 MeV/c^2; much of the remainder can be attributed to the gluons' QCBE.[67][68]

The Standard Model posits that elementary particles derive their masses from the Higgs mechanism, which is related to the Higgs boson. Physicists hope that further research into the reasons for the top quark's large mass of ~173 GeV/c^2, almost the mass of a gold atom,[67][69] might reveal more about the origin of the mass of quarks and other elementary particles.[70]

Table of properties

See also: Flavor (particle physics)

The following table summarizes the key properties of the six quarks. Flavor quantum numbers (isospin (I_3), charm (C), strangeness (S, not to be confused with spin), topness (T), and bottomness (B')) are assigned to certain quark flavors, and denote qualities of quark-based systems and hadrons. The baryon number (B) is $+\frac{1}{3}$ for all quarks, as baryons are made of three quarks. For antiquarks, the electric charge

(Q) and all flavor quantum numbers (B, I_3, C, S, T, and B') are of opposite sign. Mass and total angular momentum (J; equal to spin for point particles) do not change sign for the antiquarks.

J = total angular momentum, B = baryon number, Q = electric charge, I_3 = isospin, C = charm, S = strangeness, T = topness, B' = bottomness.

* Notation such as 4190+180

−60 denotes measurement uncertainty. In the case of the top quark, the first uncertainty is statistical in nature, and the second is systematic.

5.11.5 Interacting quarks

See also: Color confinement and Gluon

As described by quantum chromodynamics, the strong interaction between quarks is mediated by gluons, massless vector gauge bosons. Each gluon carries one color charge and one anticolor charge. In the standard framework of particle interactions (part of a more general formulation known as perturbation theory), gluons are constantly exchanged between quarks through a virtual emission and absorption process. When a gluon is transferred between quarks, a color change occurs in both; for example, if a red quark emits a red–antigreen gluon, it becomes green, and if a green quark absorbs a red–antigreen gluon, it becomes red. Therefore, while each quark's color constantly changes, their strong interaction is preserved.*[71]*[72]*[73]

Since gluons carry color charge, they themselves are able to emit and absorb other gluons. This causes *asymptotic freedom*: as quarks come closer to each other, the chromodynamic binding force between them weakens.*[74] Conversely, as the distance between quarks increases, the binding force strengthens. The color field becomes stressed, much as an elastic band is stressed when stretched, and more gluons of appropriate color are spontaneously created to strengthen the field. Above a certain energy threshold, pairs of quarks and antiquarks are created. These pairs bind with the quarks being separated, causing new hadrons to form. This phenomenon is known as *color confinement*: quarks never appear in isolation.*[72]*[75] This process of hadronization occurs before quarks, formed in a high energy collision, are able to interact in any other way. The only exception is the top quark, which may decay before it hadronizes.*[76]

Sea quarks

Hadrons, along with the *valence quarks* (q
v) that contribute to their quantum numbers, contain virtual quark–antiquark (qq) pairs known as *sea quarks* (q
s). Sea quarks form when a gluon of the hadron's color field splits; this process also works in reverse in that the annihilation of two sea quarks produces a gluon. The result is a constant flux of gluon splits and creations colloquially known as "the sea".*[77] Sea quarks are much less stable than their valence counterparts, and they typically annihilate each other within the interior of the hadron. Despite this, sea quarks can hadronize into baryonic or mesonic particles under certain circumstances.*[78]

Other phases of quark matter

Main article: QCD matter

Under sufficiently extreme conditions, quarks may become deconfined and exist as free particles. In the course of asymptotic freedom, the strong interaction becomes weaker at higher temperatures. Eventually, color confinement would be lost and an extremely hot plasma of freely moving quarks and gluons would be formed. This theoretical phase of matter is called quark–gluon plasma.*[81] The exact conditions needed to give rise to this state are unknown and have been the subject of a great deal of speculation and experimentation. A recent estimate puts the needed temperature at $(1.90 \pm 0.02) \times 10^{12}$ kelvin.*[82] While a state of entirely free quarks and gluons has never been achieved (despite numerous attempts by CERN in the 1980s and 1990s),*[83] recent experiments at the Relativistic Heavy Ion Collider have yielded evidence for liquid-like quark matter exhibiting "nearly perfect" fluid motion.*[84]

The quark–gluon plasma would be characterized by a great increase in the number of heavier quark pairs in relation to the number of up and down quark pairs. It is believed that in the period prior to $10^{*}-6$ seconds after the Big Bang (the quark epoch), the universe was filled with quark–gluon plasma, as the temperature was too high for hadrons to be stable.*[85]

Given sufficiently high baryon densities and relatively low temperatures – possibly comparable to those found in neutron stars – quark matter is expected to degenerate into a Fermi liquid of weakly interacting quarks. This liquid would be characterized by a condensation of colored quark Cooper pairs, thereby breaking the local $SU(3)_c$ symmetry. Because quark Cooper pairs harbor color charge, such a phase of quark matter would be color superconductive; that is, color charge would be able to pass through it with no resistance.*[86]

5.11.6 See also

• Color–flavor locking

- Neutron magnetic moment

- Leptons

- Preons – Hypothetical particles which were once postulated to be subcomponents of quarks and leptons

- Quarkonium – Mesons made of a quark and antiquark of the same flavor

- Quark star – A hypothetical degenerate neutron star with extreme density

- Quark–lepton complementarity – Possible fundamental relation between quarks and leptons

5.11.7   Notes

[1] Several research groups claimed to have proven the existence of tetraquarks and pentaquarks in the early 2000s. While the status of tetraquarks is still under debate, all known pentaquark candidates have previously been established as nonexistent.

[2] The main evidence is based on the resonance width of the Z0 boson, which constrains the 4th generation neutrino to have a mass greater than ~45 GeV/c^2. This would be highly contrasting with the other three generations' neutrinos, whose masses cannot exceed 2 MeV/c^2.

[3] CP violation is a phenomenon which causes weak interactions to behave differently when left and right are swapped (P symmetry) and particles are replaced with their corresponding antiparticles (C symmetry).

[4] The actual probability of decay of one quark to another is a complicated function of (amongst other variables) the decaying quark's mass, the masses of the decay products, and the corresponding element of the CKM matrix. This probability is directly proportional (but not equal) to the magnitude squared ($|V_{ij}|^2$) of the corresponding CKM entry.

[5] Despite its name, color charge is not related to the color spectrum of visible light.

5.11.8   References

[1] "Quark (subatomic particle)". *Encyclopædia Britannica*. Retrieved 2008-06-29.

[2] R. Nave. "Confinement of Quarks". *HyperPhysics*. Georgia State University, Department of Physics and Astronomy. Retrieved 2008-06-29.

[3] R. Nave. "Bag Model of Quark Confinement". *HyperPhysics*. Georgia State University, Department of Physics and Astronomy. Retrieved 2008-06-29.

[4] R. Nave. "Quarks". *HyperPhysics*. Georgia State University, Department of Physics and Astronomy. Retrieved 2008-06-29.

[5] B. Carithers, P. Grannis (1995). "Discovery of the Top Quark" (PDF). *Beam Line* (SLAC) **25** (3): 4–16. Retrieved 2008-09-23.

[6] E.D. Bloom et al. (1969). "High-Energy Inelastic *e–p* Scattering at 6° and 10°". *Physical Review Letters* **23** (16): 930–934. Bibcode:1969PhRvL..23..930B. doi:10.1103/PhysRevLett.23.930.

[7] M. Breidenbach et al. (1969). "Observed Behavior of Highly Inelastic Electron–Proton Scattering". *Physical Review Letters* **23** (16): 935–939. Bibcode:1969PhRvL..23..935B. doi:10.1103/PhysRevLett.23.935.

[8] S.S.M. Wong (1998). *Introductory Nuclear Physics* (2nd ed.). Wiley Interscience. p. 30. ISBN 0-471-23973-9.

[9] K.A. Peacock (2008). *The Quantum Revolution*. Greenwood Publishing Group. p. 125. ISBN 0-313-33448-X.

[10] B. Povh, C. Scholz, K. Rith, F. Zetsche (2008). *Particles and Nuclei*. Springer. p. 98. ISBN 3-540-79367-4.

[11] Section 6.1. in P.C.W. Davies (1979). *The Forces of Nature*. Cambridge University Press. ISBN 0-521-22523-X.

[12] M. Munowitz (2005). *Knowing*. Oxford University Press. p. 35. ISBN 0-19-516737-6.

[13] W.-M. Yao (Particle Data Group) et al. (2006). "Review of Particle Physics: Pentaquark Update" (PDF). *Journal of Physics G* **33** (1): 1–1232. arXiv:astro-ph/0601168. Bibcode:2006JPhG...33....1Y. doi:10.1088/0954-3899/33/1/001.

[14] C. Amsler (Particle Data Group) et al. (2008). "Review of Particle Physics: Pentaquarks" (PDF). *Physics Letters B* **667** (1): 1–1340. Bibcode:2008PhLB..667....1P. doi:10.1016/j.physletb.2008.07.018.
C. Amsler (Particle Data Group) et al. (2008). "Review of Particle Physics: New Charmonium-Like States" (PDF). *Physics Letters B* **667** (1): 1–1340. Bibcode:2008PhLB..667....1P. doi:10.1016/j.physletb.2008.07.018.
E.V. Shuryak (2004). *The QCD Vacuum, Hadrons and Superdense Matter*. World Scientific. p. 59. ISBN 981-238-574-6.

[15] R. Aaij et al. (LHCb collaboration) (2015). "Observation of J/ψp resonances consistent with pentaquark states in Λ0 b→J/ψK⁻ p decays". *Physical Review Letters* **115** (7). doi:10.1103/PhysRevLett.115.072001.

[16] C. Amsler (Particle Data Group) et al. (2008). "Review of Particle Physics: b′ (4th Generation)

Quarks, Searches for" (PDF). *Physics Letters B* **667** (1): 1–1340. Bibcode:2008PhLB..667....1P. doi:10.1016/j.physletb.2008.07.018.
C. Amsler (Particle Data Group) et al. (2008). "Review of Particle Physics: *t'* (4th Generation) Quarks, Searches for" (PDF). *Physics Letters B* **667** (1): 1–1340. Bibcode:2008PhLB..667....1P. doi:10.1016/j.physletb.2008.07.018.

[17] D. Decamp; Deschizeaux, B.; Lees, J.-P.; Minard, M.-N.; Crespo, J.M.; Delfino, M.; Fernandez, E.; Martinez, M. et al. (1989). "Determination of the number of light neutrino species". *Physics Letters B* **231** (4): 519. Bibcode:1989PhLB..231..519D. doi:10.1016/0370-2693(89)90704-1.
A. Fisher (1991). "Searching for the Beginning of Time: Cosmic Connection". *Popular Science* **238** (4): 70.
J.D. Barrow (1997) [1994]. "The Singularity and Other Problems". *The Origin of the Universe* (Reprint ed.). Basic Books. ISBN 978-0-465-05314-8.

[18] D.H. Perkins (2003). *Particle Astrophysics*. Oxford University Press. p. 4. ISBN 0-19-850952-9.

[19] M. Gell-Mann (1964). "A Schematic Model of Baryons and Mesons". *Physics Letters* **8** (3): 214–215. Bibcode:1964PhL.....8..214G. doi:10.1016/S0031-9163(64)92001-3.

[20] G. Zweig (1964). "An SU(3) Model for Strong Interaction Symmetry and its Breaking" (PDF). *CERN Report No.8182/TH.401.*

[21] G. Zweig (1964). "An SU(3) Model for Strong Interaction Symmetry and its Breaking: II" (PDF). *CERN Report No.8419/TH.412.*

[22] M. Gell-Mann (2000) [1964]. "The Eightfold Way: A theory of strong interaction symmetry". In M. Gell-Mann, Y. Ne'eman. *The Eightfold Way*. Westview Press. p. 11. ISBN 0-7382-0299-1.
Original: M. Gell-Mann (1961). "The Eightfold Way: A theory of strong interaction symmetry". *Synchrotron Laboratory Report CTSL-20* (California Institute of Technology).

[23] Y. Ne'eman (2000) [1964]. "Derivation of strong interactions from gauge invariance". In M. Gell-Mann, Y. Ne'eman. *The Eightfold Way*. Westview Press. ISBN 0-7382-0299-1.
Original Y. Ne'eman (1961). "Derivation of strong interactions from gauge invariance". *Nuclear Physics* **26** (2): 222. Bibcode:1961NucPh..26..222N. doi:10.1016/0029-5582(61)90134-1.

[24] R.C. Olby, G.N. Cantor (1996). *Companion to the History of Modern Science*. Taylor & Francis. p. 673. ISBN 0-415-14578-3.

[25] A. Pickering (1984). *Constructing Quarks*. University of Chicago Press. pp. 114–125. ISBN 0-226-66799-5.

[26] B.J. Bjorken, S.L. Glashow; Glashow (1964). "Elementary Particles and SU(4)". *Physics Letters* **11** (3): 255–257. Bibcode:1964PhL....11..255B. doi:10.1016/0031-9163(64)90433-0.

[27] J.I. Friedman. "The Road to the Nobel Prize". Hue University. Retrieved 2008-09-29.

[28] R.P. Feynman (1969). "Very High-Energy Collisions of Hadrons". *Physical Review Letters* **23** (24): 1415–1417. Bibcode:1969PhRvL..23.1415F. doi:10.1103/PhysRevLett.23.1415.

[29] S. Kretzer et al. (2004). "CTEQ6 Parton Distributions with Heavy Quark Mass Effects". *Physical Review D* **69** (11): 114005. arXiv:hep-ph/0307022. Bibcode:2004PhRvD..69k4005K. doi:10.1103/PhysRevD.69.114005.

[30] D.J. Griffiths (1987). *Introduction to Elementary Particles*. John Wiley & Sons. p. 42. ISBN 0-471-60386-4.

[31] M.E. Peskin, D.V. Schroeder (1995). *An introduction to quantum field theory*. Addison–Wesley. p. 556. ISBN 0-201-50397-2.

[32] V.V. Ezhela (1996). *Particle physics*. Springer. p. 2. ISBN 1-56396-642-5.

[33] S.L. Glashow, J. Iliopoulos, L. Maiani; Iliopoulos; Maiani (1970). "Weak Interactions with Lepton–Hadron Symmetry". *Physical Review D* **2** (7): 1285–1292. Bibcode:1970PhRvD...2.1285G. doi:10.1103/PhysRevD.2.1285.

[34] D.J. Griffiths (1987). *Introduction to Elementary Particles*. John Wiley & Sons. p. 44. ISBN 0-471-60386-4.

[35] M. Kobayashi, T. Maskawa; Maskawa (1973). "CP-Violation in the Renormalizable Theory of Weak Interaction". *Progress of Theoretical Physics* **49** (2): 652–657. Bibcode:1973PThPh..49..652K. doi:10.1143/PTP.49.652.

[36] H. Harari (1975). "A new quark model for hadrons". *Physics Letters B* **57B** (3): 265. Bibcode:1975PhLB...57..265H. doi:10.1016/0370-2693(75)90072-6.

[37] K.W. Staley (2004). *The Evidence for the Top Quark*. Cambridge University Press. pp. 31–33. ISBN 978-0-521-82710-2.

[38] S.W. Herb et al. (1977). "Observation of a Dimuon Resonance at 9.5 GeV in 400-GeV Proton-Nucleus Collisions". *Physical Review Letters* **39** (5): 252. Bibcode:1977PhRvL..39..252H. doi:10.1103/PhysRevLett.39.252.

[39] M. Bartusiak (1994). *A Positron named Priscilla*. National Academies Press. p. 245. ISBN 0-309-04893-1.

[40] F. Abe (CDF Collaboration) et al. (1995). "Observation of Top Quark Production in pp Collisions with the Collider Detector at Fermilab". *Physical Review Letters* **74** (14): 2626–2631. Bibcode:1995PhRvL..74.2626A. doi:10.1103/PhysRevLett.74.2626. PMID 10057978.

[41] S. Abachi (DØ Collaboration) et al. (1995). "Search for High Mass Top Quark Production in pp Collisions at $\sqrt{s}$ = 1.8 TeV". *Physical Review Letters* **74** (13): 2422–2426. Bibcode:1995PhRvL..74.2422A. doi:10.1103/PhysRevLett.74.2422.

[42] K.W. Staley (2004). *The Evidence for the Top Quark.* Cambridge University Press. p. 144. ISBN 0-521-82710-8.

[43] "New Precision Measurement of Top Quark Mass". Brookhaven National Laboratory News. 2004. Retrieved 2013-11-03.

[44] J. Joyce (1982) [1939]. *Finnegans Wake.* Penguin Books. p. 383. ISBN 0-14-006286-6.

[45] M. Gell-Mann (1995). *The Quark and the Jaguar: Adventures in the Simple and the Complex.* Henry Holt and Co. p. 180. ISBN 978-0-8050-7253-2.

[46] J. Gleick (1992). *Genius: Richard Feynman and modern physics.* Little Brown and Company. p. 390. ISBN 0-316-90316-7.

[47] J.J. Sakurai (1994). S.F Tuan, ed. *Modern Quantum Mechanics* (Revised ed.). Addison–Wesley. p. 376. ISBN 0-201-53929-2.

[48] D.H. Perkins (2000). *Introduction to high energy physics.* Cambridge University Press. p. 8. ISBN 0-521-62196-8.

[49] M. Riordan (1987). *The Hunting of the Quark: A True Story of Modern Physics.* Simon & Schuster. p. 210. ISBN 978-0-671-50466-3.

[50] F. Close (2006). *The New Cosmic Onion.* CRC Press. p. 133. ISBN 1-58488-798-2.

[51] J.T. Volk et al. (1987). "Letter of Intent for a Tevatron Beauty Factory" (PDF). Fermilab Proposal #783.

[52] G. Fraser (2006). *The New Physics for the Twenty-First Century.* Cambridge University Press. p. 91. ISBN 0-521-81600-9.

[53] "The Standard Model of Particle Physics". BBC. 2002. Retrieved 2009-04-19.

[54] F. Close (2006). *The New Cosmic Onion.* CRC Press. pp. 80–90. ISBN 1-58488-798-2.

[55] D. Lincoln (2004). *Understanding the Universe.* World Scientific. p. 116. ISBN 981-238-705-6.

[56] "Weak Interactions". *Virtual Visitor Center.* Stanford Linear Accelerator Center. 2008. Retrieved 2008-09-28.

[57] K. Nakamura et al. (2010). "Review of Particles Physics: The CKM Quark-Mixing Matrix" (PDF). *J. Phys.* G **37** (75021): 150.

[58] Z. Maki, M. Nakagawa, S. Sakata (1962). "Remarks on the Unified Model of Elementary Particles". *Progress of Theoretical Physics* **28** (5): 870. Bibcode:1962PThPh..28..870M. doi:10.1143/PTP.28.870.

[59] B.C. Chauhan, M. Picariello, J. Pulido, E. Torrente-Lujan (2007). "Quark–lepton complementarity, neutrino and standard model data predict θPMNS
$13 = 9°+1°$
$-2°$". *European Physical Journal* **C50** (3): 573–578. arXiv:hep-ph/0605032. Bibcode:2007EPJC...50..573C. doi:10.1140/epjc/s10052-007-0212-z.

[60] R. Nave. "The Color Force". *HyperPhysics.* Georgia State University, Department of Physics and Astronomy. Retrieved 2009-04-26.

[61] B.A. Schumm (2004). *Deep Down Things.* Johns Hopkins University Press. pp. 131–132. ISBN 0-8018-7971-X. OCLC 55229065.

[62] Part III of M.E. Peskin, D.V. Schroeder (1995). *An Introduction to Quantum Field Theory.* Addison–Wesley. ISBN 0-201-50397-2.

[63] V. Icke (1995). *The force of symmetry.* Cambridge University Press. p. 216. ISBN 0-521-45591-X.

[64] M.Y. Han (2004). *A story of light.* World Scientific. p. 78. ISBN 981-256-034-3.

[65] C. Sutton. "Quantum chromodynamics (physics)". *Encyclopædia Britannica Online.* Retrieved 2009-05-12.

[66] A. Watson (2004). *The Quantum Quark.* Cambridge University Press. pp. 285–286. ISBN 0-521-82907-0.

[67] K.A. Olive *et al.* (Particle Data Group), Chin. Phys. **C38**, 090001 (2014) (URL: http://pdg.lbl.gov)

[68] W. Weise, A.M. Green (1984). *Quarks and Nuclei.* World Scientific. pp. 65–66. ISBN 9971-966-61-1.

[69] D. McMahon (2008). *Quantum Field Theory Demystified.* McGraw–Hill. p. 17. ISBN 0-07-154382-1.

[70] S.G. Roth (2007). *Precision electroweak physics at electron–positron colliders.* Springer. p. VI. ISBN 3-540-35164-7.

[71] R.P. Feynman (1985). *QED: The Strange Theory of Light and Matter* (1st ed.). Princeton University Press. pp. 136–137. ISBN 0-691-08388-6.

[72] M. Veltman (2003). *Facts and Mysteries in Elementary Particle Physics.* World Scientific. pp. 45–47. ISBN 981-238-149-X.

[73] F. Wilczek, B. Devine (2006). *Fantastic Realities.* World Scientific. p. 85. ISBN 981-256-649-X.

[74] F. Wilczek, B. Devine (2006). *Fantastic Realities*. World Scientific. pp. 400ff. ISBN 981-256-649-X.

[75] T. Yulsman (2002). *Origin*. CRC Press. p. 55. ISBN 0-7503-0765-X.

[76] F. Garberson (2008). "Top Quark Mass and Cross Section Results from the Tevatron". arXiv:0808.0273 [hep-ex].

[77] J. Steinberger (2005). *Learning about Particles*. Springer. p. 130. ISBN 3-540-21329-5.

[78] C.-Y. Wong (1994). *Introduction to High-energy Heavy-ion Collisions*. World Scientific. p. 149. ISBN 981-02-0263-6.

[79] S.B. Rüester, V. Werth, M. Buballa, I.A. Shovkovy, D.H. Rischke; Werth; Buballa; Shovkovy; Rischke (2005). "The phase diagram of neutral quark matter: Self-consistent treatment of quark masses". *Physical Review D* **72** (3): 034003. arXiv:hep-ph/0503184. Bibcode:2005PhRvD..72c4004R. doi:10.1103/PhysRevD.72.034004.

[80] M.G. Alford, K. Rajagopal, T. Schaefer, A. Schmitt; Schmitt; Rajagopal; Schäfer (2008). "Color superconductivity in dense quark matter". *Reviews of Modern Physics* **80** (4): 1455–1515. arXiv:0709.4635. Bibcode:2008RvMP...80.1455A. doi:10.1103/RevModPhys.80.1455.

[81] S. Mrowczynski (1998). "Quark–Gluon Plasma". *Acta Physica Polonica B* **29**: 3711. arXiv:nucl-th/9905005. Bibcode:1998AcPPB..29.3711M.

[82] Z. Fodor, S.D. Katz; Katz (2004). "Critical point of QCD at finite T and μ, lattice results for physical quark masses". *Journal of High Energy Physics* **2004** (4): 50. arXiv:hep-lat/0402006. Bibcode:2004JHEP...04..050F. doi:10.1088/1126-6708/2004/04/050.

[83] U. Heinz, M. Jacob (2000). "Evidence for a New State of Matter: An Assessment of the Results from the CERN Lead Beam Programme". arXiv:nucl-th/0002042.

[84] "RHIC Scientists Serve Up "Perfect"Liquid". Brookhaven National Laboratory News. 2005. Retrieved 2009-05-22.

[85] T. Yulsman (2002). *Origins: The Quest for Our Cosmic Roots*. CRC Press. p. 75. ISBN 0-7503-0765-X.

[86] A. Sedrakian, J.W. Clark, M.G. Alford (2007). *Pairing in fermionic systems*. World Scientific. pp. 2–3. ISBN 981-256-907-3.

5.11.9   Further reading

- A. Ali, G. Kramer; Kramer (2011). "JETS and QCD: A historical review of the discovery of the quark and gluon jets and its impact on QCD". *European Physical Journal H* **36** (2): 245. arXiv:1012.2288. Bibcode:2011EPJH...36..245A. doi:10.1140/epjh/e2011-10047-1.

- D.J. Griffiths (2008). *Introduction to Elementary Particles* (2nd ed.). Wiley–VCH. ISBN 3-527-40601-8.

- I.S. Hughes (1985). *Elementary particles* (2nd ed.). Cambridge University Press. ISBN 0-521-26092-2.

- R. Oerter (2005). *The Theory of Almost Everything: The Standard Model, the Unsung Triumph of Modern Physics*. Pi Press. ISBN 0-13-236678-9.

- A. Pickering (1984). *Constructing Quarks: A Sociological History of Particle Physics*. The University of Chicago Press. ISBN 0-226-66799-5.

- B. Povh (1995). *Particles and Nuclei: An Introduction to the Physical Concepts*. Springer–Verlag. ISBN 0-387-59439-6.

- M. Riordan (1987). *The Hunting of the Quark: A true story of modern physics*. Simon & Schuster. ISBN 0-671-64884-5.

- B.A. Schumm (2004). *Deep Down Things: The Breathtaking Beauty of Particle Physics*. Johns Hopkins University Press. ISBN 0-8018-7971-X.

5.11.10   External links

- 1969 Physics Nobel Prize lecture by Murray Gell-Mann

- 1976 Physics Nobel Prize lecture by Burton Richter

- 1976 Physics Nobel Prize lecture by Samuel C.C. Ting

- 2008 Physics Nobel Prize lecture by Makoto Kobayashi

- 2008 Physics Nobel Prize lecture by Toshihide Maskawa

- The Top Quark And The Higgs Particle by T.A. Heppenheimer – A description of CERN's experiment to count the families of quarks.

- Bowley, Roger; Copeland, Ed. "Quarks". *Sixty Symbols*. Brady Haran for the University of Nottingham.

5.12   Quark–lepton complementarity

The **quark–lepton complementarity** (**QLC**) is a possible fundamental symmetry between quarks and leptons. First proposed in 1990 by Foot and Lew,[*][1] it assumes that leptons as well as quarks come in three "colors". Such theory may reproduce the Standard Model at low energies, and hence quark–lepton symmetry may be realized in nature.

5.12.1 Possible evidence for QLC

Recent neutrino experiments confirm that the Pontecorvo–Maki–Nakagawa–Sakata matrix U_{PMNS} contains large mixing angles. For example, atmospheric measurements of particle decay yield θPMNS
23 $\approx 45°$, while solar experiments yield θPMNS
12 $\approx 34°$. These results should be compared with θPMNS
13 which is small,[2] and with the quark mixing angles in the Cabibbo–Kobayashi–Maskawa matrix U_{CKM}. The disparity that nature indicates between quark and lepton mixing angles has been viewed in terms of a "quark–lepton complementarity" which can be expressed in the relations

$$\theta_{12}^{PMNS} + \theta_{12}^{CKM} \simeq 45° ,$$

$$\theta_{23}^{PMNS} + \theta_{23}^{CKM} \simeq 45° .$$

Possible consequences of QLC have been investigated in the literature and in particular a simple correspondence between the PMNS and CKM matrices have been proposed and analyzed in terms of a correlation matrix. The correlation matrix V_{M} is simply defined as the product of the CKM and PMNS matrices:

$$V_{\text{M}} = U_{\text{CKM}} \cdot U_{\text{PMNS}} ,$$

Unitarity implies:

$$U_{\text{PMNS}} = U_{\text{CKM}}^{\dagger} V_{\text{M}} .$$

5.12.2 Open questions

One may ask where do the large lepton mixings come from? Is this information implicit in the form of the V_{M} matrix? This question has been widely investigated in the literature, but its answer is still open. Furthermore in some Grand Unification Theories (GUTs) the direct QLC correlation between the CKM and the PMNS mixing matrix can be obtained. In this class of models, the V_M matrix is determined by the heavy Majorana neutrino mass matrix.

Despite the naive relations between the PMNS and CKM angles, a detailed analysis shows that the correlation matrix is phenomenologically compatible with a tribimaximal pattern, and only marginally with a bimaximal pattern. It is possible to include bimaximal forms of the correlation matrix V_{M} in models with renormalization effects that are relevant, however, only in particular cases with $\tan\beta > 40$ and with quasi-degenerate neutrino masses.

5.12.3 See also

- Leptoquark

5.12.4 References

[1] R. Foot, H. Lew (1990). "Quark-lepton-symmetric model". *Physical Review D* **41** (11): 3502–3505. Bibcode:1990PhRvD..41.3502F. doi:10.1103/PhysRevD.41.3502.

[2] F. P. An et al. [DAYA-BAY Collaboration], Phys. Rev. Lett. 108, 171803 (2012) [arXiv:1203.1669 [hep-ex]] http://arxiv.org/abs/arXiv:1203.1669

- B.C. Chauhan, M. Picariello, J. Pulido, E. Torrente-Lujan (2007). "Quark-lepton complementarity, neutrino and standard model data predict θPMNS 13 = (9+1 −2)°". *European Physical Journal C* **50** (3): 573–578. arXiv:hep-ph/0605032. Bibcode:2007EPJC...50..573C. doi:10.1140/epjc/s10052-007-0212-z.

- K.M. Patel (2010). "An $SO(10) \times S_4$ Model of Quark-Lepton Complementarity;". *Physics Letters B* **695**: 225. arXiv:1008.5061. Bibcode:2011PhLB..695..225P. doi:10.1016/j.physletb.2010.11.024.

5.13 Standard Model

This article is about the Standard Model of particle physics. For other uses, see Standard model (disambiguation).
This article is a non-mathematical general overview of the Standard Model. For a mathematical description, see the article Standard Model (mathematical formulation).
For the Standard Model of Big Bang cosmology, Lambda-CDM model.

The **Standard Model** of particle physics is a theory concerning the electromagnetic, weak, and strong nuclear interactions, as well as classifying all the subatomic particles known. It was developed throughout the latter half of the 20th century, as a collaborative effort of scientists around the world.[1] The current formulation was finalized in the mid-1970s upon experimental confirmation of the existence of quarks. Since then, discoveries of the top quark (1995), the tau neutrino (2000), and more recently the Higgs boson (2013), have given further credence to the Standard Model. Because of its success in explaining a wide variety of experimental results, the Standard Model is sometimes regarded as a "theory of almost everything".

Although the Standard Model is believed to be theoretically self-consistent[*][2] and has demonstrated huge and continued successes in providing experimental predictions, it does leave some phenomena unexplained and it falls short of being a complete theory of fundamental interactions. It does not incorporate the full theory of gravitation[*][3] as described by general relativity, or account for the accelerating expansion of the universe (as possibly described by dark energy). The model does not contain any viable dark matter particle that possesses all of the required properties deduced from observational cosmology. It also does not incorporate neutrino oscillations (and their non-zero masses).

The development of the Standard Model was driven by theoretical and experimental particle physicists alike. For theorists, the Standard Model is a paradigm of a quantum field theory, which exhibits a wide range of physics including spontaneous symmetry breaking, anomalies, nonperturbative behavior, etc. It is used as a basis for building more exotic models that incorporate hypothetical particles, extra dimensions, and elaborate symmetries (such as supersymmetry) in an attempt to explain experimental results at variance with the Standard Model, such as the existence of dark matter and neutrino oscillations.

5.13.1   Historical background

The first step towards the Standard Model was Sheldon Glashow's discovery in 1961 of a way to combine the electromagnetic and weak interactions.[*][4] In 1967 Steven Weinberg[*][5] and Abdus Salam[*][6] incorporated the Higgs mechanism[*][7][*][8][*][9] into Glashow's electroweak theory, giving it its modern form.

The Higgs mechanism is believed to give rise to the masses of all the elementary particles in the Standard Model. This includes the masses of the W and Z bosons, and the masses of the fermions, i.e. the quarks and leptons.

After the neutral weak currents caused by Z boson exchange were discovered at CERN in 1973,[*][10][*][11][*][12][*][13] the electroweak theory became widely accepted and Glashow, Salam, and Weinberg shared the 1979 Nobel Prize in Physics for discovering it. The W and Z bosons were discovered experimentally in 1981, and their masses were found to be as the Standard Model predicted.

The theory of the strong interaction, to which many contributed, acquired its modern form around 1973–74, when experiments confirmed that the hadrons were composed of fractionally charged quarks.

5.13.2   Overview

At present, matter and energy are best understood in terms of the kinematics and interactions of elementary particles. To date, physics has reduced the laws governing the behavior and interaction of all known forms of matter and energy to a small set of fundamental laws and theories. A major goal of physics is to find the "common ground" that would unite all of these theories into one integrated theory of everything, of which all the other known laws would be special cases, and from which the behavior of all matter and energy could be derived (at least in principle).[*][14]

5.13.3   Particle content

The Standard Model includes members of several classes of elementary particles (fermions, gauge bosons, and the Higgs boson), which in turn can be distinguished by other characteristics, such as color charge.

Fermions

The Standard Model includes 12 elementary particles of spin-½ known as fermions. According to the spin-statistics theorem, fermions respect the Pauli exclusion principle. Each fermion has a corresponding antiparticle.

The fermions of the Standard Model are classified according to how they interact (or equivalently, by what charges they carry). There are six quarks (up, down, charm, strange, top, bottom), and six leptons (electron, electron neutrino, muon, muon neutrino, tau, tau neutrino). Pairs from each classification are grouped together to form a generation, with corresponding particles exhibiting similar physical behavior (see table).

The defining property of the quarks is that they carry color charge, and hence, interact via the strong interaction. A phenomenon called color confinement results in quarks being very strongly bound to one another, forming color-neutral composite particles (hadrons) containing either a quark and an antiquark (mesons) or three quarks (baryons). The familiar proton and the neutron are the two baryons having the smallest mass. Quarks also carry electric charge and weak isospin. Hence they interact with other fermions both electromagnetically and via the weak interaction.

The remaining six fermions do not carry colour charge and are called leptons. The three neutrinos do not carry electric charge either, so their motion is directly influenced only by the weak nuclear force, which makes them notoriously difficult to detect. However, by virtue of carrying an electric charge, the electron, muon, and tau all interact electromagnetically.

Each member of a generation has greater mass than the corresponding particles of lower generations. The first generation charged particles do not decay; hence all ordinary (baryonic) matter is made of such particles. Specifically, all atoms consist of electrons orbiting around atomic nuclei, ultimately constituted of up and down quarks. Second and third generations charged particles, on the other hand, decay with very short half lives, and are observed only in very high-energy environments. Neutrinos of all generations also do not decay, and pervade the universe, but rarely interact with baryonic matter.

Gauge bosons

In the Standard Model, gauge bosons are defined as force carriers that mediate the strong, weak, and electromagnetic fundamental interactions.

Interactions in physics are the ways that particles influence other particles. At a macroscopic level, electromagnetism allows particles to interact with one another via electric and magnetic fields, and gravitation allows particles with mass to attract one another in accordance with Einstein's theory of general relativity. The Standard Model explains such forces as resulting from matter particles exchanging other particles, generally referred to as *force mediating particles*. When a force-mediating particle is exchanged, at a macroscopic level the effect is equivalent to a force influencing both of them, and the particle is therefore said to have *mediated* (i.e., been the agent of) that force. The Feynman diagram calculations, which are a graphical representation of the perturbation theory approximation, invoke "force mediating particles", and when applied to analyze high-energy scattering experiments are in reasonable agreement with the data. However, perturbation theory (and with it the concept of a "force-mediating particle") fails in other situations. These include low-energy quantum chromodynamics, bound states, and solitons.

The gauge bosons of the Standard Model all have spin (as do matter particles). The value of the spin is 1, making them bosons. As a result, they do not follow the Pauli exclusion principle that constrains fermions: thus bosons (e.g. photons) do not have a theoretical limit on their spatial density (number per volume). The different types of gauge bosons are described below.

- Photons mediate the electromagnetic force between electrically charged particles. The photon is massless and is well-described by the theory of quantum electrodynamics.

- The W+, W−, and Z gauge bosons mediate the weak interactions between particles of different flavors (all

quarks and leptons). They are massive, with the Z being more massive than the W±. The weak interactions involving the W± exclusively act on *left-handed* particles and *right-handed* antiparticles. Furthermore, the W± carries an electric charge of +1 and −1 and couples to the electromagnetic interaction. The electrically neutral Z boson interacts with both left-handed particles and antiparticles. These three gauge bosons along with the photons are grouped together, as collectively mediating the electroweak interaction.

- The eight gluons mediate the strong interactions between color charged particles (the quarks). Gluons are massless. The eightfold multiplicity of gluons is labeled by a combination of color and anticolor charge (e.g. red–antigreen).*[nb 1] Because the gluons have an effective color charge, they can also interact among themselves. The gluons and their interactions are described by the theory of quantum chromodynamics.

The interactions between all the particles described by the Standard Model are summarized by the diagrams on the right of this section.

Higgs boson

Main article: Higgs boson

The Higgs particle is a massive scalar elementary particle theorized by Robert Brout, François Englert, Peter Higgs, Gerald Guralnik, C. R. Hagen, and Tom Kibble in 1964 (see 1964 PRL symmetry breaking papers) and is a key building block in the Standard Model.*[7]*[8]*[9]*[15] It has no intrinsic spin, and for that reason is classified as a boson (like the gauge bosons, which have integer spin).

The Higgs boson plays a unique role in the Standard Model, by explaining why the other elementary particles, except the photon and gluon, are massive. In particular, the Higgs boson explains why the photon has no mass, while the W and Z bosons are very heavy. Elementary particle masses, and the differences between electromagnetism (mediated by the photon) and the weak force (mediated by the W and Z bosons), are critical to many aspects of the structure of microscopic (and hence macroscopic) matter. In electroweak theory, the Higgs boson generates the masses of the leptons (electron, muon, and tau) and quarks. As the Higgs boson is massive, it must interact with itself.

Because the Higgs boson is a very massive particle and also decays almost immediately when created, only a very high-energy particle accelerator can observe and record it. Experiments to confirm and determine the nature of the Higgs boson using the Large Hadron Collider (LHC) at CERN began in early 2010, and were performed at Fermilab's Tevatron until its closure in late 2011. Mathematical consistency of the Standard Model requires that any mechanism capable of generating the masses of elementary particles become visible at energies above 1.4 TeV;[16] therefore, the LHC (designed to collide two 7 to 8 TeV proton beams) was built to answer the question of whether the Higgs boson actually exists.[17]

On 4 July 2012, the two main experiments at the LHC (ATLAS and CMS) both reported independently that they found a new particle with a mass of about 125 GeV/c^2 (about 133 proton masses, on the order of 10^*–25 kg), which is "consistent with the Higgs boson." Although it has several properties similar to the predicted "simplest" Higgs,[18] they acknowledged that further work would be needed to conclude that it is indeed the Higgs boson, and exactly which version of the Standard Model Higgs is best supported if confirmed.[19][20][21][22][23]

On 14 March 2013 the Higgs Boson was tentatively confirmed to exist.[24]

Total particle count

Counting particles by a rule that distinguishes between particles and their corresponding antiparticles, and among the many color states of quarks and gluons, gives a total of 61 elementary particles.[25]

5.13.4 Theoretical aspects

Main article: Standard Model (mathematical formulation)

Construction of the Standard Model Lagrangian

Technically, quantum field theory provides the mathematical framework for the Standard Model, in which a Lagrangian controls the dynamics and kinematics of the theory. Each kind of particle is described in terms of a dynamical field that pervades space-time. The construction of the Standard Model proceeds following the modern method of constructing most field theories: by first postulating a set of symmetries of the system, and then by writing down the most general renormalizable Lagrangian from its particle (field) content that observes these symmetries.

The global Poincaré symmetry is postulated for all relativistic quantum field theories. It consists of the familiar translational symmetry, rotational symmetry and the inertial reference frame invariance central to the theory of special relativity. The local SU(3)×SU(2)×U(1) gauge symmetry is an internal symmetry that essentially defines the Standard

Model. Roughly, the three factors of the gauge symmetry give rise to the three fundamental interactions. The fields fall into different representations of the various symmetry groups of the Standard Model (see table). Upon writing the most general Lagrangian, one finds that the dynamics depend on 19 parameters, whose numerical values are established by experiment. The parameters are summarized in the table above (note: with the Higgs mass is at 125 GeV, the Higgs self-coupling strength $\lambda \sim 1/8$).

Quantum chromodynamics sector Main article: Quantum chromodynamics

The quantum chromodynamics (QCD) sector defines the interactions between quarks and gluons, with SU(3) symmetry, generated by T[*]a. Since leptons do not interact with gluons, they are not affected by this sector. The Dirac Lagrangian of the quarks coupled to the gluon fields is given by

$$\mathcal{L}_{QCD} = iU(\partial_\mu - ig_s G^a_\mu T^a)\gamma^\mu U + iD(\partial_\mu - ig_s G^a_\mu T^a)\gamma^\mu D.$$

G^a_μ is the SU(3) gauge field containing the gluons, γ^μ are the Dirac matrices, D and U are the Dirac spinors associated with up- and down-type quarks, and g_s is the strong coupling constant.

Electroweak sector Main article: Electroweak interaction

The electroweak sector is a Yang–Mills gauge theory with the simple symmetry group U(1)×SU(2)$_L$,

$$\mathcal{L}_{EW} = \sum_\psi \bar{\psi}\gamma^\mu \left(i\partial_\mu - g'\frac{1}{2}Y_W B_\mu - g\frac{1}{2}\vec{\tau}_L \vec{W}_\mu \right) \psi$$

where B_μ is the U(1) gauge field; Y_W is the weak hypercharge—the generator of the U(1) group; $\vec{W}_\mu$ is the three-component SU(2) gauge field; $\vec{\tau}_L$ are the Pauli matrices—infinitesimal generators of the SU(2) group. The subscript L indicates that they only act on left fermions; g' and g are coupling constants.

Higgs sector Main article: Higgs mechanism

In the Standard Model, the Higgs field is a complex scalar of the group SU(2)$_L$:

$$\varphi = \frac{1}{\sqrt{2}} \begin{pmatrix} \varphi^+ \\ \varphi^0 \end{pmatrix} ,$$

where the indices + and 0 indicate the electric charge (Q) of the components. The weak isospin (Y_W) of both components is 1.

Before symmetry breaking, the Higgs Lagrangian is:

$$\mathcal{L}_H = \varphi^\dagger \left(\partial^\mu - \frac{i}{2} \left(g' Y_W B^\mu + g \vec{\tau} \vec{W}^\mu \right) \right) \left(\partial_\mu + \frac{i}{2} \left(g' Y_W B_\mu + g \vec{\tau} \vec{W}_\mu \right) \right) \varphi - \frac{\lambda^2}{4} \left(\varphi^\dagger \varphi - v^2 \right)^2 ,$$

which can also be written as:

$$\mathcal{L}_H = \left| \left(\partial_\mu + \frac{i}{2} \left(g' Y_W B_\mu + g \vec{\tau} \vec{W}_\mu \right) \right) \varphi \right|^2 - \frac{\lambda^2}{4} \left(\varphi^\dagger \varphi - v^2 \right)^2$$

5.13.5   Fundamental forces

Main article: Fundamental interaction

The Standard Model classified all four fundamental forces in nature. In the Standard Model, a force is described as an exchange of bosons between the objects affected, such as a photon for the electromagnetic force and a gluon for the strong interaction. Those particles are called force carriers.[26]

5.13.6   Tests and predictions

The Standard Model (SM) predicted the existence of the W and Z bosons, gluon, and the top and charm quarks before these particles were observed. Their predicted properties were experimentally confirmed with good precision. To give an idea of the success of the SM, the following table compares the measured masses of the W and Z bosons with the masses predicted by the SM:

The SM also makes several predictions about the decay of Z bosons, which have been experimentally confirmed by the Large Electron-Positron Collider at CERN.

In May 2012 BaBar Collaboration reported that their recently analyzed data may suggest possible flaws in the Standard Model of particle physics.[28][29] These data show that a particular type of particle decay called "B to D-star-tau-nu" happens more often than the Standard Model says it should. In this type of decay, a particle called the B-bar meson decays into a D meson, an antineutrino and a tau-lepton. While the level of certainty of the excess (3.4 sigma) is not

enough to claim a break from the Standard Model, the results are a potential sign of something amiss and are likely to impact existing theories, including those attempting to deduce the properties of Higgs bosons.[30]

On December 13, 2012, physicists reported the constancy, over space and time, of a basic physical constant of nature that supports the *standard model of physics*. The scientists, studying methanol molecules in a distant galaxy, found the change ($\Delta\mu/\mu$) in the proton-to-electron mass ratio μ to be equal to "(0.0 ± 1.0) × 10[*]−7 at redshift z = 0.89" and consistent with "a null result".[31][32]

5.13.7   Challenges

See also: Physics beyond the Standard Model

Self-consistency of the Standard Model (currently formulated as a non-abelian gauge theory quantized through path-integrals) has not been mathematically proven. While regularized versions useful for approximate computations (for example lattice gauge theory) exist, it is not known whether they converge (in the sense of S-matrix elements) in the limit that the regulator is removed. A key question related to the consistency is the Yang–Mills existence and mass gap problem.

Experiments indicate that neutrinos have mass, which the classic Standard Model did not allow.[33] To accommodate this finding, the classic Standard Model can be modified to include neutrino mass.

If one insists on using only Standard Model particles, this can be achieved by adding a non-renormalizable interaction of leptons with the Higgs boson.[34] On a fundamental level, such an interaction emerges in the seesaw mechanism where heavy right-handed neutrinos are added to the theory. This is natural in the left-right symmetric extension of the Standard Model[35][36] and in certain grand unified theories.[37] As long as new physics appears below or around 10^{14} GeV, the neutrino masses can be of the right order of magnitude.

Theoretical and experimental research has attempted to extend the Standard Model into a Unified field theory or a Theory of everything, a complete theory explaining all physical phenomena including constants. Inadequacies of the Standard Model that motivate such research include:

- It does not attempt to explain gravitation, although a theoretical particle known as a graviton would help explain it, and unlike for the strong and electroweak interactions of the Standard Model, there is no known way of describing general relativity, the canonical theory of gravitation, consistently in terms of quantum field theory. The reason for this is, among other things, that quantum field theories of gravity generally break down before reaching the Planck scale. As a consequence, we have no reliable theory for the very early universe;

- Some consider it to be *ad hoc* and inelegant, requiring 19 numerical constants whose values are unrelated and arbitrary. Although the Standard Model, as it now stands, can explain why neutrinos have masses, the specifics of neutrino mass are still unclear. It is believed that explaining neutrino mass will require an additional 7 or 8 constants, which are also arbitrary parameters;

- The Higgs mechanism gives rise to the hierarchy problem if some new physics (coupled to the Higgs) is present at high energy scales. In these cases in order for the weak scale to be much smaller than the Planck scale, severe fine tuning of the parameters is required; there are, however, other scenarios that include quantum gravity in which such fine tuning can be avoided.[*][38]There are also issues of Quantum triviality, which suggests that it may not be possible to create a consistent quantum field theory involving elementary scalar particles.

- It should be modified so as to be consistent with the emerging "Standard Model of cosmology." In particular, the Standard Model cannot explain the observed amount of cold dark matter (CDM) and gives contributions to dark energy which are many orders of magnitude too large. It is also difficult to accommodate the observed predominance of matter over antimatter (matter/antimatter asymmetry). The isotropy and homogeneity of the visible universe over large distances seems to require a mechanism like cosmic inflation, which would also constitute an extension of the Standard Model.

- The existence of ultra-high-energy cosmic rays are difficult to explain under the Standard Model.

Currently, no proposed Theory of Everything has been widely accepted or verified.

5.13.8 See also

- Fundamental interaction:

  - Quantum electrodynamics

  - Strong interaction: Color charge, Quantum chromodynamics, Quark model

  - Weak interaction: Electroweak theory, Fermi theory of beta decay, Weak hypercharge, Weak isospin

- Gauge theory: Nontechnical introduction to gauge theory

- Generation

- Higgs mechanism: Higgs boson, Higgsless model

- J. C. Ward

- J. J. Sakurai Prize for Theoretical Particle Physics

- Lagrangian

- Open questions: BTeV experiment, CP violation, Neutrino masses, Quark matter, Quantum triviality

- Penguin diagram

- Quantum field theory

- Standard Model: Mathematical formulation of, Physics beyond the Standard Model

5.13.9 Notes and references

[1] Technically, there are nine such color–anticolor combinations. However, there is one color-symmetric combination that can be constructed out of a linear superposition of the nine combinations, reducing the count to eight.

5.13.10 References

[1] R. Oerter (2006). *The Theory of Almost Everything: The Standard Model, the Unsung Triumph of Modern Physics* (Kindle ed.). Penguin Group. p. 2. ISBN 0-13-236678-9.

[2] In fact, there are mathematical issues regarding quantum field theories still under debate (see e.g. Landau pole), but the predictions extracted from the Standard Model by current methods applicable to current experiments are all self-consistent. For a further discussion see e.g. Chapter 25 of R. Mann (2010). *An Introduction to Particle Physics and the Standard Model.* CRC Press. ISBN 978-1-4200-8298-2.

[3] Sean Carroll, Ph.D., Cal Tech, 2007, The Teaching Company, *Dark Matter, Dark Energy: The Dark Side of the Universe*, Guidebook Part 2 page 59, Accessed Oct. 7, 2013, "...Standard Model of Particle Physics: The modern theory of elementary particles and their interactions ... It does not, strictly speaking, include gravity, although it's often convenient to include gravitons among the known particles of nature..."

[4] S.L. Glashow (1961). "Partial-symmetries of weak interactions". *Nuclear Physics* **22** (4): 579–588. Bibcode:1961NucPh..22..579G. doi:10.1016/0029-5582(61)90469-2.

[5] S. Weinberg (1967). "A Model of Leptons". *Physical Review Letters* **19** (21): 1264–1266. Bibcode:1967PhRvL..19.1264W. doi:10.1103/PhysRevLett.19.1264.

[6] A. Salam (1968). N. Svartholm, ed. *Elementary Particle Physics: Relativistic Groups and Analyticity*. Eighth Nobel Symposium. Stockholm: Almquvist and Wiksell. p. 367.

[7] F. Englert, R. Brout (1964). "Broken Symmetry and the Mass of Gauge Vector Mesons". *Physical Review Letters* **13** (9): 321–323. Bibcode:1964PhRvL..13..321E. doi:10.1103/PhysRevLett.13.321.

[8] P.W. Higgs (1964). "Broken Symmetries and the Masses of Gauge Bosons". *Physical Review Letters* **13** (16): 508–509. Bibcode:1964PhRvL..13..508H. doi:10.1103/PhysRevLett.13.508.

[9] G.S. Guralnik, C.R. Hagen, T.W.B. Kibble (1964). "Global Conservation Laws and Massless Particles". *Physical Review Letters* **13** (20): 585–587. Bibcode:1964PhRvL..13..585G. doi:10.1103/PhysRevLett.13.585.

[10] F.J. Hasert et al. (1973). "Search for elastic muon-neutrino electron scattering". *Physics Letters B* **46** (1): 121. Bibcode:1973PhLB...46..121H. doi:10.1016/0370-2693(73)90494-2.

[11] F.J. Hasert et al. (1973). "Observation of neutrino-like interactions without muon or electron in the Gargamelle neutrino experiment". *Physics Letters B* **46** (1): 138. Bibcode:1973PhLB...46..138H. doi:10.1016/0370-2693(73)90499-1.

[12] F.J. Hasert et al. (1974). "Observation of neutrino-like interactions without muon or electron in the Gargamelle neutrino experiment". *Nuclear Physics B* **73** (1): 1. Bibcode:1974NuPhB..73....1H. doi:10.1016/0550-3213(74)90038-8.

[13] D. Haidt (4 October 2004). "The discovery of the weak neutral currents". *CERN Courier*. Retrieved 8 May 2008.

[14] "Details can be worked out if the situation is simple enough for us to make an approximation, which is almost never, but often we can understand more or less what is happening." from *The Feynman Lectures on Physics*, Vol 1. pp. 2–7

[15] G.S. Guralnik (2009). "The History of the Guralnik, Hagen and Kibble development of the Theory of Spontaneous Symmetry Breaking and Gauge Particles". *International Journal of Modern Physics A* **24** (14): 2601–2627. arXiv:0907.3466. Bibcode:2009IJMPA..24.2601G. doi:10.1142/S0217751X09045431.

[16] B.W. Lee, C. Quigg, H.B. Thacker (1977). "Weak interactions at very high energies: The role of the Higgs-boson mass". *Physical Review D* **16** (5): 1519–1531. Bibcode:1977PhRvD..16.1519L. doi:10.1103/PhysRevD.16.1519.

[17] "Huge $10 billion collider resumes hunt for 'God particle'". CNN. 11 November 2009. Retrieved 2010-05-04.

[18] M. Strassler (10 July 2012). "Higgs Discovery: Is it a Higgs?". Retrieved 2013-08-06.

[19] "CERN experiments observe particle consistent with long-sought Higgs boson". CERN. 4 July 2012. Retrieved 2012-07-04.

[20] "Observation of a New Particle with a Mass of 125 GeV". CERN. 4 July 2012. Retrieved 2012-07-05.

[21] "ATLAS Experiment". ATLAS. 1 January 2006. Retrieved 2012-07-05.

[22] "Confirmed: CERN discovers new particle likely to be the Higgs boson". *YouTube*. Russia Today. 4 July 2012. Retrieved 2013-08-06.

[23] D. Overbye (4 July 2012). "A New Particle Could Be Physics' Holy Grail". *New York Times*. Retrieved 2012-07-04.

[24] "New results indicate that new particle is a Higgs boson". CERN. 14 March 2013. Retrieved 2013-08-06.

[25] S. Braibant, G. Giacomelli, M. Spurio (2009). *Particles and Fundamental Interactions: An Introduction to Particle Physics*. Springer. pp. 313–314. ISBN 978-94-007-2463-1.

[26] http://home.web.cern.ch/about/physics/standard-model Official CERN website

[27] http://www.pha.jhu.edu/~{}dfehling/particle.gif

[28] "BABAR Data in Tension with the Standard Model". SLAC. 31 May 2012. Retrieved 2013-08-06.

[29] BaBar Collaboration (2012). "Evidence for an excess of B → D*(*) τ^- ν_τ decays". *Physical Review Letters* **109** (10): 101802. arXiv:1205.5442. Bibcode:2012PhRvL.109j1802L. doi:10.1103/PhysRevLett.109.101802.

[30] "BaBar data hint at cracks in the Standard Model". *e! Science News*. 18 June 2012. Retrieved 2013-08-06.

[31] J. Bagdonaite et al. (2012). "A Stringent Limit on a Drifting Proton-to-Electron Mass Ratio from Alcohol in the Early Universe". *Science* **339** (6115): 46. Bibcode:2013Sci...339...46B. doi:10.1126/science.1224898.

[32] C. Moskowitz (13 December 2012). "Phew! Universe's Constant Has Stayed Constant". Space.com. Retrieved 2012-12-14.

[33] "Particle chameleon caught in the act of changing". CERN. 31 May 2010. Retrieved 2012-07-05.

[34] S. Weinberg (1979). "Baryon and Lepton Non-conserving Processes". *Physical Review Letters* **43** (21): 1566. Bibcode:1979PhRvL..43.1566W. doi:10.1103/PhysRevLett.43.1566.

[35] P. Minkowski (1977). "$\mu \to e \gamma$ at a Rate of One Out of 10^9 Muon Decays?". *Physics Letters B* **67** (4): 421. Bibcode:1977PhLB...67..421M. doi:10.1016/0370-2693(77)90435-X.

[36] R. N. Mohapatra, G. Senjanovic (1980). "Neutrino Mass and Spontaneous Parity Nonconservation". *Physical Review Letters* **44** (14): 912–915. Bibcode:1980PhRvL..44..912M. doi:10.1103/PhysRevLett.44.912.

[37] M. Gell-Mann, P. Ramond and R. Slansky (1979). F. van Nieuwenhuizen and D. Z. Freedman, ed. *Supergravity.* North Holland. pp. 315–321. ISBN 0-444-85438-X.

[38] Salvio, Strumia (2014-03-17). "Agravity". *JHEP 1406 (2014) 080.* arXiv:1403.4226. Bibcode:2014JHEP...06..080S. doi:10.1007/JHEP06(2014)080.

5.13.11   Further reading

- R. Oerter (2006). *The Theory of Almost Everything: The Standard Model, the Unsung Triumph of Modern Physics.* Plume.

- B.A. Schumm (2004). *Deep Down Things: The Breathtaking Beauty of Particle Physics.* Johns Hopkins University Press. ISBN 0-8018-7971-X.

- "The Standard Model of Particle Physics Interactive Graphic".

Introductory textbooks

- I. Aitchison, A. Hey (2003). *Gauge Theories in Particle Physics: A Practical Introduction.* Institute of Physics. ISBN 978-0-585-44550-2.

- W. Greiner, B. Müller (2000). *Gauge Theory of Weak Interactions.* Springer. ISBN 3-540-67672-4.

- G.D. Coughlan, J.E. Dodd, B.M. Gripaios (2006). *The Ideas of Particle Physics: An Introduction for Scientists.* Cambridge University Press.

- D.J. Griffiths (1987). *Introduction to Elementary Particles.* John Wiley & Sons. ISBN 0-471-60386-4.

- G.L. Kane (1987). *Modern Elementary Particle Physics.* Perseus Books. ISBN 0-201-11749-5.

Advanced textbooks

- T.P. Cheng, L.F. Li (2006). *Gauge theory of elementary particle physics.* Oxford University Press. ISBN 0-19-851961-3. Highlights the gauge theory aspects of the Standard Model.

- J.F. Donoghue, E. Golowich, B.R. Holstein (1994). *Dynamics of the Standard Model.* Cambridge University Press. ISBN 978-0-521-47652-2. Highlights dynamical and phenomenological aspects of the Standard Model.

- L. O'Raifeartaigh (1988). *Group structure of gauge theories.* Cambridge University Press. ISBN 0-521-34785-8.

- Nagashima Y. Elementary Particle Physics: Foundations of the Standard Model, Volume 2. (Wiley 2013) 920 рапуы

- Schwartz, M.D. Quantum Field Theory and the Standard Model (Cambridge University Press 2013) 952 pages

- Langacker P. The standard model and beyond. (CRC Press, 2010) 670 pages Highlights group-theoretical aspects of the Standard Model.

Journal articles

- E.S. Abers, B.W. Lee (1973). "Gauge theories". *Physics Reports* **9**: 1–141. Bibcode:1973PhR.....9....1A. doi:10.1016/0370-1573(73)90027-6.

- M. Baak et al. (2012). "The Electroweak Fit of the Standard Model after the Discovery of a New Boson at the LHC". *The European Physical Journal C* **72** (11). arXiv:1209.2716. Bibcode:2012EPJC...72.2205B. doi:10.1140/epjc/s10052-012-2205-9.

- Y. Hayato et al. (1999). "Search for Proton Decay through $p \to \nu K^{*}$+ in a Large Water Cherenkov Detector". *Physical Review Letters* **83** (8): 1529. arXiv:hep-ex/9904020. Bibcode:1999PhRvL..83.1529H. doi:10.1103/PhysRevLett.83.1529.

- S.F. Novaes (2000). "Standard Model: An Introduction". arXiv:hep-ph/0001283 [hep-ph].

- D.P. Roy (1999). "Basic Constituents of Matter and their Interactions —A Progress Report". arXiv:hep-ph/9912523 [hep-ph].

- F. Wilczek (2004). "The Universe Is A Strange Place". *Nuclear Physics B - Proceedings Supplements* **134**: 3. arXiv:astro-ph/0401347. Bibcode:2004NuPhS.134....3W. doi:10.1016/j.nuclphysbps.2004.08.001.

5.13.12   External links

- "The Standard Model explained in Detail by CERN's John Ellis" omega tau podcast.

- "LHC sees hint of lightweight Higgs boson" "New Scientist".

- "Standard Model may be found incomplete," *New Scientist*.

- "Observation of the Top Quark" at Fermilab.

- "The Standard Model Lagrangian." After electroweak symmetry breaking, with no explicit Higgs boson.

- "Standard Model Lagrangian" with explicit Higgs terms. PDF, PostScript, and LaTeX versions.

- "The particle adventure." Web tutorial.

- Nobes, Matthew (2002) "Introduction to the Standard Model of Particle Physics" on Kuro5hin: Part 1, Part 2, Part 3a, Part 3b.

- "The Standard Model" The Standard Model on the CERN web site explains how the basic building blocks of matter interact, governed by four fundamental forces.

5.14   List of particles

This is a list of the different types of particles found or believed to exist in the whole of the universe. For individual lists of the different particles, see the list below.

5.14.1   Elementary particles

Main article: Elementary particle

Elementary particles are particles with no measurable internal structure; that is, they are not composed of other particles. They are the fundamental objects of quantum field theory. Many families and sub-families of elementary particles exist. Elementary particles are classified according to their spin. Fermions have half-integer spin while bosons have integer spin. All the particles of the Standard Model have been experimentally observed, recently including the Higgs boson.[*][1][*][2]

Fermions

Main article: Fermion

Fermions are one of the two fundamental classes of particles, the other being bosons. Fermion particles are described by Fermi–Dirac statistics and have quantum numbers described by the Pauli exclusion principle. They include the quarks and leptons, as well as any composite particles consisting of an odd number of these, such as all baryons and many atoms and nuclei.

Fermions have half-integer spin; for all known elementary fermions this is $\frac{1}{2}$. All known fermions, except neutrinos, are also Dirac fermions; that is, each known fermion has its own distinct antiparticle. It is not known whether the neutrino is a Dirac fermion or a Majorana fermion.[*][3] Fermions are the basic building blocks of all matter. They are classified according to whether they interact via the color force or not. In the Standard Model, there are 12 types of elementary fermions: six quarks and six leptons.

Quarks   Main article: Quark

Quarks are the fundamental constituents of hadrons and interact via the strong interaction. Quarks are the only known carriers of fractional charge, but because they combine in groups of three (baryons) or in groups of two with antiquarks (mesons), only integer charge is observed in nature. Their respective antiparticles are the antiquarks, which are identical except for the fact that they carry the opposite electric charge (for example the up quark carries charge $+\frac{2}{3}$, while the up antiquark carries charge $-\frac{2}{3}$), color charge, and baryon number. There are six flavors of quarks; the three positively charged quarks are called "up-type quarks" and the three negatively charged quarks are called "down-type quarks".

Leptons   Main article: Leptons

Leptons do not interact via the strong interaction. Their respective antiparticles are the antileptons which are identical, except for the fact that they carry the opposite electric charge and lepton number. The antiparticle of an electron is an antielectron, which is nearly always called a "positron" for historical reasons. There are six leptons in total; the three charged leptons are called "electron-like leptons", while the neutral leptons are called "neutrinos". Neutrinos are known to oscillate, so that neutrinos of definite flavor do not have definite mass, rather they exist in a superposition of mass eigenstates. The hypothetical heavy right-handed neutrino, called a "sterile neutrino", has been left off the list.

Bosons

Main article: Boson

Bosons are one of the two fundamental classes of particles, the other being fermions. Bosons are characterized by Bose–Einstein statistics and all have integer spins. Bosons may be either elementary, like photons and gluons, or composite, like mesons.

The fundamental forces of nature are mediated by gauge bosons, and mass is believed to be created by the Higgs field. According to the Standard Model the elementary bosons are:

The graviton is added to the list although it is not predicted by the Standard Model, but by other theories in the framework of quantum field theory. Furthermore, gravity is non-renormalizable. There are a total of eight independent gluons. The Higgs boson is postulated by the electroweak theory primarily to explain the origin of particle masses. In a process known as the "Higgs mechanism", the Higgs boson and the other gauge bosons in the Standard Model acquire mass via spontaneous symmetry breaking of the SU(2) gauge symmetry. The Minimal Supersymmetric Standard Model (MSSM) predicts several Higgs bosons. A new particle expected to be the Higgs boson was observed at the CERN/LHC on March 14, 2013, around the energy of 126.5GeV with an accuracy of close to five sigma (99.9999%, which is accepted as definitive). The Higgs mechanism giving mass to other particles has not been observed yet.

Hypothetical particles

Supersymmetric theories predict the existence of more particles, none of which have been confirmed experimentally as of 2014:

Note: just as the photon, Z boson and $W^{*}\pm$ bosons are superpositions of the B^0, W^0, W^1, and W^2 fields – the photino, zino, and $wino^{*}\pm$ are superpositions of the $bino^0$,

$wino^0$, $wino^1$, and $wino^2$ by definition.

No matter if one uses the original gauginos or this superpositions as a basis, the only predicted physical particles are neutralinos and charginos as a superposition of them together with the Higgsinos.

Other theories predict the existence of additional bosons:

Mirror particles are predicted by theories that restore parity symmetry.

"Magnetic monopole" is a generic name for particles with non-zero magnetic charge. They are predicted by some GUTs.

"Tachyon" is a generic name for hypothetical particles that travel faster than the speed of light and have an imaginary rest mass.

Preons were suggested as subparticles of quarks and leptons, but modern collider experiments have all but ruled out their existence.

Kaluza–Klein towers of particles are predicted by some models of extra dimensions. The extra-dimensional momentum is manifested as extra mass in four-dimensional spacetime.

5.14.2 Composite particles

Hadrons

Main article: Hadron

Hadrons are defined as strongly interacting composite particles. Hadrons are either:

- Composite fermions, in which case they are called baryons.

- Composite bosons, in which case they are called mesons.

Quark models, first proposed in 1964 independently by Murray Gell-Mann and George Zweig (who called quarks "aces"), describe the known hadrons as composed of valence quarks and/or antiquarks, tightly bound by the color force, which is mediated by gluons. A "sea" of virtual quark-antiquark pairs is also present in each hadron.

Baryons   See also: List of baryons

Ordinary baryons (composite fermions) contain three valence quarks or three valence antiquarks each.

- Nucleons are the fermionic constituents of normal atomic nuclei:

    - Protons, composed of two up and one down quark (uud)

    - Neutrons, composed of two down and one up quark (ddu)

- Hyperons, such as the Λ, Σ, Ξ, and Ω particles, which contain one or more strange quarks, are short-lived and heavier than nucleons. Although not normally present in atomic nuclei, they can appear in short-lived hypernuclei.

- A number of charmed and bottom baryons have also been observed.

Some hints at the existence of exotic baryons have been found recently; however, negative results have also been reported. Their existence is uncertain.

- Pentaquarks consist of four valence quarks and one valence antiquark.

Mesons   See also: List of mesons

Ordinary mesons are made up of a valence quark and a valence antiquark. Because mesons have spin of 0 or 1 and are not themselves elementary particles, they are "composite" bosons. Examples of mesons include the pion, kaon, and the J/ψ. In quantum hydrodynamic models, mesons mediate the residual strong force between nucleons.

At one time or another, positive signatures have been reported for all of the following exotic mesons but their existences have yet to be confirmed.

- A tetraquark consists of two valence quarks and two valence antiquarks;

- A glueball is a bound state of gluons with no valence quarks;

- Hybrid mesons consist of one or more valence quark-antiquark pairs and one or more real gluons.

Atomic nuclei

Atomic nuclei consist of protons and neutrons. Each type of nucleus contains a specific number of protons and a specific number of neutrons, and is called a "nuclide" or "isotope". Nuclear reactions can change one nuclide into another. See table of nuclides for a complete list of isotopes.

Atoms

Atoms are the smallest neutral particles into which matter can be divided by chemical reactions. An atom consists of a small, heavy nucleus surrounded by a relatively large, light cloud of electrons. Each type of atom corresponds to a specific chemical element. To date, 118 elements have been discovered, while only the elements 1-112,114, and 116 have received official names.

The atomic nucleus consists of protons and neutrons. Protons and neutrons are, in turn, made of quarks.

Molecules

Molecules are the smallest particles into which a non-elemental substance can be divided while maintaining the physical properties of the substance. Each type of molecule corresponds to a specific chemical compound. Molecules are a composite of two or more atoms. See list of compounds for a list of molecules.

5.14.3   Condensed matter

The field equations of condensed matter physics are remarkably similar to those of high energy particle physics. As a result, much of the theory of particle physics applies to condensed matter physics as well; in particular, there are a selection of field excitations, called quasi-particles, that can be created and explored. These include:

- Phonons are vibrational modes in a crystal lattice.

- Excitons are bound states of an electron and a hole.

- Plasmons are coherent excitations of a plasma.

- Polaritons are mixtures of photons with other quasi-particles.

- Polarons are moving, charged (quasi-) particles that are surrounded by ions in a material.

- Magnons are coherent excitations of electron spins in a material.

5.14.4   Other

- An anyon is a generalization of fermion and boson in two-dimensional systems like sheets of graphene that obeys braid statistics.

- A plekton is a theoretical kind of particle discussed as a generalization of the braid statistics of the anyon to dimension > 2.

- A WIMP (weakly interacting massive particle) is any one of a number of particles that might explain dark matter (such as the neutralino or the axion).

- The pomeron, used to explain the elastic scattering of hadrons and the location of Regge poles in Regge theory.

- The skyrmion, a topological solution of the pion field, used to model the low-energy properties of the nucleon, such as the axial vector current coupling and the mass.

- A genon is a particle existing in a closed timelike world line where spacetime is curled as in a Frank Tipler or Ronald Mallett time machine.

- A goldstone boson is a massless excitation of a field that has been spontaneously broken. The pions are quasi-goldstone bosons (quasi- because they are not exactly massless) of the broken chiral isospin symmetry of quantum chromodynamics.

- A goldstino is a goldstone fermion produced by the spontaneous breaking of supersymmetry.

- An instanton is a field configuration which is a local minimum of the Euclidean action. Instantons are used in nonperturbative calculations of tunneling rates.

- A dyon is a hypothetical particle with both electric and magnetic charges.

- A geon is an electromagnetic or gravitational wave which is held together in a confined region by the gravitational attraction of its own field energy.

- An inflaton is the generic name for an unidentified scalar particle responsible for the cosmic inflation.

- A spurion is the name given to a "particle" inserted mathematically into an isospin-violating decay in order to analyze it as though it conserved isospin.

- What is called "true muonium", a bound state of a muon and an antimuon, is a theoretical exotic atom which has never been observed.

5.14.5 Classification by speed

- A tardyon or bradyon travels slower than light and has a non-zero rest mass.

- A luxon travels at the speed of light and has no rest mass.

- A tachyon (mentioned above) is a hypothetical particle that travels faster than the speed of light and has an imaginary rest mass.

5.14.6 See also

- Acceleron

- List of baryons

- List of compounds for a list of molecules.

- List of fictional elements, materials, isotopes and atomic particles

- List of mesons

- Periodic table for an overview of atoms.

- Standard Model for the current theory of these particles.

- Table of nuclides

- Timeline of particle discoveries

5.14.7 References

[1] Observation of a new boson at a mass of 125 GeV with the CMS experiment at the LHC (2013). *arXiv:1207.7235.*

[2] Observation of a new particle in the search for the Standard Model Higgs boson with the ATLAS detector at the LHC (2012). *arXiv:1207.7214.*

[3] B. Kayser, *Two Questions About Neutrinos*, arXiv:1012.4469v1 [hep-ph] (2010).

[4] R. Maartens (2004). *Brane-World Gravity* (PDF). *Living Reviews in Relativity* **7**. p. 7. Also available in web format at http://www.livingreviews.org/lrr-2004-7.

- C. Amsler *et al.* (Particle Data Group) (2008). "Review of Particle Physics". *Physics Letters B* **667** (1–5): 1. Bibcode:2008PhLB..667....1P. doi:10.1016/j.physletb.2008.07.018. *(All information on this list, and more, can be found in the extensive, biannually-updated review by the Particle Data Group)*

5.15 Timeline of particle discoveries

This is a **timeline of subatomic particle discoveries**, including all particles thus far discovered which appear to be elementary (that is, indivisible) given the best available evidence. It also includes the discovery of composite particles and antiparticles that were of particular historical importance.

More specifically, the inclusion criteria are:

- Elementary particles from the Standard Model of particle physics that have so far been observed. The Standard Model is the most comprehensive existing model of particle behavior. All Standard Model particles including the Higgs boson have been verified, and all other observed particles are combinations of two or more Standard Model particles.

- Antiparticles which were historically important to the development of particle physics, specifically the positron and antiproton. The discovery of these particles required very different experimental methods from that of their ordinary matter counterparts, and provided evidence that *all* particles had antiparticles —an idea that is fundamental to quantum field theory, the modern mathematical framework for particle physics. In the case of most subsequent particle discoveries, the particle and its anti-particle were discovered essentially simultaneously.

- Composite particles which were the first particle discovered containing a particular elementary constituent, or whose discovery was critical to the understanding of particle physics.

5.15.1   See also

- List of mesons

- List of baryons

- List of particles

- physics

5.15.2   References

[1] Hockberger, P. E. (2002). "A history of ultraviolet photobiology for humans, animals and microorganisms" . *Photochem. Photobiol.* **76** (6): 561–579. doi:10.1562/0031-8655(2002)0760561AHOUPF2.0.CO2. ISSN 0031-8655. PMID 12511035.

[2] The ozone layer protects humans from this. Lyman, T. (1914). "Victor Schumann" . *Astrophysical Journal* **38**: 1–4. Bibcode:1914ApJ....39....1L. doi:10.1086/142050.

[3] W.C. Röntgen (1895). "Über ein neue Art von Strahlen. Vorlaufige Mitteilung" . *Sitzber. Physik. Med. Ges.* **137**: 1. as translated in A. Stanton (1896). "On a New Kind of Rays" . *Nature* **53** (1369): 274–276. Bibcode:1896Natur..53R.274.. doi:10.1038/053274b0.

[4] J.J. Thomson (1897). "Cathode Rays" . *Philosophical Magazine* **44** (269): 293–316. doi:10.1080/14786449708621070.

[5] E. Rutherford (1899). "Uranium Radiation and the Electrical Conduction Produced by it" . *Philosophical Magazine* **47** (284): 109–163. doi:10.1080/14786449908621245.

[6] P. Villard (1900). "Sur la Réflexion et la Réfraction des Rayons Cathodiques et des Rayons Déviables du Radium" . *Comptes Rendus de l'Académie des Sciences* **130**: 1010.

[7] E. Rutherford (1911). "The Scattering of α- and β- Particles by Matter and the Structure of the Atom" . *Philosophical Magazine* **21** (125): 669–688. doi:10.1080/14786440508637080.

[8] E. Rutherford (1919). "Collision of α Particles with Light Atoms IV. An Anomalous Effect in Nitrogen" . *Philosophical Magazine* **37**: 581.

[9] J. Chadwick (1932). "Possible Existence of a Neutron" . *Nature* **129** (3252): 312. Bibcode:1932Natur.129Q.312C. doi:10.1038/129312a0.

[10] E. Rutherford (1920). "Nuclear Constitution of Atoms" . *Proceedings of the Royal Society A* **97** (686): 374–400. Bibcode:1920RSPSA..97..374R. doi:10.1098/rspa.1920.0040.

[11] C.D. Anderson (1932). "The Apparent Existence of Easily Deflectable Positives" . *Science* **76** (1967): 238–9. Bibcode:1932Sci....76..238A. doi:10.1126/science.76.1967.238. PMID 17731542.

[12] S.H. Neddermeyer, C.D. Anderson (1937). "Note on the nature of Cosmic-Ray Particles" . *Physical Review* **51** (10): 884–886. Bibcode:1937PhRv...51..884N. doi:10.1103/PhysRev.51.884.

[13] M. Conversi, E. Pancini, O. Piccioni (1947). "On the Disintegration of Negative Muons" . *Physical Review* **71** (3): 209–210. Bibcode:1947PhRv...71..209C. doi:10.1103/PhysRev.71.209.

[14] C.D. Anderson (1935). "On the Interaction of Elementary Particles" . *Proceedings of the Physico-Mathematical Society of Japan* **17**: 48.

[15] G.D. Rochester, C.C. Butler (1947). "Evidence for the Existence of New Unstable Elementary Particles" . *Nature* **160** (4077): 855–857. Bibcode:1947Natur.160..855R. doi:10.1038/160855a0.

[16] The Strange Quark

[17] O. Chamberlain, E. Segrè, C. Wiegand, T. Ypsilantis (1955). "Observation of Antiprotons" . *Physical Review* **100** (3): 947–950. Bibcode:1955PhRv..100..947C. doi:10.1103/PhysRev.100.947.

[18] F. Reines, C.L. Cowan (1956). "The Neutrino" . *Nature* **178** (4531): 446–449. Bibcode:1956Natur.178..446R. doi:10.1038/178446a0.

[19] G. Danby et al. (1962). "Observation of High-Energy Neutrino Reactions and the Existence of Two Kinds of Neutrinos". *Physical Review Letters* **9** (1): 36–44. Bibcode:1962PhRvL...9...36D. doi:10.1103/PhysRevLett.9.36.

[20] R. Nave. "The Xi Baryon". Hyperphysics. Retrieved 20 June 2009.

[21] E.D. Bloom et al. (1969). "High-Energy Inelastic *e–p* Scattering at 6° and 10°". *Physical Review Letters* **23** (16): 930–934. Bibcode:1969PhRvL..23..930B. doi:10.1103/PhysRevLett.23.930.

[22] M. Breidenbach et al. (1969). "Observed Behavior of Highly Inelastic Electron-Proton Scattering". *Physical Review Letters* **23** (16): 935–939. Bibcode:1969PhRvL..23..935B. doi:10.1103/PhysRevLett.23.935.

[23] J.J. Aubert et al. (1974). "Experimental Observation of a Heavy Particle *J*". *Physical Review Letters* **33** (23): 1404–1406. Bibcode:1974PhRvL..33.1404A. doi:10.1103/PhysRevLett.33.1404.

[24] J.-E. Augustin et al. (1974). "Discovery of a Narrow Resonance in $e^+ + e^-$ Annihilation". *Physical Review Letters* **33** (23): 1406–1408. Bibcode:1974PhRvL..33.1406A. doi:10.1103/PhysRevLett.33.1406.

[25] B.J. Bjørken, S.L. Glashow (1964). "Elementary Particles and SU(4)". *Physics Letters* **11** (3): 255–257. Bibcode:1964PhL....11..255B. doi:10.1016/0031-9163(64)90433-0.

[26] M.L. Perl et al. (1975). "Evidence for Anomalous Lepton Production in $e^+ + e^-$ Annihilation". *Physical Review Letters* **35** (22): 1489–1492. Bibcode:1975PhRvL..35.1489P. doi:10.1103/PhysRevLett.35.1489.

[27] S.W. Herb et al. (1977). "Observation of a Dimuon Resonance at 9.5 GeV in 400-GeV Proton-Nucleus Collisions". *Physical Review Letters* **39** (5): 252–255. Bibcode:1977PhRvL..39..252H. doi:10.1103/PhysRevLett.39.252.

[28] D.P. Barber et al. (1979). "Discovery of Three-Jet Events and a Test of Quantum Chromodynamics at PETRA". *Physical Review Letters* **43** (12): 830–833. Bibcode:1979PhRvL..43..830B. doi:10.1103/PhysRevLett.43.830.

[29] J.J. Aubert *et al.* (European Muon Collaboration) (1983). "The ratio of the nucleon structure functions F_2^N for iron and deuterium". *Physics Letters B* **123** (3–4): 275–278. Bibcode:1983PhLB..123..275A. doi:10.1016/0370-2693(83)90437-9.

[30] G. Arnison *et al.* (UA1 collaboration) (1983). "Experimental observation of lepton pairs of invariant mass around 95 GeV/c^2 at the CERN SPS collider". *Physics Letters B* **126** (5): 398–410. Bibcode:1983PhLB..126..398A. doi:10.1016/0370-2693(83)90188-0.

[31] F. Abe *et al.* (CDF collaboration) (1995). "Observation of Top quark production in p–p Collisions with the Collider Detector at Fermilab". *Physical Review Letters* **74** (14): 2626–2631. arXiv:hep-ex/9503002. Bibcode:1995PhRvL..74.2626A. doi:10.1103/PhysRevLett.74.2626. PMID 10057978.

[32] S. Arabuchi *et al.* (D0 collaboration) (1995). "Observation of the Top Quark". *Physical Review Letters* **74** (14): 2632–2637. arXiv:hep-ex/9503003. Bibcode:1995PhRvL..74.2632A. doi:10.1103/PhysRevLett.74.2632. PMID 10057979.

[33] G. Baur et al. (1996). "Production of Antihydrogen". *Physics Letters B* **368** (3): 251–258. Bibcode:1996PhLB..368..251B. doi:10.1016/0370-2693(96)00005-6.

[34] "Physicists Find First Direct Evidence for Tau Neutrino at Fermilab" (Press release). Fermilab. 20 July 2000. Retrieved 20 March 2010.

[35] Boyle, Alan (4 July 2012). "Milestone in Higgs quest: Scientists find new particle". *MSNBC* (MSNBC). Retrieved 5 July 2012.

- V.V. Ezhela et al. (1996). *Particle Physics: One Hundred Years of Discoveries: An Annotated Chronological Bibliography*. Springer–Verlag. ISBN 1-56396-642-5.

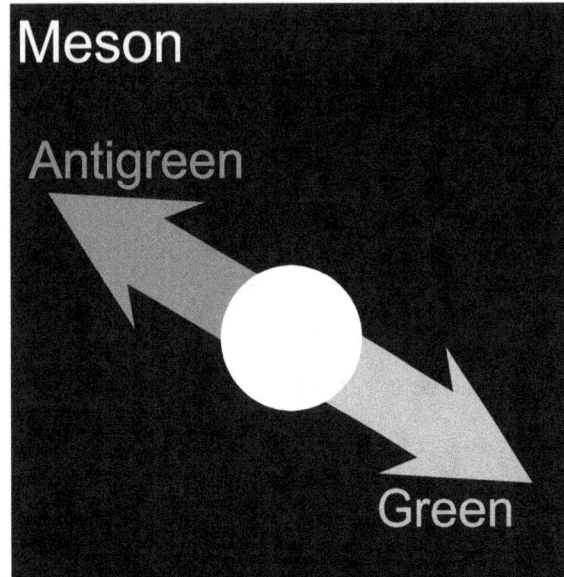

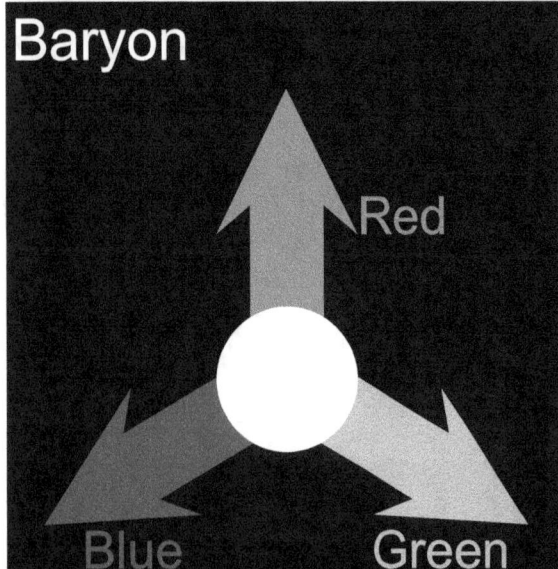

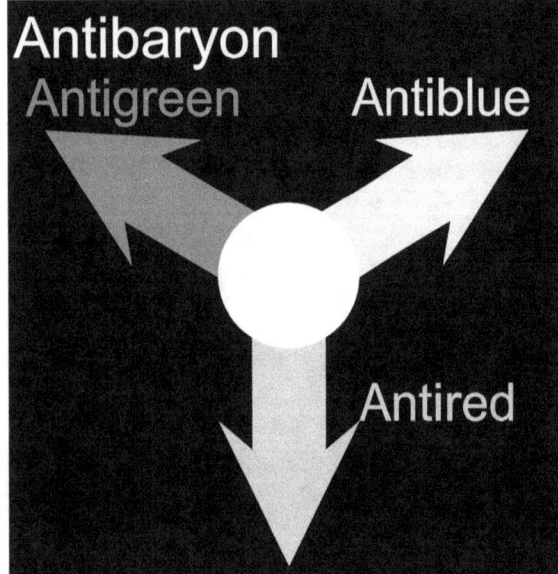

All types of hadrons have zero total color charge.

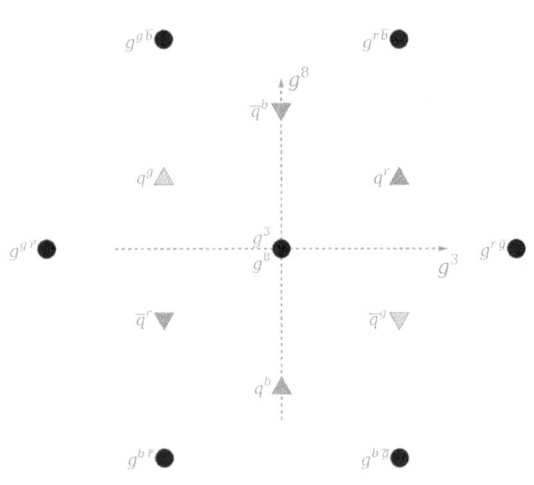

The pattern of strong charges for the three colors of quark, three antiquarks, and eight gluons (with two of zero charge overlapping).

Current quark masses for all six flavors in comparison, as balls of proportional volumes. Proton and electron (red) are shown in bottom left corner for scale

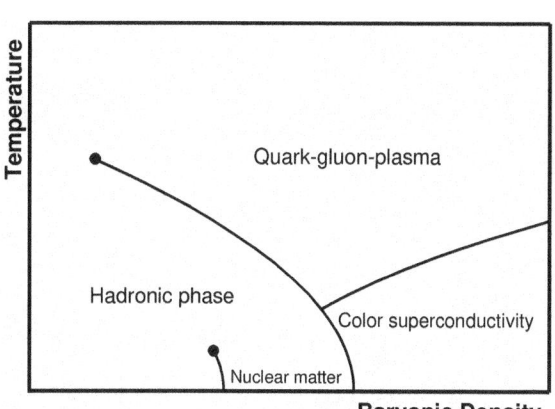

A qualitative rendering of the phase diagram of quark matter. The precise details of the diagram are the subject of ongoing research.[79]*[80]*

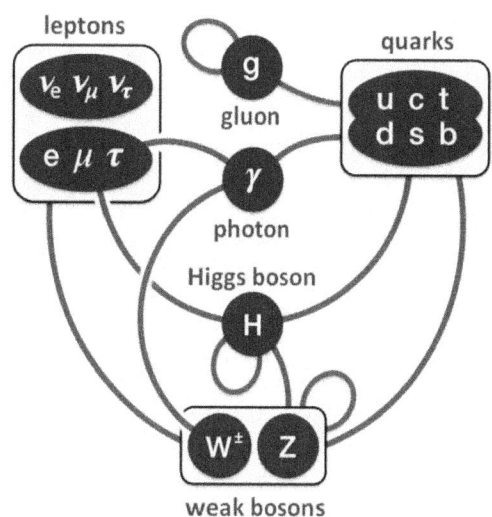

Summary of interactions between particles described by the Standard Model.

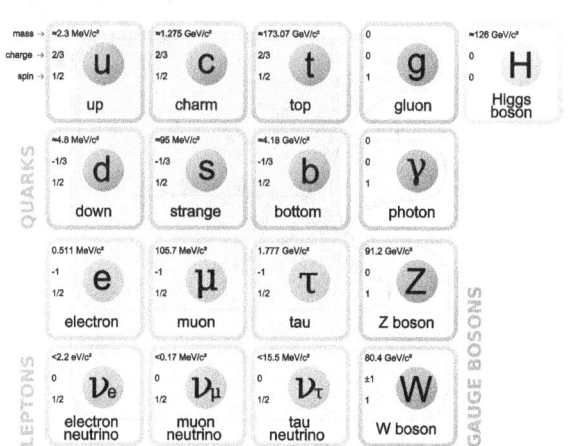

The Standard Model of elementary particles (more schematic depiction), with the three generations of matter, gauge bosons in the fourth column, and the Higgs boson in the fifth.

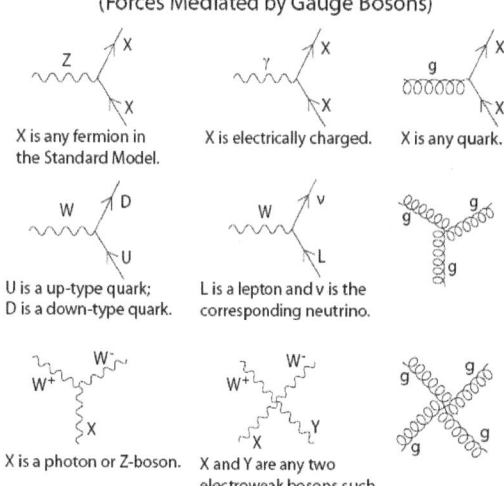

The above interactions form the basis of the standard model. Feynman diagrams in the standard model are built from these vertices. Modifications involving Higgs boson interactions and neutrino oscillations are omitted. The charge of the W bosons is dictated by the fermions they interact with; the conjugate of each listed vertex (i.e. reversing the direction of arrows) is also allowed.

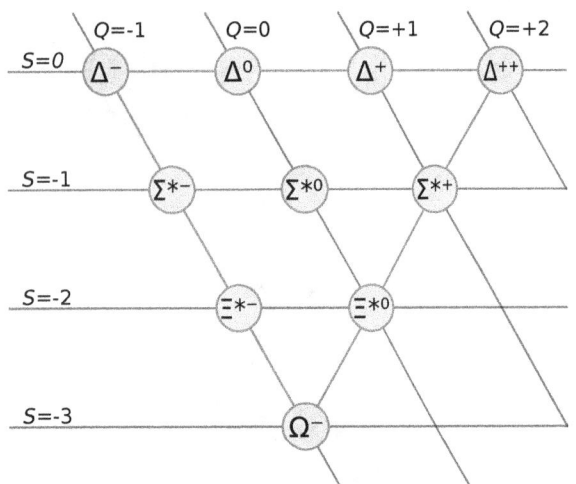

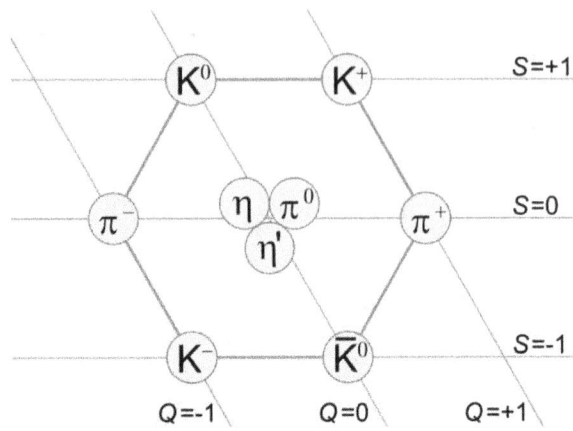

Mesons of spin 0 form a nonet

A combination of three u, d or s-quarks with a total spin of $^3/_2$ form the so-called "baryon decuplet".

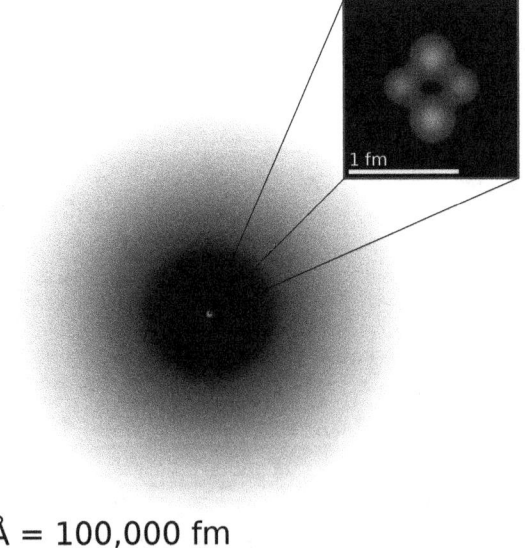

1 Å = 100,000 fm

Proton quark structure: 2 up quarks and 1 down quark. The gluon tubes or flux tubes are now known to be Y shaped.

A semi-accurate depiction of the helium atom. In the nucleus, the protons are in red and neutrons are in purple. In reality, the nucleus is also spherically symmetrical.

Chapter 6

Appendix B – Selected Biographies

6.1 Carl D. Anderson

Carl David Anderson (September 3, 1905 – January 11, 1991) was an American physicist. He is best known for his discovery of the positron in 1932, an achievement for which he received the 1936 Nobel Prize in Physics, and of the muon in 1936.

6.1.1 Biography

Anderson was born in New York City, the son of Swedish immigrants. He studied physics and engineering at Caltech (B.S., 1927; Ph.D., 1930). Under the supervision of Robert A. Millikan, he began investigations into cosmic rays during the course of which he encountered unexpected particle tracks in his (modern versions now commonly referred to as an Anderson) cloud chamber photographs that he correctly interpreted as having been created by a particle with the same mass as the electron, but with opposite electrical charge. This discovery, announced in 1932 and later confirmed by others, validated Paul Dirac's theoretical prediction of the existence of the positron. Anderson first detected the particles in cosmic rays. He then produced more conclusive proof by shooting gamma rays produced by the natural radioactive nuclide ThC" (^{208}Tl)*[1] into other materials, resulting in the creation of positron-electron pairs. For this work, Anderson shared the 1936 Nobel Prize in Physics with Victor Hess.*[2]

Also in 1936, Anderson and his first graduate student, Seth Neddermeyer, discovered the muon (or 'mu-meson', as it was known for many years), a subatomic particle 207 times more massive than the electron, but with the same negative electric charge and spin 1/2 as the electron, again in cosmic rays. Anderson and Neddermeyer at first believed that they had seen the pion, a particle which Hideki Yukawa had postulated in his theory of the strong interaction. When it became clear that what Anderson had seen was *not* the pion, the physicist I. I. Rabi, puzzled as to how the unexpected discovery could fit into any logical scheme of particle

physics, quizzically asked "Who ordered *that*?" (sometimes the story goes that he was dining with colleagues at a Chinese restaurant at the time). The muon was the first of a long list of subatomic particles whose discovery initially baffled theoreticians who could not make the confusing "zoo" fit into some tidy conceptual scheme. Willis Lamb, in his 1955 Nobel Prize Lecture, joked that he had heard it said that "the finder of a new elementary particle used to be rewarded by a Nobel Prize, but such a discovery now ought to be punished by a 10,000 dollar fine." *[3]

Anderson spent all of his academic and research career at Caltech. During World War II, he conducted research in rocketry there. He was elected a Fellow of the American Academy of Arts and Sciences in 1950.*[4] He died on January 11, 1991, and his remains were interred in the Forest Lawn, Hollywood Hills Cemetery in Los Angeles, California. His wife Lorraine died in 1984.

6.1.2 Publications

- C.D. Anderson (1933). "The Positive Electron". *Physical Review* **43** (6): 491. Bibcode:1933PhRv...43..491A. doi:10.1103/PhysRev.43.491.

- C.D. Anderson (1932). "The Apparent Existence of Easily Deflectable Positives". *Science* **76** (1967): 238–9. Bibcode:1932Sci....76..238A. doi:10.1126/science.76.1967.238. PMID 17731542.

- C.D. Anderson (1957), Technical adviser to *The Strange Case of the Cosmic Rays,* Bell System Science Film, Directed by Frank Capra, Frank Capra Productions.

6.1.3 References

[1] ThC" is a historical designation of ^{208}Tl, see Decay chains

[2] Physics 1936

149

[3] http://nobelprize.org/nobel_prizes/physics/laureates/1955/lamb-lecture.pdf

[4] "Book of Members, 1780-2010: Chapter A" (PDF). American Academy of Arts and Sciences. Retrieved 17 April 2011.

6.1.4   External links

- "Carl David Anderson". Find a Grave. Retrieved August 10, 2010.

- Weisstein, Eric W., *Anderson, Carl (1905-1991)* from ScienceWorld.

- Annotated bibliography for Carl David Anderson from the Alsos Digital Library for Nuclear Issues

- American National Biography, vol. 1, pp. 445–446.

- Oral History interview transcript with Carl D. Anderson 30 June 1966, American Institute of Physics, Niels Bohr Library and Archives

- National Academy of Sciences Biographical Memoir

6.2   Clyde Cowan

Clyde Lorrain Cowan Jr (December 6, 1919 in Detroit, Michigan – May 24, 1974 in Bethesda, Maryland) was an American physicist, the co-discoverer of the neutrino along with Frederick Reines. The discovery was made in 1956 in the neutrino experiment.[*][1] Frederick Reines received the Nobel Prize in Physics in 1995 in both their names.

6.2.1   Early life

Born the oldest of four children in Detroit, Michigan, Cowan's family moved to St. Louis, Missouri, where he began his education attending public schools. While attending the Missouri School of Mines and Metallurgy in Rolla, Missouri, Cowan was Editor-in-Chief of the Missouri Miner newspaper from 1939–1940, and graduated in 1940 with a B.S. in Chemical Engineering.

6.2.2   Military career

Cowan was a captain in the United States Army Air Forces, where he earned a bronze star in World War II.

From 1936-1940 he was in the Reserve Officers' Training Corps. Cowan joined the U.S. Army Chemical Warfare Service with the rank of Second Lieutenant when America joined World War II in 1941. In August 1942, he was transferred to Eisenhower's Eighth Air Force stationed in London, England. In 1943 he designed and built an experimental cleaning unit to be used in case of gas attack. In the following year, he joined the staff of the British Branch of the Radiation Laboratory of the Massachusetts Institute of Technology, which was located in Great Malvern, England. In 1945 he was a liaison officer with the Royal Air Force, working to expedite transmittal of technical information and equipment. He returned to the United States in 1945, and worked at Wright Patterson Air Force Base in Dayton, Ohio. He left active duty in 1946.

6.2.3   Academic career

Benefitting from the G.I. Bill, Cowan attended Washington University in St. Louis, Missouri, receiving a Masters Degree, and a Ph.D. in 1949. He then joined the staff of the Los Alamos Scientific Laboratory in New Mexico, where he met Frederick Reines.

In 1951 Reines and Cowan began their search for the neutrino. Their work was completed at the Savannah River Plant in Aiken, South Carolina, in 1956.

Cowan began his teaching career in 1957 as a Professor of Physics at George Washington University in Washington, D.C.. The following year he left GWU and joined the faculty of The Catholic University of America in Washington, D.C., a post he held until the end of his life. He also acted at various times as a consultant to the U.S. Atomic Energy Commission (AEC), US Naval Ordnance Laboratory, the United States Naval Academy, the United States Army, United Mine Workers of America, Electric Boat Co., and the Smithsonian Institution, Washington, D.C.

Cowan died in Bethesda, Maryland on May 24, 1974, and was buried in Arlington National Cemetery.

6.2.4   Family

Cowan was married in Woodford, England, January 29, 1943 to Betty Eleanor, daughter of George Henry and Mabel Jane (Mather) Dunham of Wanstead, England, and has three surviving children: Elizabeth Esthermay, who married John A. Riordon; Marian Jane, who married Charles M. Kriston; and George Langstroth, who married Justine Allen, then Kim Borkowitz. Seven other children died in infancy, and he had two adopted sons: David Lorrain (died in childhood) and Michael Lorrain. His family has blossomed to include 11 grandchildren and 8 great-grandchildren.

His grandson James Riordon, a former physicist and engineer who heads the American Physical Society media relations office, initially conceived of the distributed computing project Einstein@home, which searches gravitational wave data for signals from massive rotating objects such as pulsars. His granddaughter Barbara Riordon Maher earned her MS in Emergency Management in 2006, and is a PhD candidate in Public Health Administration. She is currently the Emergency Preparedness Coordinator for Dare County North Carolina Department of Public Health. She is a retired Maryland State Police Trooper/Flight Paramedic. Her publications include articles in plant physiology as well as medical nuclear, biological, chemical, and explosive response for the Department of Defense. She has obtained the rank of Major in the Maryland Army National Guard and serves as the Operations Officer for the JFHQ Medical Detachment. Her two sons, Joseph and Patrick, are college students seeking degrees in biology and emergency services. Joseph is currently serving in the Maryland Army National Guard as a 15P, Aviation Operations Specialist.

Cowan was a direct descendant of L. L. Langstroth, the "Father of Modern Beekeeping", and a distant relative of Katherine Drexel, a Catholic saint.

A biography can be found in The National Cyclopedia of American Biography Vol. 58 published by James T. White & Company Clifton, New Jersey, 1979.

6.2.5 References

[1] Reines, Frederick (August 1974). "Clyde L. Cowan Jr". *Physics Today* **27** (8): 68–69. Bibcode:1974PhT....27h..68R. doi:10.1063/1.3128835.

6.2.6 External links

- Neutrino history
- Cowan Reines Neutrino Experiment
- Nobel Prize in Physics 1995 Press Release
- The Neutrino with Dr. Clyde L. Cowan (Lecture on Nobel Prize–winning experiment) on YouTube

6.3 Raymond Davis, Jr.

Raymond (Ray) Davis, Jr. (October 14, 1914 – May 31, 2006) was an American chemist, physicist, and Nobel Prize in Physics laureate.[*][3]

6.3.1 Early life and education

Davis was born in Washington, D.C., where his father was a photographer for the National Bureau of Standards. He

spent several years as a choirboy to please his mother, although he could not carry a tune. He enjoyed attending the concerts at the Watergate before air traffic was loud enough to drown out the music. His brother Warren, 14 months younger than he, was his constant companion in boyhood. He graduated in chemistry from the University of Maryland in 1938. He also received a master's degree from that school and a Ph.D. from Yale University in physical chemistry in 1942.

6.3.2 Career

Davis spent most of the war years at Dugway Proving Ground, Utah observing the results of chemical weapons tests and exploring the Great Salt Lake basin for evidence of its predecessor, Lake Bonneville.

Upon his discharge from the army in 1946, Davis went to work at Monsanto's Mound Laboratory, in Miamisburg, Ohio, doing applied radiochemistry of interest to the United States Atomic Energy Commission. In 1948, he joined Brookhaven National Laboratory, which was dedicated to finding peaceful uses for nuclear power.

Davis reports that he was asked "to find something interesting to work on," and dedicated his career to the study of neutrinos, particles which had been predicted to explain the process of beta decay, but whose separate existence had not been confirmed. Davis investigated the detection of neutrinos by beta decay, the process by which a neutrino brings enough energy to a nucleus to make certain stable isotopes into radioactive ones. Since the rate for this process is very low, the number of radioactive atoms created in neutrino experiments is very small, and Davis began investigating the rates of processes other than beta decay that would mimic the signal of neutrinos. Using barrels and tanks of carbon tetrachloride as detectors, Davis characterized the rate of the production of ^{37}Argon as a function of altitude and as a function of depth underground. He deployed a detector containing chlorine atoms at the Brookhaven Reactor in 1954 and later one of the reactors at Savannah River. These experiments failed to detect a surplus of radioactive argon when the reactors were operating over when the reactors were shut down, and this was taken as the first experimental evidence that neutrinos causing the chlorine reaction, and antineutrinos produced in reactors, were distinct. Detecting neutrinos proved considerably more difficult than not detecting antineutrinos. Davis was the lead scientist behind the Homestake Experiment, the large-scale radiochemical neutrino detector which first detected evidence of neutrinos from the sun.

He shared the Nobel Prize in Physics in 2002 with Japanese physicist Masatoshi Koshiba and American Riccardo Giacconi for pioneering contributions to astrophysics, in particular for the detection of cosmic neutrinos, looking at the solar neutrino problem in the Homestake Experiment. He was 88 years old when awarded the prize.

6.3.3 Personal life

Davis met his wife Anna Torrey at Brookhaven and together they built a 21-foot wooden sailboat, the *Halcyon*. They had five children and lived in the same house in Blue Point, New York for over 50 years. He died in Blue Point, New York from Alzheimer's Disease.[*][1][*][2]

6.3.4 Honours and awards

Davis receiving the Medal of Science from President Bush, with OSTP Director Marburger on the left

- Comstock Prize in Physics of the National Academy of Sciences (1978)[*][4]

- Tom W. Bonner Prize of the American Physical Society (1988)

- W. K. H. Panofsky Prize of the American Physical Society (1992)

- Beatrice M. Tinsley Prize of the American Astronomical Society (1994)

- George Ellery Hale Prize of the American Astronomical Society (1996)

- Wolf Prize in Physics (2000)

- National Medal of Science (2001)[*][5]

- Nobel Prize in Physics (2002)

- Benjamin Franklin Medal (2003)

6.3.5 Notable works

- Davis, Raymond, Jr. (1953). "Attempt to detect the Antineutrinos from a Nuclear Reactor by the ^{37}Cl (v, e*−) ^{37}Ar Reaction". *Physical Review* **97** (3): 766. Bibcode:1955PhRv...97..766D. doi:10.1103/PhysRev.97.766. – Non-detection of antineutrinos with chlorine

- Davis, Raymond, Jr. (1964). "Solar Neutrinos II, Experimental". *Physical Review Letters* **12** (11): 303. Bibcode:1964PhRvL..12..303D. doi:10.1103/PhysRevLett.12.303. – Proposal for Homestake Experiment

- Cleveland, B. T. et al. (1998). "Measurement of the solar electron neutrino flux with the Homestake chlorine detector". *Astrophysical Journal* **496**: 505–526. Bibcode:1998ApJ...496..505C. doi:10.1086/305343. – final results of Homestake Experiment

6.3.6 Other publications

- Davis, R. Jr. & D. S. Harmer. "Solar Neutrinos", Brookhaven National Laboratory (BNL), (December 1964).

- Davis, R. Jr. "Search for Neutrinos from the Sun", Brookhaven National Laboratory (BNL), United States Department of Energy (through predecessor agency the Atomic Energy Commission), (1968).

- Davis, R. Jr. & J.C. Evans, Jr. "Report on the Brookhaven Solar Neutrino Experiment", Brookhaven National Laboratory (BNL), (September 22, 1976).

- Davis, R. Jr., Evans, J. C. & B. T. Cleveland. "Solar Neutrino Problem", Brookhaven National Laboratory (BNL), (April 28, 1978).

- Davis, R. Jr., Cleveland, B. T. & J. K. Rowley. "Variations in the Solar Neutrino Flux", Department of Astronomy and Astrophysics at University of Pennsylvania, Los Alamos National Laboratory (LANL), Brookhaven National Laboratory (BNL), (August 2, 1987).

6.3.7 References

[1] Kenneth Chang (2 June 2006). "Raymond Davis Jr., Nobelist Who Caught Neutrinos, Dies at 91". *The New York Times*. Retrieved 2007-10-10.

[2] David B. Caruso (2 June 2006). "Raymond Davis, who detected elusive solar particles, dies at 91". *The Boston Globe*. Retrieved 2007-10-10.

[3] Lande, Kenneth (October 2006). "Obituary: Raymond Davis Jr". *Physics Today* **59** (10): 78–80. Bibcode:2006PhT....59j..78L. doi:10.1063/1.2387099.

[4] "Comstock Prize in Physics". National Academy of Sciences. Retrieved 13 February 2011.

[5] National Science Foundation – The President's National Medal of Science

6.3.8 External links

- Photograph, Biography and Bibliographic Resources, from the Office of Scientific and Technical Information, United States Department of Energy

- Raymond Davis, Jr. biography at the Nobel Foundation

- Raymond Davis Jr., Brookhaven National Lab Web site

- Neutrino web at PBS NOVA

- The Raymond Davis Scholarship Society for Imaging Science and Technology

6.4 Paul Dirac

"Dirac" redirects here. For other uses, see Dirac (disambiguation).

Paul Adrien Maurice Dirac OM FRS*[2] (/dɪˈræk/ *di-RAK*; 8 August 1902 – 20 October 1984) was an English theoretical physicist who made fundamental contributions to the early development of both quantum mechanics and quantum electrodynamics. He was the Lucasian Professor of Mathematics at the University of Cambridge, a member of the Center for Theoretical Studies, University of Miami, and spent the last decade of his life at Florida State University.

Among other discoveries, he formulated the Dirac equation, which describes the behaviour of fermions and predicted the existence of antimatter. Dirac shared the Nobel Prize in Physics for 1933 with Erwin Schrödinger, "for the discovery of new productive forms of atomic theory".*[3] He also did work that forms the basis of modern attempts to reconcile general relativity with quantum mechanics.

He was regarded by his friends and colleagues as unusual in character. Albert Einstein said of him, "*This balancing on the dizzying path between genius and madness is awful*".*[4] His mathematical brilliance, however, means he is regarded as one of the most significant physicists of the 20th century.

6.4.1 Personal life

Early years

Paul Adrien Maurice Dirac was born at his parents' home in Bristol, England, on 8 August 1902,*[5] and grew up in the Bishopston area of the city.*[6] His father, Charles Adrien Ladislas Dirac, was an immigrant from Saint-Maurice, Switzerland, who worked in Bristol as a French teacher. His mother, Florence Hannah Dirac, née Holten, the daughter of a ship's captain, was born in Cornwall, England, and worked as a librarian at the Bristol Central Library. Paul had a younger sister, Béatrice Isabelle Marguerite, known as Betty, and an older brother, Reginald Charles Félix, known as Felix,*[7]*[8] who committed suicide in March 1925.*[9] Dirac later recalled: "My parents were terribly distressed. I didn't know they cared so much [...] I never knew that parents were supposed to care for their children, but from then on I knew." *[10]

Charles and the children were officially Swiss nationals until they became naturalised on 22 October 1919.*[11] Dirac's father was strict and authoritarian, although he disapproved of corporal punishment.*[12] Dirac had a strained relationship with his father, so much so that after his father's death, Dirac wrote, "I feel much freer now, and I am my own man." Charles forced his children to speak to him only in French, in order that they learn the language. When Dirac found that he could not express what he wanted to say in French, he chose to remain silent.*[13]*[14]

Education

Dirac was educated first at Bishop Road Primary School*[15] and then at the all-boys Merchant Venturers' Technical College (later Cotham School), where his father was a French teacher.*[16] The school was an institution attached to the University of Bristol, which shared grounds and staff.*[17] It emphasised technical subjects like bricklaying, shoemaking and metal work, and modern languages.*[18] This was unusual at a time when secondary education in Britain was still dedicated largely to the classics, and something for which Dirac would later express gratitude.*[17]

Dirac studied electrical engineering on a City of Bristol University Scholarship at the University of Bristol's engineering faculty, which was co-located with the Merchant Venturers' Technical College.*[19] Shortly before he completed his degree in 1921, he sat the entrance examination for St John's College, Cambridge. He passed, and was awarded a £70 scholarship, but this fell short of the amount of money required to live and study at Cambridge. Despite his having graduated with a first class honours Bachelor of Science degree in engineering, the economic climate of the post-war depression was such that he was unable to find work as an engineer. Instead he took up an offer to study for a Bachelor of Arts degree in mathematics at the University of Bristol free of charge. He was permitted to skip the first year of the course owing to his engineering degree.[*][20]

In 1923, Dirac graduated, once again with first class honours, and received a £140 scholarship from the Department of Scientific and Industrial Research.[*][21] Along with his £70 scholarship from St John's College, this was enough to live at Cambridge. There, Dirac pursued his interests in the theory of general relativity, an interest he had gained earlier as a student in Bristol, and in the nascent field of quantum physics, under the supervision of Ralph Fowler.[*][22] From 1925 to 1928 he held an 1851 Research Fellowship from the Royal Commission for the Exhibition of 1851.[*][23] He completed his PhD in June 1926 with the first thesis on quantum mechanics to be submitted anywhere.[*][24] He then continued his research in Copenhagen and Göttingen.[*][23]

Family

Paul Dirac with his wife in Copenhagen, July 1963

Dirac married Margit Wigner (Eugene Wigner's sister), in 1937. He adopted Margit's two children, Judith and Gabriel. Paul and Margit Dirac had two children together, both daughters, Mary Elizabeth and Florence Monica.

Margit, known as Manci, visited her brother in 1934 in Princeton, New Jersey, from her native Hungary and, while at dinner at the Annex Restaurant met the "lonely-looking man at the next table." This account from a Korean physi-

cist, Y. S. Kim, who met and was influenced by Dirac, also says: "It is quite fortunate for the physics community that Manci took good care of our respected Paul A. M. Dirac. Dirac published eleven papers during the period 1939–46.... Dirac was able to maintain his normal research productivity only because Manci was in charge of everything else." [*][25]

Personality

Dirac was known among his colleagues for his precise and taciturn nature. His colleagues in Cambridge jokingly defined a unit of a "dirac", which was one word per hour.[*][26] When Niels Bohr complained that he did not know how to finish a sentence in a scientific article he was writing, Dirac replied, "I was taught at school never to start a sentence without knowing the end of it." [*][27] He criticised the physicist J. Robert Oppenheimer's interest in poetry: "The aim of science is to make difficult things understandable in a simpler way; the aim of poetry is to state simple things in an incomprehensible way. The two are incompatible." [*][28]

Dirac himself wrote in his diary during his postgraduate years that he concentrated solely on his research, and stopped only on Sunday, when he took long strolls alone.[*][29]

An anecdote recounted in a review of the 2009 biography tells of Werner Heisenberg and Dirac sailing on an ocean liner to a conference in Japan in August 1929. "Both still in their twenties, and unmarried, they made an odd couple. Heisenberg was a ladies' man who constantly flirted and danced, while Dirac—'an Edwardian geek', as biographer Graham Farmelo puts it—suffered agonies if forced into any kind of socialising or small talk. 'Why do you dance?' Dirac asked his companion. 'When there are nice girls, it is a pleasure,' Heisenberg replied. Dirac pondered this notion, then blurted out: 'But, Heisenberg, how do you know beforehand that the girls are nice?'" [*][30]

According to a story told in different versions, a friend or student visited Dirac, not knowing of his marriage. Noticing the visitor's surprise at seeing an attractive woman in the house, Dirac said, "This is... this is Wigner's sister". Margit Dirac told both George Gamow and Anton Capri in the 1960s that her husband had actually said, "Allow me to present Wigner's sister, who is now my wife." [*][31][*][32]

Another story told of Dirac is that when he first met the young Richard Feynman at a conference, he said after a long silence, "I have an equation. Do you have one too?".[*][33]

After he presented a lecture at a conference, one colleague raised his hand and said "I don't understand the equation on the top-right-hand corner of the blackboard" . After a long silence, the moderator asked Dirac if he wanted to answer the question, to which Dirac replied "That was not a question, it was a comment." *[34]*[35]

Dirac was also noted for his personal modesty. He called the equation for the time evolution of a quantum-mechanical operator, which he was the first to write down, the "Heisenberg equation of motion". Most physicists speak of Fermi–Dirac statistics for half-integer-spin particles and Bose–Einstein statistics for integer-spin particles. While lecturing later in life, Dirac always insisted on calling the former "Fermi statistics". He referred to the latter as "Einstein statistics" for reasons, he explained, of "symmetry".

Religious views

Heisenberg recollected a conversation among young participants at the 1927 Solvay Conference about Einstein and Planck's views on religion between Wolfgang Pauli, Heisenberg and Dirac. Dirac's contribution was a criticism of the political purpose of religion, which was much appreciated for its lucidity by Bohr when Heisenberg reported it to him later. Among other things, Dirac said:

> I cannot understand why we idle discussing religion. If we are honest—and scientists have to be—we must admit that religion is a jumble of false assertions, with no basis in reality. The very idea of God is a product of the human imagination. It is quite understandable why primitive people, who were so much more exposed to the overpowering forces of nature than we are today, should have personified these forces in fear and trembling. But nowadays, when we understand so many natural processes, we have no need for such solutions. I can't for the life of me see how the postulate of an Almighty God helps us in any way. What I do see is that this assumption leads to such unproductive questions as why God allows so much misery and injustice, the exploitation of the poor by the rich and all the other horrors He might have prevented. If religion is still being taught, it is by no means because its ideas still convince us, but simply because some of us want to keep the lower classes quiet. Quiet people are much easier to govern than clamorous and dissatisfied ones. They are also much easier to exploit. Religion is a kind of opium that allows a nation to lull itself into wishful dreams and so forget the injustices that are being perpetrated against the people. Hence the close alliance between those two great political forces, the State and the Church. Both need the illusion

that a kindly God rewards—in heaven if not on earth—all those who have not risen up against injustice, who have done their duty quietly and uncomplainingly. That is precisely why the honest assertion that God is a mere product of the human imagination is branded as the worst of all mortal sins.*[36]

Heisenberg's view was tolerant. Pauli, raised as a Catholic, had kept silent after some initial remarks, but when finally he was asked for his opinion, said: "Well, our friend Dirac has got a religion and its guiding principle is 'There is no God and Paul Dirac is His prophet.'" Everybody, including Dirac, burst into laughter.*[37]*[38]

Later in life, Dirac's views towards the idea of God were less acerbic. As an author of an article appearing in the May 1963 edition of *Scientific American*, Dirac wrote:

> It seems to be one of the fundamental features of nature that fundamental physical laws are described in terms of a mathematical theory of great beauty and power, needing quite a high standard of mathematics for one to understand it. You may wonder: Why is nature constructed along these lines? One can only answer that our present knowledge seems to show that nature is so constructed. We simply have to accept it. One could perhaps describe the situation by saying that God is a mathematician of a very high order, and He used very advanced mathematics in constructing the universe. Our feeble attempts at mathematics enable us to understand a bit of the universe, and as we proceed to develop higher and higher mathematics we can hope to understand the universe better.*[39]

In 1971, at a conference meeting, Dirac expressed his views on the existence of God.*[40] Dirac explained that the existence of God could only be justified if an improbable event were to have taken place in the past:

> It could be that it is extremely difficult to start life. It might be that it is so difficult to start life that it has happened only once among all the planets. ...Let us consider, just as a conjecture, that the chance life starting when we have got suitable physical conditions is 10^{-100}. I don't have any logical reason for proposing this figure, I just want you to consider it as a possibility. Under those conditions...it is almost certain that life would not have started. And I feel that under those conditions it will be necessary to assume the existence of a god to start off life. I would

like, therefore, to set up this connexion between the existence of a god and the physical laws: if physical laws are such that to start off life involves an excessively small chance, so that it will not be reasonable to suppose that life would have started just by blind chance, then there must be a god, and such a god would probably be showing his influence in the quantum jumps which are taking place later on. On the other hand, if life can start very easily and does not need any divine influence, then I will say that there is no god.*[40]

Dirac did not commend himself to any definite view, but he described the possibilities for answering the question of God in a scientific manner.*[40]

Dirac's grave in Roselawn Cemetery, Tallahassee, Florida. Also buried is his wife Manci (Margit Wigner). Their daughter Mary Elizabeth Dirac, who died 20 January 2007, is buried next to them but not shown in the photograph.

Honours

Dirac shared the 1933 Nobel Prize for physics with Erwin Schrödinger "for the discovery of new productive forms of atomic theory".*[3] Dirac was also awarded the Royal Medal in 1939 and both the Copley Medal and the Max Planck Medal in 1952. He was elected a Fellow of the Royal Society in 1930,*[2] an Honorary Fellow of the American Physical Society in 1948, and an Honorary Fellow of the Institute of Physics, London in 1971. He received the inaugural J. Robert Oppenheimer Memorial Prize in 1969.*[41]*[42] Dirac became a member of the Order of Merit in 1973, having previously turned down a knighthood as he did not want to be addressed by his first name.*[30]*[43]

The commemorative marker in Westminster Abbey.

6.4.2   Career

Dirac established the most general theory of quantum mechanics and discovered the relativistic equation for the electron, which now bears his name. The remarkable notion of an antiparticle to each fermion particle – e.g. the positron as antiparticle to the electron – stems from his equation. He was the first to develop quantum field theory, which underlies all theoretical work on sub-atomic or "elementary" particles today, work that is fundamental to our understanding of the forces of nature. He proposed and investigated the concept of a magnetic monopole, an object not yet known empirically, as a means of bringing even greater symmetry to James Clerk Maxwell's equations of electromagnetism.

Death

In 1984, Dirac died in Tallahassee, Florida, and was buried at Tallahassee's Roselawn Cemetery.*[44]*[45] Dirac's childhood home in Bristol is commemorated with a blue plaque and the nearby Dirac Road is named in recognition of his links with the city. A commemorative stone was erected in a garden in Saint-Maurice, Switzerland, the town of origin of his father's family, on 1 August 1991. On 13 November 1995 a commemorative marker, made from Burlington green slate and inscribed with the Dirac equation, was unveiled in Westminster Abbey.*[44]*[46] The Dean of Westminster, Edward Carpenter, had initially refused permission for the memorial, thinking Dirac to be anti-Christian, but was eventually (over a five-year period) persuaded to relent.*[47]

Gravity

He quantised the gravitational field, and developed a general theory of quantum field theories with dynamical constraints, which forms the basis of the gauge theories and superstring theories of today. The influence and importance of his work has increased with the decades, and physicists daily use the concepts and equations that he developed.

Quantum theory

Dirac's first step into a new quantum theory was taken late in September 1925. Ralph Fowler, his research supervisor, had received a proof copy of an exploratory paper by Werner Heisenberg in the framework of the old quantum theory of Bohr and Sommerfeld, which leaned heavily on Bohr's correspondence principle but changed the equations so that they involved directly observable quantities. Fowler sent Heisenberg's paper on to Dirac, who was on vacation in Bristol, asking him to look into this paper carefully.

Dirac's attention was drawn to a mysterious mathematical relationship, at first sight unintelligible, that Heisenberg had reached. Several weeks later, back in Cambridge, Dirac suddenly recognised that this mathematical form had the same structure as the Poisson Brackets that occur in the classical dynamics of particle motion. From this thought he quickly developed a quantum theory that was based on non-commuting dynamical variables. This led him to a more profound and significant general formulation of quantum mechanics than was achieved by any other worker in this field.[*][48]

Dirac noticed an analogy between the Poisson brackets of classical mechanics and the recently proposed quantisation rules in Werner Heisenberg's matrix formulation of quantum mechanics. This observation allowed Dirac to obtain the quantisation rules in a novel and more illuminating manner. For this work, published in 1926, he received a PhD from Cambridge.

Dirac was famously not bothered by issues of interpretation in quantum theory. In fact, in a paper published in a book in his honor, he wrote: "The interpretation of quantum mechanics has been dealt with by many authors, and I do not want to discuss it here. I want to deal with more fundamental things." [*][49]

The Dirac equation

In 1928, building on 2×2 spin matrices which he discovered independently of Wolfgang Pauli's work on non-relativistic spin systems, (Abraham Pais quoted Dirac as saying "I believe I got these (matrices) independently of Pauli and possibly Pauli got these independently of me")[*][50] he proposed the Dirac equation as a relativistic equation of motion for the wave function of the electron.[*][51] This work led Dirac to predict the existence of the positron, the electron's antiparticle, which he interpreted in terms of what came to be called the *Dirac sea*.[*][52] The positron was observed by Carl Anderson in 1932. Dirac's equation also contributed to explaining the origin of quantum spin as a relativistic phenomenon.

The necessity of fermions (matter being created and destroyed in Enrico Fermi's 1934 theory of beta decay), however, led to a reinterpretation of Dirac's equation as a "classical" field equation for any point particle of spin $\hbar/2$, itself subject to quantisation conditions involving anticommutators. Thus reinterpreted, in 1934 by Werner Heisenberg, as a (quantum) field equation accurately describing all elementary matter particles – today quarks and leptons – this Dirac field equation is as central to theoretical physics as the Maxwell, Yang–Mills and Einstein field equations. Dirac is regarded as the founder of quantum electrodynamics, being the first to use that term. He also introduced the idea of vacuum polarisation in the early 1930s. This work was key to the development of quantum mechanics by the next generation of theorists, and in particular Schwinger, Feynman, Sin-Itiro Tomonaga and Dyson in their formulation of quantum electrodynamics.

Dirac's *Principles of Quantum Mechanics*, published in 1930, is a landmark in the history of science. It quickly became one of the standard textbooks on the subject and is still used today. In that book, Dirac incorporated the previous work of Werner Heisenberg on matrix mechanics and of Erwin Schrödinger on wave mechanics into a single mathematical formalism that associates measurable quantities to operators acting on the Hilbert space of vectors that describe the state of a physical system. The book also introduced the delta function. Following his 1939 article,[*][53] he also included the bra–ket notation in the third edition of his book,[*][54] thereby contributing to its universal use nowadays.

Magnetic monopoles

In 1933, following his 1931 paper on magnetic monopoles, Dirac showed that the existence of a single magnetic monopole in the universe would suffice to explain the observed quantisation of electrical charge. In 1975,[*][55] 1982,[*][56] and 2009[*][57][*][58][*][59] intriguing results suggested the possible detection of magnetic monopoles, but there is, to date, no direct evidence for their existence (see also Magnetic monopole#Searches for magnetic monopoles).

Lucasian Chair

Dirac was the Lucasian Professor of Mathematics at Cambridge from 1932 to 1969. In 1937, he proposed a speculative cosmological model based on the so-called large numbers hypothesis. During World War II, he conducted important theoretical and experimental research on uranium enrichment by gas centrifuge.

Dirac's quantum electrodynamics (QED) made predictions that were – more often than not – infinite and therefore unacceptable. A workaround known as renormalisation was developed, but Dirac never accepted this. "I must say that I am very dissatisfied with the situation" , he said in 1975, "because this so-called 'good theory' does involve neglecting infinities which appear in its equations, neglecting them in an arbitrary way. This is just not sensible mathematics. Sensible mathematics involves neglecting a quantity when it is small – not neglecting it just because it is infinitely great and you do not want it!"[*][60] His refusal to accept renormalisation resulted in his work on the subject moving increasingly out of the mainstream.

However, from his once rejected notes he managed to work on putting quantum electrodynamics on "logical foundations" based on Hamiltonian formalism that he formulated. He found a rather novel way of deriving the anomalous magnetic moment "Schwinger term" and also the Lamb shift, afresh in 1963, using the Heisenberg picture and without using the joining method used by Weisskopf and French, and by the two pioneers of modern QED, Schwinger and Feynman. That was two years before the Tomonaga–Schwinger–Feynman QED was given formal recognition by an award of the Nobel Prize for physics.

Weisskopf and French (FW) were the first to obtain the correct result for the Lamb shift and the anomalous magnetic moment of the electron. At first FW results did not agree with the incorrect but independent results of Feynman and Schwinger.[*][61] The 1963–1964 lectures Dirac gave on quantum field theory at Yeshiva University were published in 1966 as the Belfer Graduate School of Science, Monograph Series Number, 3. After having relocated to Florida to be near his elder daughter, Mary, Dirac spent his last fourteen years (of both life and physics research) at the University of Miami in Coral Gables, Florida, and Florida State University in Tallahassee, Florida.

In the 1950s in his search for a better QED, Paul Dirac developed the Hamiltonian theory of constraints[*][62] based on lectures that he delivered at the 1949 International Mathematical Congress in Canada. Dirac[*][63] had also solved the problem of putting the Tomonaga–Schwinger equation into the Schrödinger representation[*][64] and given explicit expressions for the scalar meson field (spin zero pion or pseudoscalar meson), the vector meson field (spin one rho meson), and the electromagnetic field (spin one massless boson, photon).

The Hamiltonian of constrained systems is one of Dirac's many masterpieces. It is a powerful generalisation of Hamiltonian theory that remains valid for curved space-time. The equations for the Hamiltonian involve only six degrees of freedom described by g_{rs} , p^{rs} for each point of the surface on which the state is considered. The g_{m0} ($m =$ 0, 1, 2, 3) appear in the theory only through the variables g^{r0} , $(-g^{00})^{-1/2}$ which occur as arbitrary coefficients in the equations of motion. There are four constraints or weak equations for each point of the surface $x^0 =$ constant. Three of them H_r form the four vector density in the surface. The fourth H_L is a 3-dimensional scalar density in the surface $H_L \approx 0; H_r \approx 0$ ($r = 1, 2, 3$)

In the late 1950s, he applied the Hamiltonian methods he had developed to cast Einstein's general relativity in Hamiltonian form[*][65] and to bring to a technical completion the quantisation problem of gravitation and bring it also closer to the rest of physics according to Salam and DeWitt. In 1959 he also gave an invited talk on "Energy of the Gravitational Field" at the New York Meeting of the American Physical Society later published in 1959 Phys Rev Lett 2, 368. In 1964 he published his *Lectures on Quantum Mechanics* (London:Academic) which deals with constrained dynamics of nonlinear dynamical systems including quantisation of curved spacetime. He also published a paper entitled "Quantization of the Gravitational Field" in the 1967 ICTP/IAEA Trieste Symposium on Contemporary Physics.

6.4.3 Students

Amongst his many students was John Polkinghorne, who recalls that Dirac "was once asked what was his fundamental belief. He strode to a blackboard and wrote that the laws of nature should be expressed in beautiful equations." [*][66]

6.4.4 Legacy

In 1975, Dirac gave a series of five lectures at the University of New South Wales which were subsequently published as a book, *Directions in Physics* (1978). He donated the royalties from this book to the university for the establishment of the Dirac Lecture Series. The Silver Dirac Medal for the Advancement of Theoretical Physics is awarded by the University of New South Wales to commemorate the lecture.[*][67]

Immediately after his death, two organisations of professional physicists established annual awards in Dirac's memory. The Institute of Physics, the United Kingdom's professional body for physicists, awards the Paul Dirac Medal for "outstanding contributions to theoretical (including mathematical and computational) physics" .[*][68] The first three recipients were Stephen Hawking (1987), John Stewart Bell (1988), and Roger Penrose (1989). The International Centre for Theoretical Physics awards the Dirac Medal of the ICTP each year on Dirac's birthday (8 August).[*][69]

The Dirac-Hellman Award at Florida State University was endowed by Dr Bruce P. Hellman in 1997 to reward outstanding work in theoretical physics by FSU researchers.[*][70] The Paul A.M. Dirac Science Library at Florida State University, which Manci opened in December 1989, is named in his honour, and his papers are held there.[*][71] Outside is a statue of him by Gabriella Bollobás.[*][72] The street on which the National High Magnetic Field Laboratory in Tallahassee, Florida, is located was named Paul Dirac Drive. As well as in his home town of Bristol, there is also a road named after him in Didcot Oxfordshire, Dirac Way. The BBC named a video codec, Dirac, in his honour. An asteroid discovered in 1983 was named after Dirac.[*][73]

6.4.5 Publications

- *The Principles of Quantum Mechanics* (1930): This book summarises the ideas of quantum mechanics using the modern formalism that was largely developed by Dirac himself. Towards the end of the book, he also discusses the relativistic theory of the electron (the Dirac equation), which was also pioneered by him. This work does not refer to any other writings then available on quantum mechanics.

- *Lectures on Quantum Mechanics* (1966): Much of this book deals with quantum mechanics in curved space-time.

- *Lectures on Quantum Field Theory* (1966): This book lays down the foundations of quantum field theory using the Hamiltonian formalism.

- *Spinors in Hilbert Space* (1974): This book based on lectures given in 1969 at the University of Miami, Coral Gables, Florida, USA, deals with the basic aspects of spinors starting with a real Hilbert space formalism. Dirac concludes with the prophetic words "We have boson variables appearing automatically in a theory that starts with only fermion variables, provided the number of fermion variables is infinite. There must be such boson variables connected with electrons..."

- *General Theory of Relativity* (1975): This 69-page work summarises Einstein's general theory of relativity.

6.4.6 See also

- Dirac–von Neumann axioms

- Gamma matrices

- Fermi's golden rule

- List of things named after Paul Dirac

6.4.7 References

[1] "Nobel Bio". Nobelprize.org. Retrieved 27 January 2014.

[2] Dalitz, R. H.; Peierls, R. (1986). "Paul Adrien Maurice Dirac. 8 August 1902-20 October 1984". *Biographical Memoirs of Fellows of the Royal Society* **32**: 138. doi:10.1098/rsbm.1986.0006. JSTOR 770111.

[3] "The Nobel Prize in Physics 1933". The Nobel Foundation. Retrieved 4 April 2013.

[4] Sukumar, N. (2012). *A Matter of Density: Exploring the Electron Density Concept in the Chemical, Biological, and Materials Sciences*. John Wiley & Sons. p. 27. ISBN 9781118431719. Retrieved 3 April 2013.

[5] Farmelo 2009, p. 10

[6] Farmelo 2009, pp. 18–19

[7] Kragh 1990, p. 1

[8] Farmelo 2009, pp. 10–11

[9] Farmelo 2009, pp. 77–78

[10] Farmelo 2009, p. 79

[11] Farmelo 2009, p. 34

[12] Farmelo 2009, p. 22

[13] Mehra 1972, p. 17

[14] Kragh 1990, p. 2

[15] Farmelo 2009, pp. 13–17

[16] Farmelo 2009, pp. 20–21

[17] Mehra 1972, p. 18

[18] Farmelo 2009, p. 23

[19] Farmelo 2009, p. 28

[20] Farmelo 2009, pp. 46–47

[21] Farmelo 2009, p. 53

[22] Farmelo 2009, pp. 52–53

[23] 1851 Royal Commission Archives

[24] Farmelo 2009, p. 101

[25] Kim, Y.A. (1995). "Wigner's Sisters". Retrieved 4 April 2013.

[26] Farmelo 2009, p. 89

[27] "Paul Adrien Maurice Dirac". University of St. Andrews. Retrieved 4 April 2013.

[28] Mehra 1972, pp. 17–59

[29] Kragh (1990), p. 17.

[30] McKie, Rob (1 February 2009). "Anti-matter and madness". The Guardian. Retrieved 4 April 2013.

[31] Gamow 1966, p. 121

[32] Capri 2007, p. 148

[33] Zee 2010, p. 105

[34] "A quantum leap into oddness" Review of Farmelo's The Strangest Man by Chet Raymo, Globe and Mail 2009 October 17

[35] Farmelo 2009, pp. 161–162, who attributes the story to Niels Bohr.

[36] Heisenberg 1971, pp. 85–86

[37] Heisenberg 1971, p. 87

[38] Farmelo 2009, p. 138, who says this was an old joke, pointing out in a footnote that Punch wrote in the 1850s that "There is no God, and Harriet Martineau is her prophet.

[39] Dirac, Paul (May 1963). "The Evolution of the Physicist's Picture of Nature". Scientific American. Retrieved 4 April 2013.

[40] Helge Kragh (1990). "The purest soul". Dirac: A Scientific Biography. Cambridge University Press. pp. 256–257. ISBN 9780521380898.

[41] Walter, Claire (1982). Winners, the blue ribbon encyclopedia of awards. Facts on File Inc. p. 438. ISBN 9780871963864.

[42] "Dirac Receives Miami Center Oppenheimer Memorial Prize". Physics Today (American Institute of Physics): 127. April 1969. doi:10.1063/1.3035512. Retrieved 1 March 2015.

[43] Farmelo 2009, pp. 403–404

[44] "Dirac takes his place next to Isaac Newton". Florida State University. Retrieved 4 April 2013.

[45] Paul Adrien Maurice Dirac at Find a Grave

[46] "Paul Dirac". Gisela Dirac. Retrieved 4 April 2013.

[47] Farmelo 2009, pp. 414–15

[48] "Paul Dirac: a genius in the history of physics". Cern Courier. Retrieved 13 May 2013.

[49] P. A. M. Dirac, The inadequacies of quantum field theory, in Paul Adrien Maurice Dirac, B. N. Kursunoglu and E. P. Wigner, Eds. (Cambridge University, Cambridge, 1987) p. 194

[50] Behram N. Kursunoglu and Eugene Paul Wigner (ed.). Reminiscences about a Great Physicist. Cambridge University Press. p. 98.

[51] Dirac, P. A. M. (1 February 1928). "The Quantum Theory of the Electron". Proceedings of the Royal Society of London. Series A, Containing Papers of a Mathematical and Physical Character 117 (778): 610–24. Bibcode:1928RSPSA.117..610D. doi:10.1098/rspa.1928.0023.

[52] Dirac, Paul (12 December 1933). "Theory of electrons and positrons" (PDF). Nobel Lecture. Retrieved 13 May 2013.

[53] P. A. M. Dirac (1939). "A New Notation for Quantum Mechanics". Proceedings of the Cambridge Philosophical Society 35 (3): 416. Bibcode:1939PCPS...35..416D. doi:10.1017/S0305004100021162.

[54] Gieres (2000). "Mathematical surprises and Dirac's formalism in quantum mechanics". Reports on Progress in Physics 63 (12): 1893. arXiv:quant-ph/9907069. Bibcode:2000RPPh...63.1893G. doi:10.1088/0034-4885/63/12/201.

[55] P. B. Price; E. K. Shirk; W. Z. Osborne; L. S. Pinsky (25 August 1975). "Evidence for Detection of a Moving Magnetic Monopole". Physical Review Letters (American Physical Society) 35 (8): 487–90. Bibcode:1975PhRvL..35..487P. doi:10.1103/PhysRevLett.35.487.

[56] Blas Cabrera (17 May 1982). "First Results from a Superconductive Detector for Moving Magnetic Monopoles". Physical Review Letters (American Physical Society) 48 (20): 1378–81. Bibcode:1982PhRvL..48.1378C. doi:10.1103/PhysRevLett.48.1378.

[57] "Magnetic Monopoles Detected in a Real Magnet for the First Time". Science Daily. 4 September 2009. Retrieved 13 May 2013.

[58] D.J.P. Morris, D.A. Tennant, S.A. Grigera, B. Klemke, C. Castelnovo, R. Moessner, C. Czternasty, M. Meissner, K.C. Rule, J.-U. Hoffmann, K. Kiefer, S. Gerischer, D. Slobinsky, and R.S. Perry (3 September 2009). "Dirac Strings and Magnetic Monopoles in Spin Ice $Dy_2Ti_2O_7$". Science 326 (5951): 411–4. arXiv:1011.1174. Bibcode:2009Sci...326..411M. doi:10.1126/science.1178868. PMID 19729617.

[59] S. T. Bramwell, S. R. Giblin, S. Calder, R. Aldus, D. Prabhakaran & T. Fennell (15 October 2009). "Measurement of the charge and current of magnetic monopoles in spin ice". Nature 461 (7266): 956–9. arXiv:0907.0956.

Bibcode:2009Natur.461..956B. doi:10.1038/nature08500. PMID 19829376.

[60] Kragh 1990, p. 184

[61] Schweber 1994

[62] Canad J Math 1950 vol 2, 129; 1951 vol 3, 1

[63] 1951 "The Hamiltonian Form of Field Dynamics" *Canad Jour Math*, vol 3, 1

[64] Phillips R. J. N. 1987 *Tributes to Dirac* p31 London: Adam Hilger

[65] Proc Roy Soc 1958,A vol 246, 333,Phys Rev 1959,vol 114, 924

[66] John Polkinghorne. 'Belief in God in an Age of Science' p 2

[67] "Dirac Medal awards". University of New South Wales. Retrieved 4 April 2013.

[68] "The Dirac Medal". Institute of Physics. Retrieved 24 November 2007.

[69] "The Dirac Medal". International Centre for Theoretical Physics. Retrieved 4 April 2013.

[70] "Undergraduate Awards". Florida State University. Retrieved 4 April 2013.

[71] "Paul Adrien Maurice Dirac Collection". Florida State University. Archived from the original on 15 July 2013. Retrieved 4 April 2013.

[72] Farmelo 2009, p. 417

[73] "5997 Dirac (1983 TH)". Jet Propulsion Laboratory. Retrieved 2015-01-09.

6.4.8 Sources

- Capri, Anton Z. (2007). *Quips, Quotes, and Quanta: An Anecdotal History of Physics*. Hackensack, New Jersey: World Scientific. ISBN 981-270-919-3. OCLC 214286147. Retrieved 8 June 2008.

- Crease, Robert P.; Mann, Charles C. (1986). *The Second Creation: Makers of the Revolution in Twentieth Century Physics*. New York City: Macmillan Publishing. ISBN 0-02-521440-3. OCLC 13008048.

- Farmelo, Graham (2009). *The Strangest Man: the Life of Paul Dirac*. London: Faber and Faber. ISBN 0-465-01827-0. OCLC 426938310.

- Gamow, George (1966). *Thirty Years That Shook Physics: The Story of Quantum Theory*. Garden City, New York: Doubleday. ISBN 0-486-24895-X. OCLC 11970045. Retrieved 8 June 2008.

- Heisenberg, Werner (1971). *Physics and Beyond: Encounters and Conversations*. New York City: Harper & Row. ISBN 0-06-131622-9. OCLC 115992.

- Kragh, Helge (1990). *Dirac: A Scientific Biography*. Cambridge: Cambridge University Press. ISBN 0-521-38089-8. OCLC 20013981. Retrieved 8 June 2008.

- Mehra, Jagdish (1972). "The Golden Age of Theoretical Physics: P. A. M. Dirac's Scientific Works from 1924–1933". In Wigner, Eugene Paul; Salam, Abdus. *Aspects of Quantum Theory*. Cambridge: University Press. pp. 17–59. ISBN 0-521-08600-0. OCLC 532357.

- Schweber, Silvan S. (1994). *QED and the men who made it: Dyson, Feynman, Schwinger, and Tomonaga*. Princeton, New Jersey: Princeton University Press. ISBN 0-691-03685-3. OCLC 28966591.

- Zee, A. (2010). *Quantum Field Theory in a Nutshell*. Princeton, New Jersey: Princeton University Press. ISBN 978-1-4008-3532-4. OCLC 318585662.

6.4.9 Further reading

- Brown, Helen (24 January 2009). "The Strangest Man: The Hidden Life of Paul Dirac by Graham Farmelo – review [print version: The man behind the maths]". *The Daily Telegraph (Review)*. p. 20. Retrieved 11 April 2011..

- Gilder, Louisa (13 September 2009). "Quantum Leap – Review of 'The Strangest Man: The Hidden Life of Paul Dirac by Graham Farmelo'". *The New York Times*. Retrieved 11 April 2011. Review.

6.4.10 External links

- Free online access to Dirac's classic 1920s papers from Royal Society's Proceedings A

- Annotated bibliography for Paul Dirac from the Alsos Digital Library for Nuclear Issues

- The Paul Dirac Collection at Florida State University

- Letters from Dirac (1932–36) and other papers

- Oral History interview transcript with Dirac 1 April 1962, 6, 7, 10, & 14 May 1963, American Institute of Physics, Niels Bohr Library and Archives

- O'Connor, John J.; Robertson, Edmund F., "Paul Dirac", *MacTutor History of Mathematics archive*, University of St Andrews.

- Paul Dirac at the Mathematics Genealogy Project

6.5　Masatoshi Koshiba

Masatoshi Koshiba (小柴昌俊 *Koshiba Masatoshi*, born on September 19, 1926 in Toyohashi, Aichi) is a Japanese physicist. He jointly won the Nobel Prize in Physics in 2002.

He graduated from the University of Tokyo in 1951 and received a Ph.D. in physics at the University of Rochester, New York, in 1955. From July 1955 to February 1958 he was Research Associate, Department of Physics, University of Chicago; from March 1958 to October 1963, he was Associate Professor, Institute of Nuclear Study, University of Tokyo, although from November 1959 to August 1962 he was on leave from the above as Senior Research Associate with the honorary rank of Associate Professor and as the Acting Director, Laboratory of High Energy Physics and Cosmic Radiation, Department of Physics, University of Chicago. At the University of Tokyo he became Associate Professor in March 1963 and then Professor in March 1970 in the Department of Physics, Faculty of Science, and Emeritus Professor there in 1987. From 1987 to 1997, Koshiba taught at Tokai University. In 2002, he jointly won the Nobel Prize in Physics "for pioneering contributions to astrophysics, in particular for the detection of cosmic neutrinos". (The other shares of that year's Prize were awarded to Raymond Davis Jr. & Riccardo Giacconi of the U.S.A.)[1]

He is now Senior Counselor of International Center for Elementary Particle Physics (ICEPP) and Emeritus Professor of University of Tokyo.

Koshiba's award-winning work centred on neutrinos, subatomic particles that had long perplexed scientists. Since the 1920s it had been suspected that the Sun shines because of nuclear fusion reactions that transform hydrogen into helium and release energy. Later, theoretical calculations indicated that countless neutrinos must be released in these reactions and, consequently, that Earth must be exposed to a constant flood of solar neutrinos. Because neutrinos interact weakly with matter, however, only one in a trillion is stopped on its way to Earth. Neutrinos thus developed a reputation as being undetectable.

In the 1980s, Koshiba, drawing on the work done by Raymond Davis Jr, constructed an underground neutrino detector in a zinc mine in Japan. Called Kamiokande II, it was an enormous water tank surrounded by electronic detectors to sense flashes of light produced when neutrinos interacted with atomic nuclei in water molecules. Koshiba was able to confirm Davis's results—that the Sun produces neutrinos and that fewer neutrinos were found than had been expected (a deficit that became known as the solar neutrino problem). In 1987 Kamiokande also detected neutrinos from a supernova explosion outside the Milky Way. After building a larger, more sensitive detector named Super-Kamiokande, which became operational in 1996, Koshiba found strong evidence for what scientists had already suspected—that neutrinos, of which three types are known, change from one type into another in flight; this resolves the solar neutrino problem, since early experiments could only detect one type, not all three.

In 2003, he was awarded the Benjamin Franklin Medal in Physics.

Koshiba is a member of the Board of Sponsors of The Bulletin of the Atomic Scientists.

He is a foreign fellow of Bangladesh Academy of Sciences [2]

In commemoration of the Nobel Prize-winning by Masatoshi Koshiba, *Koshiba* hall was established at the University of Tokyo.[3]

6.5.1　Publications

- Koshiba, M.; Fukuda, Y et al. (1998). "Evidence for Oscillation of Atmospheric Neutrinos". *Physical Review Letters* **81** (8): 1562. arXiv:hep-ex/9807003. Bibcode:1998PhRvL..81.1562F. doi:10.1103/PhysRevLett.81.1562.

- Koshiba, M.; Fukuda, Y et al. (1999). "Constraints on Neutrino Oscillation Parameters from the Measurement of Day-Night Solar Neutrino Fluxes at Super-Kamiokande". *Physical Review Letters* **82** (9): 1810. arXiv:hep-ex/9812009. Bibcode:1999PhRvL..82.1810F. doi:10.1103/PhysRevLett.82.1810.

6.5.2　See also

- Kamioka Observatory

- Institute for Cosmic Ray Research

6.5.3　References

[1] The Nobel Prize in Physics 2002.

[2] List of Fellows of Bangladesh Academy of Sciences

[3] 寺崎昌男 2007 『東京大学の歴史大学制度の先駆け』 講談社

6.5.4 External links

- Prof. Koshiba has won the Nobel prize.

- Nobelprize.org Birography

- Photograph, Biography and Bibliographic Resources, from the Office of Scientific and Technical Information, United States Department of Energy

- Masatoshi Koshiba, Autobiography in English

- Freeview video 'An Interview with Masatoshi Koshiba' by the Vega Science Trust

6.6 Leon M. Lederman

Leon Max Lederman (born July 15, 1922) is an American experimental physicist who received, along with Martin Lewis Perl, the Wolf Prize in Physics in 1982, for their research on quarks and leptons, and the Nobel Prize for Physics in 1988, along with Melvin Schwartz and Jack Steinberger, for their research on neutrinos. He is Director Emeritus of Fermi National Accelerator Laboratory (Fermilab) in Batavia, Illinois, USA. He founded the Illinois Mathematics and Science Academy, in Aurora, Illinois in 1986, and has served in the capacity of Resident Scholar since 1998.*[2] In 2012, he was awarded the Vannevar Bush Award for his extraordinary contributions to understanding the basic forces and particles of nature.*[3]

6.6.1 Early life and career

Lederman was born in New York City, New York, the son of Minna (née Rosenberg) and Morris Lederman, a laundryman.*[4] Lederman graduated from the James Monroe High School in the South Bronx. He received his bachelor's degree from the City College of New York in 1943, and received a Ph.D. from Columbia University in 1951. He then joined the Columbia faculty and eventually became Eugene Higgins Professor of Physics. In 1960, on leave from Columbia, he spent some time at CERN in Geneva as a Ford Foundation Fellow.*[5] He took an extended leave of absence from Columbia in 1979 to become director of Fermilab. Resigning from Columbia (and retiring from Fermilab) in 1989 to teach briefly at the University of Chicago, he then moved to the physics department of the Illinois Institute of Technology, where he currently serves as the Pritzker Professor of Science. In 1991, Lederman became President of the American Association for the Advancement of Science.

Lederman is also one of the main proponents of the "Physics First" movement. Also known as "Right-side Up Science" and "Biology Last," this movement seeks to rearrange the current high school science curriculum so that physics precedes chemistry and biology.

A former president of the American Physical Society, Lederman also received the National Medal of Science, the Wolf Prize and the Ernest O. Lawrence Medal. Lederman serves as President of the Board of Sponsors of The Bulletin of the Atomic Scientists. He also served on the board of trustees for Science Service, now known as Society for Science & the Public, from 1989 to 1992, and is a member of the JASON defense advisory group.*[6]

Among his achievements are the discovery of the muon neutrino in 1962 and the bottom quark in 1977. These helped establish his reputation as among the top particle physicists.

In 1977, a group of physicists led by Leon Lederman announced that a particle with a mass of about 6.0 GeV was being produced by the Fermilab particle accelerator. The particle's initial name was the greek letter Upsilon (Υ). After taking further data, the group discovered that this particle did not actually exist, and the "discovery" was named "Oops-Leon" as a pun on the original name (mispronounced /ˈjuːpsɪlɒn/) and Lederman's first name.

As the director of Fermilab and subsequent Nobel physics prizewinner, Leon Lederman was a very prominent early supporter – some sources say the architect*[7] or proposer*[8] – of the Superconducting Super Collider project, which was endorsed around 1983, and was a major proponent and advocate throughout its lifetime.*[9]*[10] Lederman later wrote his 1993 popular science book *The God Particle: If the Universe Is the Answer, What Is the Question?* – which sought to promote awareness of the significance of such a project – in the context of the project's last years and the changing political climate of the 1990s.*[11] The increasingly moribund project was finally shelved that same year after some $2 billion of expenditure.*[7]

In 1988, Lederman received the Nobel Prize for Physics along with Melvin Schwartz and Jack Steinberger "for the neutrino beam method and the demonstration of the doublet structure of the leptons through the discovery of the muon neutrino" .*[2] Lederman also received the National Medal of Science (1965), the Elliott Cresson Medal for Physics (1976), the Wolf Prize for Physics (1982) and the Enrico Fermi Award (1992).

In 1995, he received the Chicago History Museum "Making History Award" for Distinction in Science Medicine and Technology.

Lederman was an early supporter of Science Debate 2008, an initiative to get the then-candidates for president, Barack Obama and John McCain, to debate the nation's top science policy challenges. In October 2010, Lederman participated in the USA Science and Engineering Festival's Lunch with a Laureate program where middle and high school students got to engage in an informal conversation with a Nobel Prize-winning scientist over a brown-bag lunch.[12] Lederman was also a member of the USA Science and Engineering Festival's Advisory Board [13] and CRDF Global.

6.6.2   Personal life

Lederman was born in New York to a family of Jewish immigrants from Russia.[14] His father operated a hand laundry while encouraging Leon to pursue his education. He went to elementary school in New York City, continuing on to college and his doctorate in the city.[15]

In his book,*The God Particle: If the Universe Is the Answer, What Is the Question?*, Lederman writes that, although he was a chemistry major, he became fascinated with physics, because of the clarity of the logic and the unambiguous results from experimentation. His best friend during his college years, Martin Klein, convinced him of "the splendors of physics during a long evening over many beers." After that conversation he became resolute and unwavering regarding his desire to pursue physics. When he joined the Army with a B.S. in Chemistry, he was determined to become a physicist following his service.[16]

After three years in the U.S. Army during World War II, he took up physics at Columbia University, and received his Masters in 1948. Lederman began his Ph.D research working with Columbia's Nevis synchro-cyclotron,[17] which was the most powerful particle accelerator in the world at that time.[16] Dwight D. Eisenhower, then the president of Columbia University, and future president of the United States, cut the ribbon dedicating the synchro-cyclotron in June 1950.[18] These atom smashers were just coming of age at this time and created the new discipline of particle physics.[16]

After receiving his Ph.D and then becoming a faculty member at Columbia University he was promoted to full professor in 1958.[19]

In "The God Particle" he once wrote "The history of atomism is one of reductionism – the effort to reduce all the operations of nature to a small number of laws governing a small number of primordial objects."[20] And this was the quest he undertook. This book shows that he pursued the quark, and hopes to find the Higgs boson. The top quark, which he and other physicists realized must exist according to the standard model, was, in fact, produced at Fermilab not long after this book was published.[21]

He is known for his sense of humor in the physics community.[22] On August 26, 2008 Lederman was video-recorded by a science focused organization called Scien-

Central, on the street in a major U.S. city, answering questions from passersby.[23] He answered questions such as "What is the strong force?" and "What happened before the Big Bang?".

He has three children with his first wife, Florence Gordon, and now lives with his second wife, Ellen (Carr), in Driggs, Idaho.[24]

He is an atheist.[25][26]

6.6.3   Publications

- *The God Particle: If the Universe Is the Answer, What Is the Question?* by Leon M. Lederman, Dick Teresi (ISBN 0-385-31211-3)

- *From Quarks to the Cosmos* by Leon Lederman and David N. Schramm (ISBN 0-7167-6012-6)

- *Portraits of Great American Scientists* Leon M. Lederman, et al. (ISBN 1-57392-932-8)

- *Symmetry and the Beautiful Universe* Leon M. Lederman and Christopher T. Hill (ISBN 1-59102-242-8)

- *What We'll Find Inside the Atom* by Leon Lederman is an essay he wrote for the September 15, 2008 issue of Newsweek

- "Quantum Physics for Poets" Leon M. Lederman and Christopher T. Hill (ISBN 978-1616142339)

6.6.4   Honorary degrees and awards

- Election to the National Academy of Sciences, 1965.

-

- U.S. National Medal of Science, 1965.

-

- Member, American Academy of Arts and Sciences, 13 May 1970.

-

- Townsend Harris Medal, Alumni Association of the City College of New York, 1973.

-

- Elliot Cresson Prize of the Franklin Institute, 1976.

-

- President Jimmy Carter's Committee on the National Medal of Science, 21 March 1979.

-

- Wolf Foundation Prize in Physics, Israel, 1982.

-

- Doctor of Science, University of Chicago, Chicago, Illinois, 10 June 1983.

-

- Doctor of Humane Letters and Science, IIT, Chicago, Illinois, 17 May 1987.

-

- Doctor of Science, Lake Forest College, Lake Forest, Illinois, 7 May 1988.

-

- Honorary Degree of Doctor of Science, Carnegie Mellon University, Pittsburgh, Pennsylvania, 15 May 1988.

-

- Department of Energy Distinguished Associate Award, May 1988.

-

- Doctor of Science, City College of New York, New York, New York, 8 June 1988.

-

- Nobel Prize in Physics, December 1988.

-

- La laurea honoris causa in Fisica, Universita' degli Studi di Pisa, 21 March 1989.

-

- Member, American Philosophical Society, 21 April 1989.

-

- Doctor of Science, honoris causae, Aurora University, Aurora, Illinois, 20 May 1989.

-

- Doctor of Humane Letters, Columbia College, Chicago, Illinois, 3 June 1989.

-

- Citation on the occasion of the dedication of IMSA (founded 1985) to the State of Illinois in honor of Leon Lederman and Gov. James Thompson, IMSA, Aurora, Illinois, 10 June 1989.

-

- Doctor of Humane Letters, Rush University, Chicago, Illinois, 10 June 1989.

-

- Doctor of Science, University of Illinois, Chicago, Illinois, 11 June 1989.

-

- Doctor en Filosofia – Fisica, Honoris Causa, Universidad de Guanajuato, Guanajuato, Mexico, 2 August 1989.

-

- Honorary Degree, Academia Nacional de Ciencias Exactas, Fisicas y Naturales, Buenos Aires, Argentina, 3 November 1989.

-

- Doctor of Philosophy of Physics, honoris causa, University of Guanajuato, Guanajuato, Mexico, 1989.

-

- Appointment as member of the Secretary of Energy Advisory Board, U.S. Department of Energy, Washington, DC, 15 February 1990.

-

- Laureate of the Lincoln Academy, State of Illinois, Springfield, Illinois, 21 April 1990.

-

- Doctor of Science, Bradley University, Peoria, Illinois, 19 May 1990.

-

- Doctor of Science, Columbia University, New York, New York, 26 September 1990.

-

- Public Affairs Committee Award, American Chemical Society, 22 March 1991.

-

- 1991 William Proctor Prize, Sigma Xi, The Scientific Research Society, 1991.

-
- Doctor of Humane Letters, honoris causa, Mount Sinai School of Medicine, The City University of New York, New York City, New York, 11 May 1992.

-
- Doctor of Science, Adelphi University, Long Island, New York, 17 May 1992.

-
- Doctor of Science, honoris causa, Saint Xavier University, Chicago, Illinois, 23 May 1992.

-
- Doctor of Science – honoris causa, University of Southampton, Southampton, United Kingdom, 10 July 1992.

-
- Doctor of Science, honoris causa, Drury College, Springfield, Missouri, 14 October 1992.

-
- Enrico Fermi Prize of the U.S. Department of Energy, 1992.

-
- Doctor of Humane Letters, State University of New York, Geneseo, New York, 15 May 1993.

-
- Doctor of Science, honoris causa, Case-Western Reserve University, Cleveland, Ohio, 23 May 1993.

-
- President's Medal, The City College, The City University of New York, New York, New York, 28 May 1993.

-
- Doctor of Science, honoris causa, Marywood College, Scranton, Pennsylvania, 15 October 1993.

-
- Doctor of Humane Letters, honoris causa, Illinois Benedictine College, Lisle, Illinois, 21 May 1994.

-
- Appointment as a Tetelman Fellow at Jonathan Edwards College, Yale University, New Haven, Connecticut, 7 November 1994.

-
- Doctor of Humane Letters, University of Dallas, Dallas, Texas, 15 November 1994.

-
- The first Enrico Fermi History Maker Award, for distinction in Science, Medicine and Technology, Chicago Historical Society, Chicago, Illinois, 8 June 1995.

-
- Doctor of Humane Letters, DePaul University, Chicago, Illinois, 11 June 1995.

-
- Ordem Nacional do Merito Cientifico, Brasilia, Brazil, 13 June 1995.

-
- Universitario de la Universidad Nacional de San Antonio Abad Del Cusco, Cusco, Peru, Doctor, honoris causa, 16 August 1995.

-
- University of Notre Dame, Notre Dame, Indiana, Doctor of Science, honoris causa, 18 May 1997.

-
- Doctor of Science, Bethany College, Bethany, Virginia, 24 May 1997.

-
- Doctor of Science, Illinois State University, Normal, Illinois, 17 May 1998.

-
- Doctor of Science, honoris causa, Clark University, Worcester, Massachusetts, 17 May 1998.

-
- Doctor of Science, Pennsylvania State University, University Park, Pennsylvania, December 1998.

-
- Doctor, honoris causa, Institute for High Energy Physics, Protvino, Russia, 15 July 1998.

-
- Member, President Bill Clinton's Commission on White House Fellowships, 13 April 1999.

-

- Diploma, Miembro Correspondiente, La Academia Mexicana de Ciencias, October 1999.

-

- 1999 Medallion, Division of Particles and Fields, Mexican Physical Society, Mérida, Yucatán, Mexico, 11 November 1999.

-

- Honorary Professor, Beijing Normal University, Beijing, China, 5 November 2000.

-

- Doctor of Humane Letters, Roosevelt University, Chicago, Illinois, 21 January 2001.

-

- Doctor of Science, Florida Institute of Technology, Melbourne, Florida, May 2004.

-

- Doctor of Public Service, The George Washington University, Washington, D.C., 16 May 2004.

-

- Doctor of Science Education, honoris causa, The Ohio State University, Columbus, Ohio, 12 December 2004.

-

- Doctor of Humane Letters, Honoris Causa, Dominican University, River Forest, Illinois, 6 May 2006.

-

- Doctor of Science, Honoris Causa, Gustavus Adolphus College, St. Peter, Minnesota, 24 October 2008.

-

- Vannevar Bush Prize, 2012.*[27]

6.6.5 See also

- List of Jewish Nobel laureates

6.6.6 References

[1] American Scientists - Charles W. Carey - Google Books

[2] Lederman, Leon M. (1988). Frängsmyr, Tore; Ekspång, Gösta, eds. "The Nobel Prize in Physics 1988: Leon M. Lederman, Melvin Schwartz, Jack Steinberger". *Nobel Lectures, Physics 1981–1990* (Singapore: World Scientific Publishing Co.). Retrieved 22 May 2012.

[3] National Science Board - Honorary Awards - Vannevar Bush Award Recipients

[4] Fermilab: Physics, the Frontier, and Megascience - Lillian Hoddeson, Adrienne W. Kolb, Catherine Westfall - Google Books

[5] "CERN affiliated article by Lederman". Springer. Retrieved 11 June 2015.

[6] Horgan, John (April 16, 2006). "Rent-a-Genius". The New York Times.

[7] ASCHENBACH, JOY (1993-12-05). "No Resurrection in Sight for Moribund Super Collider : Science: Global financial partnerships could be the only way to salvage such a project. But some feel that Congress delivered a fatal blow." . *Los Angeles Times*. Retrieved 16 January 2013. Disappointed American physicists are anxiously searching for a way to salvage some science from the ill-fated superconducting super collider ... "We have to keep the momentum and optimism and start thinking about international collaboration," said Leon M. Lederman, the Nobel Prize-winning physicist who was the architect of the super collider plan

[8] Lillian Hoddeson; Adrienne Kolb. "Vision to reality: From Robert R. Wilson's frontier to Leon M. Lederman's Fermilab". arXiv:1110.0486. Lederman also planned what he saw as Fermilab's next machine, the Superconducting SuperCollider (SSC)

[9] Abbott, Charles (June 1987). "Illinois Issues journal, June 1987". p. 18. Lederman, who considers himself an unofficial propagandist for the super collider, said the SSC could reverse the physics brain drain in which bright young physicists have left America to work in Europe and elsewhere. (direct link to article:)

[10] Kevles, Dan. "Good-bye to the SSC" (PDF). *California Institute of Technology* "Engineering & Science". 58 no. 2 (Winter 1995): 16–25. Retrieved 16 January 2013. Lederman, one of the principal spokesmen for the SSC, was an accomplished high-energy experimentalist who had made Nobel Prize-winning contributions to the development of the Standard Model during the 1960s (although the prize itself did not come until 1988). He was a fixture at congressional hearings on the collider, an unbridled advocate of its merits []

[11] Calder, Nigel (2005). *Magic Universe:A Grand Tour of Modern Science*. pp. 369–370. The possibility that the next

big machine would create the Higgs became a carrot to dangle in front of funding agencies and politicians. A prominent American physicist, Leon lederman, advertised the Higgs as The God Particle in the title of a book published in 1993 ...Lederman was involved in a campaign to persuade the US government to continue funding the Superconducting Super Collider... the ink was not dry on Lederman's book before the US Congress decided to write off the billions of dollars already spent

[12]   Archived December 30, 2010 at the Wayback Machine

[13] USA Science and Engineering Festival - Advisors

[14] Humes, Edward (2006), *Over Here: How the G.I. Bill Transformed the American Dream*, Houghton Mifflin Harcourt, p. 275, ISBN 9780151007103.

[15] Lederman, Leon M. "Leon M. Lederman – Autobiography." http://nobelprize.org/nobel_prizes/physics/laureates/1988/lederman-autobio.html Retrieved 06 April 2009. Published 1988, The Nobel Foundation.

[16] The God Particle: If the Universe is the Answer, What is the Question – page 5
by Leon Lederman with Dick Teresi (copyright 1993) Houghton Mifflin Company

[17] see **High Energy Physics** section on this page: http://www.columbia.edu/cu/physics/news/DeptHistory/index/

[18] Eisenhower at ribbon cutting for the dedication of the Nevis synchro-cyclotron http://www.columbia.edu/cu/physics/images/columbia-cyclotron.jpeg

[19] The God Particle: If the Universe is the Answer, What is the Question – page 296
by Leon Lederman with Dick Teresi (copyright 1993) Houghton Mifflin Company

[20] The God Particle: If the Universe is the Answer, What is the Question – page 87
by Leon Lederman with Dick Teresi (copyright 1993) Houghton Mifflin Company

[21] Observation of Top Quark Production in [anti-p][and][p] Collisions with the Collider Detector at Fermilab http://prola.aps.org/abstract/PRL/v74/i14/p2626_1

[22] The God Particle: If the Universe is the Answer, What is the Question – page 17 – by Leon Lederman with Dick Teresi (copyright 1993) Houghton Mifflin Company

[23] Street Corner Science with Leon Lederman, http://blogs.discovermagazine.com/cosmicvariance/2008/08/26/street-corner-science-with-leon-lederman

[24] http://411.info/people/Idaho/Driggs/Lederman-Leon/132837548.html." http://nobelprize.org/nobel_prizes/physics/laureates/1988/lederman-autobio.html Retrieved 29.7.2008. Published 1988, The Nobel Foundation.

[25] Dan Falk (2005). "What About God?". *Universe on a T-Shirt: The Quest for the Theory of Everything*. Arcade Publishing. p. 195. ISBN 9781559707336. "Physics isn't a religion. If it were, we'd have a much easier time raising money." - Leon Lederman

[26] Babu Gogineni (July 10, 2012). "It's the Atheist Particle, actually". Postnoon News. Retrieved 10 July 2012. Leon Lederman is himself an atheist and he regrets the term, and Peter Higgs who is an atheist too, has expressed his displeasure, but the damage has been done!

[27] Fermilab History and Archives Project I Leon M. Lederman Honorary Degrees and Awards

6.6.7   External links

- Education, Politics, Einstein and Charm *The Science Network* interview with Leon Lederman

- Biography and Bibliographic Resources, from the Office of Scientific and Technical Information, United States Department of Energy

- Fermilab's Leon M. Lederman webpage

- The Nobel Prize in Physics 1988

- Video Interview with Lederman from the Nobel Foundation

- Leon M. Lederman – Autobiography

- Timeline of Nobel Prize Winners in Physics webpage for Leon Max Lederman

- *Story of Leon* by Leon Lederman

- Honeywell – Nobel Interactive Studio

- 1976 Cresson Medal recipient from The Franklin Institute.

- Scientific publications of Leon M. Lederman on INSPIRE-HEP

6.7   Louis Michel

Louis Michel was a French mathematical physicist at the Institut des Hautes Études Scientifiques (IHÉS).*[1] He was born in Roanne, near Loire, on 4 May 1923 and died in Bures-sur-Yvette on 30 December 1999.

6.7.1 Biography

Michel completed his studies at the École Polytechnique in Paris. After the World War II, he was in Manchester, where he worked on weak interactions. Back in France, he was teaching in Lille and Orsay before creating the Centre de Physique Théorique of[École Polytechnique. In 1962 he became a permanent professor at IHÉS in Bures-sur-Yvette, where he remained until his retirement, and as an emeritus professor until his death.

Louis Michel was President of the Société Française de Physique between 1978 and 1980, and a member of the French Academy of Sciences since 1979. In 1982 he was awarded the Wigner Medal.

His scientific activities in the domain of theoretical physics encompassed many fields, from elementary particles and High Energy Physics to Crystals, and provided pioneering insights in spontaneous symmetry breaking in many contexts. His name is associated to the Bargmann–Michel–Telegdi equation describing spin evolution in a magnetic field,[*][2] the theory of phase transitions as a symmetry-breaking,[*][3] the Michel–Radicati theory for the SU(3) octet,[*][4][*][5] and more generally his geometric theory of spontaneous symmetry breaking,[*][6][*][7][*][8] and to several results in crystallography.[*][8]

After his death, the IHÉS created the Louis Michel Chairs for distinguished long-term visitors to honour his memory.

6.7.2 References

[1] Cipra, B. A. (8 January 1998). "A Gem of a Definition". *Scientific American*. Retrieved 2007-11-19.

[2] Bargman, V.; Michel, L.; Telegdi, V. (1959). "Precession of the Polarization of Particles Moving in a Homogeneous Electromagnetic Field". *Physical Review Letters* **2** (10): 435. Bibcode:1959PhRvL...2..435B. doi:10.1103/PhysRevLett.2.435.

[3] Michel, L. (1953). "Selection rules imposed by charge conjugation". *Il Nuovo Cimento* **10** (3): 10. doi:10.1007/BF02786202.

[4] Michel, L.; Radicati, L. (1971). "Properties of the breaking of hadronic internal symmetry". *Annals of Physics* **66** (2): 758. Bibcode:1971AnPhy..66..758M. doi:10.1016/0003-4916(71)90079-0.

[5] Michel, L.; Radicati, L. (1973). "The geometry of the octet" (PDF). *Annales de l'Institut Henri Poincaré A* **18**: 185. MR 325036. Zbl 0267.22019.

[6] Michel, L. (1971). "Points critiques des fonctions invariantes sur une G-variété" (PDF). *Comptes Rendus de l'Académie des Sciences de Paris* **272**: 433.

[7] Michel, L. (1980). "Symmetry defects and broken symmetry. Configurations Hidden Symmetry". *Reviews of Modern Physics* **52** (3): 618. Bibcode:1980RvMP...52..617M. doi:10.1103/RevModPhys.52.617.

[8] Michel, L.; Kim, J. S.; Zak, J.; Zhilinskii, B. (2001). "Symmetry, invariants, topology". *Physics Reports* **341**: 7. Bibcode:2001PhR...341....7M. doi:10.1016/S0370-1573(00)00087-9.

6.7.3 Further reading

* Kim, J. S.; Zak, J.; Zhilinskii, B. (2001). "Obituary". *Physics Reports* **341**: 5. Bibcode:2001PhR...341....5K. doi:10.1016/S0370-1573(00)00086-7.

6.8 Wolfgang Pauli

This article is about the Austrian-Swiss physicist. For the German physicist, see Wolfgang Paul.

Wolfgang Ernst Pauli (25 April 1900 – 15 December 1958) was an Austrian-born Swiss theoretical physicist and one of the pioneers of quantum physics. In 1945, after having been nominated by Albert Einstein, Pauli received the Nobel Prize in Physics for his "decisive contribution through his discovery of a new law of Nature, the exclusion principle or Pauli principle." The discovery involved spin theory, which is the basis of a theory of the structure of matter.

6.8.1 Biography

Early years

Pauli was born in Vienna to a chemist Wolfgang Joseph Pauli (*né* Wolf Pascheles, 1869–1955) and his wife Bertha Camilla Schütz. His middle name was given in honor of his godfather, physicist Ernst Mach. Pauli's paternal grandparents were from prominent Jewish families of Prague; his great-grandfather was the Jewish publisher Wolf Pascheles.[*][4] Pauli's father converted from Judaism to Roman Catholicism shortly before his marriage in 1899. Pauli's mother, Bertha Schütz, was raised in her own mother's Roman Catholic religion; her father was Jewish writer Friedrich Schütz. Pauli was raised as a Roman Catholic, although eventually he and his parents left the Church.[*][5] He is considered to have been a deist and a mystic.[*][6][*][7]

Pauli attended the Döblinger-Gymnasium in Vienna, graduating with distinction in 1918. Only two months after graduation, he published his first paper, on Albert Einstein's theory of general relativity. He attended the Ludwig-Maximilians University in Munich, working under Arnold Sommerfeld,[2] where he received his PhD in July 1921 for his thesis on the quantum theory of ionized diatomic hydrogen (H+ 2).[1][8]

Sommerfeld asked Pauli to review the theory of relativity for the *Encyklopädie der mathematischen Wissenschaften* (*Encyclopedia of Mathematical Sciences*). Two months after receiving his doctorate, Pauli completed the article, which came to 237 pages. It was praised by Einstein; published as a monograph, it remains a standard reference on the subject to this day.

Wolfgang Pauli lecturing

Pauli spent a year at the University of Göttingen as the assistant to Max Born, and the following year at the Institute for Theoretical Physics in Copenhagen, which later became the Niels Bohr Institute in 1965. From 1923 to 1928, he was a lecturer at the University of Hamburg. During this period, Pauli was instrumental in the development of the modern theory of quantum mechanics. In particular, he formulated the exclusion principle and the theory of non-relativistic spin.

In 1928, he was appointed Professor of Theoretical Physics at ETH Zurich in Switzerland where he made significant scientific progress. He held visiting professorships at the

University of Michigan in 1931, and the Institute for Advanced Study in Princeton in 1935. He was awarded the Lorentz Medal in 1931.

At the end of 1930, shortly after his postulation of the neutrino and immediately following his divorce in November, Pauli had a severe breakdown. He consulted psychiatrist and psychotherapist Carl Jung who, like Pauli, lived near Zurich. Jung immediately began interpreting Pauli's deeply archetypal dreams,[9] and Pauli became one of the depth psychologist's best students. He soon began to criticize the epistemology of Jung's theory scientifically, and this contributed to a certain clarification of the latter's thoughts, especially about the concept of synchronicity. A great many of these discussions are documented in the Pauli/Jung letters, today published as *Atom and Archetype*. Jung's elaborate analysis of more than 400 of Pauli's dreams is documented in *Psychology and Alchemy*.

The German annexation of Austria in 1938 made him a German citizen, which became a problem for him in 1939 after the outbreak of World War II. In 1940, he tried in vain to obtain Swiss citizenship, which would have allowed him to remain at the ETH.[10]

Pauli moved to the United States in 1940, where he was employed as a professor of theoretical physics at the Institute for Advanced Study. In 1946, after the war, he became a naturalized citizen of the United States and subsequently returned to Zurich, where he mostly remained for the rest of his life. In 1949, he was granted Swiss citizenship.

In 1958, Pauli was awarded the Max Planck medal. In that same year, he fell ill with pancreatic cancer. When his last assistant, Charles Enz, visited him at the Rotkreuz hospital in Zurich, Pauli asked him: "Did you see the room number?" It was number 137. Throughout his life, Pauli had been preoccupied with the question of why the fine structure constant, a dimensionless fundamental constant, has a value nearly equal to 1/137. Pauli died in that room on 15 December 1958.[11]

Scientific research

Pauli made many important contributions in his career as a physicist, primarily in the field of quantum mechanics. He seldom published papers, preferring lengthy correspondences with colleagues such as Niels Bohr and Werner Heisenberg, with whom he had close friendships. Many of his ideas and results were never published and appeared only in his letters, which were often copied and circulated by their recipients.

Pauli proposed in 1924 a new quantum degree of freedom (or quantum number) with two possible values, in order to resolve inconsistencies between observed molecular spectra

Niels Bohr, Werner Heisenberg, and Wolfgang Pauli, ca. *1935*

and the developing theory of quantum mechanics. He formulated the Pauli exclusion principle, perhaps his most important work, which stated that no two electrons could exist in the same quantum state, identified by four quantum numbers including his new two-valued degree of freedom. The idea of spin originated with Ralph Kronig. George Uhlenbeck and Samuel Goudsmit one year later identified Pauli's new degree of freedom as electron spin.

In 1926, shortly after Heisenberg published the matrix theory of modern quantum mechanics, Pauli used it to derive the observed spectrum of the hydrogen atom. This result was important in securing credibility for Heisenberg's theory.

Pauli introduced the 2 × 2 Pauli matrices as a basis of spin operators, thus solving the nonrelativistic theory of spin. This work is sometimes said to have influenced Paul Dirac in his creation of the Dirac equation for the relativistic electron, though Dirac stated that he invented these same matrices himself independently at the time, without Pauli's influence. Dirac invented similar but larger (4x4) spin matrices for use in his relativistic treatment of fermionic spin.

In 1930, Pauli considered the problem of beta decay. In a letter of 4 December to Lise Meitner *et al.*, beginning, "Dear radioactive ladies and gentlemen", he proposed the existence of a hitherto unobserved neutral particle with a small mass, no greater than 1% the mass of a proton, in order to explain the continuous spectrum of beta decay. In 1934, Enrico Fermi incorporated the particle, which he called a neutrino, into his theory of beta decay. The neutrino was first confirmed experimentally in 1956 by Frederick Reines and Clyde Cowan, two and a half years before Pauli's death. On receiving the news, he replied by telegram: "Thanks for message. Everything comes to him who knows how to wait. Pauli." [12]

In 1940, he re-derived the spin-statistics theorem, a criti-cal result of quantum field theory which states that particles with half-integer spin are fermions, while particles with integer spin are bosons.

In 1949, he published a paper on Pauli–Villars regularization: regularization is the term for techniques which modify infinite mathematical integrals to make them finite during calculations, so that one can identify whether the intrinsically infinite quantities in the theory (mass, charge, wavefunction) form a finite and hence calculable set which can be redefined in terms of their experimental values, which criterion is termed renormalization, and which removes infinities from quantum field theories, but also importantly allows the calculation of higher order corrections in perturbation theory.

Pauli made repeated criticisms of the modern synthesis of evolutionary biology,[13][14] and his contemporary admirers point to modes of epigenetic inheritance as supportive of his arguments.[15]

Personality and reputation

Wolfgang Pauli, ca. 1945

The Pauli effect was named after the anecdotal bizarre ability of his to break experimental equipment simply by being in the vicinity. Pauli was aware of his reputation and was delighted whenever the Pauli effect manifested. These strange occurrences were in line with his investigations into the legitimacy of parapsychology, particularly his collaboration with C. G. Jung on the concept of synchronicity.

Regarding physics, Pauli was famously a perfectionist. This extended not just to his own work, but also to the work of his colleagues. As a result, he became known in the physics community as the "conscience of physics," the critic to whom his colleagues were accountable. He could be scathing in his dismissal of any theory he found lacking, often labelling it *ganz falsch*, utterly false.

However, this was not his most severe criticism, which he reserved for theories or theses so unclearly presented as to be untestable or unevaluatable and, thus, not properly belonging within the realm of science, even though posing as such. They were worse than wrong because they could not be proven wrong. Famously, he once said of such an unclear paper: *It is not even wrong!"*[2]

His supposed remark when meeting another leading physicist, Paul Ehrenfest, illustrates this notion of an arrogant Pauli. The two met at a conference for the first time. Ehrenfest was familiar with Pauli's papers and was quite impressed with them. After a few minutes of conversation, Ehrenfest remarked, "I think I like your Encyclopedia article [on relativity theory] better than I like you," to which Pauli shot back, "That's strange. With me, regarding you, it is just the opposite." *[16] The two became very good friends from then on.

A somewhat warmer picture emerges from this story which appears in the article on Dirac:

"Werner Heisenberg [in *Physics and Beyond*, 1971] recollects a friendly conversation among young participants at the 1927 Solvay Conference, about Einstein and Planck's views on religion. Wolfgang Pauli, Heisenberg, and Dirac took part in it. Dirac's contribution was a poignant and clear criticism of the political manipulation of religion, that was much appreciated for its lucidity by Bohr, when Heisenberg reported it to him later. Among other things, Dirac said: "I cannot understand why we idle discussing religion. If we are honest – and as scientists honesty is our precise duty – we cannot help but admit that any religion is a pack of false statements, deprived of any real foundation. The very idea of God is a product of human imagination. [...] I do not recognize any religious myth, at least because they contradict one another. [...]" Heisenberg's view was tolerant. Pauli had kept silent, after some initial remarks. But when finally he was asked for his opinion, jokingly he said: "Well, I'd say that also our friend Dirac has got a religion and the first commandment of this religion is 'God does not exist and Paul Dirac is his prophet'". Everybody burst into laughter, including Dirac.

Many of Pauli's ideas and results were never published and appeared only in his letters, which were often copied and circulated by their recipients. Pauli may have been unconcerned that much of his work thus went uncredited, but when it came to Heisenberg's world-renowned 1958 lecture at Göttingen on their joint work on a unified field theory, and the press release calling Pauli a mere "assistant to Professor Heisenberg", Pauli became offended, shooting back several times at CERN and elsewhere by denouncing Heisenberg's physics prowess. The deterioration between them resulted in Heisenberg ignoring Pauli's funeral, and writing in his autobiography that Pauli's criticisms were overwrought.*[17] Pauli was elected a Foreign Member of the Royal Society (ForMemRS) in 1953.*[2] In 1958 he became foreign member of the Royal Netherlands Academy of Arts and Sciences.*[18]

Personal life

In May 1929, Pauli left the Roman Catholic Church. In December of that year, he married Käthe Margarethe Deppner. The marriage was an unhappy one, ending in divorce in 1930 after less than a year. He married again in 1934 to Franziska Bertram (1901-1987). They had no children.

6.8.2 Bibliography

by Pauli

- Pauli, Wolfgang; Jung, C.G. (1955). *The Interpretation of Nature and the Psyche*. Ishi Press. ISBN 4-87187-713-2.

- Pauli, Wolfgang (1981). *Theory of Relativity*. New York: Dover Publications. ISBN 0-486-64152-X.

- Pauli, Wolfgang; Jung, C.G. (2001). ed. C.A. Meier, ed. *Atom and Archetype, The Pauli/Jung Letters, 1932–1958*. Princeton, New Jersey: Princeton University Press. ISBN 978-0-691012-07-0.

about Pauli

- Enz, Charles P. (2002). *No Time to be Brief, A scientific biography of Wolfgang Pauli*. Oxford Univ. Press.

- Enz, Charles P. (1995). "Rationales und Irrationales im Leben Wolfgang Paulis". In ed. H. Atmanspacher et al. *Der Pauli-Jung-Dialog*. Berlin: Springer-Verlag.

- Fischer, Ernst Peter (2004). *Brücken zum Kosmos. Wolfgang Pauli – Denkstoffe und Nachtträume zwischen Kernphysik und Weltharmonie*. Libelle. ISBN 978-3-909081-44-8.

- Gieser, Suzanne (2005). *The Innermost Kernel. Depth Psychology and Quantum Physics. Wolfgang Pauli's Dialogue with C.G. Jung*. Springer Verlag.

- Jung, C.G. (1980). *Psychology and Alchemy*. Princeton, New Jersey: Princeton Univ. Press.

- Keve, Tom (2000). *Triad: the physicists, the analysts, the kabbalists*. London: Rosenberger & Krausz.

- Lindorff, David (1994). *Pauli and Jung: The Meeting of Two Great Minds*. Quest Books.

- Pais, Abraham (2000). *The Genius of Science*. Oxford: Oxford University Press.

- Enz, P.; von Meyenn, Karl (editors); Schlapp, Robert (translator) (1994). *Wolfgang Pauli – Writings on physics and philosophy*. Berlin: Springer Verlag. ISBN 978-3-540-56859-9.

- Laurikainen, K. V. (1988). *Beyond the Atom – The Philosophical Thought of Wolfgang Pauli*. Berlin: Springer Verlag. ISBN 0-387-19456-8.

- Casimir, H. B. G. (1983). *Haphazard Reality: Half a Century of Science*. New York: Harper & Row. ISBN 0-06-015028-9.

- Casimir, H. B. G. (1992). *Het toeval van de werkelijkheid: Een halve eeuw natuurkunde*. Amsterdam: Meulenhof. ISBN 90-290-9709-4.

- Miller, Arthur I. (2009). *Deciphering the Cosmic Number: The Strange Friendship of Wolfgang Pauli and Carl Jung*. New York: W.W. Norton & Co. ISBN 978-0-393-06532-9.

- Remo, F. Roth: *Return of the World Soul, Wolfgang Pauli, C.G. Jung and the Challenge of Psychophysical Reality [unus mundus], Part 1: The Battle of the Giants*. Pari Publishing, 2011, ISBN 978-88-95604-12-1.

- Remo, F. Roth: *Return of the World Soul, Wolfgang Pauli, C.G. Jung and the Challenge of Psychophysical Reality [unus mundus], Part 2: A Psychophysical Theory*. Pari Publishing, 2012, ISBN 978-88-95604-16-9.

6.8.3 References

[1] Wolfgang Pauli at the Mathematics Genealogy Project

[2] Peierls, Rudolf (1960). "Wolfgang Ernst Pauli 1900-1958". *Biographical Memoirs of Fellows of the Royal Society* (Royal Society) **6**. doi:10.1098/rsbm.1960.0014.

[3] Gerald E. Brown and Chang-Hwan Lee (2006): *Hans Bethe and His Physics*, World Scientific, ISBN 981-256-610-4, p. 338

[4] Ernst Mach and Wolfgang Pauli's ancestors in Prague

[5] "Jewish Physicists". Retrieved 2006-09-30.

[6] Charles Paul Enz (2002). *No Time to Be Brief: A Scientific Biography of Wolfgang Pauli*. Oxford University Press. ISBN 9780198564799. At the same time Pauli writes on 11 October 1957 to the science historian Shmuel Sambursky whom he had met on his trip to Israel (see Ref. [7], p. 964): 'In opposition to the monotheist religions – but in unison with the mysticism of all peoples, including the Jewish mysticism – I believe that the ultimate reality is not personal.'

[7] Werner Heisenberg (2007). *Physics and Philosophy: The Revolution in Modern Science*. HarperCollins. pp. 214–215. ISBN 9780061209192. Wolfgang shared my concern. ..."Einstein's conception is closer to mine. His God is somehow involved in the immutable laws of nature. Einstein has a feeling for the central order of things. He can detect it in the simplicity of natural laws. We may take it that he felt this simplicity very strongly and directly during his discovery of the theory of relativity. Admittedly, this is a far cry from the contents of religion. I don't believe Einstein is tied to any religious tradition, and I rather think the idea of a personal God is entirely foreign to him."

[8] Pauli, Wolfgang Ernst (1921). *Über das Modell des Wasserstoff-Molekülions* (PhD thesis). Ludwig-Maximilians-Universität München.

[9] Varlaki, P.; Nadai L.; Bokor, J. (2008). "Number Archetypes and Background Control Theory Concerning the Fine Structure Constant" (PDF). *Acta Polytechnica Hungarica* **5** (2). Retrieved 2009-02-12.

[10] Charles Paul Enz: *No Time to be Brief: A Scientific Biography of Wolfgang Pauli*, first published 2002, reprinted 2004, ISBN 0-19-856479-1, p. 338

[11] "By a 'cabalistic' coincidence, Wolfgang Pauli died in room 137 of the Red-Cross hospital at Zurich on 15 December 1958." - Of Mind and Spirit, Selected Essays of Charles Enz, Charles Paul Enz, World Scientific, 2009, ISBN 978-981-281-900-0, pg.95.

[12] Enz, Charles; Meyenn, Karl von (1994). *Wolfgang Pauli, A Biographical Introduction. Writings on Physics and Philosophy* (Springer-Verlag). p. 19.

[13] Pauli, W. (1954). "Naturwissenschaftliche und erken-ntnistheoretische Aspekte der Ideen vom Unbewussten". *Dialectica* **8** (4): 283–301. doi:10.1111/j.1746-8361.1954.tb01265.x.

[14] Atmanspacher, H.; Primas, H. (2006). "Pauli's ideas on mind and matter in the context of contemporary science" (PDF). *Journal of Consciousness Studies* **13** (3): 5–50. Retrieved 2009-02-12.

[15] Conference on Wolfgang Pauli's Philosophical Ideas and Contemporary Science organised by ETH May 20–25, 2007. The abstract of a paper discussing this by Richard Jorgensen is here

[16] *The Historical Development of Quantum Theory*, By Jagdish Mehra, Helmut Rechenberg, page 488, Springer (December 28, 2000), ISBN 978-0-387-95175-1, citing Oskar Klein.

[17] Arthur I. Miller (10 Dec 2009). "The strange friendship of Pauli and Jung – Part 6" (flv). *CERN*. University College London. pp. 4–6:00,8:10–8:50. ...a press release that read, most offensively to Pauli, 'Professor Heisenberg and his assistant W. Pauli...

[18] "Wolfgang Ernst Pauli (1900 - 1958)". Royal Netherlands Academy of Arts and Sciences. Retrieved 26 July 2015.

6.8.4 External links

- Publications by and about Wolfgang Pauli in the catalogue Helveticat of the Swiss National Library

- Wolfgang Pauli at the official Nobel Prize site

- Pauli bio at the University of St Andrews, Scotland

- Wolfgang Pauli bio at "Nobel Prize Winners"

- Wolfgang Pauli, Carl Jung and Marie-Louise von Franz

- Photos of Wolfgang Pauli at the Emilio Segrè Visual Archives, American Institute of Physics

- Virtual walk-through exhibition of the life and times of Pauli

- Annotated bibliography for Wolfgang Pauli from the Alsos Digital Library for Nuclear Issues

- Pauli Archives at CERN Document Server

- Virtual exhibition at ETH-Bibliothek, Zurich

- Key Participants: Wolfgang Pauli – *Linus Pauling and the Nature of the Chemical Bond: A Documentary History*

6.9 Martin Lewis Perl

Martin Lewis Perl (June 24, 1927 – September 30, 2014) was an American physicist who won the Nobel Prize in Physics in 1995 for his discovery of the tau lepton.

6.9.1 Life and career

Perl was born in New York City, New York. His parents, Fay (née Resenthal), a secretary and bookkeeper, and Oscar Perl, a stationery salesman who founded a printing and advertising company, were Jewish emigrants to the US from the Polish area of Russia.[1]

Perl is a 1948 chemical engineering graduate of Brooklyn Polytechnic Institute (now known as NYU-Poly) in Brooklyn. After graduation, Perl worked for the General Electric Company, as a chemical engineer in a factory producing electron vacuum tubes. To learn about how the electron tubes worked, Perl signed up for courses in atomic physics and advanced calculus at Union College in Schenectady, New York, which led to his growing interest in physics, and eventually to becoming a graduate student in physics in 1950.[1]

He received his Ph.D. from Columbia University in 1955, where his thesis advisor was I.I. Rabi. Perl's thesis described measurements of the nuclear quadrupole moment of sodium, using the atomic beam resonance method that Rabi had won the Nobel Prize in Physics for in 1944.[1]

Following his Ph.D., Perl spent 8 years at the University of Michigan, where he worked on the physics of strong interactions, using bubble chambers and spark chambers to study the scattering of pions and later neutrons on protons.[1] While at Michigan, Perl and Lawrence W. Jones served as co-advisors to Samuel C. C. Ting, who earned the Nobel Prize in Physics in 1976.

Seeking a simpler interaction mechanism to study, Perl started to consider electron and muon interactions.[2] He had the opportunity to start planning experimental work in this area when he moved in 1963 to the Stanford Linear Accelerator Center (SLAC), then being built in California. He was particularly interested in understanding the muon: why it should interact almost exactly like the electron but be 206.8 times heavier, and why it should decay through the route that it does. Perl chose to look for answers to these questions in experiments on high-energy charged leptons. In addition, he considered the possibility of finding a third generation of lepton through electron-positron collisions. He died after a heart attack[3] at Stanford University Hospital on September 30, 2014 at the age of 87.[4]

Discovery of the tau particle

The tau lepton (τ, also called the tau particle, tauon or simply tau) is an elementary particle similar to the electron, with negative electric charge and a spin of $\frac{1}{2}$, but with 3477 times the mass. Together with the electron, the muon, and the three neutrinos, it is classified as a lepton.

The tau was first detected in a series of experiments between 1974 and 1977 by Perl with his colleagues at the SLAC-LBL group.[*][5] Their equipment consisted of SLAC's then-new e+–e– colliding ring, called SPEAR, and the LBL magnetic detector. They could detect and distinguish between leptons, hadrons and photons. SPEAR was able to collide electrons and positrons at higher energies than had previously been possible, initially at up to 4.8 GeV and eventually at 8 GeV, energies high enough to lead to the production of a tau/antitau pair.[*][2] The tau has a lifetime of only $2.9 \times 10^{*}{-}13$ s and so these particles decayed within a few millimetres of the collision.[*][6] Hence Perl and his coworkers did not detect the tau directly, but rather discovered anomalous events where they detected either an electron and a muon, or a positron and an antimuon:

> "*We have discovered 64 events of the form*
>
> e+ + e– $\rightarrow$ e$\pm$ + $\mu\mp$ + at least two undetected particles
>
> *for which we have no conventional explanation.*"

The need for at least two undetected particles was shown by the inability to conserve energy and momentum with only one. However, no other muons, electrons, photons, or hadrons were detected. It was proposed that this event was the production and subsequent decay of a new particle pair:

$$e+ + e- \rightarrow \tau+ + \tau- \rightarrow e\pm + \mu\mp + 4\nu$$

This was difficult to verify, because the energy to produce the $\tau+\tau-$ pair is similar to the threshold for D meson production. Work done at DESY-Hamburg, and with the Direct Electron Counter (DELCO) at SPEAR, subsequently established the mass and spin of the tau.

The symbol τ was derived from the Greek $\tau\rho\acute{\iota}\tau o\nu$ (*triton*, meaning "third" in English), since it was the third charged lepton discovered.[*][7]

Nobel Prize and later career

Perl won the Nobel Prize in 1995 jointly with Frederick Reines. The prize was awarded "for pioneering experimental contributions to lepton physics". Perl received half

"for the discovery of the tau lepton" while Reines received his share "for the detection of the neutrino".[*][8]

He joined University of Liverpool as a visiting professor.[*][9] He served on the board of advisors of Scientists and Engineers for America, an organization focused on promoting sound science in American government. In 2009, Perl received an honorary doctorate from the University of Belgrade.[*][10]

6.9.2 See also

- List of Jewish Nobel laureates

6.9.3 References

[1] "Martin L. Perl - Biographical". Nobel Media AB. 1995. Retrieved 2013-12-28.

[2] Martin L. Perl (1995). "Reflections on the Discovery of the Tau Lepton". Retrieved 2013-12-28.

[3] http://www.nytimes.com/2014/10/04/science/martin-perl-physicist-who-discovered-electrons-long-lost-brother-dies-at-87.html

[4] http://news.stanford.edu/news/2014/october/martin-perl-obit-100114.html

[5] Perl, M. L.; Abrams, G.; Boyarski, A.; Breidenbach, M.; Briggs, D.; Bulos, F.; Chinowsky, W.; Dakin, J. et al. (1975). "Evidence for Anomalous Lepton Production in e+e– Annihilation". *Physical Review Letters* **35** (22): 1489. Bibcode:1975PhRvL..35.1489P. doi:10.1103/PhysRevLett.35.1489.

[6] "The Nobel Prize in Physics 1995 - Press Release". Nobel Media AB. 1995. Retrieved 2014-01-01.

[7] M.L. Perl (1977). "Evidence for, and properties of, the new charged heavy lepton" (PDF). In T. Thanh Van (ed.). *Proceedings of the XII Rencontre de Moriond*. SLAC-PUB-1923.

[8] "Nobel Prize in Physics, 1995". 1995. Retrieved 2013-12-28.

[9] "Professor Martin Perl joins University of Liverpool". BBC. 3 December 2011. Retrieved 3 December 2011.

[10] "Promovisani počasni doktori Beogradskog univerziteta - RADIO-TELEVIZIJA VOJVODINE". Rtv.rs. 2009-10-20. Retrieved 2011-02-17.

6.9.4 External links

- Nobel autobiography

- Nobel Prize press release, explaining the significance of Perl's work

- Biography and Bibliographic Resources, from the Office of Scientific and Technical Information, United States Department of Energy

- Personal blog: Reflections on Physics

- U.S. Patent 5943075 Universal fluid droplet ejector (Martin Lewis Perl)

- U.S. Patent 5975682 Two-dimensional fluid droplet arrays generated using a single nozzle (Martin Lewis Perl)

6.10   Bruno Pontecorvo

Bruno Pontecorvo (Russian: Брýно Макси́мович Понтекóрво, *Bruno Maksimovich Pontekorvo*; 22 August 1913 – 24 September 1993) was an Italian nuclear physicist, an early assistant of Enrico Fermi and then the author of numerous studies in high energy physics, especially on neutrinos. According to Oleg Gordievsky (the highest-ranking KGB officer ever to defect)[*][1] and Pavel Sudoplatov (former deputy director of Foreign Intelligence for the Soviet Union),[*][2] Pontecorvo was also a Soviet agent.[*][3] Convinced communist, he defected to the Soviet Union in 1950, where he continued his research on the decay of the muon and on neutrinos. The prestigious Pontecorvo Prize was instituted in his memory in 1995.

6.10.1   Early life and education

Pontecorvo was born in Marina di Pisa into a wealthy non-observant Italian Jewish family. After attending the first two years of Engineering at the University of Pisa, at only 18 he was admitted to the third year of Physics at the University of Rome *La Sapienza*. There he soon became one of the closest (and the youngest) assistants of Fermi and one of the so-called Via Panisperna boys (as Fermi's group of scientists is often called, after the name of the street where the Institute of Physics of Rome University was then situated). Fermi described Pontecorvo as "scientifically one of the brightest men with whom I have come in contact in my scientific career".[*][4]

In 1934 he contributed to Fermi's famous experiment showing the properties of slow neutrons that led the way to the discovery of nuclear fission.

Bruno Pontecorvo was the older brother of Gillo Pontecorvo, the director of *The Battle of Algiers*, as well as Guido Pontecorvo, a geneticist, and Poli Pontecorvo, an engineer who worked on radar after World War II.

6.10.2   Early career

In 1936 he moved to Paris to work in the laboratory of Irène and Frédéric Joliot-Curie on the effects of collisions of neutrons with protons and on the electromagnetic transitions among isomers. During this period he was influenced by the ideas of socialism to which he remained loyal for the rest of his life. In Paris, in 1938, he formed a relationship with Marianne Nordblom, a young student of French Literature, and their first son was born during that year.

Carlo Franzinetti (Left) and Bruno Pontecorvo (Right)

Pontecorvo was unable to return to Italy because of the fascist regime's racial laws against the Jews. He remained in Paris until the Nazis entered the city, then fled with his family to Spain and shortly after to the United States, where he had found employment with an oil company in Tulsa, Oklahoma. While at the oil company he developed a technology and an instrument for *well logging*, based on the properties of neutrons. This technology may be considered the first practical application of the *Via Panisperna boys* discovery of slow neutrons.

He was not called upon to participate in the Manhattan Project in the USA for the construction of the atomic bomb, possibly because of his committed socialist beliefs. But in 1943 he was invited to join the associated Montreal Laboratory in Canada, where he concentrated on reactor design, cosmic rays, neutrinos and the decay of muons.

In 1948, after he obtained British citizenship, he was invited by John Cockcroft to contribute to the British atomic bomb project at AERE, Harwell where he joined the Nuclear Physics Division under Egon Bretscher. In 1950 he was appointed to the chair of physics at the University of Liverpool which he was due to take up in January 1951.

6.10.3 Defection

However, on 31 August 1950, in the middle of a holiday in Italy, he abruptly left Rome for Stockholm with his wife and three sons without informing friends or relatives. The next day he was helped by Soviet agents to enter the Soviet Union from Finland. His abrupt disappearance caused much concern to many of the western intelligence services, especially those of Britain and the USA who were worried about the escape of atomic secrets to the Soviet Union after the then recent case of Klaus Fuchs. But as was pointed out immediately, Pontecorvo had had only limited access to "secret subjects" and even later no allegation of spying or of transferring of secrets to the Soviets has ever been made against him.

In the USSR Pontecorvo was welcomed with honor and given a number of privileges reserved only to the Soviet nomenklatura. He worked until his death in what is now the Joint Institute for Nuclear Research (JINR) in Dubna, concentrating entirely on theoretical studies of high energy particles and continuing his research on neutrinos and decay of muons. In recognition of his research he was awarded the Stalin Prize in 1953, membership of the Soviet Academy of Sciences in 1958 and two Orders of Lenin. In 1955 he appeared in public at a press conference where he explained to the world the motivations of his choice to leave the West and work in the USSR. Pontecorvo did not leave the Soviet Union for many years, the first trip being in 1978 when he travelled to Italy.

6.10.4 Personal life

Pontecorvo was brother of film director Gillo Pontecorvo and geneticist Guido Pontecorvo. He was a great-uncle of Flavio Pontecorvo, the electronics engineer. He had one wife: Marianna Nordblom (born in Sweden) with whom he had three children, and a long-term relationship with Rodam Amiredzhibi (born in Georgia, Soviet Union).

6.10.5 Death

He died in Dubna in 1993, afflicted by Parkinson's disease. Half of his ashes are now buried in the Protestant Cemetery in Rome, and another half in Dubna, Russia, according to his will.

6.10.6 Legacy

In 1995, in recognition of his scientific merits, the prestigious Pontecorvo Prize has been instituted by the Joint Institute for Nuclear Research. The prize, awarded annually to an individual scientist, recognizes "the most significant investigations in elementary particle physics", as acknowledged by the international scientific community.

The scientific work of Bruno Pontecorvo is full of formidable intuitions, some of which have represented milestones in modern physics. These include:

- the intuition of how to detect anti-neutrinos generated in nuclear reactors (methodology used by Frederick Reines who was awarded for this the Nobel prize in 1995);

- the prediction that neutrinos associated with electrons are different from those associated with muons (for experimental verification of this another Nobel prize was awarded to J. Steinberger, L. Lederman and M. Schwartz in 1988);

- the idea that neutrinos may convert into other types of neutrinos, a phenomenon known as neutrino oscillation.

This last idea was proposed in 1957 and developed in subsequent years by Pontecorvo, until 1967 where it was given its modern form. A hint for this phenomenon was first seen with solar neutrinos in 1968 (see Solar neutrino problem); the existence of the oscillations was finally established by the Super-Kamiokande experiment in 1998 and later confirmed by other experiments. However it has not yet been recognized by a Nobel prize (the prize awarded to Masatoshi Koshiba and Ray Davis in 2002 was for neutrino astronomy).

In 2006 Moscow historical society Moskultprog has unveiled an artistic plaque celebrating Pontecorvo's Moscow house at 9 Tverskaya.*[5]

6.10.7 Selected publications

- "Neutron Well Logging – A New Geological Method Based on Nuclear Physics". *Oil and Gas Journal* **40**: 32–33. 1941.

- Pages in the Development of Neutrino Physics, Usp.Fiz.Nauk 141, 1983, 675 [English ed. Sov. Phys. Usp. 26, 1983, 1087]

- B. Pontecorvo, "Mesonium and anti-mesonium", Sov. Phys. JETP 6 429 (1957)

6.10.8 See also

- PMNS matrix

- Solar neutrino problem

- Sudbury Neutrino Observatory

- Super-Kamiokande

6.10.9   References

[1] de Lisle, Leanda. "Pinkos and patriots". *The Guardian* (30 January 2001). Retrieved 22 March 2011.

[2] Stout, David (28 September 1996). "Pavel Sudoplatov, 89, Dies; Top Soviet Spy Who Accused Oppenheimer". *The New York Times*. Retrieved 22 March 2011.

[3] Andrew, Christopher M.; Oleg Gordievsky (1990). *KGB: the inside story of its foreign operations from Lenin to Gorbachev*. HarperCollins. pp. 317–318, 379. ISBN 0-06-016605-3.; Sudoplatov, Pavel; Anatoli Sudoplatov; Jerrold L. Schecter; Leona P. Schecter (1995). *Special tasks: the memoirs of an unwanted witness, a Soviet spymaster*. Little, Brown. p. 3. ISBN 0-316-82115-2.

[4] Wellerstein, Alex (20 February 2015). "Physicist. Defector. Spy?". *Science* **347** (6224): 833. doi:10.1126/science.aaa3654.

[5] В Москве появилась неофициальная мемориальная доска Бруно Понтекорво. Regnum.ru. 14 June 2006.

6.10.10   Further reading

- Close, Frank. *Half-Life: The Divided Life of Bruno Pontecorvo, Physicist or Spy* (Basic Books; 2015) 377 pages

- Mafai, Miriam (1992). *Il lungo freddo: Storia di Bruno Pontecorvo, lo scienziato che scelse l'URSS*. Milan.

- Turchetti, Simone. "Atomic secrets and governmental lies: nuclear science, politics and security in the Pontecorvo case." *British Journal for the History of Science* (2003) 36#4 pp: 389–415. online

- Turchetti, Simone. *The Pontecorvo Affair: a cold war defection and nuclear physics* (University of Chicago Press, 2012)

6.10.11   External links

- Biography / Scientific Works / Popular Articles / About B. Pontecorvo / Photoalbum (in English and Russian)

- 1950s news of Pontecorvo's disappearance from the BBC archive

- Confessions of an atom spy: Forty years after Bruno Pontecorvo, a British scientist, went to work for Moscow, he tells Charles Richards in Rome why he changed sides

- Annotated bibliography of Bruno Pontecorvo from the Alsos Digital Library for Nuclear Issues

6.11   Frederick Reines

Frederick Reines (*RYE-ness*);[1] (March 16, 1918 – August 26, 1998) was an American physicist. He was awarded the 1995 Nobel Prize in Physics for his co-detection of the neutrino with Clyde Cowan in the neutrino experiment. He may be the only scientist in history "so intimately associated with the discovery of an elementary particle and the subsequent thorough investigation of its fundamental properties".[2]

A graduate of the Stevens Institute of Technology and New York University, Reines joined the Manhattan Project's Los Alamos Laboratory in 1944, working in the Theoretical Division in Richard Feynman's group. He became a group leader there in 1946. He participated in a number of nuclear tests, culminating in his becoming the director of the Operation Greenhouse test series in the Pacific in 1951.

In the early 1950s, working in Hanford and Savannah River Sites, Reines and Cowan developed the equipment and procedures with which they first detected the supposedly undetectable neutrinos in June 1956. Reines dedicated the major part of his career to the study of the neutrino's properties and interactions, which work would influence study of the neutrino for many researchers to come. This included the detection of neutrinos created in the atmosphere by cosmic rays, and the 1987 detection of neutrinos emitted from Supernova SN1987A, which inaugurated the field of neutrino astronomy.

6.11.1   Early life

Frederick Reines was born in Paterson, New Jersey, one of four children of Gussie (Cohen) and Israel Reines. His parents were Jewish emigrants from the same town in Russia, but only met in New York City, where they were later married. He had an older sister, Paula, who became a doctor, and two older brothers, David and William, who became lawyers. He said that his "early education was strongly influenced" by his studious siblings. He was the great-nephew of the Rabbi Yitzchak Yaacov Reines, the founder of Mizrachi, a religious Zionist movement.[3]

The family moved to Hillburn, New York, where his father ran the general store, and he spent much of his childhood. He was an Eagle Scout. Looking back, Reines said: "My early childhood memories center around this typical American country store and life in a small American town, including Independence Day July celebrations marked by fireworks and patriotic music played from a pavilion bandstand." *[4]

Reines sang in a chorus, and as a soloist. For a time he considered the possibility of a singing career, and was instructed by a vocal coach from the Metropolitan Opera who provided lessons for free because the family did not have the money for them.*[4] The family later moved to North Bergen, New Jersey, residing on Kennedy Boulevard and 57th Street. Because North Bergen did not have a high school,*[5] he attended Union Hill High School in Union Hill, New Jersey,*[4]*[5] from which he graduated in 1935.*[5]

From an early age, Reines exhibited an interest in science, and liked creating and building things. He later recalled that:

> The first stirrings of interest in science that I remember occurred during a moment of boredom at religious school, when, looking out of the window at twilight through a hand curled to simulate a telescope, I noticed something peculiar about the light; it was the phenomenon of diffraction. That began for me a fascination with light.*[4]

Ironically, Reines excelled in literary and history courses, but received average or low marks in science and math in his freshman year of high school, though he improved in those areas by his junior and senior years through the encouragement of a teacher who gave him a key to the school laboratory. This cultivated a love of science by his senior year. In response to a question seniors were asked about what they wanted to do for a yearbook quote, he responded: "To be a physicist extraordinaire." *[4]

Reines was accepted into the Massachusetts Institute of Technology, but chose instead to attend Stevens Institute of Technology in Hoboken, New Jersey, where he earned his Bachelor of Science (B.S.) degree in mechanical engineering in 1939, and his Master of Science (M.S.) degree in mathematical physics in 1941, writing a thesis on "A Critical Review of Optical Diffraction Theory" .*[3] He married Sylvia Samuels on August 30, 1940.*[3] They had two children, Robert and Alisa.*[4] He then entered New York University, where he earned his Doctor of Philosophy (Ph.D.) in 1944. He studied cosmic rays there under Serge A. Korff,*[4] but wrote his thesis under the supervision of Richard D. Present*[3] on "Nuclear fission and the liquid drop model of the nucleus" .*[6] Publication of the thesis

was delayed until after the end of World War II; it appeared in Physical Review in 1946.*[3]*[7]

6.11.2 Los Alamos Laboratory

Operation Greenhouse – Dog shot

In 1944 Richard Feynman recruited Reines to work in the Theoretical Division at the Manhattan Project's Los Alamos Laboratory, where he would remain for the next fifteen years.*[4] He joined Feynman's T-4 (Diffusion Problems) Group, which was part of Hans Bethe's T (Theoretical) Division. Diffusion was an important aspect of critical mass calculations.*[3] In June 1946, he became a group leader, heading the T-1 (Theory of Dragon) Group. An outgrowth of the "tickling the Dragon's tail" experiment, the Dragon was a machine that could attain a critical state for short bursts of time, which could be used as a research tool or power source.*[8]

Reines participated in a number of nuclear tests, and writing reports on their results. These included Operation Crossroads at Bikini Atoll in 1946, Operation Sandstone at Eniwetok Atoll in 1948, and Operation Ranger and Operation Buster–Jangle at the Nevada Test Site. In 1951 he was the director of Operation Greenhouse series of nuclear tests in the Pacific. This saw the first American tests of boosted fission weapons, an important step towards thermonuclear weapons. He studied the effects of nuclear blasts, and co-authored a paper with John von Neumann on Mach stem formation, an important aspect of an air blast wave.*[3]*[4]

In spite or perhaps because of his role in these nuclear tests, Reines was concerned about the dangers of radioactive pollution from atmospheric nuclear tests, and became an advocate of underground nuclear testing. In the wake of the Sputnik crisis, he participated in John Archibald Wheeler's Project 137, which evolved into JASON. He was also a delegate at the Atoms for Peace Conference in Geneva in 1958.[3][4]

6.11.3 Discovery of the neutrino and the inner workings of stars

The neutrino was a subatomic particle first proposed theoretically by Wolfgang Pauli on December 4, 1930, to explain undetected energy that escaped during beta decay when neutron decayed into a proton and an electron so that the law of conservation of energy was not violated. Enrico Fermi renamed it the neutrino, Italian for "little neutron",[9] and in 1934, proposed his theory of beta decay which explained that the electrons emitted from the nucleus were created by the decay of a neutron into a proton, an electron, and a neutrino:[10][11]

$$n0 \rightarrow p+ + e- + ve$$

The neutrino accounted for the missing energy, but Fermi's theory described a particle with little mass and no electric charge that would be difficult to observe directly. In a 1934 paper, Rudolf Peierls and Hans Bethe calculated that neutrinos could easily pass through the Earth, and concluded "there is no practically possible way of observing the neutrino."[12] In 1951, at the conclusion of the Greenhouse test series, Reines received permission from the head of T Division, J. Carson Mark, for a leave in residence to study fundamental physics. Reines and his colleague Clyde Cowan decided to see if they could detect neutrinos. "So why did we want to detect the free neutrino?" he later explained, "Because everybody said, you couldn't do it."[13]

According to Fermi's theory, there was also a corresponding reverse reaction, in which a neutrino combines with a proton to create a neutron and a positron:[13]

$$\overset{\nu}{e} + p+ \rightarrow n0 + e+$$

The positron would soon be annihilated by an electron and produce two 0.51 MeV gamma rays, while the neutron would be captured by a proton and release a 2.2 MeV gamma ray. This would produce a distinctive signature that could be detected. They then realised that by adding cadmium salt to their liquid scintillator to enhance the neutron capture reaction, resulting in a 9 MeV burst of gamma rays.[14] For a neutrino source, they proposed using an atomic bomb. Permission for this was obtained from the laboratory director, Norris Bradbury. Work began on digging a shaft for the experiment when J. M. B. Kellogg convinced them to use a nuclear reactor instead of a bomb.

Although a less intense source of neutrinos, it had the advantage in allowing for multiple experiments to be carried out over a long period of time.[3][14]

In 1953, the made their first attempts using one of the large reactors at the Hanford nuclear site in what is now known as the Cowan–Reines neutrino experiment. Their detector now included 300 litres (66 imp gal; 79 US gal) of scintillating fluid and 90 photomultiplier tubes, but the effort was frustrated by background noise from cosmic rays. With encouragement from John A. Wheeler, they tried again in 1955, this time using one of the newer, larger 700 MW reactors at the Savannah River Site that emitted a high neutrino flux of 1.2×10^{12} / cm^2 sec. They also had a convenient, well-shielded location 11 metres (36 ft) from the reactor and 12 metres (39 ft) underground.[13] On June 14, 1956, they were able to send Pauli a telegram announcing that the neutrino had been found.[15] When Bethe was informed that he had been proven wrong, he said he said, "Well, you shouldn't believe everything you read in the papers." [13]

Supernova SN1987A (the bright object in the center), as seen through the Hubble Space Telescope

From then on Reines dedicated the major part of his career to the study of the neutrino's properties and interactions, which work would influence study of the neutrino for future researchers to come.[16] Cowan left Los Alamos in 1957 to teach at George Washington University, ending their collaboration.[3] On the basis of his work in first detecting the neutrino, Reines became the head of the physics department of Case Western Reserve University from 1959 to 1966. At Case, he led a group that was the first to detect neutrinos created in the atmosphere by cosmic rays.[14] Reines had a booming voice, and had been a singer since childhood. During this time, besides performing his duties as a research supervisor and chairman of the physics department, Reines sang in the Cleveland Orchestra Chorus under the direction of Robert Shaw in performances with George Szell and the Cleveland Orchestra.[17]

In 1966, Reines took most of his neutrino research team with him when he left for the new University of California, Irvine (UCI), becoming its first dean of physical sciences. At UCI, Reines extended the research interests of some of his graduate students into the development of medical radiation detectors, such as for measuring total radiation delivered to the whole human body in radiation therapy.[17]

Reines had prepared for the possibility of measuring the distant events of a supernova explosion. Supernova explosions are rare, but Reines thought he might be lucky enough to see one in his lifetime, and be able to catch the neutrinos streaming from it in his specially-designed detectors. During his wait for a supernova to explode, he put signs on some of his large neutrino detectors, calling them "Supernova Early Warning Systems" .[17] In 1987, neutrinos emitted from Supernova SN1987A were detected by the Irvine–Michigan–Brookhaven (IMB) Collaboration. which used an 8,000 ton Cherenkov detector located in a salt mine near Cleveland.[18] Normally, the detectors recorded only a few background events each day. The supernova registered 19 events in just ten seconds.[13] This discovery is regarded as inaugurating the field of neutrino astronomy.[18]

In 1995, Reines was honored, along with Martin L. Perl with the Nobel Prize in Physics for his work with Cowan in first detecting the neutrino. Unfortunately, Cowan had died in 1974, and the Nobel Prize is not awarded posthumously.[16] Reines also received many other awards, including the J. Robert Oppenheimer Memorial Prize in 1981,[19] the National Medal of Science in 1985, the Bruno Rossi Prize in 1989, the Michelson–Morley Award in 1990, the Panofsky Prize in 1992, and the Franklin Medal in 1992. He was elected a member of the National Academy of Sciences in 1980 and a foreign member of the Russian Academy of Sciences in 1994.[3] He remained dean of physical sciences at UCI until 1974, and became a professor emeritus in 1988, but he continued teaching until 1991, and remained on UCI's faculty until his death.[20]

6.11.4 Death

Reines died after a long illness at the University of California, Irvine Medical Center in Orange, California,[1] on August 26, 1998.[3] He was survived by his wife and children.[1] His papers are in the UCI Libraries.[21] Reines Hall at UCI was named in his honor.[22]

Frederick Reines Hall at the University of California, Irvine. The building houses the Physics and Astronomy Department and part of the Chemistry Department.

6.11.5 Publications

- Reines, F. & C. L. Cowan, Jr. "On the Detection of the Free Neutrino" , Los Alamos National Laboratory (LANL) (through predecessor agency Los Alamos Scientific Laboratory), United States Department of Energy (through predecessor agency the Atomic Energy Commission), (August 6, 1953).

- Reines, F., Cowan, C. L. Jr., Carter, R. E., Wagner, J. J. & M. E. Wyman. "The Free Antineutrino Absorption Cross Section. Part I. Measurement of the Free Antineutrino Absorption Cross Section. Part II. Expected Cross Section from Measurements of Fission Fragment Electron Spectrum" , Los Alamos National Laboratory (LANL) (through predecessor agency Los Alamos Scientific Laboratory), United States Department of Energy (through predecessor agency the Atomic Energy Commission), (June 1958).

- Reines, F., Gurr, H. S., Jenkins, T. L. & J. H. Munsee. "Neutrino Experiments at Reactors" , University of California-Irvine, Case Western Reserve University, United States Department of Energy (through predecessor agency the Atomic Energy Commission), (September 9, 1968).

- Roberts, A., Blood, H., Learned, J. & F. Reines. "Status and Aims of the DUMAND Neutrino Project: the Ocean as a Neutrino Detector" , Fermi National Accelerator Laboratory (FNAL), United States Department of Energy (through predecessor agency the Energy Research and Development Administration), (July 1976).

- Reines, F. (1991). *Neutrinos and Other Matters: Selected Works of Frederick Reines.* Teaneck, N.J.: World Scientific. ISBN 978-981-02-0392-4.

6.11.6   Notes

[1] Wilford, John Noble (August 28, 1998). "Frederick Reines Dies at 80; Nobelist Discovered Neutrino". *The New York Times.* Retrieved February 18, 2015.

[2] Schultz, Jonas; Sobel, Hank. "Frederick Reines and the Neutrino". University of California, Irvine School of Physical Sciences. Archived from the original on February 20, 2014.

[3] Kropp, William; Schultz, Jonas; Sobel, Henry (2009). *Frederick Reines 1918-1998 A Biographical Memoir* (PDF). Washington D.C.: National Academy of Sciences. Retrieved March 17, 2010.

[4] "The Nobel Prize in Physics 1995". Nobel Foundation. Retrieved March 23, 2012.

[5] Pope, Gennarose (March 25, 2012). "Bridge of troubled Kennedy Boulevard". *The Union City Reporter.* p. 12.

[6] "Nuclear fission and the liquid drop model of the nucleus". New York University. Retrieved February 18, 2015.

[7] Present, R. D.; Reines, F.; Knipp, J. K. (October 1946). "The Liquid Drop Model for Nuclear Fission". *Physical Review* (American Physical Society) **70** (7-8): 557–558. Bibcode:1946PhRv...70..557P. doi:10.1103/PhysRev.70.557.2.

[8] Truslow & Smith 1961, pp. 56-59.

[9] Close 2012, pp. 15–18.

[10] Fermi, E. (1968). Wilson, Fred L. (trans.). "Fermi's Theory of Beta Decay" (PDF). *American Journal of Physics* **36**. Bibcode:1968AmJPh..36.1150W. doi:10.1119/1.1974382. Retrieved January 20, 2013.

[11] Close 2012, pp. 22–25.

[12] "The Neutrino". *Nature* (133): 532–532. April 7, 1934. Bibcode:1934Natur.133..532B. doi:10.1038/133532a0. ISSN 0028-0836.

[13] Reines, Frederick (December 8, 1995). "The Neutrino: From Poltergeist to Particle" (PDF). Nobel Foundation. Retrieved February 20, 2015. Nobel Prize lecture

[14] Lubkin, Gloria B. "Nobel Prize in Physics goes to Frederick Reines for the Detection of the Neutrino" (PDF). *Physics Today* **54** (2): 17–19. ISSN 0031-9228. Archived from the original (PDF) on December 17, 2008.

[15] Close 2012, pp. 37–41.

[16] Close 2012, p. 42.

[17] "In Memoriam, 1998. Frederick Reines, Physics; Radiological Sciences: Irvine". University of California. Retrieved February 19, 2015.

[18] Schultz, Jonas; Sobel, Hank. "Frederick Reines and the Neutrino". Archived from the original on February 20, 2014.

[19] "Frederick Reines wins Oppenheimer Prize". *Physics Today* (American Institute of Physics): 94. May 1981. Bibcode:1981PhT....34R..94.. doi:10.1063/1.2914589. Retrieved March 1, 2015.

[20] "The Passing of Frederick Reines, Physics Nobel Laureate in 1995". University of California, Irvine. Archived from the original on November 2, 2013.

[21] "Guide to the Frederick Reines Papers". Retrieved February 18, 2015 – via California Digital Library.

[22] Benjamin, Marisa. "Frederick Reines Hall at UC Irvine". About.com. Retrieved February 18, 2015.

6.11.7   References

- Close, Frank E. (2012). *Neutrino.* Oxford: Oxford University Press. ISBN 9780199574599. OCLC 840096946.

- Truslow, Edith C.; Smith, Ralph Carlisle (1961). *Manhattan District history, Project Y, the Los Alamos story, Volume II: August 1945 to December 1946* (PDF). Los Angeles: Tomash Publishers. ISBN 978-0-938228-08-0. Retrieved February 20, 2014. Originally published as Los Alamos Report LAMS-2532

6.11.8   External links

- Biography and Bibliographic Resources, from the Office of Scientific and Technical Information, United States Department of Energy

- The Neutrino: From Poltergeist to Particle (Nobel lecture)

- Guide to the Frederick Reines Papers. Special Collections and Archives, The UC Irvine Libraries, Irvine, California.

6.12   Shoichi Sakata

Dr. **Shoichi Sakata** (坂田昌一 *Sakata Shōichi*, 18 January 1911, near Hiroshima – 16 October 1970) was a Japanese physicist who was internationally known for theoretical work on the structure of the atom.*[1] He proposed the Sakata model, which was an early precursor to the quark model.

After the end of World War II, he joined other physicists in campaigning for the peaceful uses of atomic energy.*[1]

6.12.1 Career

Between 1929 and 1933 Sakata studied physics in Tokyo under Yoshio Nishina and later at the Kyoto Imperial University under Hideki Yukawa, the first Japanese Nobel laureate. He first met Yukawa at Rikagaku Kenkyūsho in Ōsaka, a private research foundation started by Yukawa. Here he worked with him from 1937 on meson theory and in 1939 accompanied him to Kyoto University where Yukawa was a lecturer. Sakata was appointed professor at Nagoya University in 1942 and remained there until his death.

Sakata was a leading Japanese researcher in elementary particles in the 1950s and 1960s, and became well-known outside Japan for his 1956 model of hadrons, later termed the Sakata model, which proposed that the fundamental building blocks of all strongly interacting particles are the proton, the neutron and the lambda baryon. For example, the positively charged pion is made out of a proton and an antineutron. Aside from the integer charges, the proton, neutron, and lambda have the same properties as the up quark, down quark, and strange quark respectively, explaining the model's success.

Sakata's model was superseded by the quark model, due to Murray Gell-Mann and George Zweig, which made the constituents fractionally charged and rejected the idea that they could be identified with observed particles. This leads to the Gell-Mann–Nishijima formula and the eightfold way, which provides the most correct fundamental description. Still, within Japan, integer charged quark models parallel to Sakata's were used until the 1970s, and are still used as effective descriptions in certain domains.

Sakata's model was used in Harry J. Lipkin's book *"Lie Groups for Pedestrians"* (1965). In 1960, with his Nagoya University associates, he expanded his model to include leptons. Shortly thereafter he developed the Neutrino mixing matrix, a precursor to the currently accepted Neutrino oscillation.*[2] In the early 1960s there was already evidence of a second neutrino type.

6.12.2 Influences

The 2008 physics Nobel laureates Yoichiro Nambu, Toshihide Maskawa and Makoto Kobayashi, who received their awards for work on symmetry breaking, all came under his tutelage and influence.*[3] The Nagoya Model was the inspiration for the later Cabibbo–Kobayashi–Maskawa

matrix of 1973, which specifies the mismatch of quantum states of quarks, when they propagate freely and when they take part in weak interactions. Physicists however, generally attribute the introduction of a third generation of quarks (the "top" and "bottom" quarks) into the Standard Model of the elementary particles to that 1973 paper by Kobayashi and Maskawa.

Kent Staley (2004) describes the historical background to their paper, emphasizing the largely forgotten role of theorists at Nagoya University and the "Nagoya model" they developed. Several of the authors of the Nagoya model embraced the philosophy of dialectical materialism, and he discusses the role that such metaphysical commitments play in physical theorizing. Both theoretical and experimental developments that generated great interest in Japan, and ultimately stimulated Kobayashi and Masukawa's 1973 work, went almost entirely unnoticed in the U.S. The episode exemplifies both the importance of untestable "themata" in developing new theories, and the difficulties that may arise, when two parts of a research community work in relative isolation from one another.*[4]

6.12.3 Missed out on Nobel Prize*[5]

Shoichi Sakata's "Sakata model" inspired Murray Gell-Mann and George Zweig's quark model, but 1969 prize was only awarded to Murray Gell-Mann. Afterward, Ivar Waller, the member of Nobel Committee for Physics was sorry that Sakata had not received a prize.

In September 1970, Hideki Yukawa politely wrote to Waller informing him that Sakata had been ill when the nomination was written; since then, his condition had worsened significantly. Three weeks later, Sakata died. Yukawa informed Waller that a prize to Sakata would have brought him much honor and encouragement. He, then, in the name of leading Japanese particle physicists, asked to know what the Nobel committee thought of Sakata's merits, for that would perhaps bring them consolation.

6.12.4 Honors

- Asahi Prize 1948

- Imperial Prize of the Japan Academy 1950

- Order of the Sacred Treasure (瑞宝章 *Zuihōshō*) 1970

6.12.5 Notes

[1] Nussbaum, Louis-Frédéric. (2005). "*Sakata Shōichi*" in Japan Encyclopedia, *p. 812*, p. 812, at Google Books; n.b.,

Louis-Frédéric is pseudonym of Louis-Frédéric Nussbaum, *see* Deutsche Nationalbibliothek Authority File.

[2] Ziro MAKI, Masami NAKAGAWA and Shoichi SAKATA; *Remarks on the Unified Model of Elementary Particles.* In Progress of Theoretical Physics, Vol. 28, No. 5 (November 1962).

[3] Asia News & Thailand News

[4] Kent W. Staley; *Lost Origins of the Third Generation of Quarks: Theory, Philosophy*: Pages 210-229 in Physics in Perspective (PIP), Birkhäuser, Basel (2004). ISSN 1422-6944

[5] Robert Marc Friedman, *The Politics of Excellence: Behind the Nobel Prize in Science.* New York: Henry Holt & Company (October 2001)

6.12.6   References

- Nussbaum, Louis-Frédéric and Käthe Roth. (2005). *Japan encyclopedia.* Cambridge: Harvard University Press. ISBN 978-0-674-01753-5; OCLC 58053128

6.12.7   External links

- Theoretical Physics and Dialectics of Nature - June 1947

- Philosophy and Methodology of Present-Day Science - 1968

- Engels' "Dialektik der Natur" - July 1969

- CP Violation and Flavour Mixing

6.13   Melvin Schwartz

Melvin Schwartz (November 2, 1932 – August 28, 2006) was an American physicist. He shared the 1988 Nobel Prize in Physics with Leon M. Lederman and Jack Steinberger for their development of the neutrino beam method and their demonstration of the doublet structure of the leptons through the discovery of the muon neutrino.[*][2]

6.13.1   Biography

He grew up in New York City in the Great Depression and went to the Bronx High School of Science. His interest in physics began there at the age of 12.

He earned his B.A. (1953) and Ph.D. (1958) at Columbia University, where Nobel laureate I. I. Rabi was the head of the physics department. Schwartz became an assistant professor at Columbia in 1958. He was promoted to associate professor in 1960 and full professor in 1963. Tsung-Dao Lee, a Columbia colleague who had recently won the Nobel prize at age 30, inspired the experiment for which Schwartz received his Nobel. Schwartz and his colleagues performed the experiments which led to their Nobel Prize in the early 1960s, when all three were on the Columbia faculty. The experiment was carried out at the nearby Brookhaven National Laboratory.

In 1966, after 17 years at Columbia, he moved west to Stanford University, where SLAC, a new accelerator, was just being completed. There, he was involved in research investigating the charge asymmetry in the decay of long-lived neutral kaons and another project which produced and detected relativistic hydrogen-like atoms made up of a pion and a muon.

In the 1970s he founded and became president of Digital Pathways. In 1991, he became Associate Director of High Energy and Nuclear Physics at Brookhaven National Laboratory. At the same time, he rejoined the Columbia faculty as Professor of Physics. He became I. I. Rabi Professor of Physics in 1994 and retired as Rabi Professor Emeritus in 2000. He spent his retirement years in Ketchum, Idaho, and died August 28, 2006 at a Twin Falls, Idaho, nursing home after struggling with Parkinson's disease and hepatitis C.

6.13.2   Publications

- Samios, N. P., Plano, R., Prodell, A., Schwartz, M. and J. Steinberger. "The Parity of the Neutral Pion and the Decay pi{sup 0} Yields 2e{sup +} + 2e{sup -}", Nevis Cyclotron Laboratory, Columbia University, United States Department of Energy (through predecessor agency the Atomic Energy Commission), Office of Naval Research, (January 1962).

- Lee, T. D., Robinson, H., Schwartz, M. and R. Cool. "Intensity of Upward Muon Flux Due to Cosmic-Ray Neutrinos Produced in the Atmosphere", Nevis Cyclotron Laboratory, Columbia University, United States Department of Energy (through predecessor agency the Atomic Energy Commission), (June 1963).

- Franzini, P., Leontic, B., Rahm, D., Samios, N. and M. Schwartz. "Search for Massive Particles Produced in Interactions at 30 BeV", Brookhaven National Laboratory, Columbia University, United States Department of Energy (through predecessor agency the Atomic Energy Commission), (January 1965).

- Schwartz M. Principles of Electrodynamics. (October 1987).

6.13.3 References

[1] http://www.nasonline.org/publications/
biographical-memoirs/memoir-pdfs/schwartz-melvin.pdf

[2] Samios, Nicholas P. (December 2006). "Obituary: Melvin Schwartz". *Physics Today* **59** (12): 75–76. Bibcode:2006PhT....59l..75S. doi:10.1063/1.2435691.

6.13.4 External links

- Photograph, Biography and Bibliographic Resources, from the Office of Scientific and Technical Information, United States Department of Energy

- 1988 Nobel Physics winners

- Nobel autobiography

6.14 Jack Steinberger

Hans Jakob "Jack"Steinberger (born May 25, 1921) is a physicist currently residing near Geneva, Switzerland. He co-discovered the muon neutrino, along with Leon Lederman and Melvin Schwartz, for which they were given the 1988 Nobel Prize in Physics.

6.14.1 Life

Steinberger was born in the city of Bad Kissingen in Bavaria, Germany, on 1921. The rise of the Nazi party in Germany, with its open anti-Semitism, prompted his parents, Berta and Ludwig Steinberger, who was a cantor and religious teacher,[1] to send him out of the country.

Steinberger emigrated to the United States at the age of 13, making the trans-Atlantic trip with his brother Herbert. Barnett Farroll cared for him as a foster child, the connection was made by Jewish charities in the United States. During this period, Steinberger attended New Trier Township High School, in Winnetka, Illinois.

Steinberger studied chemical engineering at Armour Institute of Technology (now Illinois Institute of Technology) but left after his scholarship ended to help supplement his family's income. He obtained a bachelor's degree in Chemistry from the University of Chicago, in 1942. Shortly thereafter, he joined the Signal Corps at MIT. With the help of the G.I. Bill, he returned to graduate studies at the University of Chicago in 1946, where he studied under Edward Teller and Enrico Fermi. His Ph.D. thesis concerned the energy spectrum of electrons emitted in muon decay; his results showed that this was a three-body decay, and implied the participation of two neutral particle in the

decay (later identified as the electron (ν_e) and muon (ν_μ) neutrinos) rather than one.

As an atheist and a humanist, Steinberger is a Humanist Laureate in the International Academy of Humanism.[2][3]

He is the father of Ned Steinberger, founder of the eponymous company for headless guitars and basses.

6.14.2 Early career

After receiving his doctorate, Steinberger attended the Institute for Advanced Study in Princeton for a year. In 1949 he published a calculation of the lifetime of the neutral pion,[4] which anticipated the study of anomalies in quantum field theory.

Following Princeton, Steinberger went to the Radiation Lab at the University of California at Berkeley, where he performed an experiment which demonstrated the production of neutral pions and their decay to photon pairs. This experiment utilized the 330 MeV synchrotron and the newly invented scintillation counters.[5] Despite this and other achievements, he was asked to leave the Radiation Lab at Berkeley due to his refusal to sign the so-called *Non-Communist Oath*.

Steinberger accepted a faculty position at Columbia in 1950. The newly commissioned meson beam at Nevis Labs provided the tool for several important experiments. Measurements of the production cross section of pions on various nuclear targets showed that the pion has odd parity.[6] A direct measurement of the production of pions on a liquid hydrogen target, then not a common tool, provided the data needed to show that the pion has spin zero. The same target was used to observe the relative rare decay of neutral pions to a photon, an electron and a positron. A related experiment measured the mass difference between the charged and neutral pions based on the angular correlation between the neutral pions produced when the negative pion is captured by the proton in the hydrogen nucleus.[7] Other important experiments studied the angular correlation between electron-positron pairs in neutral pion decays, and established the rare decay of a charged pion to an electron and neutrino; the latter required use of a liquid-hydrogen bubble chamber.[8]

6.14.3 Investigations of strange particles

During 1954–1955, Steinberger contributed to the development of the bubble chamber with the construction of a 15 cm device for use with the Cosmotron at Brookhaven National Laboratory. The experiment used a pion beam to produce pairs of hadrons with strange quarks in order to elucidate the puzzling production and decay properties of these particles.[*][9] Somewhat later, in 1956, a 30 cm chamber outfitted with three cameras was used in the discovery of the neutral Sigma hyperon and a measurement of its mass.[*][10] This observation was important for confirming the existence of the SU(3) flavor symmetry which hypothesizes the existence of the strange quark.

An important characteristic of the weak interaction is its violation of parity symmetry. This characteristic was established through the measurement of the spins and parities of many hyperons. Steinberger and his collaborators contributed several such measurements using large (75 cm) liquid-hydrogen bubble chambers and separated hadron beams at Brookhaven. One example is the measurement of the invariant mass distribution of electron-positron pairs produced in the decay of Sigma-zero hyperons to Lambda-zero hyperons.[*][11]

6.14.4 Neutrinos and the weak neutral current

In the 1960s, the emphasis in the study of the weak interaction shifted from strange particles to neutrinos. Leon Lederman, Steinberger and Schwartz built large spark chambers at Nevis Lab and exposed them in 1961 to neutrinos produced in association with muons in the decays of charged pions and kaons. They used the Alternating Gradient Synchrotron (AGS) at Brookhaven, and obtained a number of convincing events in which muons were produced, but no electrons.[*][12] This result, for which they received the Nobel Prize in 1988, proved the existence of a type of neutrino associated with the muon, distinct from the neutrino produced in beta decay.

6.14.5 Study of CP violation

The CP violation (charge conjugation and parity) was established in the neutral kaon system in 1964. Steinberger recognized that the phenomenological parameter epsilon (ε) which quantifies the degree of CP violation could be measured in interference phenomena (See CP violation). In collaboration with Carlo Rubbia, he performed an experiment while on sabbatical at CERN during 1965 which demonstrated robustly the expected interference effect, and also measured precisely the difference in mass of the short-lived and long-lived neutral kaon masses.[*][13][*][14]

Back in the United States, Steinberger conducted an experiment at Brookhaven to observe CP violation in the semileptonic decays of neutral kaons. The charge asymmetry relates directly to the epsilon parameter, which was thereby measured precisely.[*][15] This experiment also allowed the deduction of the phase of epsilon, and confirmed that CPT is a good symmetry of nature.

6.14.6 CERN

In 1968, Steinberger left Columbia University and accepted a position as a department director at CERN. He constructed an experiment there utilizing multi-wire proportional chambers (MWPC), recently invented by Georges Charpak. The MWPC's, augmented by micro-electronic amplifiers, allowed much larger samples of events to be recorded. Several results for neutral kaons were obtained and published in the early 1970s, including the observation of the rare decay of the neutral kaon to a muon pair, the time-dependence of the asymmetry for semi-leptonic decays, and a more precise measurement of the neutral kaon mass difference. A new era in experimental technique was opened.

These new techniques proved crucial for the first demonstration of direct CP-violation. The NA31 experiment at CERN was built in the early 1980s using the CERN SPS 400 GeV proton synchrotron. Aside from banks of MWPC's and a hadron calorimeter, it featured a liquid argon electromagnetic calorimeter with exceptional spatial and energy resolution. NA31 showed that direct CP violation is real.[*][16]

6.14.7 Nobel Prize

Jack Steinberger was awarded the Nobel Prize in Physics in the year 1988, "for the neutrino beam method and the demonstration of the doublet structure of the leptons through the discovery of the muon neutrino" .[*][17] He shares this prize with Leon M. Lederman and Melvin Schwartz. At the time, all three experimenters were at Columbia University.

The experiment used charged pion beams generated with the *Alternating Gradient Synchrotron* (AGS) at Brookhaven National Laboratory. The pions decayed to muons which were detected in front of a steel wall; the neutrinos were detected in spark chambers installed behind the wall. The coincidence of muons and neutrinos demonstrated that a second kind of neutrino was created in association with muons. Subsequent experiments proved this neutrino to be distinct from the first kind (electron-type). Steinberger, Lederman and Schwartz published their work in Physical Review Letters in 1962.[*][12]

He gave his Nobel medal to New Trier High School in Winnetka, Illinois (USA), of which he is an alumnus.

6.14.8 See also

- List of Jewish Nobel laureates

6.14.9 References

[1] http://www.nobelprize.org/nobel_prizes/physics/laureates/1988/steinberger-autobio.html

[2] The International Academy of Humanism at the website of the Council for Secular Humanism. Retrieved 18 October 2007. Some of this information is also at the International Humanist and Ethical Union website

[3] Istva´n Hargittai, Magdolna Hargittai (2006). *Candid Science VI: More Conversations with Famous Scientists*. Imperial College Press. p. 749. ISBN 9781860948855. Jack Steinberger: "I'm now a bit anti-Jewish since my last visit to the synagogue, but my atheism does not necessarily reject religion."

[4] J. Steinberger (1949). "On the use of subtraction fields and the lifetimes of some types of meson decay". *Physical Review* **76** (8): 1180. Bibcode:1949PhRv...76.1180S. doi:10.1103/PhysRev.76.1180.

[5] J. Steinberger, W. K. H. Panofsky and J. Steller (1950). "Evidence for the production of neutral mesons by photons". *Physical Review* **78** (6): 802. Bibcode:1950PhRv...78..802S. doi:10.1103/PhysRev.78.802.

[6] C. Chedester, P. Isaacs, A. Sachs and J. Steinberger (1951). "Total cross-sections of π-mesons on protons and several other nuclei". *Physical Review* **82** (6): 958. Bibcode:1951PhRv...82..958C. doi:10.1103/PhysRev.82.958.

[7] W. Chinkowsky and J. Steinberger (1954). "The mass difference of neutral and negative π mesons". *Physical Review* **93** (3): 586. Bibcode:1954PhRv...93..586C. doi:10.1103/PhysRev.93.586.

[8] G. Impeduglia, R. Plano, A. Prodell, N. Samios, M. Schwartz and J. Steinberger (1958). "β decay of the pion". *Physical Review Letters* **1** (7): 249. Bibcode:1958PhRvL...1..249I. doi:10.1103/PhysRevLett.1.249.

[9] R. Budde, M. Chretien, J. Leitner, N.P. Samios, M. Schwartz and J. Steinberger (1956). "Properties of heavy unstable particles produced by 1.3 BeV π^- mesons". *Physical Review* **103** (6): 1827. Bibcode:1956PhRv..103.1827B. doi:10.1103/PhysRev.103.1827.

[10] R. Plano, N. Samios, M. Schwartz and J. Steinberger (1957). "Demonstration of the existence of the Σ^0 hyperon and a measurement of its mass". *Il Nuovo Cimento* **5**: 216. doi:10.1007/BF02812828.

[11] C. Alff-Steinberger et al. (1963). *Siena 1963 Conference Report*: 205. Missing or empty |title= (help)

[12] G. Danby, J.-M. Gaillard, K. Goulianos, L. M. Lederman, N. B. Mistry, M. Schwartz, J. Steinberger (1962). "Observation of high-energy neutrino reactions and the existence of two kinds of neutrinos". *Physical Review Letters* **9**: 36. Bibcode:1962PhRvL...9...36D. doi:10.1103/PhysRevLett.9.36.

[13] C. Alff-Steinberger et al. (1966). "K_S and K_L interference in the $\pi^++\pi^-$ decay mode, CP invariance and the K_S–K_L mass difference". *Physics Letters* **20** (2): 207. Bibcode:1966PhL.....20..207A. doi:10.1016/0031-9163(66)90937-1.

[14] C. Alff-Steinberger et al. (1966). "Further results from the interference of K_S and K_L in the $\pi^++\pi^-$ decay modes". *Physics Letters* **21** (5): 595. Bibcode:1966PhL.....21..595A. doi:10.1016/0031-9163(66)91312-6.

[15] S. Bennett, D. Nygren, H. Saal, J. Steinberger and J. Sutherland (1967). "Measurement of the charge asymmetry in the decay K0
L $\to$ $\pi\pm+e\mp+\nu$". *Physical Review Letters* **19** (17): 993. Bibcode:1967PhRvL..19..993B. doi:10.1103/PhysRevLett.19.993.

[16] H. Burkhardt et al. (1988). "First evidence for direct CP violation". *Physics Letters B* **206**: 169. Bibcode:1988PhLB..206..169B. doi:10.1016/0370-2693(88)91282-8.

[17] Anthony, Katarina (11 July 2011). "In conversation with Nobel laureate Jack Steinberger". *CERN Bulletin* (28-29).

6.14.10 Publications

- Steinberger, J. & A. S. Bishop. "The Detection of Artificially Produced Photomesons with Counters", Radiation Laboratory, University of California-Berkeley, United States Department of Energy (through predecessor agency the Atomic Energy Commission), (March 8, 1950).

- Steinberger, J., W. K. H. Panofsky & J. Steller. "Evidence for the Production of Neutral Mesons by Photons", Radiation Laboratory, University of California-Berkeley, United States Department of Energy (through predecessor agency the Atomic Energy Commission), (April 1950).

- Panofsky, W. K. H., J. Steinberger & J. Steller. "Further Results on the Production of Neutral Mesons by Photons", Radiation Laboratory, University of California-Berkeley, United States Department of Energy (through predecessor agency the Atomic Energy Commission), (October 1, 1950).

- Steinberger, J. "Experimental Survey of Strange Particle Decays" , Columbia University, Nevis Laboratories, United States Department of Energy (through predecessor agency the Atomic Energy Commission), (June 1964).

6.14.11 External links

- Biography and Bibliographic Resources, from the Office of Scientific and Technical Information, United States Department of Energy

- Autobiography (at the Nobel Prize web site)

- official web site for the Nobel Prize in Physics

- CERN web site for Jack Steinberger

- Scientific publications of Jack Steinberger on INSPIRE-HEP

6.15   J. J. Thomson

This article is about the Nobel laureate and physicist. For the moral philosopher, see Judith Jarvis Thomson.

Sir Joseph John "J. J."Thomson, OM, FRS[1] (/ˈtɒmsən/; 18 December 1856 – 30 August 1940) was an English physicist. He was elected as a fellow of the Royal Society of London[2] and appointed to the Cavendish Professorship of Experimental Physics at the Cambridge University's Cavendish Laboratory in 1884.[3]

In 1897, Thomson showed that cathode rays were composed of previously unknown negatively charged particles, which he calculated must have bodies much smaller than atoms and a very large value for their charge-to-mass ratio.[3] Thus he is credited with the discovery and identification of the electron; and with the discovery of the first subatomic particle. Thomson is also credited with finding the first evidence for isotopes of a stable (non-radioactive) element in 1913, as part of his exploration into the composition of canal rays (positive ions). His experiments to determine the nature of positively charged particles, with Francis William Aston, were the first use of mass spectrometry and led to the development of the mass spectrograph.[3]

Thomson was awarded the 1906 Nobel Prize in Physics for the discovery of the electron and for his work on the conduction of electricity in gases.[4] Seven of his students, and his son George Paget Thomson, also became Nobel Prize winners.

6.15.1   Biography

Joseph John Thomson was born 18 December 1856 in Cheetham Hill, Manchester, Lancashire, England. His mother, Emma Swindells, came from a local textile family. His father, Joseph James Thomson, ran an antiquarian bookshop founded by a great-grandfather. He had a brother two years younger than he was, Frederick Vernon Thomson.[5]

His early education was in small private schools where he demonstrated outstanding talent and interest in science. In 1870 he was admitted to Owens College at the unusually young age of 14. His parents planned to enroll him as an apprentice engineer to Sharp-Stewart & Co, a locomotive manufacturer, but these plans were cut short when his father died in 1873.[5]

He moved on to Trinity College, Cambridge in 1876. In 1880, he obtained his BA in mathematics (Second Wrangler in the Tripos[6] and 2nd Smith's Prize).[7] He applied for and became a Fellow of Trinity College as of 1881.[8] Thomson received his MA (with Adams Prize) in 1883.[7]

Thomson was elected a Fellow of the Royal Society[1] on 12 June 1884 and served as President of the Royal Society from 1915 to 1920.

On 22 December 1884 Thomson was chosen to become Cavendish Professor of Physics at the University of Cambridge.[3] The appointment caused considerable surprise, given that candidates such as Richard Glazebrook were older and more experienced in laboratory work. Thomson was known for his work as a mathematician, where he was recognized as an exceptional talent.[9]

In 1890, Thomson married Rose Elisabeth Paget, daughter of Sir George Edward Paget, KCB, a physician and then Regius Professor of Physic at Cambridge. They had one son, George Paget Thomson, and one daughter, Joan Paget Thomson.

He was awarded a Nobel Prize in 1906, "in recognition of the great merits of his theoretical and experimental investigations on the conduction of electricity by gases." He was knighted in 1908 and appointed to the Order of Merit in 1912. In 1914 he gave the Romanes Lecture in Oxford on "The atomic theory" . In 1918 he became Master of Trinity College, Cambridge, where he remained until his death. Joseph John Thomson died on 30 August 1940 and was buried in Westminster Abbey, close to Sir Isaac Newton.

One of Thomson's greatest contributions to modern science was in his role as a highly gifted teacher. One of his students was Ernest Rutherford, who later succeeded him as Cavendish Professor of Physics. In addition to Thomson himself, seven of his research assistants and his son won

Nobel Prizes in physics. His son won the Nobel Prize in 1937 for proving the wavelike properties of electrons.

6.15.2 Career

Early work

Thomson's prize-winning master's work, *Treatise on the motion of vortex rings*, shows his early interest in atomic structure.[4] In it, Thomson mathematically described the motions of William Thomson's vortex theory of atoms.[9]

Thomson published a number of papers addressing both mathematical and experimental issues of electromagnetism. He examined the electromagnetic theories of light of James Clerk Maxwell, introduced the concept of electromagnetic mass of a charged particle, and demonstrated that a moving charged body would apparently increase in mass.[9]

Much of his work in mathematical modelling of chemical processes can be thought of as early computational chemistry.[3] In further work, published in book form as *Applications of dynamics to physics and chemistry* (1888), Thomson addressed the transformation of energy in mathematical and theoretical terms, suggesting that all energy might be kinetic.[9] His next book, *Notes on recent researches in electricity and magnetism* (1893), built upon Maxwell's *Treatise upon electricity and magnetism*, and was sometimes referred to as "the third volume of Maxwell".[4] In it, Thomson emphasized physical methods and experimentation and included extensive figures and diagrams of apparatus, including a number for the passage of electricity through gases.[9] His third book, *Elements of the mathematical theory of electricity and magnetism* (1895)[10] was a readable introduction to a wide variety of subjects, and achieved considerable popularity as a textbook.[9]

A series of four lectures, given by Thomson on a visit to Princeton University in 1896, were subsequently published as *Discharge of electricity through gases* (1897). Thomson also presented a series of six lectures at Yale University in 1904.[4]

Discovery of the electron

Several scientists, such as William Prout and Norman Lockyer, had suggested that atoms were built up from a more fundamental unit, but they envisioned this unit to be the size of the smallest atom, hydrogen. Thomson, in 1897, was the first to suggest that one of the fundamental units was more than 1,000 times smaller than an atom, suggesting the subatomic particle now known as the electron. Thomson discovered this through his explorations on the properties of cathode rays. Thomson made his suggestion on 30 April

1897 following his discovery that cathode rays (at the time known as Lenard rays) could travel much further through air than expected for an atom-sized particle.[11] He estimated the mass of cathode rays by measuring the heat generated when the rays hit a thermal junction and comparing this with the magnetic deflection of the rays. His experiments suggested not only that cathode rays were over 1,000 times lighter than the hydrogen atom, but also that their mass was the same in whichever type of atom they came from. He concluded that the rays were composed of very light, negatively charged particles which were a universal building block of atoms. He called the particles "corpuscles", but later scientists preferred the name electron which had been suggested by George Johnstone Stoney in 1891, prior to Thomson's actual discovery.[12]

In April 1897, Thomson had only early indications that the cathode rays could be deflected electrically (previous investigators such as Heinrich Hertz had thought they could not be). A month after Thomson's announcement of the corpuscle, he found that he could reliably deflect the rays by an electric field if he evacuated the discharge tube to a very low pressure. By comparing the deflection of a beam of cathode rays by electric and magnetic fields he obtained more robust measurements of the mass to charge ratio that confirmed his previous estimates.[13] This became the classic means of measuring the charge and mass of the electron.

Thomson believed that the corpuscles emerged from the atoms of the trace gas inside his cathode ray tubes. He thus concluded that atoms were divisible, and that the corpuscles were their building blocks. In 1904 Thomson suggested a model of the atom, hypothesizing that it was a sphere of positive matter within which electrostatic forces determined the positioning of the corpuscles.[3] To explain the overall neutral charge of the atom, he proposed that the corpuscles were distributed in a uniform sea of positive charge. In this "plum pudding" model the electrons were seen as embedded in the positive charge like plums in a plum pudding (although in Thomson's model they were not stationary, but orbiting rapidly).[14][15]

Isotopes and mass spectrometry

In 1912, as part of his exploration into the composition of the streams of positively charged particles then known as canal rays, Thomson and his research assistant F. W. Aston channelled a stream of neon ions through a magnetic and an electric field and measured its deflection by placing a photographic plate in its path.[5] They observed two patches of light on the photographic plate (see image on right), which suggested two different parabolas of deflection, and concluded that neon is composed of atoms of two different atomic masses (neon-20 and neon-22), that is to

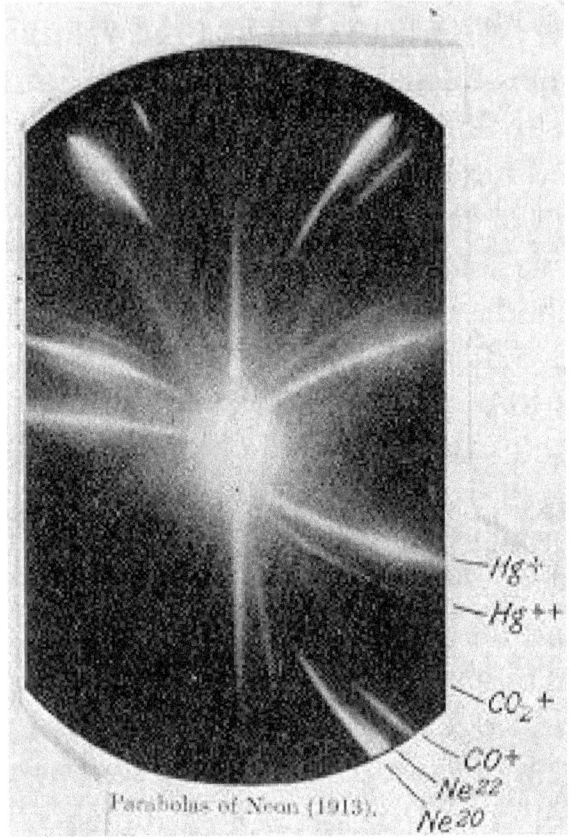

In the bottom right corner of this photographic plate are markings for the two isotopes of neon: neon-20 and neon-22.

say of two isotopes.*[16] This was the first evidence for isotopes of a stable element; Frederick Soddy had previously proposed the existence of isotopes to explain the decay of certain radioactive elements.

J.J. Thomson's separation of neon isotopes by their mass was the first example of mass spectrometry, which was subsequently improved and developed into a general method by F. W. Aston and by A. J. Dempster.*[3]

6.15.3   Experiments with cathode rays

Earlier, physicists debated whether cathode rays were immaterial like light ("some process in the aether") or were "in fact wholly material, and ... mark the paths of particles of matter charged with negative electricity", quoting Thomson.*[13] The aetherial hypothesis was vague,*[13] but the particle hypothesis was definite enough for Thomson to test.

Experiments on the magnetic deflection of cathode rays

Thomson first investigated the magnetic deflection of cathode rays. Cathode rays were produced in the side tube

on the left of the apparatus and passed through the anode into the main bell jar, where they were deflected by a magnet. Thomson detected their path by the fluorescence on a squared screen in the jar. He found that whatever the material of the anode and the gas in the jar, the deflection of the rays was the same, suggesting that the rays were of the same form whatever their origin.*[17]

Experiment to show that cathode rays were electrically charged

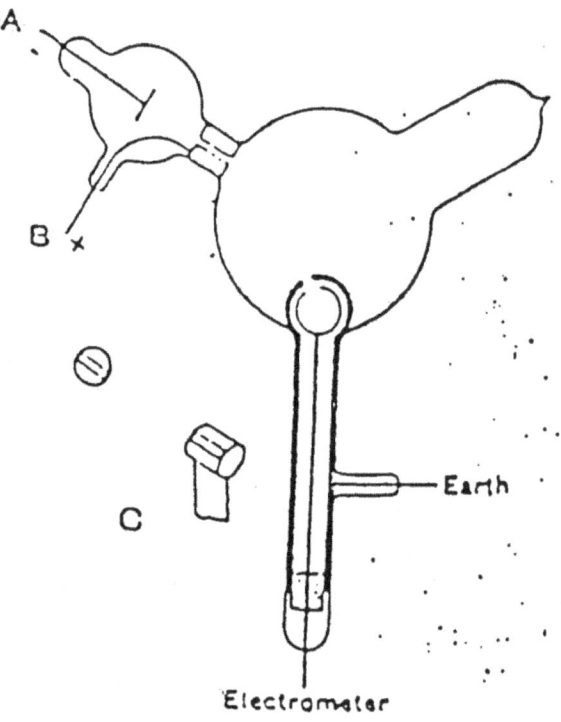

The cathode ray tube by which J.J. Thomson demonstrated that cathode rays could be deflected by a magnetic field, and that their negative charge was not a separate phenomenon.

While supporters of the aetherial theory accepted the possibility that negatively charged particles are produced in Crookes tubes, they believed that they are a mere by-product and that the cathode rays themselves are immaterial. Thomson set out to investigate whether or not he could actually separate the charge from the rays.

Thomson constructed a Crookes tube with an electrometer set to one side, out of the direct path of the cathode rays. Thomson could trace the path of the ray by observing the phosphorescent patch it created where it hit the surface of the tube. Thomson observed that the electrometer registered a charge only when he deflected the cathode ray to it with a magnet. He concluded that the negative charge and the rays were one and the same.*[11]

Experiment to show that cathode rays could be deflected electrically

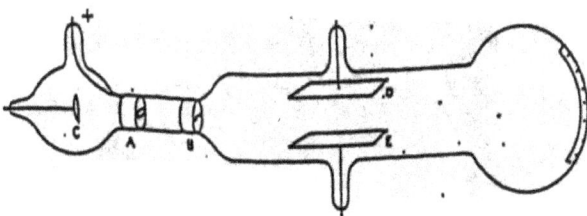

Thomson's illustration of the Crookes tube by which he observed the deflection of cathode rays by an electric field (and later measured their mass to charge ratio). Cathode rays were emitted from the cathode C, passed through slits A (the anode) and B (grounded), then through the electric field generated between plates D and E, finally impacting the surface at the far end.

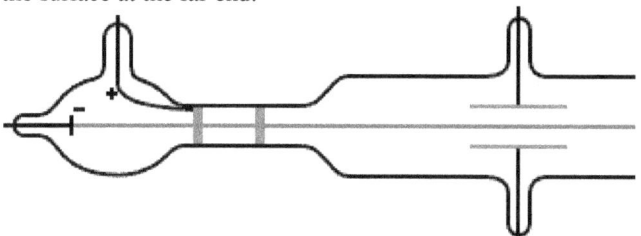

The cathode ray (blue line) was deflected by the electric field (yellow).

In May–June 1897, Thomson investigated whether or not the rays could be deflected by an electric field.[*][5] Previous experimenters had failed to observe this, but Thomson believed their experiments were flawed because their tubes contained too much gas.

Thomson constructed a Crookes tube with a better vacuum. At the start of the tube was the cathode from which the rays projected. The rays were sharpened to a beam by two metal slits – the first of these slits doubled as the anode, the second was connected to the earth. The beam then passed between two parallel aluminium plates, which produced an electric field between them when they were connected to a battery. The end of the tube was a large sphere where the beam would impact on the glass, created a glowing patch. Thomson pasted a scale to the surface of this sphere to measure the deflection of the beam. Note that any electron beam would collide with some residual gas atoms within the Crookes tube, thereby ionizing them and producing electrons and ions in the tube (space charge); in previous experiments this space charge electrically screened the externally applied electric field. However, in Thomson's Crookes tube the density of residual atoms was so low that the space charge from the electrons and ions was insufficient to electrically screen the externally applied electric field, which

permitted Thomson to successfully observe electrical deflection.

When the upper plate was connected to the negative pole of the battery and the lower plate to the positive pole, the glowing patch moved downwards, and when the polarity was reversed, the patch moved upwards.

Experiment to measure the mass to charge ratio of cathode rays

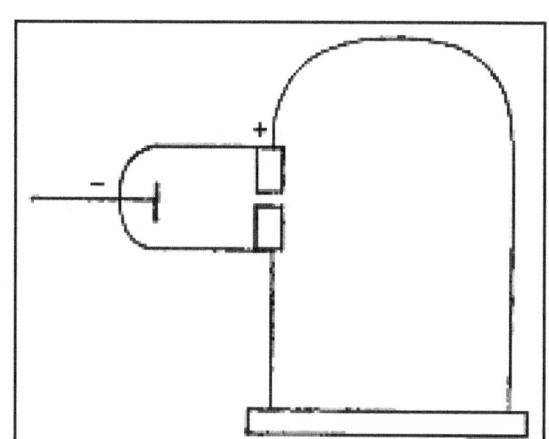

In his classic experiment, Thomson measured the mass-to-charge ratio of the cathode rays by measuring how much they were deflected by a magnetic field and comparing this with the electric deflection. He used the same apparatus as in his previous experiment, but placed the discharge tube between the poles of a large electromagnet. He found that the mass to charge ratio was over a thousand times *lower* than that of a hydrogen ion ($H^{*}+$), suggesting either that the particles were very light and/or very highly charged.[*][13] Significantly, the rays from every cathode yielded the same mass-to-charge ratio. This is in contrast to anode rays (now known to arise from positive ions emitted by the anode), where the mass-to-charge ratio varies from anode-to-anode. Thomson himself remained critical of what his work established, in his Nobel Prize acceptance speech referring to "corpuscles" rather than "electrons".

Thomson's calculations can be summarised as follows (notice that we reproduce here Thomson's original notations, using F instead of E for the Electric field and H instead of B for the magnetic field):

The electric deflection is given by $\Theta = Fel/mv^2$ where Θ is the angular electric deflection, F is applied electric intensity, e is the charge of the cathode ray particles, l is the length of the electric plates, m is the mass of the cathode ray particles and v is the velocity of the cathode ray particles.

The magnetic deflection is given by φ = Hel/mv where φ is the angular magnetic deflection and H is the applied magnetic field intensity.

The magnetic field was varied until the magnetic and electric deflections were the same, when $\Theta = \varphi$ and Fel/mv^2= Hel/mv. This can be simplified to give m/e = H^{2}l/FΘ. The electric deflection was measured separately to give Θ and H, F and l were known, so m/e could be calculated.

Conclusions

> As the cathode rays carry a charge of negative electricity, are deflected by an electrostatic force as if they were negatively electrified, and are acted on by a magnetic force in just the way in which this force would act on a negatively electrified body moving along the path of these rays, I can see no escape from the conclusion that they are charges of negative electricity carried by particles of matter.
> —J. J. Thomson[13]

As to the source of these particles, Thomson believed they emerged from the molecules of gas in the vicinity of the cathode.

> If, in the very intense electric field in the neighbourhood of the cathode, the molecules of the gas are dissociated and are split up, not into the ordinary chemical atoms, but into these primordial atoms, which we shall for brevity call corpuscles; and if these corpuscles are charged with electricity and projected from the cathode by the electric field, they would behave exactly like the cathode rays.
> —J. J. Thomson[18]

Thomson imagined the atom as being made up of these corpuscles orbiting in a sea of positive charge; this was his plum pudding model. This model was later proved incorrect when his student Ernest Rutherford showed that the positive charge is concentrated in the nucleus of the atom.

Other work

In 1905, Thomson discovered the natural radioactivity of potassium.[19]

In 1906, Thomson demonstrated that hydrogen had only a single electron per atom. Previous theories allowed various numbers of electrons.[20][21]

6.15.4   Awards and recognition

Plaque commemorating J. J. Thomson's discovery of the electron outside the old Cavendish Laboratory in Cambridge

- Adams Prize (1882)

- Royal Medal (1894)

- Hughes Medal (1902)

- Nobel Prize for Physics (1906)

- Elliott Cresson Medal (1910)

- Copley Medal (1914)

- Franklin Medal (1922)

In 1991, the thomson (symbol: Th) was proposed as a unit to measure mass-to-charge ratio in mass spectrometry in his honour.[22]

J J Thomson Avenue, on the University of Cambridge campus, is named after Thomson.[23]

In November 1927, J.J. Thomson opened the Thomson building, named in his honor, in the Leys School, Cambridge.[24]

6.15.5   Notes

[1] Rayleigh (1941). "Joseph John Thomson. 1856-1940". *Obituary Notices of Fellows of the Royal Society* **3** (10): 586–609. doi:10.1098/rsbm.1941.0024.

[2] Thomson, Sir George Paget. *Sir J.J. Thomson, British Physicist*. Encyclopædia Brittanica. Retrieved 11 February 2015.

[3] "Joseph John Thomson". Chemical Heritage Foundation. Retrieved 18 November 2013.

[4] "J.J. Thomson - Biographical". *The Nobel Prize in Physics 1906*. The Nobel Foundation. Retrieved 11 February 2015.

[5] Davis & Falconer, *J.J. Thomson and the Discovery of the Electron*

[6] Grayson, Mike. "The Early Life of J.J. Thomson: Computational Chemistry and Gas Discharge Experiments". *Profiles in Chemistry*. Chemical Heritage Foundation. Retrieved 11 February 2015.

[7] "Thomson, Joseph John (THN876JJ)". *A Cambridge Alumni Database*. University of Cambridge.

[8] *The Victoria University Calendar for the Session 1881-2*. 1882. p. 184. Retrieved 11 February 2015.

[9] Kim, Dong-Won (2002). *Leadership and creativity : a history of the Cavendish Laboratory, 1871 - 1919*. Dordrecht: Kluwer Acad. Publ. ISBN 9781402004759. Retrieved 11 February 2015.

[10] Mackenzie, A. Stanley (1896). "Review: *Elements of the Mathematical Theory of Electricity and Magnetism* by J. J. Thomson" (PDF). *Bull. Amer. Math. Soc.* **2** (10): 329–333. doi:10.1090/s0002-9904-1896-00357-8.

[11] J.J. Thomson (1897) "Cathode Rays", *The Electrician* 39, 104

[12] Falconer (2001) "Corpuscles to electrons"

[13] Thomson, J. J. (7 August 1897). "Cathode Rays". *Philosophical Magazine*. 5 **44**: 293. doi:10.1080/14786449708621070. Retrieved 4 August 2014.

[14] Mellor, Joseph William (1917), *Modern Inorganic Chemistry*, Longmans, Green and Company, p. 868, According to J. J. Thomson's hypothesis, atoms are built of systems of rotating rings of electrons.

[15] Dahl (1997), p. 324: "Thomson's model, then, consisted of a uniformly charged sphere of positive electricity (the pudding), with discrete corpuscles (the plums) rotating about the center in circular orbits, whose total charge was equal and opposite to the positive charge."

[16] See:
- J.J. Thomson (1912) "Further experiments on positive rays," *Philosophical Magazine*, series 6, **24** (140): 209–253.
- J.J. Thomson (1913) "Rays of positive electricity," *Proceedings of the Royal Society* A, **89**: 1–20.

[17] Thomson (8 February 1897)'On the cathode rays', Proceedings of the Cambridge Philosophical Society, 9, 243

[18] *Cathode rays* Philosophical Magazine, 44, 293 (1897)

[19] Thomson, J. J. (1905). "On the emission of negative corpuscles by the alkali metals". *Philosophical Magazine*. Series 6 **10** (59): 584–590. doi:10.1080/14786440509463405.

[20] Hellemans, Alexander; Bunch, Bryan (1988). *The Timetables of Science*. Simon & Schuster. p. 411. ISBN 0671621300.

[21] Thomson, J. J. (June 1906). "On the Number of Corpuscles in an Atom". *Philosophical Magazine* **11**: 769–781. doi:10.1080/14786440609463496. Archived from the original on 19 December 2007. Retrieved 4 October 2008.

[22] Cooks, R. G.; A. L. Rockwood (1991). "The 'Thomson'. A suggested unit for mass spectroscopists". *Rapid Communications in Mass Spectrometry* **5** (2): 93.

[23] "Cambridge Physicist is streets ahead". 2002-07-18. Retrieved 2014-07-31.

[24] "Opening of the New Science Building: Thomson". 2005-12-01. Retrieved 2015-01-10.

6.15.6 References

- Thomson, George Paget. (1964) *J.J. Thomson: Discoverer of the Electron*. Great Britain: Thomas Nelson & Sons, Ltd.

- 1883. *A Treatise on the Motion of Vortex Rings: An essay to which the Adams Prize was adjudged in 1882, in the University of Cambridge*. London: Macmillan and Co., pp. 146. Recent reprint: ISBN 0-543-95696-2.

- 1888. *Applications of Dynamics to Physics and Chemistry*. London: Macmillan and Co., pp. 326. Recent reprint: ISBN 1-4021-8397-6.

- 1893. *Notes on recent researches in electricity and magnetism: intended as a sequel to Professor Clerk-Maxwell's 'Treatise on Electricity and Magnetism'*. Oxford University Press, pp.xvi and 578. 1991, Cornell University Monograph: ISBN 1-4297-4053-1.

- 1921 (1895). *Elements Of The Mathematical Theory Of Electricity And Magnetism*. London: Macmillan and Co. Scan of 1895 edition.

- *A Text book of Physics in Five Volumes*, co-authored with J.H. Poynting: (1) Properties of Matter, (2) Sound, (3) Heat, (4) Light, and (5) Electricity and Magnetism. Dated 1901 and later, and with revised later editions.

- Navarro, Jaume, 2005, "Thomson on the Nature of Matter: Corpuscles and the Continuum," *Centaurus* 47(4): 259–82.

- Downard, Kevin, 2009. "J.J. Thomson Goes to America" J. Am. Soc. Mass Spectrom. 20(11): 1964–1973.

- Dahl, Per F., "*Flash of the Cathode Rays: A History of J.J. Thomson's Electron*". Institute of Physics Publishing. June 1997. ISBN 0-7503-0453-7

- J.J. Thomson (1897) "Cathode Rays", *The Electrician* 39, 104, also published in *Proceedings of the Royal Institution* 30 April 1897, 1–14—first announcement of the "corpuscle" (before the classic mass and charge experiment)

- J.J. Thomson (1897), *Cathode rays*, *Philosophical Magazine*, 44, 293—The classic measurement of the electron mass and charge

- J.J. Thomson (1912), "Further experiments on positive rays" *Philosophical Magazine*, 24, 209–253—first announcement of the two neon parabolae

- J.J. Thomson (1913), *Rays of positive electricity*, *Proceedings of the Royal Society*, A 89, 1–20—Discovery of neon isotopes

- J.J. Thomson, "On the Structure of the Atom: an Investigation of the Stability and Periods of Oscillation of a number of Corpuscles arranged at equal intervals around the Circumference of a Circle; with Application of the Results to the Theory of Atomic Structure," *Philosophical Magazine* Series 6, Volume 7, Number 39, pp. 237–265. This paper presents the classical "plum pudding model" from which the Thomson Problem is posed.

- The Master of Trinity at Trinity College, Cambridge

- J.J. Thomson, *The Electron in Chemistry: Being Five Lectures Delivered at the Franklin Institute*, Philadelphia (1923).

- Davis, Eward Arthur & Falconer, Isobel. *J.J. Thomson and the Discovery of the Electron*. 1997. ISBN 978-0-7484-0696-8

- Falconer, Isobel (1988) "J.J. Thomson's Work on Positive Rays, 1906–1914" *Historical Studies in the Physical and Biological Sciences* 18(2) 265–310

- Falconer, Isobel (2001) "Corpuscles to Electrons" in J Buchwald and A Warwick (eds) *Histories of the Electron*, Cambridge, Mass: MIT Press, pp. 77–100

6.15.7   External links

- Media related to Joseph John Thomson at Wikimedia Commons

- Works written by or about J. J. Thomson at Wikisource

- The Discovery of the Electron

- The Nobel Prize in Physics 1906

- Annotated bibliography for Joseph J. Thomson from the Alsos Digital Library for Nuclear Issues

- Essay on Thomson life and religious views

- The Cathode Ray Tube site

- Nobel Prize acceptance lecture (1906)

- Thomson's discovery of the isotopes of Neon

- Photos of some of Thomson's remaining apparatus at the Cavendish Laboratory Museum

- Works by J. J. Thomson at Project Gutenberg

- Works by or about J. J. Thomson at Internet Archive

6.16   Lincoln Wolfenstein

Lincoln Wolfenstein (February 10, 1923, Cleveland, Ohio - March 27, 2015, Oakland, California) was an American particle physicist who studied the weak interaction. Wolfenstein was born in 1923 and obtained his PhD in 1949 from the University of Chicago.[1] He retired from Carnegie Mellon University in 2000 after being a faculty member for 52 years. Despite being retired, he continued to come into work nearly every day.[2][3][4]

Wolfenstein was a particle phenomenologist, a theorist who focused primarily on connecting theoretical physics to experimental observations. In 1978, he noted that the presence of electrons in Earth and Solar matter could affect neutrino propagation. This work led to an eventual understanding of the MSW effect, which acts to enhance neutrino oscillation in matter. Wolfenstein received the 2005 Bruno Pontecorvo Prize from The Scientific Council of the Joint Institute for Nuclear Research (JINR), for his pioneering work on the MSW effect.

In 1992, Wolfenstein was awarded the American Physical Society's J.J. Sakurai Prize for Theoretical Particle Physics for "his many contributions to the theory of weak interactions, particularly CP violation and the properties of neutrinos" .[5][6]

6.16.1   See also

- Cabibbo–Kobayashi–Maskawa matrix (Wolfenstein parameters)

- Carnegie Mellon University

6.16.2 References

[1] Lincoln Wolfenstein at the Mathematics Genealogy Project

[2]

[3] Biography at APS

[4] Belculfine, Lexi (4 April 2015). "Obituary: Lincoln Wolfenstein / Internationally known physicist who taught at CMU". Pittsburgh Post-Gazette. Retrieved 12 May 2015.

[5] American Physical Society - J. J. Sakurai Prize Winners

[6] 1992 J. J. Sakurai Prize

6.16.3 External links

- Lincoln Wolfenstein's profile at Carnegie Mellon University

Chapter 7

Appendix C – Selected facilities and experiments

7.1 Cowan–Reines neutrino experiment

The **Cowan–Reines neutrino experiment** was performed by Clyde L. Cowan and Frederick Reines in 1956. This experiment confirmed the existence of the antineutrino—a neutrally charged subatomic particle with very low mass.

7.1.1 Background

During the 1910s and 1920s, through the study of electron spectra from the nuclear beta decay, it became apparent that, in addition to an electron, another particle with very small mass and with no electric charge is emitted in the beta-decay but not observed. The observed electron energy spectrum was continuous. Assuming energy conservation, this is only possible if the beta decay is a three-body rather than a two-body decay: the latter produces monochromatic peak rather than a continuous energy spectrum. This and other reasons led Wolfgang Pauli to postulate the existence of the neutrino in 1930.

7.1.2 Potential for experiment

Via the inverse beta decay, the predicted electron antineutrino (ν
e), should interact with a proton (p) to produce a neutron (n) and positron (e+) – the antimatter counterpart of the electron.

$$\nu e + p \rightarrow n + e+$$

The positron quickly finds an electron, and they annihilate each other. The two resulting gamma rays (γ) are detectable. The neutron can be detected by its capture on an appropriate nucleus, releasing a gamma ray. The coincidence of both events - positron annihilation and neutron capture - gives a unique signature of an antineutrino interaction.

Most hydrogen atoms bound in water molecules have a single proton for a nucleus. Those protons serve as a target for the antineutrinos from a reactor. For heavier nuclei, with several protons and neutrons, the interaction mechanism is more complicated and is not always well described by considering the constituent protons as free.

7.1.3 Setup

Cowan and Reines used a nuclear reactor, as advised by Los Alamos physics division leader J.M.B. Kellogg,[1] as a source of a neutrino flux of 5×10^{13} neutrinos per second per square centimeter;[2] far higher than any attainable flux from other radioactive sources.

The neutrinos then interacted (as shown above) with protons in two tanks of water, creating neutrons and positrons. Each positron created a pair of gamma rays when it annihilated with an electron. The gamma rays were detected by sandwiching the water tanks between tanks filled with liquid scintillator. The scintillator material gives off flashes of light in response to the gamma rays, and these light flashes are detected by photomultiplier tubes.

This experiment was not conclusive enough, so they devised a second layer of certainty. They detected the neutrons by placing cadmium chloride in the tank. Cadmium is a highly effective neutron absorber and gives off a gamma ray when it absorbs a neutron.

$$n + 108Cd \rightarrow 109mCd \rightarrow 109Cd + \gamma$$

The arrangement was such that the gamma ray from the cadmium would be detected 5 microseconds after the gamma ray from the positron, if it were truly produced by a neutrino.

7.1.4 Results

They performed the experiment preliminarily at Hanford Site, but later moved the experiment to the Savannah River Plant in South Carolina near Aiken where they had better shielding against cosmic rays. This shielded location was 11 m from the reactor and 12 m underground.

They used two tanks with a total of about 200 liters of water with about 40 kg of dissolved $CdCl_2$. The water tanks were sandwiched between three scintillator layers which contained 110 five-inch (127 mm) photomultiplier tubes.

After months of data collection, they had accumulated data on about three neutrinos per hour in their detector. To be absolutely sure that they were seeing neutrino events from the detection scheme described above, they shut down the reactor to show that there was a difference in the number of detected events.

They had predicted a cross-section for the reaction to be about $6\times10^{*}-44$ cm^2 and their measured cross-section was $6.3\times10^{*}-44$ cm^2. Their results were published in the July 20, 1956 issue of Science.*[3]*[4]

Clyde Cowan died in 1974; Frederick Reines was honored with the Nobel Prize in 1995 for his work on neutrino physics.*[5]

7.1.5 See also

- Homestake Experiment (a contemporary experiment which detected neutrinos from nuclear fusion in the solar core)

- The Neutrino with Dr. Clyde L. Cowan (Lecture on Nobel Prize winning experiment)

7.1.6 References

[1] "The Reines-Cowan Experiments: Detecting the Poltergeist" (PDF). *Los Alamos Science* **25**: 3. 1997.

[2] Griffiths, David J. (1987). *Introduction to Elementary Particles*. John Wiley & Sons. ISBN 0-471-60386-4.

[3] C. L Cowan Jr., F. Reines, F. B. Harrison, H. W. Kruse, A. D McGuire (July 20, 1956). "Detection of the Free Neutrino: a Confirmation". *Science* **124** (3212): 103–4. Bibcode:1956Sci...124..103C. doi:10.1126/science.124.3212.103. PMID 17796274.

[4] Winter, Klaus (2000). *Neutrino physics*. Cambridge University Press. p. 38ff. ISBN 978-0-521-65003-8. This source reproduces the 1956 paper.

[5] "The Nobel Prize in Physics 1995". The Nobel Foundation. Retrieved 201-06-29. Check date values in: |accessdate= (help)

7.1.7 Further reading

- Cowan and Reines Neutrino Experiment

- Decay of the Neutron

- Beta Decay

- Electron Neutrinos and Antineutrinos

- Subatomic particles

- Cowan & Reines Experiments: Poltergeist, Hanford, Savannah River

7.2 DESY

The **Deutsches Elektronen-Synchrotron** (english *German Electron Synchrotron*) commonly referred to by the abbreviation **DESY**, is a national research center in Germany that operates particle accelerators used to investigate the structure of matter. It conducts a broad spectrum of interdisciplinary scientific research in three main areas: particle and high energy physics; photon science; and the development, construction and operation of particle accelerators. Its name refers to its first project, an electron synchrotron. DESY is publicly financed by the Federal Republic of Germany, the States of Germany, and the German Research Foundation (DFG). DESY is a member of the Helmholtz Association and operates at sites in Hamburg and Zeuthen.

7.2.1 Functions

DESY's function is to conduct fundamental research. It specializes in:

- **Particle accelerator** development, construction and operation.

- **Particle physics** research to explore the fundamental characteristics of matter and forces, including astroparticle physics

- **Photon science** research in surface physics, material science, chemistry, molecular biology, geophysics and medicine through the use of synchrotron radiation and free electron lasers

In addition to operating its own large accelerator facilities, DESY also provides consulting services to research initiatives, institutes and universities. It is closely involved in major international projects such as the European X-Ray Free-Electron Laser, the Large Hadron Collider in Geneva, the IceCube Neutrino Observatory at the South Pole and the International Linear Collider.*[1]

7.2.2   Sites

DESY operates in two locations. The primary location is in a suburb of Hamburg. In 1992, DESY expanded to a second site in Zeuthen near Berlin.

Main entrance of the DESY campus in Hamburg.

Hamburg

The DESY Hamburg site is located in the suburb Bahrenfeld, west of the city. Most of DESY's research in high energy physics with elementary particles has been taking place here since 1960. The site is bounded by the ring of the former PETRA particle accelerator (since 2007 PETRA III, a synchrotron source) and part of the larger HERA (Hadron Elektron Ring Anlage) ring. Besides these accelerators there is also the free electron laser FLASH, and its offspring XFEL, which is under construction since 2009. This project is meant to secure DESY's future place among the top research centers of the world.

Zeuthen

Following German reunification, the DESY expanded to a second site. The former Institute for High Energy Physics (German: *Institut für Hochenergiephysik IfH*) in Zeuthen, southeast of Berlin, was the high energy physics laboratory of the German Democratic Republic and belonged to the Academy of Sciences of the GDR. The institute was merged with DESY on 1 January 1992.

7.2.3   Employees and training

DESY employs approximately 2000 people, of which 650 are scientists, working in the fields of accelerator operation, research and development. Staff is distributed on the two sites as follows.

- Hamburg: 1800 employees, of which 600 are scientists

- Zeuthen: 200 employees, of which 50 are scientists

It also trains more than 100 apprentices in commercial and technical vocations, and more than 700 undergraduates, graduates, and postdocs. DESY also hosts 3000 scientists from over 40 countries annually.*[2]

7.2.4   Budget and financing

The research center has an annual budget around € 192 million. Of this, approximately € 173 million is budgeted for the Hamburg site and € 19 million for Zeuthen. The primary source of financing is the Federal Ministry for Education and Research (German: *Bundesministerium für Bildung und Forschung*) with 10% support coming from the German States of Hamburg and Brandenburg. Individual experiments and projects at the accelerators are financed by the participating German and foreign institutes, which in turn are often publicly financed. Special projects are funded by the German Research Foundation.*[2]

7.2.5   International Co-operation

Segment of a particle accelerator at DESY

2500 external scientists used the DESY facilities for research with photons at PETRA III and FLASH in 2012.*[3]

The International Project HERA

The construction of the accelerator HERA was one of the first really internationally financed projects of this magnitude. Beforehand the construction of scientific facilities was always financed by the country in which it is located. Only the costs for the experiments were carried by the conducting national or foreign institutes. But due to the enormous scope of the HERA project many international facilities consented to already help with the construction. All in all more than 45 institutes and 320 corporations participated with donations of money and/or materials in the construction of the facility, more than 20% of the costs were carried by foreign institutions.

Following the example of HERA, many scientific projects of a large scale are financed jointly by several states. By now this model is established and international cooperation is pretty common with the construction of those facilities.

7.2.6 Particle Accelerators, Facilities and Experiments at DESY

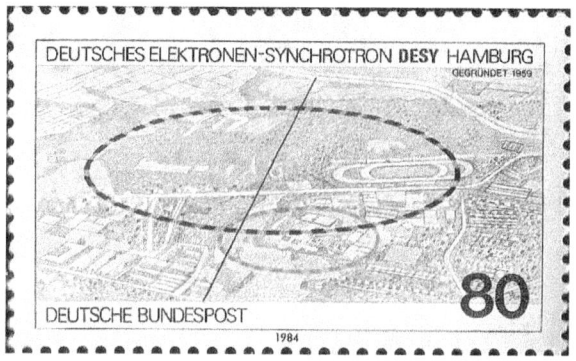

German 1984 postal stamp - 25th anniversary of DESY's foundation

DESY's accelerators were not built all at once, but rather were added one by one to meet the growing demand of the scientists for higher and higher energies to gain more insight into particle structures. In the course of the construction of new accelerators the older ones were converted to pre-accelerators or to sources for synchrotron radiation for laboratories with new research tasks (for example for HASYLAB).

Nowadays, after the shutdown of the accelerator HERA in 2007, DESY's most important facilities are the high intensity source for synchrotron radiation, PETRA III, the synchrotron-research lab HASYLAB, the free-electron laser FLASH (previously called VUV-FEL), and the test facility for the planned European XFEL. The development of the different facilities will be described chronologically in the following section.

DESY

The construction of the first particle accelerator **DESY** (*D*eutsches *E*lektronen *S*ynchrotron, "German Electron Synchrotron") began in 1960. At that time it was the biggest facility of this kind and was able to accelerate electrons to 7.4 GeV. On 1 January 1964 the first electrons were accelerated in the synchrotron and the research on elementary particles began.

The international attention first focused on DESY in 1966 due to its contribution to the validation of quantum electrodynamics, which was achieved with results from the accelerator. In the following decade DESY established itself as a center of excellence for the development and operation of high-energy accelerators.

The synchrotron radiation, which comes up as a side effect, was first used in 1967 for absorption measurements. For the arising spectrum there had not been any conventional radiation sources beforehand. The European Molecular Biology Laboratory (EMBL) made use of the possibilities that arose with the new technology and in 1972 established a permanent branch at DESY with the aim of analyzing the structure of biological molecules by means of synchrotron radiation.

The electron-synchrotron DESY II and the proton-synchrotron DESY III were taken into operation in 1987 and 1988 respectively as pre-accelerators for HERA.

DORIS III

DORIS (**Do**ppel-**Ri**ng-**S**peicher, "double-ring storage"), built between 1969 and 1974, was DESY's second circular accelerator and its first storage ring with a circumference of nearly 300 m. Constructed as an electron-positron storage ring, one could conduct collision-experiments with electrons and their antiparticles at energies of 3.5 GeV per beam. In 1978 the energy of the beams was raised to 5 GeV each.

With evidence of the "excited charmonium states" DORIS made an important contribution to the process of proving the existence of heavy quarks. In the same year there were the first tests of X-ray lithography at DESY, a procedure that was later refined to X-ray depth lithography.

In 1987 the ARGUS detector of the DORIS storage ring was the first place where the conversion of a B-meson into its antiparticle, the anti-B-meson was observed. From this one could conclude that it was possible, for the second-heaviest quark - the bottom-quark - under certain circumstances to convert into a different quark. One could also conclude from this that the unknown sixth quark - the top quark - had to possess a huge mass. The top quark was found eventually in 1995 at the Fermilab in the USA.

After the commissioning of HASYLAB in 1980 the synchrotron radiation, which was generated at DORIS as a byproduct, was used for research there. While in the beginning DORIS was used only ⅓ of the time as a radiation source, from 1993 on the storage-ring solely served that purpose under the name DORIS III. In order to achieve more intense and controllable radiation, DORIS was upgraded in 1984 with wigglers and undulators. By means of a special array of magnets the accelerated positrons could now be brought onto a slalom course. By this the intensity of the emitted synchrotron radiation was increased a hundredfold in comparison to conventional storage ring systems.

DORIS III provided 33 photon beamlines, where 44 instruments are operated in circulation. The overall beam time per year amounts to 8 to 10 months. It was finally shut down in favour of it successor PETRA III in the end of 2012.

OLYMPUS   The former site of ARGUS in DORIS is now the location of the OLYMPUS experiment, which began installation in 2010.[*][4] OLYMPUS uses the toroidal magnet and pair of drift chambers from the MIT-Bates BLAST experiment along with refurbished time-of-flight detectors and multiple luminosity monitoring systems. OLYMPUS measures the positron-proton to electron-proton cross section ratio to precisely determine the size of two-photon exchange in elastic ep scattering. This may help resolve the proton form factor discrepancy between recent measurements made using polarization techniques and ones using the Rosenbluth separation method.[*][5] OLYMPUS began taking data in early 2012.

PETRA II

Main article: PETRA

PETRA   (**P**ositron-**E**lektron-**T**andem-**R**ing-Anlage, "positron-electron tandem-ring facility") was built between 1975 and 1978. At the time of its construction it was the biggest storage ring of its kind and still is DESY's second largest synchrotron after HERA. PETRA originally served for research on elementary particles. The discovery of the gluon, the carrier particle of the strong nuclear force, in 1979 is counted as one of the biggest successes. PETRA can accelerate electrons and positrons to 19 GeV.

Research at PETRA lead to an intensified international use of the facilities at DESY. Scientists from China, England, France, Israel, the Netherlands, Norway and the USA participated in the first experiments at PETRA alongside many German colleagues.

In 1990 the facility was taken into operation under the name PETRA II as a pre-accelerator for protons and electrons/positrons for the new particle accelerator HERA. In March 1995, PETRA II was equipped with undulators to create greater amounts of synchrotron radiation with higher energies, especially in the X-ray part of the spectrum. Since then PETRA serves HASYLAB as a source of high-energy synchrotron radiation and for this purpose possesses three test experimental areas. Positrons are accelerated to up to 12 GeV nowadays.

PETRA III Max von Laue hall at DESY campus in Hamburg.

PETRA III

PETRA III is the third incarnation for the PETRA storage ring operating a regular user programme as the most brilliant storage ring based X-ray source worldwide since August 2010. The accelerator produces a particle energy of 6 GeV.[*][6]

HASYLAB

The HASYLAB (**Ha**mburger **S**ynchrotronstrahlungs**lab**or, "Hamburg Synchrotron radiation Laboratory") is used for research with synchrotron radiation at DESY. It was opened in 1980 with 15 experimental areas (today there are 42). The laboratory adjoins to the storage ring DORIS in order to be able to use the generated synchrotron radiation for its research. While in the beginning DORIS served only one third of the time as a radiation source for HASYLAB, since 1993 all its running time is available for experiments with synchrotron radiation. On top of the 42 experimental areas DORIS provides, there are also three test experimental

View inside the PETRA III Max von Laue hall at DESY campus in Hamburg.

The ARGUS detector at DESY

areas available for experiments with high-energy radiation generated with the storage ring PETRA.

After the upgrade of DORIS with the first wigglers, which produced far more intense radiation, the first Mössbauer spectrum acquired by means of synchrotron radiation was recorded at HASYLAB in 1984.

In 1985 the development of more advanced X-ray technology made it possible to bring to light the structure of the influenza virus. In the following year researchers at HASYLAB were the first to successfully make the attempt of exciting singular grid oscillations in solid bodies. Thus it was possible to conduct analyses of elastic materials, which were possible prior to this only with nuclear reactors via neutron scattering.

In 1987 the workgroup for structural molecular biology of the Max Planck Society founded a permanent branch at HASYLAB. It uses synchrotron radiation to study the structure of ribosomes.

Nowadays many national and foreign groups of researchers conduct their experiments at HASYLAB: All in all 1900 scientists participate in the work. On the whole the spectrum of the research ranges from fundamental research to experiments in physics, material science, chemistry, molecular biology, geology and medicine to industrial cooperations.

One example is OSRAM, which since recently uses HASYLAB to study the filaments of their light bulbs. The gained insights helped to notably increase the life span of the lamps in certain fields of application.

In addition researchers at HASYLAB analysed among other things minuscule impurities in silicon for computer chips, the way catalysators work, the microscopic properties of materials and the structure of protein molecules.

HERA

Main article: Hadron Elektron Ring Anlage

HERA (**H**adron-**E**lektron-**R**ing-**A**nlage, "Hadron Electron Ring Facility") was DESY's largest synchrotron and storage ring, with a circumference of 6336 metres. The construction of the subterranean facility began in 1984 and was an international task: In addition to Germany, 11 further countries participated in the development of HERA. The accelerator began operation on November 8, 1990 and the first two experiments started taking data in 1992. HERA was mainly used to study the structure of protons and the properties of quarks. It was closed on June 30, 2007.[*][7]

HERA was the only accelerator in the world that was able to collide protons with either electrons or positrons. To make this possible, HERA used mainly superconducting magnets, which was also a world first. At HERA, it was possible to study the structure of protons up to 30 times more accurately than before. The resolution covered structures 1/1000 of the proton in size. In the years to come, there were made many discoveries concerning the composition of protons from quarks and gluons.

HERA's tunnels run 10 to 25 metres below ground level and have an inner diameter of 5.2 metres. For the construction, the same technology was used as for the construction of subway tunnels. Two circular particle accelerators run inside the tube. One accelerated electrons to energies of 27.5 GeV, the other one protons to energies of 920 GeV in the opposite direction. Both beams completed their circle nearly at the speed of light, making approximately 47 000 revolutions per second.

At two places of the ring the electron and the proton beam could be brought to collision. In the process, electrons or positrons are scattered at the constituents of the protons, the quarks. The products of these particle collisions, the scattered lepton and the quarks, which are produced by the fragmentation of the proton, were registered in huge detectors. In addition to the two collision zones, there are two more interaction zones. All four zones are placed in big subterranean halls. A different international group of researchers were at work in each hall. These groups developed, constructed and run house-high, complex measurement devices in many years of cooperative work and evaluate enormous amounts of data.

The experiments in the four halls will be presented in the following section:

H1 Main article: H1 (particle detector)

H1 is a universal detector for the collision of electrons and protons and was located in DESY's HERA-Hall North. It had been active since 1992, measured 12 m × 10 m × 15 m and weighed 2,800 tons.

It was designed for the decryption of the inner structure of the proton, the exploration of the strong interaction as well as the search for new kinds of matter and unexpected phenomena in particle physics.

ZEUS Main article: ZEUS

ZEUS is like H1 a detector for electron-proton collisions and was located in HERA-Hall South. Built in 1992, it measured 12 m × 11 m × 20 m and weighs 3600 tons.

Its tasks resemble H1's.

HERA-B Main article: HERA-B

HERA-B was an experiment in HERA-Hall West which collected data from 1999 to February 2003. By using HERA's proton beam, researchers at HERA-B conducted experiments on heavy quarks. It measured 8 m × 20 m × 9 m and weighed 1 000 tons.

HERMES Main article: HERMES experiment

The HERMES experiment in HERA-Hall East was taken into operation in 1995. HERA's longitudinally polarised

electron beam was used for the exploration of the spin structure of nucleons. For this purpose the electrons were scattered at energies of 27.5 GeV at an internal gas target. This target and the detector itself were designed especially with a view to spin polarised physics. It measured 3.5 m × 8 m × 5 m and weighed 400 tons.

FLASH II experimental hall at DESY campus in Hamburg.

FLASH

FLASH (**F**ree-electron -**LAS**er in **H**amburg) is a superconducting linear accelerator with a free electron laser for radiation in the vacuum-ultraviolet and soft X-ray range of the spectrum. It originated from the TTF (TESLA Test Facility), which was built in 1997 to test the technology that was to be used in the planned linear collider TESLA, a project which was replaced by the ILC (International Linear Collider). For this purpose the TTF was enlarged from 100 m to 260 m.

At FLASH technology for the future-project European XFEL is tested as well as for the ILC. Five test experimental areas have been in use since the commissioning of the facility in 2004.

European XFEL

Main article: European x-ray free electron laser

The European x-ray free electron laser (European XFEL) is an X-ray laser currently under construction. It is a European project in collaboration with DESY and is as of 2012 slated to be operational by the end of 2015.[8] The 3.4 km long tunnel will contain an 2.1 km long superconducting linear accelerator where electrons will be accelerated to an energy of up to 17.5 GeV.[9] It will produce extremely short and powerful X-ray flashes that have many applications. Construction of the tunnels was completed in summer 2012.[8]

Center for Free-Electron Laser Science, CFEL, at DESY campus in Hamburg.

Further Accelerators

In addition to the larger ones, there are also several smaller particle accelerators which serve mostly as pre-accelerators for PETRA and HERA. Among these are the linear accelerators LINAC I (operated from 1964 to 1991 for electrons), LINAC II (operated since 1969 for positrons) and LINAC III (operated since 1988 as a pre-accelerator for protons for HERA).

7.2.7 Plans for the future

DESY is involved in the project International Linear Collider (ILC). This project consists of a 30-kilometer-long linear accelerator. An international consortium decided to build it with the technology originally developed for the TESLA project. There has been no final decision on where to build the accelerator but Japan is the most likely candidate.

7.2.8 References

[1] DESY About retrieved 24-May-2012.

[2] DESY Portrait retrieved 24-May-2012.

[3] All Beamtime and Operating Schedules retrieved 23-Sep-2013.

[4] "OLYMPUS - Deutsches Elektronen-Synchrotron DESY". Retrieved 16 August 2012.

[5] "OLYMPUS Collaboration". Retrieved 16 August 2012.

[6] "PETRA III". Hasylab, Desy. Retrieved 2012-09-04.

[7] Last run of HERA. Read September 30th 2007.

[8] "European X-Ray Free-Electron Laser tunnel construction completed".

[9] "European XFEL facts & figures". Retrieved 2009-11-27.

7.2.9 External links

- Official website

7.3 DONUT

This page is for the Fermilab experiment. For other uses, see Doughnut (disambiguation).

DONUT (**D**irect **O**bservation of the **NU Tau**, E872) was

DONUT Detector

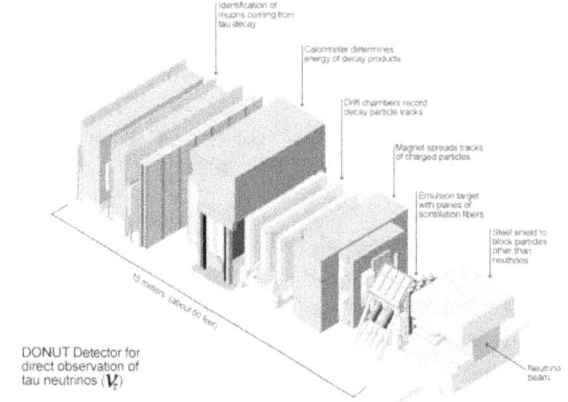

Schematic overview of the DONUT detector

an experiment at Fermilab dedicated to the search for tau neutrino interactions. The detector operated during a few months in the summer of 1997, and successfully detected the tau neutrino.[1] It confirmed the existence of the last lepton predicted by the Standard Model.[2] The data from the experiment was also used to put an upper limit on the tau neutrino magnetic moment[3] and measure its interaction cross section.[4]

7.3.1 Principle

In DONUT, protons accelerated by the Tevatron were used to produce tau neutrinos via decay of charmed mesons. After eliminating as many unwanted background particles as possible by a system of magnets and bulk matter (mostly iron and concrete), the beam passed through several sheets of nuclear emulsion. In very rare cases one of the neutrinos would interact in the detector, producing electrically charged particles which left visible tracks in the emulsion and could be electronically registered by a system of scintillators and drift chambers.[1]

Using the electronic information, possible neutrino interactions were identified and selected for further analysis. This meant photographically developing the emulsion sheets so any traces left by particles passing through them would show up as a small black dot. By connecting these dots across subsequent sheets, the path that each particle had taken was reconstructed and likely neutrino interactions identified. The characteristic properties of tau neutrino interactions were that several tracks suddenly appeared without any leading up to them and that one of those tracks would show a "kink" after a few millimeters, indicating decay of a tau lepton.*[1]

7.3.2 Result

In July 2000, the DONUT collaboration announced the first observation of tau neutrino interactions. Even though this result was based on only four events, the signal was far in excess of the expected background (< 0.2 events) and is therefore valid. Its significance lies in the fact that the tau neutrino had so far remained the only particle of the Standard Model that had not been directly observed except for the Higgs boson.*[2]

Other than the result itself, DONUT also allowed validation of new techniques for high energy neutrino detection, notably the *Emulsion Cloud Chamber*, in which nuclear emulsion sheets are interspersed with layers of iron, leading to an increase in the number of interactions.

7.3.3 References

[1] K. Kodama *et al.*   (DONUT Collaboration) (2001). "Observation of tau neutrino interactions".   *Physics Letters B* **504** (3): 218.   arXiv:hep-ex/0012035. Bibcode:2001PhLB..504..218D.   doi:10.1016/S0370-2693(01)00307-0.

[2] "Physicists Find First Direct Evidence for Tau Neutrino at Fermilab" (Press release). Fermilab. 20 July 2000.

[3] K. Kodama *et al.* (DONUT Collaboration) (2008). "A first measurement of the interaction cross section of the tau neutrino". *Physical Review D* **78** (5): 052002. arXiv:0711.0728.   Bibcode:2008PhRvD..78e2002K. doi:10.1103/PhysRevD.78.052002.

[4] R. Schwienhorst *et al.* (DONUT Collaboration) (2001). "A new upper limit for the tau-neutrino magnetic moment" .   *Physics Letters B* **513**: 23.   arXiv:hep-ex/0102026. Bibcode:2001PhLB..513...23D.   doi:10.1016/S0370-2693(01)00746-8.

7.3.4 External links

- DONUT home page

7.4 Fermilab

Fermi National Accelerator Laboratory (**Fermilab**), located just outside Batavia, Illinois, near Chicago, is a United States Department of Energy national laboratory specializing in high-energy particle physics. Since 2007, Fermilab has been operated by the Fermi Research Alliance, a joint venture of the University of Chicago, Illinois Institute of Technology and the Universities Research Association (URA). Fermilab is a part of the Illinois Technology and Research Corridor.

Fermilab's Tevatron was a landmark particle accelerator; at 3.9 miles (6.3 km) in circumference, it was the world's second-largest energy particle accelerator (after CERN's Large Hadron Collider, which is 27 km in circumference), until it was shut down in 2011. In 1995, the discovery of the top quark was announced by researchers who used the Tevatron's CDF and DØ detectors.

In addition to high-energy collider physics, Fermilab hosts smaller fixed-target and neutrino experiments, such as MiniBooNE and MicroBooNE (Mini Booster Neutrino Experiment and Micro Booster Neutrino Experiment), SciBooNE (SciBar Booster Neutrino Experiment) and MINOS (Main Injector Neutrino Oscillation Search). The MiniBooNE detector is a 40-foot (12 m) diameter sphere that contains 800 tons of mineral oil lined with 1,520 phototube detectors.   An estimated 1 million neutrino events are recorded each year. SciBooNE is the newest neutrino experiment at Fermilab; it sits in the same neutrino beam as MiniBooNE but has fine-grained tracking capabilities. The MINOS experiment uses Fermilab's NuMI (Neutrinos at the Main Injector) beam, which is an intense beam of neutrinos that travels 455 miles (732 km) through the Earth to the Soudan Mine in Minnesota.

In the public realm, Fermilab hosts many cultural events: not only public science lectures and symposia, but also classical and contemporary music concerts, folk dancing and arts galleries. The site is open from dawn to dusk to visitors who present valid photo identification.

Asteroid 11998 Fermilab is named in honor of the laboratory.

7.4.1 History

Weston, Illinois, was a community next to Batavia voted out of existence by its village board in 1966 to provide a site for Fermilab.*[2]

The laboratory was founded in 1967 as the **National Accelerator Laboratory**; it was renamed in honor of Enrico Fermi in 1974. The laboratory's first director was Robert Rathbun Wilson, under whom the laboratory opened ahead

Robert Rathbun Wilson Hall

of time and under budget. Many of the sculptures on the site are of his creation. He is the namesake of the site's high-rise laboratory building, whose unique shape has become the symbol for Fermilab and which is the center of activity on the campus.

After Wilson stepped down in 1978 to protest the lack of funding for the lab, Leon M. Lederman took on the job. It was under his guidance that the original accelerator was replaced with the Tevatron, an accelerator capable of colliding proton and an antiproton at a combined energy of 1.96 TeV. Lederman stepped down in 1989 and remains Director Emeritus. The science education center at the site was named in his honor.

The later directors include:

- John Peoples, 1989 to 1999

- Michael S. Witherell, July 1999 to June 2005

- Piermaria Oddone, July 2005 to July 2013[*][3]

- Nigel Lockyer, September 2013 to the present[*][4]

Fermilab continues to participate in the work in the LHC; it serves as a Tier 1 site in the Worldwide LHC Computing Grid.[*][5]

7.4.2 Accelerators

Current state

As of 2014, the first stage in the acceleration process (pre-accelerator injector) takes place in two ion sources which turn hydrogen gas into H^*- ions. The gas is introduced into a container lined with molybdenum electrodes, each a matchbox-sized, oval-shaped cathode and a surrounding anode, separated by 1 mm and held in place by glass ceramic

insulators. A magnetron generates a plasma to form the ions near the metal surface. The ions are accelerated by the source to 35 keV and matched by low energy beam transport (LEBT) into the radio-frequency quadrupole (RFQ) which applies a 750 keV electrostatic field giving the ions their second acceleration. At the exit of RFQ, the beam is matched by medium energy beam transport (MEBT) into the entrance of the linear accelerator (linac).[*][6]

The next stage of acceleration is linear particle accelerator (linac). This stage consists of two segments. The first segment has 5 vacuum vessel for drift tubes, operating at 201 MHz. The second stage has 7 side-coupled cavities, operating at 805 MHz. At the end of linac, the particles are accelerated to 400 MeV, or about 70% of the speed of light.[*][7][*][8] Immediately before entering the next accelerator, the H^*- ions pass through a carbon foil, becoming H^*+ ions (protons).[*][9]

The resulting protons then enter the booster ring, a 468 m-circumference circular accelerator whose magnets bend beams of protons around a circular path. The protons travel around the Booster about 20,000 times in 33 milliseconds, adding energy with each revolution until they leave the Booster accelerated to 8 GeV.[*][9]

The final acceleration is applied by the Main Injector, which is the smaller of the two rings in the last picture below (foreground). Completed in 1999, it has become Fermilab's "particle switchyard" in that it can route protons to any of the experiments installed along the beam lines after accelerating them to 120 GeV. Until 2011, the Main Injector provided protons to the antiproton ring and the Tevatron for further acceleration but now provides the last push before the particles reach the beam line experiments.

- Two ion sources at the center with two high-voltage electronics cabinets next to them[*][1]

- Beam direction right to left: RFQ (silver), MEBT (green), first drift tube linac (blue)[*][1]

- A 7835 power amplifier that is used at the first stage of linac[*][2]

- A 12 MW klystron used at the second stage of linac[*][2]

- A cutaway view of the 805 MHz side-couple cavities[*][3]

- Booster ring[*][4]

- Fermilab's accelerator rings

1. ^ [*]*a* [*]*b* "35 years of H- ions at Fermilab" (PDF). *Fermilab*. Retrieved 12 August 2015.

2. ^ Cite error: The named reference slideshow was invoked but never defined (see the help page).

3. ^ May, Michael P.; Fritz, James R.; Jurgens, Thomas G.; Miller, Harold W.; Olson, James; Snee, Daniel (1990). "Mechanical Construction of the 805 MHz Side Couple Cavities for the Fermilab Linac Upgrade" (PDF). *Proceedings of the Linear Accelerator Conference 1990, Albuquerque, New Mexico, USA*. Retrieved 13 August 2015.

4. ^ "Wilson Hall & vicinity" . *Fermilab*. Retrieved 12 August 2015.

Proton improvement plan

In recognizing higher demands of proton beams to support new experiments, Fermilab started an initiative to enhance their accelerators. The project started in 2011 and will continue for many years.[*][10] The project has two phases called Proton Improvement Plan (PIP) and Proton Improvement Plan-II (PIP-II).[*][11]

PIP (2011–2018)   The overall goals of PIP are to increase the repetition rate of the Booster beam from 7 Hz to 15 Hz and replace old hardware to increase reliability of the operation.[*][11] Before the start of the PIP project, a replacement of the pre-accelerator injector was underway. The replacement of almost 40-year-old Cockcroft–Walton generators to RFQ started in 2009 and completed in 2012. At the linac stage, the analog beam position monitor (BPM) modules were replaced with digital boards in 2013. A replacement of Linac vacuum pumps and related hardware is expected to be completed in 2015. A study on the replacement of 201-MHz drift tubes is still ongoing. At the boosting stage, a major component of the PIP is to upgrade the Booster ring to 15-Hz operation. The Booster has 19 radio frequency stations. Originally, the Booster stations were operating without solid-state drive system which was acceptable for 7-Hz, but not for 15-Hz operation. A demonstration project in 2004 converted one of the stations to solid state drive prior to the PIP project. As part of the project, the remaining stations were successfully converted to solid state in 2013. Another major part of the PIP project is to refurbish and replace 40-year-old Booster cavities. Many cavities have been refurbished and tested to operate at 15 Hz repetition rate. The completion of cavity refurbishment is expected to be completed in 2015 and the repetition rate can be gradually increased to 15-Hz operation from that point. A longer term upgrade task is to replace the Booster cavities with a new design. The research and development of the new cavities is underway. The completion of the Booster cavity replacement is expected to be in 2018.[*][10]

Prototypes of SRF cavities to be used in the last segment of PIP-II linac[*][12]

PIP-II   The goals of PIP-II include a plan to delivery 1.2 MW of proton beam power from the Main Injector to the Deep Underground Neutrino Experiment target at 120 GeV and the power near 1 MW at 60 GeV with a possibility to extend the power to 2 MW in the future. The plan should also support the current 8 GeV experiments including Mu2e, g-2, and other short-baseline neutrino experiments. These require an upgrade to the linac to inject to the Booster with 800 MeV. The first option is to add 400 MeV "afterburner" superconducting linac at the tail end of the existing 400 MeV. This requires moving the existing linac up 50 metres (160 ft). However, there are many technical issues with this approach. The preferred option is to build a new 800 MeV superconducting linac to inject to the Booster ring. The new linac site will be located on top of a small portion of Tevatron near the Booster ring in order to take advantage of existing electrical and water, and cryogenic infrastructure. The PIP-II linac will have low energy beam transport line (LEBT), radio frequency quadrupole (RFQ), and medium energy beam transport line (MEBT) operated at the room temperature at with a 162.5 MHz and energy increasing from 0.03 MeV. The first segment of linac will be operated at 162.5 MHz and energy increased up to 11 MeV. The second segment of linac will be operated at 325 MHz and energy increased up to 177 MeV. The last segment of linac will be operated at 650 MHz and will have the final energy level of 800 MeV.[*][13]

Project X

Main article: Project X (Accelerator)

Project X is a long range plan to bring accelerators at Fermilab campus to new frontiers. The plan for accelerators focuses on two of the three frontiers that are long-term plan of Fermilab. In the intensity frontier, the new high-intensity accelerators will support experiments that require intense particle beam to understand particles such as neutrinos, muons, kaons and nuclei. In the energy frontier, the accelerators will support detection of new particles and forces with potential future projects such as multi-TeV Muon Collider. The immediate plan of Project X is to focus on the intensity frontier. The project is broken down into 3 stages. Stage one includes upgrade to existing facilities to support immediate experiments. This stage has translated into work done in the Proton Improvement Plan. Stage two includes delivery of three concurrent beam levels: 2.9 MW at 3 GeV; 50–200 kW at 8 GeV and 2.3 MW at 60–120 GeV. Stage three is to build next generation accelerators as the front end to the energy frontier based on international collaboration in projects such as Neutrino Factory and Muon Collider.[*][14]

7.4.3 Experiments

- Cryogenic Dark Matter Search (CDMS)

- COUPP: Chicagoland Observatory for Underground Particle Physics

- Dark Energy Survey (DES)

- Deep Underground Neutrino Experiment (DUNE), formerly known as Long Baseline Neutrino Experiment (LBNE)

- Holometer interferometer

- MiniBooNE: Mini Booster Neutrino Experiment

- MicroBooNE: Micro Booster Neutrino Experiment

- MINOS: Main Injector Neutrino Oscillation Search

- MINERvA: Main INjector ExpeRiment with vs on As

- MIPP: Main Injector Particle Production

- Mu2e: Muon-to-Electron Conversion Experiment

- Muon g-2

- NOvA: NuMI Off-axis v_e Appearance

- SELEX: SEgmented Large-X baryon spectrometer EXperiment, run to study charmed baryons

- Sciboone: SciBar Booster Neutrino Experiment

- SeaQuest

Interior of Wilson Hall

7.4.4 Architecture

Fermilab's first director, Wilson, insisted that the site's aesthetic complexion not be marred by a collection of concrete block buildings. The design of the administrative building (Wilson Hall) harkens back to St. Pierre's Cathedral in Beauvais, France. Several of the buildings and sculptures within the Fermilab reservation represent various mathematical constructs as part of their structure.

The Archimedean Spiral is the defining shape of several pumping stations as well as the building housing the MINOS experiment. The reflecting pond at Wilson Hall also showcases a 32-foot-tall (9.8 m) hyperbolic obelisk, designed by Wilson. Some of the high-voltage transmission lines carrying power through the laboratory's land are built to echo the Greek letter π. One can also find structural examples of the DNA double-helix spiral and a nod to the geodesic sphere.

Wilson's sculptures on the site include *Tractricious*, a freestanding arrangement of steel tubes near the Industrial Complex constructed from parts and materials recycled from the Tevatron collider, and the soaring *Broken Symmetry*, which greets those entering the campus via the Pine Street entrance.[*][15] Crowning the Ramsey Auditorium is a representation of the Möbius strip with a diameter of more than 8 feet (2.4 m). Also scattered about the access roads and village are a massive hydraulic press and old magnetic containment channels, all painted blue.

7.4.5 Current developments

Fermilab is dismantling the CDF (Collider Detector at Fermilab) and DØ (D0 experiment) facilities, and has been approved to continue moving forward with MINOS, NOvA, G-2, and Liquid Argon Test Facility.

LBNE

Fermilab has been approved and currently stands to become the world leader in Neutrino physics through its Long Baseline Neutrino Experiment (LBNE). Other leaders are CERN, which leads in Accelerator physics with the Large Hadron Collider (LHC), and Japan, which has been approved to build and lead the International Linear Collider (ILC).

"Over 350 people from over 60 institutions participate in the Long-Baseline Neutrino Experiment (LBNE), working together to plan and develop both the experimental facilities and the physics program. LBNE is expected to be fully constructed and ready for operations in 2022.

LBNE plans a world-class program in neutrino physics that will measure fundamental physical parameters to high precision and explore physics beyond the Standard Model. The measurements LBNE makes will greatly increase our understanding of neutrinos and their role in the universe, thereby better elucidating the nature of matter and antimatter.

LBNE will send the world's highest-intensity neutrino beam 800 miles through the Earth's mantle to a large detector, a multi-kiloton volume of target material instrumented such that it can record interactions between neutrinos and the target material. Neutrinos are harmless and can pass right through matter, only very rarely colliding with other matter particles. Therefore, no tunnel is needed; the vast majority of the neutrinos will pass through the mantle's material, and in turn, right through the detector. The experiment will thus need to collect data for a decade or two since neutrinos interact so rarely.

Fermilab, in Batavia, IL, is the host laboratory and the site of LBNE's future beamline, and the Sanford Underground Research Facility (SURF), in Lead, SD, is the site selected to house the massive far detector. The term "baseline" refers to the distance between the neutrino source and the detector.

Why neutrinos: Neutrinos, astonishingly abundant yet not well understood, may provide the key to answering some of the most fundamental questions about the nature of our universe. The discovery that neutrinos are not massless, as previously thought, has opened a first crack in the highly successful Standard Model of Particle Physics. Neutrinos may play a key role in solving the mystery of how the universe came to consist only of matter rather than antimatter."

g−2

Transportation of the 600-ton magnet to Fermilab

Muon g-2 building (white and orange) that hosts the magnet

"In the summer of 2013, the Muon g−2 team successfully transported a 50-foot-wide electromagnet from Brookhaven National Laboratory in Long Island, New York, to Fermilab in one piece. The move took 35 days and traversed 3,200 miles over land and sea."

"Muon g−2 (pronounced gee minus two) will use Fermilab's powerful accelerators to explore the interactions of short-lived particles known as muons with a strong magnetic field in "empty" space. Scientists know that even in a vacuum, space is never empty. Instead, it is filled with an invisible sea of virtual particles that—in accordance with the laws of quantum physics—pop in and out of existence for incredibly short moments of time. Scientists can test the presence and nature of these virtual particles with particle beams traveling in a magnetic field."

Particle discovery

On September 3, 2008, the discovery of a new particle, the bottom Omega baryon ($\Omega-$
b) was announced at the DØ experiment of Fermilab. It is made up of two strange quarks and a bottom quark. This discovery helps to complete the "periodic table of the baryons" and offers insight into how quarks form matter.[*][16]

7.4.6 Wildlife at Fermilab

In 1967, Wilson brought five American Bison to the site, a bull and four cows, and an additional 21 were provided by the Illinois Department of Conservation. Some fearful locals believed at first that the bison were introduced in order to serve as an alarm if and when radiation at the laboratory reached dangerous levels, but they were assured by Fermilab that this claim had no merit. Today, the herd is a popular attraction that draws many visitors[*][17] and the grounds are also a sanctuary for other local wildlife populations.[*][18]

Working with the Forest Preserve District of DuPage County, Fermilab has introduced Barn owls to selected structures around the grounds.

7.4.7 See also

- Big Science

- Center for the Advancement of Science in Space—operates the US National Laboratory on the ISS.

- CERN

- Fermi Linux LTS

- Scientific Linux

- Stanford Linear Accelerator Center

7.4.8 References

[1] "DOE Budget Report" (PDF). Retrieved 2014-12-27.

[2] Fermilab. "Before Weston". Retrieved 2009-11-25.

[3] "Fermilab director Oddone announces plan to retire next year". *The Beacon-News*. August 2, 2012. Retrieved 10 July 2013.

[4] "New Fermilab director named". *Crain's Chicago Business*. June 21, 2013. Retrieved 10 July 2013.

[5] National Science Foundation. "The US and LHC Computing". Retrieved 2011-01-11.

[6] Carneiro, J.P. (13 Nov 2014). "Transmission efficiency measurement at the FNAL 4-rod RFQ (FERMILAB-CONF-14-452-APC)" (PDF). *27th International Linear Accelerator Conference (LINAC14)*. Retrieved 12 August 2015.

[7] "Fermilab Linac Slide Show Description". *Fermilab*. Retrieved 12 August 2015.

[8] Kubik, Donna (2005). *Fermilab* (PDF). Retrieved 12 August 2015.

[9] "Accelerator". *Fermilab*. Retrieved 12 August 2015.

[10] "FNAL – The Proton Improvement Plan (PIP)". *Proceedings of IPAC2014, Dresden, Germany* (PDF). p. 3409–3411. ISBN 978-3-95450-132-8. Retrieved 15 August 2015.

[11] Holmes, Steve (16 December 2013). *MegaWatt Proton Beams for Particle Physics at Fermilab* (PDF). Fermilab. Retrieved 15 August 2015.

[12] "02 Proton and Ion Accelerators and Applications". *Proceedings of LINAC2014, Geneva, Switzerland* (PDF). September 2014. pp. 171–173. ISBN 978-3-95450-142-7. Retrieved 16 August 2015.

[13] *Proton Improvement Plan-II* (PDF). Fermilab. 12 December 2013. Retrieved 15 August 2015.

[14] *A Fermilab Plan for Discovery* (PDF). 2011. Retrieved 18 August 2015.

[15] "About Fermilab - The Fermilab Campus". 2005-12-01. Retrieved 2007-02-27.

[16] "Fermilab physicists discover "doubly strange" particle". Fermilab. 9 September 2008.

[17] Fermilab (30 December 2005). "Safety and the Environment at Fermilab". Retrieved 2006-01-06.

[18] http://www.fnal.gov/pub/about/campus/ecology/wildlife/ retrieved 3/30/2013

7.4.9 External links

- Fermi National Accelerator Laboratory

  - *Fermilab Today* Daily newsletter
  - Other Fermilab online publications
  - Fermilab Virtual Tour
  - Architecture at the Fermilab campus

Coordinates: 41°49′55″N 88°15′26″W / 41.83194°N 88.25722°W

7.5   Homestake experiment

The **Homestake experiment** (sometimes referred to as the **Davis experiment**) was an experiment headed by astrophysicists Raymond Davis, Jr. and John N. Bahcall in the late 1960s. Its purpose was to collect and count neutrinos emitted by nuclear fusion taking place in the Sun. Bahcall did the theoretical calculations and Davis designed the experiment. After Bahcall calculated the rate at which the detector should capture neutrinos, Davis's experiment turned up only one third of this figure. The experiment was the first to successfully detect and count solar neutrinos, and the discrepancy in results essentially created the solar neutrino problem. The experiment operated continuously from 1970 until 1994. The University of Pennsylvania took it over in 1984. The discrepancy between the predicted and measured rates of neutrino detection was later found to be due to neutrino "flavour" oscillations.

7.5.1   Methodology

The experiment took place in the Homestake Gold Mine in Lead, South Dakota. Davis placed 1,478 meters (4,850 feet) underground a 380 cubic meter (100,000 gallon) tank of perchloroethylene, a common dry-cleaning fluid. A big target deep underground was needed to prevent interference from cosmic rays, taking into account the very small probability of a successful neutrino capture, and, therefore, very low effect rate even with the huge mass of the target. Perchloroethylene was chosen because it is rich in chlorine. Upon interaction with an electron neutrino, a chlorine−37 atom transforms into a radioactive isotope of argon−37, which can then be extracted and counted. The reaction of neutrino capture is

$$\nu_e + {}^{37}Cl \longrightarrow {}^{37}Ar + e^-.$$

The reaction threshold is 0.814 MeV, i.e. the neutrino should have at least this energy to be captured by the chlorine-37 nucleus.

Every few weeks, Davis bubbled helium through the tank to collect the argon that had formed. A small (few cubic cm) gas counter was filled by the collected few tens atoms of argon-37 (together with the stable argon) to detect its decays. In such way, Davis was able to determine how many neutrinos had been captured.[*][1][*][2]

7.5.2   Conclusions

Davis's figures were consistently very close to one-third of Bahcall's calculations. The first response from the scientific community was that either Bahcall or Davis had made a mistake. Bahcall's calculations were checked repeatedly, with no errors found. Davis scrutinized his own experiment and insisted there was nothing wrong with it. The Homestake experiment was followed by other experiments with the same purpose, such as Kamiokande in Japan, SAGE in the former Soviet Union, GALLEX in Italy, Super Kamiokande, also in Japan, and SNO (Sudbury Neutrino Observatory) in Ontario, Canada. SNO was the first detector able to detect neutrino oscillation, solving the solar neutrino problem. The results of the experiment, published in 2001, revealed that of the three "flavours" between which neutrinos are able to oscillate, Davis's detector was sensitive to only one. After it had been proven that his experiment was sound, Davis shared the 2002 Nobel Prize in Physics. Among those sharing the prize was Masatoshi Koshiba of Japan, who worked on the Kamiokande and the Super Kamiokande.

7.5.3   See also

- Cowan–Reines neutrino experiment (a contemporary experiment by Reines and Cowan which discovered the antineutrino)

7.5.4   References

[1] Martin, B.R.; Shaw, G (1999). *Particle Physics (2nd ed.)*. Wiley. p. 265. ISBN 0-471-97285-1.

[2] B. T. Cleveland et al. (1998). "Measurement of the Solar Electron Neutrino Flux with the Homestake Chlorine Detector". *Astrophysical Journal* **496**: 505–526. Bibcode:1998ApJ...496..505C. doi:10.1086/305343.

- Raymond Davis Jr.'s Solar Neutrino Experiments (at BNL.gov)

Coordinates: 44°21′12″N 103°44′39″W / 44.35333°N 103.74417°W

7.6   Kamioka Observatory

The **Kamioka Observatory, Institute for Cosmic Ray Research** (神岡宇宙素粒子研究施設 *Kamioka Uchū Soryūshi Kenkyū Shisetsu*) is a neutrino physics laboratory located underground in the Mozumi Mine of the Kamioka Mining and Smelting Co. near the Kamioka section of the city of Hida in Gifu Prefecture, Japan. A set of groundbreaking neutrino experiments have taken place at the observatory over the past two decades. All of the experiments have been very large and have contributed substantially to the advancement of particle physics, in particular to the study of neutrino astronomy and neutrino oscillation.

7.6.1 Past experiments

KamiokaNDE

A model of KamiokaNDE

The first of the Kamioka experiments was named KamiokaNDE for **Kamioka Nucleon Decay Experiment**. It was a large water Čerenkov detector designed to search for proton decay. To observe the decay of a particle with a lifetime as long as a proton an experiment must run for a long time and observe an enormous number of protons. This can be done most cost effectively if the target (the source of the protons) and the detector itself are made of the same material. Water is an ideal candidate because it is inexpensive, easy to purify, stable, and can detect relativistic charged particles through their production of Čerenkov radiation. A proton decay detector must be buried deep underground or in a mountain because the background from cosmic ray muons in such a large detector located on the surface of the Earth would be far too large. The muon rate in the KamiokaNDE experiment was about 0.4 events per second, roughly five orders of magnitude smaller than what it would have been if the detector had been located at the surface.*[1]

The distinct pattern produced by Čerenkov radiation allows for particle identification, an important tool both understanding the potential proton decay signal and for rejecting backgrounds. The ID is possible because the sharpness of the edge of the ring depends on the particle producing the radiation. Electrons (and therefore also gamma rays) produce fuzzy rings due to the multiple scattering of the low mass electrons. Minimum ionizing muons, in contrast produce very sharp rings as their heavier mass allows them to propagate directly.

Construction of Kamioka Underground Observatory (the predecessor of the present Kamioka Observatory, Institute for Cosmic Ray Research, University of Tokyo) began in 1982 and was completed in April, 1983. The detector was a cylindrical tank which contained 3,000 tons of pure water and had about 1,000 50 cm diameter photomultiplier tubes (PMTs) attached to the inner surface. The size of the outer detector was 16.0 m in height and 15.6 m in diameter. The detector failed to observe proton decay, but set what was then the world's best limit on the lifetime of the proton.

When pronounced in Japanese, the name of the project, *kamiokande*, can—among other meanings—be understood to mean 神を噛んで (*kami wo kande*), which roughly translates to *bite into God*.*[2]

Kamiokande-II

The **Kamiokande-II** experiment was a major step forward from KamiokaNDE, and made a significant number of important observations.

Solar Neutrinos In the 1930s, Hans Bethe and Carl Friedrich von Weizsäcker had hypothesized that the source of the sun's energy was fusion reactions in its core. While this hypothesis was widely accepted for decades there was no way of observing the sun's core and directly testing the hypothesis. Ray Davis's Homestake Experiment was the first to detect solar neutrinos, strong evidence that the nuclear theory of the sun was correct. Over a period of decades the Davis experiment consistently observed only about 1/3 the number of neutrinos predicted by the Standard Solar Models of his colleague and close friend John Bahcall. Because of the great technical difficulty of the experiment and its reliance on radiochemical techniques rather than real time direct detection many physicists were suspicious of his result.

It was realized that a large water Čerenkov detector could be an ideal neutrino detector, for several reasons. First, the enormous volume possible in a water Čerenkov detector can overcome the problem of the very small cross section of the 5-15 MeV solar neutrinos. Second, water Čerenkov detectors offer real time event detection. This meant that Individual neutrino-electron interaction candidate events could be studied on an event-by-event basis, starkly different from the month-to-month observation required in radiochemical experiments. Third, in the neutrino-electron scattering interaction the electron recoils in roughly the direction that the neutrino was travelling (similar to the motion of billiard balls), so the electrons "point back" to the sun. Fourth, neutrino-electron scattering is an elastic process, so the energy distribution of the neutrinos can be studied, further testing the solar model. Fifth, the characteristic "ring" produced by Čerenkov radiation allows discrimination of the signal against backgrounds. Finally, since a water Čerenkov experiment would use a different target, interaction process, detector technology, and location it would be a very complementary test of Davis's results.

It was clear that KamiokaNDE could be used to perform a fantastic and novel experiment, but a serious problem needed to be overcome first. The presence of radioactive backgrounds in KamiokaNDE meant that the detector had an energy threshold of tens of MeV. The signals produced by proton decay and atmospheric neutrino interactions are considerably larger than this, so the original KamiokaNDE detector had not needed to be particularly aggressive about its energy threshold or resolution. The problem was attacked in two ways. The participants of the KamiokaNDE experiment designed and built new purification systems for the water to reduce the radon background, and instead of constantly cycling the detector with "fresh" mine water they kept the water in the tank allowing the radon to decay away. A group from the University of Pennsylvania joined the collaboration and supplied new electronics with greatly superior timing capabilities. The extra information provided by the electronics further improved the ability to distinguish the neutrino signal from radioactive backgrounds. One further improvement was the expansion of the cavity, and the installation of an instrumented "outer detector". The extra water provided shielding from gamma rays from the surrounding rock, and the outer detector provided a veto for cosmic ray muons.[*][1]

With the upgrades completed the experiment was renamed **Kamiokande-II**, and started data taking in 1985. The experiment spent several years fighting the radon problem, and started taking "production data" in 1987. Once 450 days of data had been accumulated the experiment was able to see a clear enhancement in the number of events which pointed away from sun over random directions.[*][1] The directional information was the smoking gun signature of solar neutrinos, demonstrating directly for the first time that the sun is a source of neutrinos. The experiment continued to take data for many years and eventually found the solar neutrino flux to be about 1/2 that predicted by solar models. This was in conflict with both the solar models and Davis's experiment, which was ongoing at the time and continued to observe only 1/3 of the predicted signal. This conflict between the flux predicted by solar theory and the radiochemical and water Čerenkov detectors became known as

the solar neutrino problem.

Atmospheric neutrinos   The flux of atmospheric neutrinos is considerably smaller than that of the solar neutrinos, but because the reaction cross sections increase with energy they are detectable in a detector of Kamiokande-II's size. The experiment used a "ratio of ratios" to compare the ratio of electron to muon flavor neutrinos to the ratio predicted by theory (this technique is used because many systematic errors cancel each other out). This ratio indicated a deficit of muon neutrinos, but the detector was not large enough to obtain the statistics necessary to call the result a discovery. This result came to be known as the **atmospheric neutrino deficit**.

Supernova 1987A   The Kamiokande-II experiment happened to be running at a particularly fortuitous time, as a supernova took place while the detector was online and taking data. With the upgrades that had taken place the detector was sensitive enough to observe the thermal neutrinos produced by Supernova 1987A, which took place roughly 160,000 light years away in the Large Magellanic Cloud. The neutrinos arrived at Earth in February 1987, and the Kamiokande-II detector observed 11 events.

Nucleon decay   Kamiokande-II continued KamiokaNDE's search for proton decay and again failed to observe it. The experiment once again set a lower-bound on the half-life of the proton.

Nobel Prize   For his work directing the Kamioka experiments, and in particular for the first-ever detection of astrophysical neutrinos Masatoshi Koshiba was awarded the Nobel Prize in Physics in 2002. Raymond Davis Jr. and Riccardo Giacconi were co-winners of the prize.

K2K

Main article: K2K

The **KEK To Kamioka** experiment[*][3] used accelerator neutrinos to verify the oscillations observed in the atmospheric neutrino signal with a well controlled and understood beam. A neutrino beam was directed from the KEK accelerator to Super Kamiokande. The experiment found oscillation parameters which were consistent with those measured by Super-K.

7.6.2   Current experiments

Super Kamiokande

Main article: Super Kamiokande

By the 1990s particle physicists were starting to suspect that the solar neutrino problem and atmospheric neutrino deficit had something to do with neutrino oscillation. The **Super Kamiokande** detector was designed to test the oscillation hypothesis for both solar and atmospheric neutrinos. The Super-Kamiokande detector is massive, even by particle physics standards. It consists of 50,000 tons of pure water surrounded by about 11,200 photomultiplier tubes. The detector was again designed as a cylindrical structure, this time 41.4 m tall and 39.3 m across. The detector was surrounded with a considerably more sophisticated outer detector which could not only act as a veto for cosmic muons but actually help in their reconstruction.

Super-Kamiokande started data taking in 1996 and has made several important measurements. These include precision measurement of the solar neutrino flux using the elastic scattering interaction, the first very strong evidence for atmospheric neutrino oscillation, and a considerably more stringent limit on proton decay.

Super Kamiokande-II   On November 12, 2001, several thousand photomultiplier tubes in the Super-Kamiokande detector imploded, apparently in a chain reaction as the shock wave from the concussion of each imploding tube cracked its neighbours. The detector was partially restored by redistributing the photomultiplier tubes which did not implode, and by adding protective acrylic shells that it was hoped would prevent another chain reaction from recurring. The data taken after the implosion is referred to as the **Super Kamiokande-II** data.

Super Kamiokande-III   In July 2005, preparation began to restore the detector to its original form by reinstalling about 6,000 new PMTs. It was finished in June 2006. Data taken with the newly restored machine will be called the **SuperKamiokande-III** dataset.

KamLAND

Main article: Kamioka Liquid Scintillator Antineutrino Detector

The KamLAND experiment is a liquid scintillator detector designed to detect reactor antineutrinos. KamLAND is a complementary experiment to the Sudbury Neutrino Observatory because while the SNO experiment has good sensitivity to the solar mixing angle but poor sensitivity to the squared mass difference, KamLAND has very good sensitivity to the squared mass difference with poor sensitivity to the mixing angle. The data from the two experiments may be combined as long as CPT is a valid symmetry of our universe. The KamLAND experiment is located in the original KamiokaNDE cavity.

T2K

The **Tokai To Kamioka** long baseline experiment started in 2009. It is making a precision measurement of the atmospheric neutrino oscillation parameters and is helping ascertain the value of θ_{13}. It uses a neutrino beam directed at the Super Kamiokande detector from the Japanese Hadron Facility 50 GeV (currently 30 GeV) proton synchrotron such that the neutrinos travel a total distance of 295 km.

In 2013 T2K observed for the first time the neutrino oscillations in the appearance channel: transformation of muon neutrinos to electron neutrinos.[4] In 2014 the collaboration provided the first constraints on the value of CP violating phase, together with the most precise measurement of the mixing angle θ_{23}.[5]

7.6.3   Future experiments

KAGRA

Main article: KAGRA

The **KA**mioka **GRA**vitational wave detector (formerly LCGT, the Large-scale Cryogenic Gravitational Wave Telescope) was approved in 2010. It will have two sets of 3 km long laser interferometers, and will have a planned sensitivity to detect coalescing binary neutron stars at hundreds of Mpc distance. The tunnels were completed in March 2014,[6] and the experiment is likely to be operational by 2018.

Hyper-Kamiokande

There are proposals[7] to build a detector ten times larger than Super Kamiokande, and this project is known by the name **Hyper-Kamiokande**. As of December 2010, construction of Hyper-Kamiokande was projected to begin around 2014.[8] As of January 2015, it is expected to begin construction in 2018 and start observation in 2025.[9]

7.6.4   See also

- MINOS

- Supernova Early Warning System

7.6.5 References

[1] Nakahata, Masayuki. "Kamiokande and Super-Kamiokande" (PDF). Association of Asia Pacific Physical Societies. Retrieved 2014-04-08.

[2] "Google Translate: " 神を嚙んで" from Japanese to English". Retrieved 13 July 2011.

[3] "Long Baseline neutrino oscillation experiment, from KEK to Kamioka (K2K)". Retrieved 2008-09-10.

[4] Abe, A. "Observation of Electron Neutrino Appearance in a Muon Neutrino Beam". *Phys.Rev.Lett.* **112**: 061802. doi:10.1103/PhysRevLett.112.061802.

[5] Abe, K. "Measurements of neutrino oscillation in appearance and disappearance channels by the T2K experiment with 6.6×10^{20} protons on target". *Phys.Rev.* D **91**: 072010. doi:10.1103/PhysRevD.91.072010.

[6] "Excavation of KAGRA's 7 km Tunnel Now Complete" (Press release). University of Tokyo. 31 March 2014. Retrieved 2015-06-07.

[7] Abe, K.; Aihara, H.; Fukuda, Y.; Hayato, Y.; Huang, K.; Ichikawa, A. K.; Ikeda, M.; Inoue, K.; Ishino, H.; Itow, Y.; Kajita, T.; Kameda, J.; Kishimoto, Y.; Koga, M.; Koshio, Y.; Lee, K. P.; Minamino, A.; Miura, M.; Moriyama, S.; Nakahata, M.; Nakamura, K.; Nakaya, T.; Nakayama, S.; Nishijima, K.; Nishimura, Y.; Obayashi, Y.; Okumura, K.; Sakuda, M.; Sekiya, H. (2011). "Letter of Intent: The Hyper-Kamiokande Experiment --- Detector Design and Physics Potential ---". arXiv:1109.3262 [hep-ex].

[8] Masato Shiozawa, "Hyper-Kamiokande design", 15 December 2010 (accessed 27 August 2011).

[9] Normile, Dennis (6 February 2015). "Japanese neutrino physicists think really big". *Science* (American Association for the Advancement of Science) **347** (6222): 598. doi:10.1126/science.347.6222.598. PMID 25657225. Retrieved 8 February 2015.

7.6.6 External links

- The official Super-Kamiokande home page
- American Super-K home page
- Official report on the Super-K accident (in PDF format)
- T2K website

Coordinates: 36°25.6′N   137°18.7′E   /   36.4267°N 137.3117°E (Mt. Ikeno)

7.7 Large Electron–Positron Collider

"LEP" redirects here. For other uses, see LEP (disambiguation).

The **Large Electron–Positron Collider** (**LEP**) was one of the largest particle accelerators ever constructed.

It was built at CERN, a multi-national centre for research in nuclear and particle physics near Geneva, Switzerland. LEP was a circular collider with a circumference of 27 kilometres built in a tunnel roughly 100 m (300 ft) underground and passing through Switzerland and France. It was used from 1989 until 2000. Around 2001 it was dismantled to make way for the LHC, which re-used the LEP tunnel. To date, LEP is the most powerful accelerator of leptons ever built.

7.7.1 Collider Background

LEP was a circular lepton collider – the most powerful such ever built. For context, modern colliders can be generally categorized based on their shape (circular or linear) and on what types of particles they accelerate and collide (leptons or hadrons). Leptons are point particles and are relatively light. Because they are point particles, their collisions are clean and amenable to precise measurements; however, because they are light, the collisions cannot reach the same energy that can be achieved with heavier particles. Hadrons are composite particles (composed of quarks) and are relatively heavy; protons, for example, have a mass 2000 times greater than electrons. Because of their higher mass, they can be accelerated to much higher energies, which is the key to directly observing new particles or interactions that are not predicted by currently accepted theories. However, hadron collisions are very messy (there are often lots of unrelated tracks, for example, and it is not straightforward to determine the energy of the collisions), and therefore more challenging to analyze and less amenable to precision measurements.

The shape of the collider is also important. High energy physics colliders collect particles into bunches, and then collide the bunches together. However, only a very tiny fraction of particles in each bunch actually collide. In circular colliders, these bunches travel around a roughly circular shape in opposite directions and therefore can be collided over and over. This enables a high rate of collisions and facilitates collection of a large amount of data, which is important for precision measurements or for observing very rare decays. However, the energy of the bunches is limited due to losses from synchrotron radiation. In linear colliders, particles move in a straight line and therefore do not suffer from synchrotron radiation, but bunches cannot be re-used and it is therefore more challenging to collect large amounts of data.

As a circular lepton collider, LEP was well suited for precision measurements of the electroweak interaction at energies that were not previously achievable.

7.7.2 History

When the LEP collider started operation in August 1989 it accelerated the electrons and positrons to a total energy of 45 GeV each to enable production of the Z boson, which has a mass of 91 GeV.*[1] The accelerator was upgraded later to enable production of a pair of W bosons, each having a mass of 80 GeV. LEP collider energy eventually topped at 209 GeV at the end in 2000. At a Lorentz factor (= particle energy/rest mass = [104.5 GeV/0.511 MeV]) of over 200,000, LEP still holds the particle accelerator speed record, extremely close to the limiting speed of light. At the end of 2000, LEP was shut down and then dismantled in order to make room in the tunnel for the construction of the Large Hadron Collider (LHC).

7.7.3 Operation

An old RF cavity from **LEP**, *now on display at the Microcosm exhibit at CERN*

The Super Proton Synchrotron (an older ring collider) was used to accelerate electrons and positrons to nearly the speed of light. These are then injected into the ring. As in all ring colliders, the LEP's ring consists of many magnets which force the charged particles into a circular trajectory (so that they stay inside the ring), RF accelerators which

accelerate the particles with radio frequency waves, and quadrupoles that focus the particle beam (i.e. keep the particles together). The function of the accelerators is to increase the particles' energies so that heavy particles can be created when the particles collide. When the particles are accelerated to maximum energy (and focused to so-called bunches), an electron and a positron bunch is made to collide with each other at one of the collision points of the detector. When an electron and a positron collide, they annihilate to a virtual particle, either a photon or a Z boson. The virtual particle almost immediately decays into other elementary particles, which are then detected by huge particle detectors.

7.7.4 Detectors

The Large Electron–Positron Collider had four detectors, built around the four collision points within underground halls. Each was the size of a small house and was capable of registering the particles by their energy, momentum and charge, thus allowing physicists to infer the particle reaction that had happened and the elementary particles involved. By performing statistical analysis of this data, knowledge about elementary particle physics is gained. The four detectors of LEP were called Aleph, Delphi, Opal, and L3. They were built differently to allow for complementary experiments.

ALEPH

Main article: ALEPH experiment

ALEPH stands for *Apparatus for* **LEP PH***ysics at CERN*. The detector determined the mass of the W-boson and Z-boson to within one part in a thousand. The number of families of particles with light neutrinos was determined to be 2.982±0.013, which is consistent with the standard model value of 3. The running of the quantum chromodynamics (QCD) coupling constant was measured at various energies and found to run in accordance with perturbative calculations in QCD.*[2]

DELPHI

Main article: DELPHI experiment

DELPHI stands for **DE***tector with Lepton, Photon and* **Hadron I***dentification*.

OPAL

Main article: OPAL experiment

OPAL stands for *Omni-Purpose Apparatus for LEP*. The name of the experiment was a pun since some of the founding members of the scientific collaboration which first proposed the design had previously worked on the JADE detector at DESY in Hamburg.*[3] OPAL was a general-purpose detector designed to collect a broad range of data. Its data were used to make high precision measurements of the Z boson lineshape, perform detailed tests of the Standard Model, and place limits on new physics. The detector was dismantled in 2000 to make way for LHC equipment. The lead glass blocks from the OPAL barrel electromagnetic calorimeter are currently being re-used in the large-angle photon veto detectors at the NA62 experiment at CERN.

L3

Main article: L3 experiment

L3 was another LEP experiment.*[4] Its enormous octagonal magnet return yoke remained in place in the cavern and became part of the ALICE detector for the LHC.

7.7.5 Results

The results of the LEP experiments allowed precise values of many quantities of the Standard Model—most importantly the mass of the Z boson and the W boson (which were discovered in 1983 at an earlier CERN collider [the Intersecting Storage Rings project]) to be obtained—and so confirm the Model and put it on a solid basis of empirical data.

7.7.6 A not quite discovery of the Higgs boson

Near the end of the scheduled run time, data suggested tantalizing but inconclusive hints that the Higgs particle of a mass around 115 GeV might have been observed, a sort of Holy Grail of current high-energy physics. The run-time was extended for a few months, to no avail. The strength of the signal remained at 1.7 standard deviations which translates to the 91% confidence level, much less than the confidence expected by particle physicists to claim a discovery, and was at the extreme upper edge of the detection range of the experiments with the collected LEP data. There was a proposal to extend the LEP operation by another year in order to seek confirmation, which would have delayed the

start of the LHC. However, the decision was made to shut down LEP and progress with the LHC as planned.

For years, this observation was the only hint of a Higgs Boson; subsequent experiments until 2010 at the Tevatron had not been sensitive enough to confirm or refute these hints.*[5] Beginning in July 2012, however, the ATLAS and CMS experiments at LHC presented evidence of a Higgs particle around 125 GeV,*[6] and strongly excluded the 115 GeV region.

7.7.7 See also

- Electron–positron annihilation

- Large Hadron Collider

7.7.8 References

[1] http://sl-div.web.cern.ch/sl-div/history/lep_doc.html CERN 1990 historical reference with much information on the design issues and details of LEP.

[2] "Welcome to ALEPH". Retrieved 2011-09-14.

[3] "The OPAL Experiment at LEP 1989–2000". Retrieved 2011-09-14.

[4] "L3 Homepage". Retrieved 2011-09-14.

[5] CDF Collaboration, D0 Collaboration, Tevatron New Physics, Higgs Working Group (2010-06-26). "Combined CDF and D0 Upper Limits on Standard Model Higgs-Boson Production with up to 6.7 fb*−1 of Data". arXiv:1007.4587 [hep-ex].

[6] http://home.web.cern.ch/about/updates/2013/03/new-results-indicate-new-particle-higgs-boson

7.7.9 External links

- LEP Working Groups

- The LEP Collider from Design to Approval and Commissioning excerpts from the John Adams memorial lecture delivered at CERN on 26 November 1990

- A short but good (though slightly outdated) overview (with nice photographs) about LEP and related subjects can be found in this online booklet of the British *Particle Physics and Astronomy Research Council*.

7.8 Lawrence Berkeley National Laboratory

The **Lawrence Berkeley National Laboratory** (LBNL or LBL), commonly referred to as **Berkeley Lab**, is a United States national laboratory located in the Berkeley Hills near Berkeley, California that conducts scientific research on behalf of the United States Department of Energy (DOE). It is managed and operated by the University of California,[*][2] The laboratory overlooks University of California, Berkeley's main campus.

7.8.1 History

The laboratory was founded in 1931 as the **Radiation Laboratory** of the University of California, associated with the Physics Department, on August 26 by Ernest Lawrence. It centered physics research around his new instrument, the cyclotron, a type of particle accelerator for which he won the Nobel Prize in Physics in 1939.[*][4] Throughout the 1930s, Lawrence pushed to create larger and larger machines for physics research, courting private philanthropists for funding. After the laboratory was scooped on a number of fundamental discoveries that they felt they ought to have made, the "cyclotroneers" began to collaborate more closely with the department's theoretical physicists, led by Robert Oppenheimer. The lab moved to its site on the hill above campus in 1940 as its machines, specifically the 184-inch (4.67 m) cyclotron, became too large for the university grounds.[*][4]

The laboratory developed the Calutrons used at Y-12 in Oak Ridge, Tennessee to create enriched uranium for the first atomic bombs.

Lawrence courted government as his sponsor in the early years of the Manhattan Project, the American effort to produce the first atomic bomb during World War II, and along with Applied Physics Laboratory at Johns Hopkins (which helped develop the proximity fuse), and the MIT Radiation Laboratory (which helped to develop radar) ushered in the era of "Big Science". Lawrence's lab helped contribute to what has been judged to be the three most valuable technology developments of the war (the atomic bomb, proximity fuse, and radar). Using the newly created 184-inch cyclotron as a mass spectrometer, Lawrence and his colleagues developed the principle behind the electromagnetic enrichment of uranium, which was put to use in the calutrons (named after the university) at the massive Y-12 facility in Oak Ridge, Tennessee. The cyclotron was finished in November 1946; the Manhattan Project shut down two months later.

After the war, Lawrence sought to maintain strong government and military ties at his lab, which became incorporated into the new system of Atomic Energy Commission (AEC) (now Department of Energy (DOE)) National Laboratories, but in the early 1950s set out that the lab's purpose would be primarily non-classified research. For security purposes, classified weapons research was assigned to the more isolated locations, Los Alamos National Laboratory (LANL) (established during the war) in New Mexico and the new UC Radiation Laboratory at Livermore (today's Lawrence Livermore National Laboratory (LLNL)). Livermore, about an hour southeast of Berkeley, was established at a former naval air station in 1952 by Lawrence and Edward Teller from what was originally a splinter from the original Radiation Laboratory. Some weapons-related and collaborative research continued at Berkeley Lab until the 1970s, however.

Shortly after the death of Lawrence in August 1958, the UC Radiation Laboratory (both branches) was renamed the **Lawrence Radiation Laboratory** and the Berkeley location became the **Lawrence Berkeley Laboratory** in 1971,[*][5] although many continued to call it the "Rad Lab." Gradually, another shortened form came into common usage, "**LBL**". Its formal name was amended to Ernest Orlando Lawrence Berkeley National Laboratory in 1995, when "National" was added to the names of all DOE labs. "Ernest Orlando" was later dropped to shorten the name. Today, the lab is commonly referred to as "**Berkeley Lab**".[*][6]

Lab Directors

(2009–present): Paul Alivisatos
(2004–2008): Steven Chu
(1989–2004): Charles Shank
(1980–1989): David Shirley
(1973–1980): Andrew Sessler
(1958–1972): Edwin McMillan

(1931–1958): Ernest Lawrence

Alvarez Physics Memos

The Alvarez Physics Memos are a set of informal working papers of the large group of physicists, engineers, computer programmers, and technicians led by Luis W. Alvarez from the early 1950s until his death in 1988. Over 1700 memos are available on-line, hosted by the Laboratory.*[7]

7.8.2   Science mission

From the 1950s through the present, Berkeley Lab has maintained its status as a major international center for physics research, and has also diversified its research program into almost every realm of scientific investigation. The Laboratory's 14 scientific divisions are organized within the areas of Computing Sciences, General Sciences, Energy and Environmental Sciences, Life Sciences, and Photon Sciences. Many research projects are staffed and supported by multiple divisions, with computational and engineering integrated across the biosciences, general sciences, and energy sciences. The scientific divisions include: earth sciences, genomics, life sciences, chemical sciences, environmental energy technologies, materials science, physical biosciences, computational research, accelerator and fusion research, engineering, nuclear science, nuclear medicine and physics.

Berkeley Lab has six main science thrusts: soft x-ray science for discovery, climate change and environmental sciences, matter and force in the universe, energy efficiency and sustainable energy, computational science and networking, and biological science for energy research. It was Lawrence's belief that scientific research is best done through teams of individuals with different fields of expertise, working together. His teamwork concept is a Berkeley Lab tradition that continues today.

Additionally Berkeley Lab is host to six major National User Facilities: the Advanced Light Source, the National Center for Electron Microscopy, National Energy Research Scientific Computing Center, the Energy Sciences Network, the Molecular Foundry, and the Joint Genome Institute (JGI).

The Joint Genome Institute, in Walnut Creek, was founded in 1997 to unite the expertise and resources in genome mapping, DNA sequencing, technology development, and information sciences pioneered at the three genome centers at Berkeley Lab, Livermore, and Los Alamos. Today the JGI's partner laboratories include Berkeley Lab, LLNL, LANL, as well as Oak Ridge (ORNL), Pacific Northwest National Laboratory (PNNL), and the HudsonAlpha Insti-

tute for Biotechnology (formerly associated with the Stanford Human Genome Center). The JGI workforce draws most heavily from Berkeley Lab and LLNL.

The laboratory also manages the Department of Energy's high speed research network, ESnet.*[8]

Berkeley Lab is the lead partner in the Joint BioEnergy Institute (JBEI), located in Emeryville, California. Other partners are the Sandia National Laboratories, the University of California (UC) campuses of Berkeley and Davis, the Carnegie Institution for Science, and the Lawrence Livermore National Laboratory (LLNL). JBEI's primary scientific mission is to advance the development of the next generation of biofuels – liquid fuels derived from the solar energy stored in plant biomass. JBEI is one of three new U.S. Department of Energy (DOE) Bioenergy Research Centers (BRCs).

7.8.3   Operations and governance

The site consists of 76 buildings (owned by the U.S. Department of Energy) located on 200 acres (0.81 km^2) owned by the university in the Berkeley Hills. Altogether, it has some 4,000 UC employees, of whom about 800 are students. Each year, the Lab also hosts more than 3,000 participating guests. There are approximately two dozen DOE employees stationed at the laboratory to provide federal oversight of Berkeley Lab's work for the DOE.

The laboratory director is appointed by the university regents and reports to the university president. In 2009, Paul Alivisatos was appointed interim director on January 21 to replace Steven Chu, who was sworn in as Secretary of Energy in the new Obama administration.*[9] Ten months later, Alivisatos was officially named Berkeley Lab's seventh director by the regents on November 19. Although Berkeley Lab is governed by UC independently of the Berkeley campus, the two entities are closely interconnected: more than 200 Berkeley Lab researchers hold joint appointments as UC Berkeley faculty and more than 500 UC Berkeley graduate students conduct research at Berkeley Lab.

The Lab's budget for fiscal year 2011 was $735 million, plus $101 million from the American Recovery and Reinvestment Act.

7.8.4   Scientific achievements, inventions, and discoveries

Notable scientific accomplishments at the Lab since World War II include the observation of the antiproton, the discovery of several transuranic elements, and the discovery of the accelerating universe.

Since its inception, 12 researchers associated with Berkeley Lab (Ernest Lawrence, Glenn T. Seaborg, Edwin M. McMillan, Owen Chamberlain, Emilio G. Segrè, Donald A. Glaser, Melvin Calvin, Luis W. Alvarez, Yuan T. Lee, Steven Chu, George F. Smoot and Saul Perlmutter) have been awarded the Nobel Prize.

Fifty-seven Berkeley Lab scientists are members of the U.S. National Academy of Sciences (NAS), one of the highest honors for a scientist in the United States. Thirteen Berkeley Lab scientists have won the National Medal of Science, the nation's highest award for lifetime achievement in fields of scientific research. Eighteen Berkeley Lab engineers have been elected to the National Academy of Engineering, and three Berkeley Lab scientists have been elected into the Institute of Medicine.

Elements discovered by Berkeley Lab physicists include astatine, neptunium, plutonium, curium, americium, berkelium*, californium*, einsteinium, fermium, mendelevium, nobelium, lawrencium*, dubnium, and seaborgium*. Those elements listed with asterisks (*) are named after the University, Professors Lawrence and Seaborg. Seaborg was the principal scientist involved in their discovery. The element technetium was discovered after Ernest Lawrence gave Emilio Segrè a molybdenum strip from the Berkeley Lab cyclotron.*[10]

Some inventions and discoveries to come out of Berkeley Lab include: "smart" windows with embedded electrodes that enable window glass to respond to changes in sunlight, synthetic genes for antimalaria and anti-AIDS superdrugs based on breakthroughs in synthetic biology, electronic ballasts for more efficient lighting, Home Energy Saver, the web's first do-it-yourself home energy audit tool, a pocket-sized DNA sampler called the PhyloChip, and the Berkeley Darfur Stove, which uses one-quarter as much firewood as traditional cook stoves. One of Berkeley Lab's most notable breakthroughs is the discovery of dark energy. During the 1980s and 1990s Berkeley Lab physicists and astronomers formed the Supernova Cosmology Project (SCP), using Type Ia supernovae as "standard candles" to measure the expansion rate of the universe. Their successful methods inspired competition, with the result that early in 1998 both the SCP and the High-Z Supernova Search Team announced the surprising discovery that expansion is accelerating; the cause was soon named dark energy.

Networking tools libpcap,*[11] tcpdump, and traceroute were developed by the Network Engineering Group staff at the Laboratory.

As of July 2012, LBNL holds the world record for the most powerful laser.

7.8.5 Scandal

The fabricated evidence used to claim the creation of ununoctium and ununhexium (now livermorium) by Victor Ninov, a researcher employed at Berkeley Lab, led to the retraction of two articles and was one of the big scandals in physics in 2002.*[12]

7.8.6 See also

- Center for the Advancement of Science in Space—operates the US National Laboratory on the ISS.

- List of Advanced Scientific Computing Research Leadership Computing Challenge allocations

- Lawrence Hall of Science

- Lawrence Livermore National Laboratory

- Los Alamos National Laboratory

7.8.7 References

[1] About Berkeley Lab

[2] University of California | Office of the President (accessed 2013-07-15).

[3] Nobel Prizes affiliated with Berkeley Lab

[4] "Lawrence and His Laboratory: Chapter 1: A New Lab for a New Science". Lbl.gov. Retrieved 2009-07-12.

[5] "Ernest Lawrence and M. Stanley Livingston". American Physical Society. Retrieved May 9, 2014.

[6] "Ernest Lawrence's Cyclotron". Lbl.gov. 1958-08-27. Retrieved 2009-07-12.

[7] "Alvarez Physics Memos".

[8] "Department of Energy's ESnet Wins 2009 Excellence.Gov Award for Effectively Leveraging Technology « Berkeley Lab News Center". Newscenter.lbl.gov. 2009-04-16. Retrieved 2009-07-12.

[9] "University of California - UC Newsroom | UC appoints Paul Alivisatos interim director of Berkeley Lab". Universityofcalifornia.edu. 2009-01-22. Retrieved 2009-07-12.

[10] "Chemical Elements Discovered at Lawrence Berkeley National Laboratory". Lbl.gov. 1999-06-07. Retrieved 2009-07-12.

[11] Jacobson, Van. "Libpcap". Securityfocus.com. Retrieved 2009-07-12.

[12] Dalton, Rex (2002). "The stars who fell to Earth" (PDF). *Nature* **420**: 728–9. doi:10.1038/420728a. PMID 12490902. Retrieved 21 April 2012.

7.8.8   External links

- Official website

  - YouTube channel

- University of California Office of Laboratory Management

- The Rad Lab - Ernest Lawrence and the Cyclotron: American Institute of Physics web exhibit

- *Lawrence and His Laboratory: A Historian's View of the Lawrence Years* by J. L. Heilbron, Robert W. Seidel, and Bruce R. Wheaton.

- A Century of Physics at Berkeley: Seedtime for "Big Science", 1930-1950

- SPIE Video: Paul Alivisatos: Berkeley Lab director navigates uncertain times with a focus on research

Coordinates: 37°52′34″N   122°14′49″W / 37.876°N 122.247°W

7.9   Lawrence Livermore National Laboratory

●

LLNL

Location in the United States

Lawrence Livermore National Laboratory (**LLNL**) is a federal research facility in Livermore, California, founded by the University of California in 1952. A Federally Funded Research and Development Center (FFRDC), it is primarily funded by the United States Department of Energy (DOE) and managed and operated by Lawrence Livermore National Security, LLC (LLNS), a partnership of the University of California, Bechtel, Babcock & Wilcox, URS, and Battelle Memorial Institute in affiliation with the Texas A&M University System. The laboratory was honored in 2012 by having the synthetic chemical element livermorium named after it.

7.9.1   Background

Aerial view of Lawrence Livermore National Laboratory

LLNL is self-described as "a premier research and development institution for science and technology applied to national security." [1] Its principal responsibility is ensuring the safety, security and reliability of the nation's nuclear weapons through the application of advanced science, engineering and technology. The Laboratory also applies its special expertise and multidisciplinary capabilities to preventing the proliferation and use of weapons of mass destruction, bolstering homeland security and solving other nationally important problems, including energy and environmental security, basic science and economic competitiveness.

LLNL is home to many unique facilities and a number of the most powerful computer systems in the world, according to the TOP500 list, including Blue Gene/L, the world's fastest computer from 2004 until Los Alamos National Laboratory's IBM Roadrunner supercomputer surpassed it in 2008. On June 18, 2012, LLNL re-took the lead on the latest edition of the list of the world's Top 500 supercomputers with IBM Sequoia, a 16.32 petaflops system packing more than 1.5 million custom Power cores. It is based on the same IBM BlueGene/Q architecture used in three other top ten systems which also were the most power efficient on the list. Since 1978, LLNL has received a total of 118 R&D 100 Awards, including five in 2007.[2] The awards are given annually by the editors of *R&D Magazine* to the most innovative ideas of the year.

The Laboratory is located on a one-square-mile (2.6 km^2) site at the eastern edge of Livermore. It also operates a 7,000 acres (28 km^2) remote experimental test site, called Site 300, situated about 15 miles (24 km) southeast of the main lab site. LLNL has an annual budget of about $1.5 billion and a staff of roughly 5,800 employees.

7.9.2 Origins

LLNL was established 63 years ago in 1952 as the **University of California Radiation Laboratory at Livermore**, an offshoot of the existing UC Radiation Laboratory at Berkeley. It was intended to spur innovation and provide competition to the nuclear weapon design laboratory at Los Alamos in New Mexico, home of the Manhattan Project that developed the first atomic weapons. Edward Teller and Ernest Lawrence,*[3] director of the Radiation Laboratory at Berkeley, are regarded as the co-founders of the Livermore facility.

The new laboratory was sited at a former naval air station of World War II. It was already home to several UC Radiation Laboratory projects that were too large for its location in the hills above the Berkeley campus, including one of the first experiments in the magnetic approach to confined thermonuclear reactions (i.e. fusion). About half an hour southeast of Berkeley, the Livermore site provided much greater security for classified projects than an urban university campus.

Lawrence tapped 32-year-old Herbert York, a former graduate student of his, to run Livermore. Under York, the Lab had four main programs: Project Sherwood (the Magnetic Fusion Program), Project Whitney (the weapons design program), diagnostic weapon experiments (both for the Los Alamos and Livermore laboratories), and a basic physics program. York and the new lab embraced the Lawrence "big science" approach, tackling challenging projects with physicists, chemists, engineers, and computational scientists working together in multidisciplinary teams.

Lawrence died in August 1958 and shortly after, the university's board of regents named both laboratories for him, as the **Lawrence Radiation Laboratory**.

Historically, the Berkeley and Livermore laboratories have had very close relationships on research projects, business operations and staff. The Livermore Lab was established initially as a branch of the Berkeley Laboratory. The Livermore Lab was not officially severed administratively from the Berkeley Lab until 1971. To this day, in official planning documents and records, Lawrence Berkeley National Laboratory is designated as Site 100, Lawrence Livermore National Lab as Site 200, and LLNL's remote test location as Site 300.*[4]

The laboratory was renamed **Lawrence Livermore Laboratory (LLL)** in 1971. On October 1, 2007 LLNS assumed management of LLNL from the University of California, which had exclusively managed and operated the Laboratory since its inception 55 years before. The laboratory was honored in 2012 by having the synthetic chemical element livermorium named after it. The LLNS takeover of the Laboratory has been controversial. In May 2013,

an Alameda County jury awarded over $2.7 million to five former Laboratory employees who were among 430 employees LLNS laid off during 2008.*[5] The jury found that LLNS breached a contractual obligation to terminate the employees only for "reasonable cause." *[6] The five plaintiffs also have pending age discrimination claims against LLNS, which will be heard by a different jury in a separate trial.*[7] There are 125 co-plaintiffs awaiting trial on similar claims against LLNS.*[8] The May 2008 layoff was the first layoff at the Laboratory in nearly 40 years.*[7]

On March 14, 2011, the City of Livermore officially expanded the city's boundaries to annex LLNL and move it within the city limits. The unanimous vote by the Livermore City Council expanded Livermore's southeastern boundaries to cover 15 land parcels covering 1,057 acres (4.28 km^2) that comprise the LLNL site. Prior to this, the site was in an unincorporated area of Alameda County. The LLNL campus continues to be owned by the federal government.

7.9.3 Nuclear weapons projects

From its inception, Livermore focused on innovative weapon design concepts; as a result, its first three nuclear tests were unsuccessful. The lab persevered and its subsequent designs proved increasingly successful. In 1957, the Livermore Lab was selected to develop the warhead for the Navy's Polaris missile. This warhead required numerous innovations to fit a nuclear warhead into the relatively small confines of the missile nosecone.*[9]

During the Cold War, scores of Livermore-designed warheads entered service. These were used in missiles ranging in size from the Lance surface-to-surface tactical missile to the megaton-class Spartan antiballistic missile. Over the years, LLNL designed the following warheads: W27 (Regulus cruise missile; 1955; joint with Los Alamos), W38 (Atlas/Titan ICBM; 1959), B41 (B52 bomb; 1957), W45 (Little John/Terrier missiles; 1956), W47 (Polaris SLBM; 1957), W48 (155-mm howitzer; 1957), W55 (submarine rocket; 1959), W56 (Minuteman ICBM; 1960), W58 (Polaris SLBM; 1960), W62 (Minuteman ICBM; 1964), W68 (Poseidon SLBM; 1966), W70 (Lance missile; 1969), W71 (Spartan missile; 1968), W79 (8-in. artillery gun; 1975), W82 (155-mm howitzer; 1978), B83 (modern strategic bomb; 1979), and W87 (Peacekeeper/MX ICBM; 1982). The W87, and the B83 are the only LLNL designs still in the U.S. nuclear stockpile.*[10]*[11]*[12]

With the collapse of the Soviet Union in 1991 and the end of the Cold War, the United States began a moratorium on nuclear testing and development of new nuclear weapon designs. To sustain existing warheads for the indefinite future, a science-based Stockpile Stewardship Program (SSP) was defined that emphasized the development and application of greatly improved technical capabilities to assess the safety, security, and reliability of existing nuclear warheads without the use of nuclear testing. Confidence in the performance of weapons, without nuclear testing, is maintained through an ongoing process of stockpile surveillance, assessment and certification, and refurbishment or weapon replacement.

With no new designs of nuclear weapons, the warheads in the U.S. stockpile must continue to function far past their original expected lifetimes. As components and materials age, problems can arise. Stockpile Life Extension Programs can extend system lifetimes, but they also can introduce performance uncertainties and require maintenance of outdated technologies and materials. Because there is concern that it will become increasingly difficult to maintain high confidence in the current warheads for the long term, the Department of Energy/National Nuclear Security Administration initiated the Reliable Replacement Warhead (RRW) Program. RRW designs could reduce uncertainties, ease maintenance demands, and enhance safety and security. In March 2007, the LLNL design was chosen for the Reliable Replacement Warhead.[13] Since that time, Congress has not allocated funding for any further development of the RRW.

The Livermore Action Group organized many mass protests, from 1981 to 1984, against nuclear weapons which were being produced by the Lawrence Livermore National Laboratory. Peace activists Ken Nightingale and Eldred Schneider were involved.[14] On June 22, 1982, more than 1,300 anti-nuclear protesters were arrested in a non-violent demonstration.[15] More recently, there has been an annual protest against nuclear weapons research at Lawrence Livermore. In August 2003, 1,000 people protested at Livermore Labs against "new-generation nuclear warheads".[16] In the 2007 protest, 64 people were arrested.[17] More than 80 people were arrested in March 2008 while protesting at the gates.[18] 31 people were arrested in August 2013 during a protest marking the 68th anniversary of the atomic bombings of Hiroshima and Nagasaki, including famous whistle blower and author of the Pentagon Papers, Daniel Ellsberg.[19]

In the 1980s, Lawrence's widow petitioned the Regents of the University of California on several occasions to remove her husband's name from the Livermore laboratory, due to its focus on nuclear weapons.[20][21][22][23] She outlived her husband by more than 44 years and died in Walnut Creek at the age of 92 in January 2003.[24]

7.9.4 Plutonium research

LLNL conducts research into the properties and behavior of plutonium to learn how plutonium performs as it ages and how it behaves under high pressure (e.g., with the impact of high explosives). Plutonium has seven temperature-dependent solid allotropes. Each possesses a different density and crystal structure. Alloys of plutonium are even more complex; multiple phases can be present in a sample at any given time. Experiments are being conducted at LLNL and elsewhere to measure the structural, electrical and chemical properties of plutonium and its alloys and to determine how these materials change over time. Such measurements will enable scientists to better model and predict plutonium's long-term behavior in the aging stockpile.[25]

The Lab's plutonium research is conducted in a specially designed, ultra-safe, and highly secure facility called the SuperBlock. Work with highly enriched uranium is also conducted here. In March 2008, the National Nuclear Security Administration (NNSA) presented its preferred alternative for the transformation of the nation's nuclear weapons complex. Under this plan, LLNL would be a center of excellence for nuclear design and engineering, a center of excellence for high explosive research and development, and a science magnet in high-energy-density (i.e., laser) physics. In addition, most of its special nuclear material would be removed and consolidated at a more central, yet-to-be-named site.[26]

On September 30, 2009, the NNSA announced that about two thirds of the special nuclear material (e.g., plutonium) at LLNL requiring the highest level of security protection had been removed from LLNL. The move was part of NNSA's efforts initiated in October 2006 to consolidate special nuclear material at five sites by 2012, with significantly reduced square footage at those sites by 2017. The federally mandated project intends to improve security and reduce security costs, and is part of NNSA's overall effort to transform the Cold War era "nuclear weapons" enterprise into a 21st-century "nuclear security" enterprise. The original date to remove all high-security nuclear material from LLNL, based on equipment capability and capacity, was 2014. NNSA and LLNL developed a timeline to remove this material as early as possible, accelerating the target completion date to 2012.[27]

7.9.5 Global security program

The Lab's work in global security aims to reduce and mitigate the dangers posed by the spread or use of weapons of mass destruction and by threats to energy and environmental security. Livermore has been working on global security and homeland security for decades, predating both the collapse of the Soviet Union in 1991 and the September 11, 2001, terrorist attacks. LLNL staff have been heavily involved in the cooperative nonproliferation programs with Russia to secure at-risk weapons materials and assist former weapons workers in developing peaceful applications and self-sustaining job opportunities for their expertise and technologies.*[28]*[29] In the mid-1990s, Lab scientists began efforts to devise improved biodetection capabilities, leading to miniaturized and autonomous instruments that can detect biothreat agents in a few minutes instead of the days to weeks previously required for DNA analysis.*[30]*[31]

Today, Livermore researchers address the full spectrum of threats – radiological/nuclear, chemical, biological, explosives, and cyber. They combine physical and life sciences, engineering, computations, and analysis to develop technologies that solve real-world problems. Activities are grouped into five programs:

- **Nonproliferation**. Preventing the spread of materials, technology and expertise related to weapons of mass destruction (WMD) and detecting WMD proliferation activities worldwide.*[32]

- **Domestic security**: Anticipating, innovating and delivering technological solutions to prevent and mitigate devastating high-leverage attacks on U.S. soil.*[33]*[34]*[35]*[36]

- **Defense**: Developing and demonstrating new concepts and capabilities to help the Department of Defense prevent and deter harm to the nation, its citizens and its military forces.*[37]*[38]

- **Intelligence**: Working at the intersection of science, technology and analysis to provide insight into the threats to national security posed by foreign entities.*[39]

- **Energy and environmental security**: Furnishing scientific understanding and technological expertise to devise energy and environmental solutions at global, regional and local scales.*[40]*[41]

7.9.6   Other programs

LLNL supports capabilities in a broad range of scientific and technical disciplines, applying current capabilities to existing programs and developing new science and technologies to meet future national needs.

- The LLNL chemistry, materials, and life science research focuses on chemical engineering, nuclear chemistry, materials science, and biology and bio-nanotechnology.

- Physics thrust areas include condensed matter and high-pressure physics, optical science and high energy density physics, medical physics and biophysics, and nuclear, particle and accelerator physics.

- In the area of energy and environmental science, Livermore' s emphasis is on carbon and climate, energy, water and the environment, and the national nuclear waste repository.

- The LLNL engineering activities include micro- and nanotechnology, lasers and optics, biotechnology, precision engineering, nondestructive characterization, modeling and simulation, systems and decision science, and sensors, imaging and communications.

- The LLNL is very strong in computer science, with thrust areas in computing applications and research, integrated computing and communications systems, and cyber security.

Lawrence Livermore National Laboratory has worked out several energy technologies in the field of coal gasification, shale oil extraction, geothermal energy, advanced battery research, solar energy, and fusion energy. Main oil shale processing technologies worked out by the Lawrence Livermore National Laboratory are LLNL HRS (hot-recycled-solid), LLNL RISE (*in situ* extraction technology) and LLNL radiofrequency technologies.*[42]

7.9.7   Key accomplishments

Over its 60-year history, Lawrence Livermore has made many scientific and technological achievements, including:*[43]

- Critical contributions to the U.S. nuclear deterrence effort through the design of nuclear weapons to meet military requirements and, since the mid-1990s, through the Stockpile Stewardship Program, by which the safety and reliability of the enduring stockpile is ensured without underground nuclear testing.

- Design, construction, and operation of a series of ever larger, more powerful, and more capable laser systems, culminating in the 192-beam National Ignition Facility (NIF), completed in 2009.

- Advances in particle accelerator and fusion technology, including magnetic fusion, Free-electron lasers, accelerator mass spectrometry, and inertial confinement fusion.

- Breakthroughs in high-performance computing, including the development of novel concepts for massively parallel computing and the design and application of computers that can carry out hundreds of trillions of operations per second.

- Development of technologies and systems for detecting nuclear, radiological, chemical, biological, and explosive threats to prevent and mitigate WMD proliferation and terrorism.

- Development of extreme ultraviolet lithography (EUVL) for fabricating next-generation computer chips.

- First-ever detection of massive compact halo objects (MACHOs), a suspected but previously undetected component of dark matter.

- Advances in genomics, biotechnology, and biodetection, including major contributions to the complete sequencing of the human genome though the Joint Genome Institute and the development of rapid PCR (polymerase chain reaction) technology that lies at the heart of today's most advanced DNA detection instruments.

- Development and operation of the National Atmospheric Release Advisory Center (NARAC), which provides real-time, multi-scale (global, regional, local, urban) modeling of hazardous materials released into the atmosphere.

- Development of highest resolution global climate models and contributions to the International Panel on Climate Change which, together with former vice president Al Gore, was awarded the 2007 Nobel Peace Prize.

- Co-discoverers of new superheavy elements 113, 114, 115, 116, 117, and 118.

- Invention of new healthcare technologies, including a microelectrode array for construction of an artificial retina, a miniature glucose sensor for the treatment of diabetes, and a compact proton therapy system for radiation therapy.

On July 17, 2009 LLNL announced that the Laboratory had captured eight R&D 100 Awards – more than it had ever received in the annual competition. The previous LLNL record of seven awards was reached five times – in 1987, 1988, 1997, 1998 and 2006.

Also known as the "Oscars of invention", the awards are given each year for the development of cutting-edge scientific and engineering technologies with commercial potential.

The awards raises LLNL's total to 129 since 1978. The winning technologies were:

- GeMini Spectrometer

- Artificial Retina —Restoring Sight to the Blind

- The ROSE compiler framework

- The Babel Middleware

- The FemtoScope: A Time Microscope

- ROSE: Making Compiler Technology Accessible to all Programmers

- Land Mine Locator: Eradicating the Aftermath of War

- Laser Beam Centering and Pointing System

- Spectral Sentry —Protecting High-Intensity Lasers from Bandwidth-Related Damage

- Precision Robotic Assembly Machine —for Building Nuclear Fusion Ignition Targets

7.9.8   Unique facilities

- Biosecurity and Nanoscience Laboratory. Researchers apply advances in nanoscience to develop novel technologies for the detection, identification, and characterization of harmful biological pathogens (viruses, spores, and bacteria) and chemical toxins.

- Center for Accelerator Mass Spectrometry: LLNL's Center for Accelerator Mass Spectrometry (CAMS) develops and applies a wide range of isotopic and ion-beam analytical tools used in basic research and technology development, addressing a spectrum of scientific needs important to the Laboratory, the university community, and the nation. CAMS is the world's most versatile and productive accelerator mass spectrometry facility, performing more than 25,000 AMS measurement operations per year.

- High Explosives Applications Facility and Energetic Materials Center: At HEAF, teams of scientists, engineers, and technicians address nearly all aspects of high explosives: research, development and testing, material characterization, and performance and safety tests. HEAF activities support the Laboratory's Energetic Materials Center, a national resource for research and development of explosives, pyrotechnics, and propellants.

- National Atmospheric Release Advisory Center: NARAC is a national support and resource center for planning, real-time assessment, emergency response, and detailed studies of incidents involving a wide variety of hazards, including nuclear, radiological, chemical, biological, and natural atmospheric emissions.

- National Ignition Facility: This 192-beam, stadium-size laser system will be used to compress fusion targets to conditions required for thermonuclear burn. Experiments at NIF will study physical processes at conditions that exist only in the interior of stars and in exploding nuclear weapons (see National Ignition Facility and photon science).

- Superblock: This unique high-security facility houses modern equipment for research and engineering testing of nuclear materials and is the place where plutonium expertise is developed, nurtured, and applied. Research on highly enriched uranium also is performed here.

- Terascale Simulation Facility: LLNL' s Terascale Simulation Facility houses one of the world' s most powerful computers, Sequoia. Sequoia occupied the No. 1 position on the Top500 list in June 2012;[*][44] the current system achieves a Linpack benchmark performance of 16.32 PFlop/s (Petaflops, or quadrillions of calculations per second). Another Blue Gene class machine, Dawn, was installed to act as a developmental testbed for multi-petaflop computing.[*][45]

- Titan Laser: Titan is a combined nanosecond-long pulse and ultrashort-pulse (subpicosecond) laser, with hundreds of joules of energy in each beam. This petawatt-class laser is used for a range of high-energy density physics experiments, including the science of fast ignition for inertial confinement fusion energy.

7.9.9 Largest computers

Throughout its history, LLNL has been a leader in computers and scientific computing. Even before the Livermore Lab opened its doors, E.O. Lawrence and Edward Teller recognized the importance of computing and the potential of computational simulation. Their purchase of one of the first UNIVAC computers, set the precedent for LLNL's history of acquiring and exploiting the fastest and most capable supercomputers in the world. A succession of increasingly powerful and fast computers have been used at the Lab over the years:

- 1953 Remington-Rand UNIVAC 1 (Universal Automatic Computer)

- 1954 IBM 701

- 1956 IBM 704

- 1958 IBM 709

- 1960 IBM 7090

- 1960 Remington-Rand LARC (Livermore Advanced Research Computer)

- 1961 IBM 7030 (Stretch)

- 1963 IBM 7094

- 1963 CDC 1604

- 1963 CDC 3600

- 1964 CDC 6600

- 1969 CDC 7600

- 1974 CDC STAR 100

- 1978 Cray-1

- 1984 Cray X-MP

- 1985 Cray-2

- 1989 Cray Y-MP

- 1992 BBN Butterfly

- 1994 Meiko CS-2

- 1995 Cray C90

- 1995 Cray T3D

- 1996 IBM ASCI Blue Pacific

- 2000 IBM ASCI White

- 2004 Thunder

- 2005 IBM Blue Gene/L

- 2005 ASC Purple

- 2006 Zeus

- 2006 Rhea

- 2006 Atlas

- 2007 Minos

- 2012 IBM Sequoia

The November 2007 release of the 30th TOP500 list of the 500 most powerful computer systems in the world, has LLNL's Blue Gene/L computer in first place for the seventh consecutive time. Five other LLNL computers are in the top 100. The November 2008 release of the TOP500 list places the Blue Gene/L supercomputer behind the Pleiades supercomputer in NASA/Ames Research Center, the Jaguar supercomputer in Oak Ridge National Laboratory, and the IBM Roadrunner supercomputer in Los Alamos National Laboratory. Currently, the Blue Gene/L computer can sustain 478.2 trillion operations per second, with a peak of 596.4 trillion operations per second.

On June 22, 2006, researchers at LLNL announced that they had devised a scientific software application that sustained 207.3 trillion operations per second. The record performance was made at LLNL on Blue Gene/L, the world's fastest supercomputer with 131,072 processors. The record was a milestone in the evolution of predictive science, a field in which researchers use supercomputers to answer questions about such subjects as: materials science simulations, global warming, and reactions to natural disasters.

LLNL has a long history of developing computing software and systems. Initially, there was no commercially available software, and computer manufacturers considered it the customer's responsibility to develop their own. Users of the early computers had to write not only the codes to solve their technical problems, but also the routines to run the machines themselves. Today, LLNL computer scientists focus on creating the highly complex physics models, visualization codes, and other unique applications tailored to specific research requirements. A great deal of software also has been written by LLNL personnel to optimize the operation and management of the computer systems, including operating system extensions such as CHAOS (Linux Clustering) and resource management packages such as SLURM.*[46] The Peloton procurements in late 2006 (Atlas and other computers) were the first in which a commercial resource management package, Moab, was used to manage the clusters.*[47]

7.9.10 Livermore Valley Open Campus (LVOC)

In August 2009 a joint venture was announced between Sandia National Laboratories/California campus and LLNL to create an open, unclassified research and development space called the Livermore Valley Open Campus (LVOC). The motivation for the LVOC stems from current and future national security challenges that require increased coupling to the private sector to understand threats and deploy solutions in areas such as high performance computing, energy and environmental security, cyber se-

curity, economic security, and non-proliferation.

The LVOC is modeled after research and development campuses found at major industrial research parks and other U.S. Department of Energy laboratories with campus-like security, a set of business and operating rules devised to enhance and accelerate international scientific collaboration and partnerships with U.S. government agencies, industry and academia. Ultimately, the LVOC will consist of an approximately 110-acre parcel along the eastern edge of the Livermore Laboratory and Sandia sites, and will house additional conference space, collaboration facilities and a visitor's center to support educational and research activities.

Objectives of LVOC

- Enhance the two laboratories' national security missions by substantially increasing engagement with the private sector and academic community.

- Stay at the forefront of the science, technology and engineering fields.

- Ensure a quality future workforce by expanding opportunities for open engagement of the broader scientific community.

Initial research areas for the LVOC:

- High Performance Computing

- CyberSecurity Science

- Sandia's Combustion Research Facility and Transportation Research

- High Energy Density Physics

- Climate and Energy Research

The architecture of the LVOC is planned in stages; first steps including:

- A new High Performance Computing Innovation Center at LLNL

- Open access to Sandia's Combustion Research Facility

- Planning for an unclassified computing center

7.9.11 Sponsors

LLNL's principal sponsor is the Department of Energy/National Nuclear Security Administration (DOE/NNSA) Office of Defense Programs, which supports its stockpile stewardship and advanced scientific computing programs. Funding to support LLNL's global security and homeland security work comes from the DOE/NNSA Office of Defense Nuclear Nonproliferation as well as the Department of Homeland Security. LLNL also receives funding from DOE' s Office of Science, Office of Civilian Radioactive Waste Management, and Office of Nuclear Energy. In addition, LLNL conducts work-for-others research and development for various Defense Department sponsors, other federal agencies, including NASA, Nuclear Regulatory Commission (NRC), National Institutes of Health, and Environmental Protection Agency, a number of California State agencies, and private industry.

7.9.12 Budget

For Fiscal Year 2009 LLNL spent $1.497 billion[48] on research and laboratory operations activities:

Research/Science Budget:

- National Ignition Facility - $301.1 million
- Nuclear Weapon Deterrent (Safety/Security/Reliability) - $227.2 million
- Advance Simulation and Computing - $221.9 million
- Nonproliferation - $152.2 million
- Department of Defense - $125.9 million
- Basic and Applied Science - $86.6 million
- Homeland Security - $83.9 million
- Energy - $22.4 million

Site Management/Operations Budget:

- Safeguards/Security - $126.5 million
- Facility Operations - $118.2 million
- Environmental Restoration - $27.3 million

7.9.13 Directors

The LLNL Director is appointed by the Board of Governors of Lawrence Livermore National Security, LLC (LLNS) and reports to the board. The Laboratory Director also serves as the President of LLNS. Over the course of its history, the following eminent scientists have served as LLNL Director:

- 1952–1958 Herbert York
- 1958–1960 Edward Teller
- 1960–1961 Harold Brown
- 1961–1965 John S. Foster, Jr.
- 1965–1971 Michael M. May
- 1971–1988 Roger E. Batzel
- 1988–1994 John H. Nuckolls
- 1994–2002 C. Bruce Tarter
- 2002–2006 Michael R. Anastasio
- 2006–2011 George H. Miller[49]
- 2011–2013 Penrose C. Albright[50]
- 2013–2014 Bret Knapp, acting director [51]
- 2014–present William H. Goldstein[52]

7.9.14 Organization

The LLNL Director is supported by a senior executive team consisting of the Deputy Director, the Deputy Director for Science and Technology, Principal Associate Directors, and other senior executives who manage areas/functions directly reporting to the Laboratory Director.

The Directors Office is organized into these functional areas/offices:

- Chief Information Office
- Contractor Assurance and Continuous Improvement
- Environment, Safety and Health
- Government and External Relations
- Independent Audit and Oversight
- Office of General Counsel
- Prime Contract Management Office
- Quality Assurance Office
- Security Organization
- LLNS, LLC Parent Oversight Office

The Laboratory is organized into four principal directorates, each headed by a Principal Associate Director:

- Global Security

- Weapons and Complex Integration

- National Ignition Facility and Photon Science

- Operations and Business

  - Business
  - Facilities & Infrastructure
  - Institutional Facilities Management
  - Integrated Safety Management System Project Office
  - Nuclear Operations
  - Planning and Financial Management
  - Staff Relations
  - Strategic Human Resources Management

Three other directorates are each headed by an Associate Director who reports to the LLNL Director:

- Computation

- Engineering

- Physical & Life Sciences

7.9.15 Corporate management

The LLNL Director reports to the Lawrence Livermore National Security, LLC (LLNS) Board of Governors, a group of key scientific, academic, national security and business leaders from the LLNS partner companies that jointly own and control LLNL. The LLNS Board of Governors has a total of 16 positions, with six of these Governors constituting an Executive Committee. All decisions of the Board are made by the Governors on the Executive Committee. The other Governors are advisory to the Executive Committee and do not have voting rights.

The University of California is entitled to appoint three Governors to the Executive Committee, including the Chair. Bechtel is also entitled to appoint three Governors to the Executive Committee, including the Vice Chair. One of the Bechtel Governors must be a representative of Babcock and Wilcox (B&W) or the Washington Division of URS Corporation (URS), who is nominated jointly by B&W and URS each year, and who must be approved and appointed by Bechtel. The Executive Committee has a seventh Governor who is appointed by Battelle; they are non-voting and advisory to the Executive Committee. The remaining Board positions are known as Independent Governors (also referred to as Outside Governors), and are selected from among individuals, preferably of national stature, and can not be employees or officers of the partner companies.

The University of California-appointed Chair has tie-breaking authority over most decisions of the Executive Committee. The Board of Governors is the ultimate governing body of LLNS and is charged with overseeing the affairs of LLNS in its operations and management of LLNL.

LLNS managers and employees who work at LLNL, up to and including the President/Laboratory Director, are generally referred to as Laboratory Employees. All Laboratory Employees report directly or indirectly to the LLNS President. While most of the work performed by LLNL is funded by the federal government, Laboratory employees are paid by LLNS which is responsible for all aspects of their employment including providing health care benefits and retirement programs.

Within the Board of Governors, authority resides in the Executive Committee to exercise all rights, powers, and authorities of LLNS, excepting only certain decisions that are reserved to the parent companies. The LLNS Executive Committee is free to appoint officers or other managers of LLNS and LLNL, and may delegate its authorities as it deems appropriate to such officers, employees, or other representatives of LLNS/LLNL. The Executive Committee may also retain auditors, attorneys, or other professionals as necessary. For the most part the Executive Committee has appointed senior managers at LLNL as the primary officers of LLNS. As a practical matter most operational decisions are delegated to the President of LLNS, who is also the Laboratory Director. The positions of President/Laboratory Director and Deputy Laboratory Director are filled by joint action of the Chair and Vice Chair of the Executive Committee, with the University of California nominating the President/Laboratory Director and Bechtel nominating the Deputy Laboratory Director.*[53]

The current LLNS Chairman is Norman J. Pattiz - founder and chairman of Westwood One, America's largest radio network, and he also currently serves on the Board of Regent of the University of California. The Vice Chairman is J. Scott Ogilvie - president of Bechtel Systems & Infrastructure, Inc., he serves on the Board of Directors of Bechtel Group, Inc. (BGI) and on the BGI Audit Committee.*[54]

The Board of Governors uses the following committees to oversee the management and operations of LLNL by LLNS:

- Business and Operations

- Ethics and Audit

- Mission

- Nominations and Compensation

- Nuclear Weapons Complex Integration

- Safeguards and Security

- Science and Technology

7.9.16   See also

- Center for the Advancement of Science in Space—operates the US National Laboratory on the ISS.

- Dielectric wall accelerator

- Top 100 US Federal Contractors

- List of articles associated with nuclear issues in California

7.9.17   Footnotes

[1] "Missions & Programs". Lawrence Livermore National Laboratory. February 13, 2008. Retrieved March 19, 2008.

[2] "R&D 2007 Award Index of Winners". R&D Technologies & Strategies for Research & Development. August 16, 2008. Retrieved May 20, 2008.

[3] http://education.llnl.gov/archives/ *Multimedia timeline of EO Lawrence*

[4] "Science and Technology Review (September 1998)". "A Short History of the Laboratory at Livermore".

[5] Jeff Garberson (May 16, 2013). "$2.7 Million Awarded to Former Lab Employees". *The Independent.*

[6] Max Taves (May 10, 2013). "Laid Off Lab Workers Awarded $2.8 Million". *The Recorder.*

[7] Bob Egelko (May 13, 2013). "Livermore lab jury awards $2.7 million". *SF Chronicle.*

[8] Todd Jacobson (May 17, 2013). "Five Former Livermore Workers Receive $2.7 Million in Layoff Case". *Nuclear Weapons & Materials Monitor.*

[9] "Global Security" (April 27, 2005). ["http://www.globalsecurity.org/wmd/intro/miniaturization.htm" "Weapons of Mass Destruction: Miniaturization"]. Retrieved June 3, 2008.

[10] James N. Gibson (October 14, 2006). "Complete List of All U.S. Nuclear Weapons". *The Nuclear Weapon Archive.* Retrieved March 19, 2008.

[11] "U.S. Nuclear Weapon Enduring Stockpile". *The Nuclear Weapon Archive.* August 31, 2007. Retrieved March 19, 2008.

[12] "Nuclear Weapons Stockpile Stewardship". Lawrence Livermore National Laboratory. February 13, 2008. Retrieved March 19, 2008.

[13] Scott Lindlaw (March 2, 2007). "Bush Administration Picks Lawrence Livermore Warhead Design". *The San Francisco Chronicle.* Archived from the original on March 12, 2007. Retrieved March 19, 2008.

[14] Barbara Epstein. Political protest and cultural revolution: nonviolent direct action in the 1970s and 1980s University of California Press, 1993. pp. 125-133.

[15] 1,300 Arrested in California Anti-nuclear Protest

[16] Diana Walsh (August 11, 2003). "Nuclear Protest Blooms Again at Lab: 1,000 in Livermore Demonstrate Against New Buster Bomb". *SF Chronicle.*

[17] Police arrest 64 at California anti-nuclear protest *Reuters*, April 6, 2007.

[18] "Scores arrested during protest at Livermore Lab". *Oakland Tribune.* March 22, 2008.

[19] "Daniel Ellsberg among 31 arrested at Livermore lab during Hiroshima Day protests". *San Jose Mercury News.* August 6, 2013.

[20] "University rejects widow's request". *Ocala Star-Banner* (Ocala, FL). Associated Press. July 16, 1983. p. 15A.

[21] Savage, David G. (September 7, 1985). "Physicist's widow asks that husband's name be removed from weapons lab". *Los Angeles Times.* Retrieved May 9, 2014.

[22] Lawrence, Mary B. (October 1986). "So they say:". *The Scientist.* excerpts. Retrieved May 9, 2014.

[23] "Name change". *Milwaukee Journal.* Associated Press. June 8, 1987. p. 2A.

[24] Yarris, Lynn (January 8, 2003). "Lab mourns death of Molly Lawrence, widow of Ernest O. Lawrence". Lawrence Berkeley National Laboratory. Retrieved May 9, 2014.

[25] "Plutonium Up Close...Way Close". Lawrence Livermore National Laboratory. Retrieved May 20, 2008.

[26] "Lawrence Livermore National Laboratory Fact Sheet for NNSA Complex Transformation– Preferred Alternative" (PDF). Lawrence Livermore National Laboratory. Archived from the original (PDF) on August 20, 2008. Retrieved May 20, 2008.

[27] NNSA Press Release, September 30, 2009, NNSA Ships Additional Special Nuclear Material from Lawrence Livermore National Laboratory as Part of Deinventory Project

[28] "Science and Technology Review, Lawrence Livermore National Laboratory" (November 2007). ["https://www.llnl.gov/str/Nov07/bissani.html" "Scientists without Borders"].

[29] Science and Technology Review, Lawrence Livermore National Laboratory (December 2007). "Out of Harms Way"
.

[30] Science and Technology Review, Lawrence Livermore National Laboratory (November 2007). "Characterizing Virulent Pathogens".

[31] Science and Technology Review, Lawrence Livermore National Laboratory (September 2007). "Assessing the Threat of Bioterrorism".

[32] Science and Technology Review, Lawrence Livermore National Laboratory (August 2008). "Antineutrino Detectors Improve Reactor Safeguards".

[33] Science and Technology Review, Lawrence Livermore National Laboratory (January 2007). "Identifying the Source of Stolen Nuclear Materials".

[34] Science and Technology Review, Lawrence Livermore National Laboratory (October 2007). "Mobile Mapping for Radioactive Materials".

[35] Science and Technology Review, Lawrence Livermore National Laboratory (March 2007). "On the Leading Edge of Atmospheric Predictions".

[36] Science and Technology Review, Lawrence Livermore National Laboratory (May 2006). "Protecting our Nation's Livestock".

[37] Science and Technology Review, Lawrence Livermore National Laboratory. "Simulating Warfare Is No Video Game".

[38] Science and Technology Review, Lawrence Livermore National Laboratory. "Leveraging Science and Technology in the National Interest".

[39] Science and Technology Review, Lawrence Livermore National Laboratory (July 2002). "Knowing the Enemy, Anticipating the Threat".

[40] Science and Technology Review, Lawrence Livermore National Laboratory (June 2007). "Setting a World Driving Record with Hydrogen".

[41] Science and Technology Review, Lawrence Livermore National Laboratory (March 2007). "Climate and Agriculture: Change Begets Change".

[42] Burnham, Alan K.; McConaghy, James R. (October 16, 2006). *Comparison of the acceptability of various oil shale processes* (PDF). 26th Oil shale symposium. Golden, Colorado: Lawrence Livermore National Laboratory. pp. 2; 17. UCRL-CONF-226717. Retrieved May 27, 2007.

[43] "Lawrence Livermore National Laboratory". Lawrence Livermore National Laboratory. Retrieved May 20, 2008.

[44] TOP500 Supercomputer Sequoia

[45] IBM Press Release: 20 Petaflop Sequoia Supercomputer

[46] "Linux at Livermore". Lawrence Livermore National Laboratory. Retrieved February 28, 2007.

[47] "Peloton Capability Cluster". Lawrence Livermore National Laboratory. Retrieved February 28, 2007.

[48] FY2009 LLNL Annual Report

[49] George Miller to step down as Laboratory director

[50] UC Newsroom - Livermore Lab director named

[51] LANL's Knapp to take over as Livermore acting director

[52] William H. Goldstein named director of LLNL

[53] An Introductory Guide to UC's Ties to LANS LLC and LLNS LLC and their Management of the Weapons Labs at Los Alamos and Livermore, prepared by the UC Academic Council and University Counsel

[54] LLNS Board of Governors

7.9.18 References

- *Nuclear Rites: A Weapons Laboratory at the End of the Cold War*, by Hugh Gusterson, University of California Press, Berkeley, 1996 (ISBN 0-520-21373-4)

- *The Stockpile Stewardship and Management Program: Maintaining Confidence in the Safety and Reliability of the Enduring U.S. Nuclear Weapon Stockpile* U.S. Department of Energy, Office of Defense Programs. May 1995.

- *Preparing for the 21st Century: 40 Years of Excellence.* Lawrence Livermore National Laboratory. Report UCRL-AR-108618. 1992.

7.9.19 External links and sources

- Lawrence Livermore National Laboratory (official website)

- Lawrence Livermore National Security, a Limited Liability Corporation (official website)

- "History of the Laboratory". LLNL. July 14, 2002. Archived from the original on October 12, 2006.

- Serving the Nation for Fifty Years: 1952 - 2002 Lawrence Livermore National Laboratory [LLNL], Fifty Years of Accomplishments (PDF) (Report). LLNL. 2002. UCRL-AR-148833.

- LLNL Industrial Partnerships and Commercialization (IPAC) (official website)

- University of California Office of Laboratory Management (official website)

- Society of Professionals, Scientists and Engineers (Union representing UC Scientists and Engineers at LLNL)

- University of California LLNL Retiree Group (Legal Defense Fund for UC Retirees from LLNL)

- Annotated bibliography for Livermore from the Alsos Digital Library for Nuclear Issues

Coordinates: 37°41′N 121°43′W / 37.69°N 121.71°W

MINOS service building at Fermilab, the entrance to the underground MINOS hall that hosts the near detector.[5]

7.10 MINOS

For other uses, see Minos (disambiguation).
MINOS (or **Main Injector Neutrino Oscillation**

Front face of the MINOS far detector. On the left is the control room and on the right is a mural by Joseph Giannetti.

Search) is a particle physics experiment designed to study the phenomena of neutrino oscillations, first discovered by a Super-Kamiokande (Super-K) experiment in 1998. Neutrinos produced by the NuMI ("Neutrinos at Main Injector") beamline at Fermilab near Chicago are observed at two detectors, one very close to where the beam is produced (the *near detector*), and another much larger detector 735 km away in northern Minnesota (the *far detector*).

The MINOS experiment started detecting neutrinos from the NuMI beam in February 2005. On 30 March 2006, the MINOS collaboration announced that the analysis of the initial data, collected in 2005, is consistent with neutrino oscillations, with the oscillation parameters which are consistent with Super-K measurements.[1] MINOS received the last neutrinos from the NUMI beam line at midnight on 30 April 2012.[2][3] It has been upgraded to MINOS+ which started taking data in 2013 for 3 years.[4]

7.10.1 Detectors

There are two detectors in the experiment.

- The near detector is similar to the far detector in design, but smaller in size with a mass of 980 tons (t). It is located at Fermilab, a few hundred meters away from the graphite target which the protons interact with, and approximately 100 meters underground. The commissioning of the near detector was completed in December 2004, and it is now fully operational.

- The far detector has a mass of 5.4 kt. It is located in the Soudan mine in Northern Minnesota at a depth of 716 meters. The far detector has been fully operational since summer 2003, and has been taking cosmic ray and atmospheric neutrino data since early in its construction.

Both MINOS detectors are steel-scintillator sampling calorimeters made out of alternating planes of magnetized steel and plastic scintillators. The magnetic field causes the path of a muon produced in a muon neutrino interaction to bend, making it possible to distinguish interactions with neutrinos from those with antineutrinos. This feature of the MINOS detectors allows MINOS to search for CPT-violation with atmospheric neutrinos and anti-neutrinos.

7.10.2 Neutrino beam

To produce the NuMI beamline, 120 GeV Main Injector proton pulses hit a water-cooled graphite target. The resulting interactions of protons with the target material produce pions and kaons, which are focused by a system of magnetic horns. The neutrinos from subsequent decays of pions and kaons form the neutrino beam. Most of these are

NuMI Target Hall (left), the starting point of the NuMI tunnel with the Main Injector in the background.[6]

muon neutrinos, with a small electron neutrino contamination. Neutrino interactions in the near detector are used to measure the initial neutrino flux and energy spectrum. Because they are weakly interacting and therefore usually pass through matter, the vast majority of the neutrinos travel through the near detector and the 734 km of rock, then through the far detector and off into space. On the way toward Soudan, about 20% of the muon neutrinos oscillate into other flavors.

7.10.3   Physics goals and results

MINOS measures the difference in neutrino beam composition and energy distribution in the near and far detectors with the aim of producing precision measurements of the neutrino squared mass difference and mixing angle. In addition, MINOS looks for the appearance of electron neutrinos in the far detector, and will either measure or set a limit on the oscillation probability of muon neutrinos into electron neutrinos.

On 29 July 2006, the MINOS collaboration published a paper giving their initial measurements of oscillation parameters as judged from muon neutrino disappearance. These are: $\Delta m2$
$23 = 2.74 + 0.44$
$-0.26 \times 10^{*-3}$ eV^2/c^4 and $\sin^2(2\theta_{23}) > 0.87$ (68% confidence limit).[7][8]

In 2008 MINOS released a further result using over twice the previous data (3.36×10^{20} protons-on-target; this includes the first data set). This is the most precise measurement of Δm^2. The results are: $\Delta m2$
$23 = 2.43 + 0.13$
$-0.13 \times 10^{*-3}$ eV^2/c^4 and $\sin^2(2\theta_{23}) > 0.90$ (90% confidence limit).[9]

In 2011, the above results were updated again, using a more

than double data sample (exposure of 7.25×10^{20} protons on target) and improved analysis methodology. The results are: $\Delta m2$
$23 = 2.32 + 0.12$
$-0.08 \times 10^{*-3}$ eV^2/c^4 and $\sin^2(2\theta_{23}) > 0.90$ (90% confidence limit).[10]

In 2010 and 2011, MINOS reported results according to which there is a difference in the disappearance and consequently the masses between antineutrinos and neutrinos, which would violate CPT symmetry. [11][12][13] However, after additional data were evaluated in 2012, MINOS reported that this gap has closed and no excess is there any more.[14][15]

Cosmic ray results from the MINOS far detector have shown that there is a strong correlation between high energy cosmic rays measured and the temperature of the stratosphere. This is the first time, daily variations in secondary cosmic rays from an underground muon detector are shown to be associated with planetary–scale meteorological phenomena in the stratosphere such as the Sudden stratospheric warming [16] as well as the change in seasons.[17] The MINOS far detector is also able to observe a reduction in cosmic rays caused by the Sun and the Moon[18]

Time of flight of neutrinos

Main article: Measurements of neutrino speed

In 2007 an experiment with the MINOS detectors found the speed of 3 GeV neutrinos to be 1.000051(29) c at 68% confidence level, and at 99% confidence level a range between 0.999976 c to 1.000126 c. The central value was higher than the speed of light; however, the uncertainty was great enough that the result also did not rule out speeds less than or equal to light at this high confidence level.[19][20]

After the detectors for the project were upgraded in 2012, MINOS corrected their initial result and found agreement with the speed of light, with the difference in the arrival times of −0.0006% (±0.0012%) between neutrinos and light. Further measurements are going to be conducted.[21]

7.10.4   References

[1] "MINOS experiment sheds light on mystery of neutrino disappearance" (Press release). 30 March 2006. Retrieved 2009-08-03.

[2] "MINOS Run Period Run Subrun Ranges (MRPRSR)". Retrieved 4 November 2012.

[3] de Jong, Jeffrey (12 September 2012). "'Final' MINOS Results" (PDF). Retrieved 13 December 2012.

[4] Tzanankos, G et al. (2011). "MINOS+: a Proposal to FNAL to run MINOS with the medium energy NuMI beam" (PDF). *FERMILAB-PROPOSAL-1016.*

[5] Basu, Paroma (30 March 2006). "Physicists Say Multimillion Dollar Experiment Advancing Smoothly". *Wisconsin Online.* Retrieved 14 August 2015.

[6] "Site map of NuMI/MINOS". *Fermilab.* Retrieved 14 August 2015.

[7] D.G. Michael et al. (2006). "Observation of muon neutrino disappearance with the MINOS detectors in the NuMI neutrino beam". *Physical Review Letters* **97** (19): 191801. arXiv:hep-ex/0607088. Bibcode:2006PhRvL..97s1801M. doi:10.1103/PhysRevLett.97.191801. PMID 17155614.

[8] P. Adamson et al. (2008). "Study of muon neutrino disappearance using the Fermilab Main Injector neutrino beam". *Physical Review D* **77** (7): 072002. arXiv:0711.0769. Bibcode:2008PhRvD..77g2002A. doi:10.1103/PhysRevD.77.072002.

[9] P. Adamson et al. (2008). "Measurement of neutrino oscillations with the MINOS detectors in the NuMI beam". *Physical Review Letters* **101** (13): 131802. arXiv:0806.2237. Bibcode:2008PhRvL.101m1802A. doi:10.1103/PhysRevLett.101.131802. PMID 18851439.

[10] P. Adamson et al. (2011). "Measurement of the neutrino mass splitting and flavor mixing by MINOS". *Physical Review Letters* **106** (18): 181801. arXiv:1103.0340. Bibcode:2011PhRvL.106r1801A. doi:10.1103/PhysRevLett.106.181801.

[11] "New measurements from Fermilab's MINOS experiment suggest a difference in a key property of neutrinos and antineutrinos". Fermilab press release. June 14, 2010. Retrieved 14 December 2011.

[12] MINOS Collaboration (2011). "First Direct Observation of Muon Antineutrino Disappearance". *Physical Review Letters* **107** (2): 021801. arXiv:1104.0344. Bibcode:2011PhRvL.107b1801A. doi:10.1103/PhysRevLett.107.021801.

[13] MINOS Collaboration (2011). "Search for the disappearance of muon antineutrinos in the NuMI neutrino beam". *Physical Review D* **84** (7): 071103. arXiv:1108.1509. Bibcode:2011PhRvD..84g1103A. doi:10.1103/PhysRevD.84.071103.

[14] "Fermilab experiment announces world's best measurement of key property of neutrinos". Fermilab press release. June 5, 2012. Retrieved June 20, 2012.

[15] MINOS Collaboration (2012). "An improved measurement of muon antineutrino disappearance in MINOS". *Physical Review Letters* **108** (19): 191801. arXiv:1202.2772. Bibcode:2012PhRvL.108s1801A. doi:10.1103/PhysRevLett.108.191801.

[16] Osprey, S.; Barnett, J.; Smith, J.; the MINOS Collaboration (7 March 2009). "Sudden stratospheric warmings seen in MINOS deep underground muon data". *Geophysical Research Letters* **36** (5). Bibcode:2009GeoRL..36.5809O. doi:10.1029/2008GL036359.

[17] Adamson, P. et al. (1 January 2010). "Observation of muon intensity variations by season with the MINOS far detector". *Physical Review D* **81** (1). arXiv:0909.4012. Bibcode:2010PhRvD..81a2001A. doi:10.1103/PhysRevD.81.012001.

[18] Adamson, P. et al. "Observation in the MINOS far detector of the shadowing of cosmic rays by the sun and moon". *Astroparticle Physics* **34** (6): 457–466. arXiv:1008.1719. Bibcode:2011APh....34..457A. doi:10.1016/j.astropartphys.2010.10.010.

[19] P. Adamson et al. (MINOS Collaboration) (2007). "Measurement of neutrino velocity with the MINOS detectors and NuMI neutrino beam". *Physical Review D* **76** (7). arXiv:0706.0437. Bibcode:2007PhRvD..76g2005A. doi:10.1103/PhysRevD.76.072005.

[20] D. Overbye (22 September 2011). "Tiny neutrinos may have broken cosmic speed limit". *New York Times.* That group found, although with less precision, that the neutrino speeds were consistent with the speed of light.

[21] "MINOS reports new measurement of neutrino velocity". Fermilab today. June 8, 2012. Retrieved June 8, 2012.

Coordinates: 47°49′12″N 92°14′30″W / 47.82000°N 92.24167°W

7.10.5 External links

- NuMI and MINOS

7.11 MiniBooNE

MiniBooNE is an experiment at Fermilab designed to observe neutrino oscillations (BooNE is an acronym for the Booster Neutrino Experiment). A neutrino beam consisting primarily of muon neutrinos is directed at a detector filled with 800 tons of mineral oil and lined with 1,280 photomultiplier tubes. An excess of electron neutrino events in the detector would support the neutrino oscillation interpretation of the LSND (Liquid Scintillator Neutrino Detector) result.

It started collecting data in 2002[*][1] and was still running in 2012.[*][2]

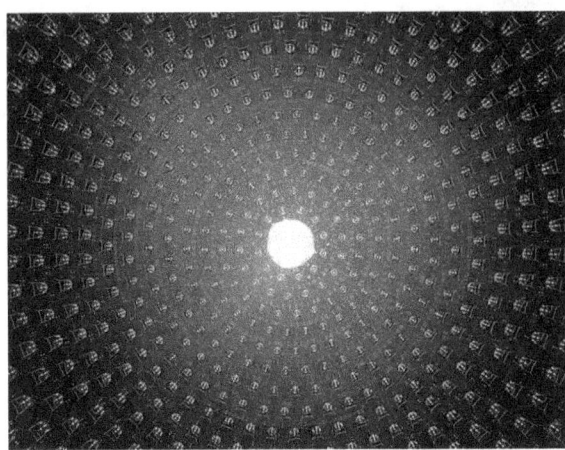

The interior of the MiniBooNE detector.

7.11.1 History and motivation

Experimental observation of solar neutrinos and atmospheric neutrinos provided evidence for neutrino oscillations, implying that neutrinos have masses. Data from the LSND experiment at Los Alamos National Laboratory are controversial since they are not compatible with the oscillation parameters measured by other neutrino experiments in the framework of the Standard Model. Either there must be an extension to the Standard Model, or one of the experimental results must have a different explanation. Moreover, the KARMEN experiment in Karlsruhe[*][3] examined a [low energy] region similar to the LSND experiment, but saw no indications of neutrino oscillations. This experiment was less sensitive than LSND, and both could be right.

Cosmological data can provide an indirect but rather model-dependent bound to the mass of sterile neutrinos, such as the $m_s < 0.26$ eV (0.44 eV) at 95% (99.9%) confidence limit given by Dodelson et al..[*][4] However, cosmological data can be accommodated within models with different assumptions, such as that by Gelmini et al.[*][5]

MiniBooNE was designed to unambiguously verify or refute the LSND controversial result in a controlled environment.

2007 The first results came in late March 2007, and showed no evidence for muon neutrino to electron neutrino oscillations in the LSND [low energy] region, refuting a simple 2-neutrino oscillation interpretation of the LSND results.[*][6] More advanced analyses of their data are currently being undertaken by the MiniBooNE collaboration; early indications are pointing towards the existence of the sterile neutrino,[*][7] an effect interpreted by some physicists to be hinting of the existence of the bulk[*][8] or Lorentz violation.[*][9]

2008 Some members of MiniBooNE have formed a new collaboration with outside scientists and proposed a new experiment (called MicroBooNE) designed to further investigate this.[*][10]

7.11.2 References

[1] "MiniBooNE website" .

[2] "Progress in Delivering Beam to MiniBooNE" .

[3] "KARMEN experiment" (Press release). 3 August 2011.

[4] S. Dodelson; A. Melchiorri; A. Slosar (2006). "Is cosmology compatible with sterile neutrinos?". *Physical Review Letters* **97** (4): 04301. arXiv:astro-ph/0511500. Bibcode:2006PhRvL..97d1301D. doi:10.1103/PhysRevLett.97.041301.

[5] G. Gelmini; S. Palomares-Ruiz & S. Pascoli (2004). "Low reheating temperature and the visible sterile neutrino" . *Physical Review Letters* **93** (8): 081302. arXiv:astro-ph/0403323. Bibcode:2004PhRvL..93h1302G. doi:10.1103/PhysRevLett.93.081302. PMID 15447171.

[6] A. A. Aguilar-Arevalo; et al. (MiniBooNE Collaboration) (2007). "A Search for Electron Neutrino Appearance at the $\Delta m^2 \sim 1$ eV^2 Scale" . *Physical Review Letters* **98** (23): 231801. arXiv:0704.1500. Bibcode:2007PhRvL..98w1801A. doi:10.1103/PhysRevLett.98.231801.

[7] M. Alpert (August 2007). "Dimensional Shortcuts" . *Scientific American*. Retrieved 2007-07-23.

[8] H. Päs; S. Pakvasa; T.J. Weiler (2007). "Shortcuts in extra dimensions and neutrino physics" . *AIP Conference Proceedings* **903**: 315. arXiv:hep-ph/0611263. doi:10.1063/1.2735188.

[9] T. Katori; V.A. Kostelecky; R. Tayloe (2006). "Global three-parameter model for neutrino oscillations using Lorentz violation" . *Physical Review D* **74** (10): 105009. arXiv:hep-ph/0606154. Bibcode:2006PhRvD..74j5009K. doi:10.1103/PhysRevD.74.105009.

[10] M. Alpert (September 2008). "Fermilab Looks for Visitors from Another Dimension" . *Scientific American*. Retrieved 2008-09-23.

7.11.3 External links

- MiniBooNe first results press release and arXiv:0704.1500

- MiniBooNE website

  - MiniBooNE publications

  - Experiment details

- Overview of MiniBooNE for Mineral Oil Suppliers

- An informal discussion of the experiment and initial results

- Experiment Nixes Fourth Neutrino (April 2007 Scientific American)

- Dimensional Shortcuts - evidence for sterile neutrino; (August 2007; Scientific American)

7.12 SLAC National Accelerator Laboratory

"SLAC" redirects here. For other uses, see SLAC (disambiguation).

SLAC National Accelerator Laboratory, originally named **Stanford Linear Accelerator Center**,[*][2][*][3] is a United States Department of Energy National Laboratory operated by Stanford University under the programmatic direction of the U.S. Department of Energy Office of Science and located in Menlo Park, California.

The SLAC research program centers on experimental and theoretical research in elementary particle physics using electron beams and a broad program of research in atomic and solid-state physics, chemistry, biology, and medicine using synchrotron radiation.

7.12.1 History

The entrance to SLAC in Menlo Park.

Founded in 1962 as the Stanford Linear Accelerator Center, the facility is located on 426 acres (1.72 square kilometers) of Stanford University-owned land on Sand Hill Road in Menlo Park, California—just west of the University's main campus. The main accelerator is 2 miles long—the longest linear accelerator in the world—and has been operational since 1966.

Nobel Prize

Research at SLAC has produced three Nobel Prizes in Physics:

- 1976: The charm quark—see J/ψ meson[*][4]

- 1990: Quark structure inside protons and neutrons[*][5]

- 1995: The tau lepton[*][6]

SLAC's meeting facilities also provided a venue for the Homebrew Computer Club and other pioneers of the home computer revolution of the late 1970s and early 1980s.

In 1984 the laboratory was named an ASME National Historic Engineering Landmark and an IEEE Milestone.[*][7]

SLAC developed and, in December 1991, began hosting the first World Wide Web server outside of Europe.[*][8]

In the early-to-mid 1990s, the Stanford Linear Collider (SLC) investigated the properties of the Z boson using the Stanford Large Detector.

As of 2005, SLAC employs over 1,000 people, some 150 of which are physicists with doctorate degrees, and serves over 3,000 visiting researchers yearly, operating particle accelerators for high-energy physics and the Stanford Synchrotron Radiation Laboratory (SSRL) for synchrotron light radiation research, which was "indispensable" in the research leading to the 2006 Nobel Prize in Chemistry awarded to Stanford Professor Roger D. Kornberg.[*][9]

In October 2008, the Department of Energy announced that the Center's name would be changed to SLAC National Accelerator Laboratory. The reasons given include a better representation of the new direction of the lab and the ability to trademark the laboratory's name. Stanford University had legally opposed the Department of Energy's attempt to trademark "Stanford Linear Accelerator Center".[2][10]

In March 2009 it was announced that the SLAC National Accelerator Laboratory was to receive $68.3 Million in Recovery Act Funding to be disbursed by Department of Energy's Office of Science.[11]

7.12.2 Components

SLAC 1.9 mile (3 kilometer) long Klystron Gallery above the beamline Accelerator

Accelerator

Part of the SLAC beamline

The main accelerator is an RF linear accelerator that can accelerate electrons and positrons up to 50 GeV. At 2.0 miles (about 3.2 kilometers) long, the accelerator is the longest linear accelerator in the world, and is claimed to be

"the world's straightest object." [12] The main accelerator is buried 30 feet (about 10 meters) below ground[13] and passes underneath Interstate Highway 280. The above-ground klystron gallery atop the beamline is the longest building in the United States.

SLC pit and detector

Stanford Linear Collider

The Stanford Linear Collider was a linear accelerator that collided electrons and positrons at SLAC.[14] The center of mass energy was about 90 GeV, equal to the mass of the Z boson, which the accelerator was designed to study. Grad student Barrett D. Milliken discovered the first Z event on 12 April 1989 while poring over the previous day's computer data from the Mark II detector.[15] The bulk of the data was collected by the SLAC Large Detector, which came online in 1991. Although largely overshadowed by the Large Electron-Positron Collider at CERN, which began running in 1989, the highly polarized electron beam at SLC (close to 80%[16]) made certain unique measurements possible, such as parity violation in Z Boson-b quark coupling.

Presently no beam enters the south and north arcs in the machine, which leads to the Final Focus, therefore this section is mothballed to run beam into the PEP2 section from the beam switchyard.

SLAC Large Detector

The SLAC Large Detector (SLD) was the main detector for the Stanford Linear Collider. It was designed primarily to detect Z bosons produced by the accelerator's electron-positron collisions. The SLD operated from 1992 to 1998.

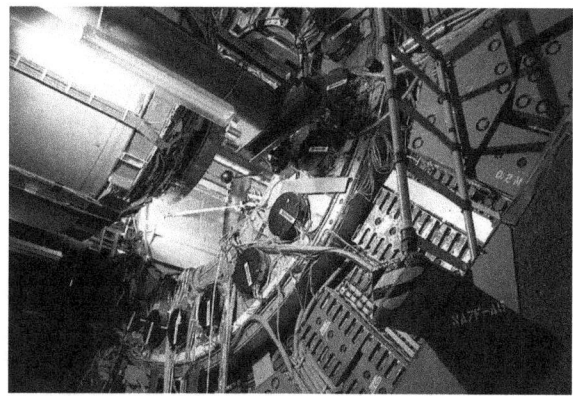

Inside view of the SLD

PEP

PEP (Positron-Electron Project) began operation in 1980, with center-of-mass energies up to 29 GeV. At its apex, PEP had five large particle detectors in operation, as well as a sixth smaller detector. About 300 researchers made used of PEP. PEP stopped operating in 1990, and PEP-II began construction in 1994.[*][17]

PEP-II

From 1999 to 2008, the main purpose of the linear accelerator was to inject electrons and positrons into the PEP-II accelerator, an electron-positron collider with a pair of storage rings 1.4 miles (2.2 km) in circumference. PEP-II was host to the BaBar experiment, one of the so-called B-Factory experiments studying charge-parity symmetry.

Stanford Synchrotron Radiation Lightsource

Main article: Stanford Synchrotron Radiation Lightsource

The Stanford Synchrotron Radiation Lightsource (SSRL) is a synchrotron light user facility located on the SLAC campus. Originally built for particle physics, it was used in experiments where the J/ψ meson was discovered. It is now used exclusively for materials science and biology experiments which take advantage of the high-intensity synchrotron radiation emitted by the stored electron beam to study the structure of molecules. In the early 1990s, an independent electron injector was built for this storage ring, allowing it to operate independently of the main linear accelerator.

Fermi Gamma-ray Space Telescope

Fermi Gamma-ray Space Telescope

Main article: Fermi Gamma-ray Space Telescope

SLAC plays a primary role in the mission and operation of the Fermi Gamma-ray Space Telescope, launched in August 2008. The principle scientific objectives of this mission are:

- To understand the mechanisms of particle acceleration in AGNs, pulsars, and SNRs.

- To resolve the gamma-ray sky: unidentified sources and diffuse emission.

- To determine the high-energy behavior of gamma-ray bursts and transients.

- To probe dark matter and fundamental physics.

KIPAC

Main article: Kavli Institute for Particle Astrophysics and Cosmology

The Kavli Institute for Particle Astrophysics and Cosmology (KIPAC) is partially housed on the grounds of SLAC, in addition to its presence on the main Stanford campus.

PULSE

The Stanford PULSE Institute (PULSE) is a Stanford Independent Laboratory located in the Central Laboratory at SLAC. PULSE was created by Stanford in 2005 to help Stanford faculty and SLAC scientists develop ultrafast x-ray research at LCLS. PULSE research publications can be viewed here.

LCLS

The Linac Coherent Light Source (LCLS) is a free electron laser facility located at SLAC. The LCLS is partially a reconstruction of the last 1/3 of the original linear accelerator at SLAC, and can deliver extremely intense x-ray radiation for research in a number of areas. It achieved first lasing in April 2009.[18]

Aerial photo of the Stanford Linear Accelerator Center, with detector complex at the right (east) side

The laser produces hard X-rays, 10^9 times the relative brightness of traditional synchrotron sources and is the most powerful x-ray source in the world. LCLS enables a variety of new experiments and provides enhancements for existing experimental methods. Often, x-rays are used to take "snapshots" of objects on the nearly atomic level before obliterating samples. The laser's wavelength, ranging from 200 to 2000 electron volts (eV)[19] is similar to the width of an atom, providing extremely detailed images for objects previously unattainable.[20] Additionally, the laser is capable of capturing images with a "shutter speed" measured in femtoseconds, or million-billionths of a second, necessary because the intensity of the beam is often high enough so that the sample explodes on the femtosecond timescale.[21]

LCLS-II

The LCLS-II project is to provide a major upgrade to LCLS by adding two new X-ray laser beams. The new system will utilize the 500 metres (1,600 ft) of existing tunnel to add new superconducting accelerator at 4 GeV and two undulators. The advancement from the discoveries using this new capabilities may include new drugs, next-generation computers, and new materials.[22]

FACET

In 2012, the first two thirds (~2 km) of the original SLAC LINAC were re-commissioned for a new user facility, the Facility for Advanced Accelerator Experimental Tests (FACET). This new facility is capable of delivering 23 GeV, 3 nC electron (and positron) beams with short bunch lengths and small spot sizes, ideal for beam-driven Plasma Acceleration studies. [23]

NLCTA

The Next Linear Collider Test Accelerator (NLCTA) is a 60-120 MeV high-brightness electron beam linear accelerator used for experiments on advanced beam manipulation and acceleration techniques. It is located at SLAC's end station B. A list of relevant research publications can be viewed here.

7.12.3   Other discoveries

- SLAC has also been instrumental in the development of the klystron, a high-power microwave amplification tube.

- There is active research on plasma acceleration with recent successes such as the doubling of the energy of 42 GeV electrons in a meter-scale accelerator.

- There was a *Paleoparadoxia* found at the SLAC site, and its skeleton can be seen at a small museum there in the Breezeway.[24]

- The SSRL facility was used to reveal hidden text in the Archimedes Palimpsest. X-rays from the synchrotron radiation lightsource caused the iron in the original ink to glow, allowing the researchers to photograph the original document that a Christian monk had scrubbed off.[25]

7.12.4   See also

- Accelerator physics

- Beamline

- CERN

- Cyclotron

- Dipole magnet

- Electromagnetism

- List of particles

- List of United States college laboratories conducting basic defense research

- Particle beam

- Particle physics

- Quadrupole magnet

- Spallation Neutron Source

- Wolfgang Panofsky (1961–84, SLAC Director; Professor, Stanford University)

7.12.5 References

[1] Labs at a glance - SLAC http://science.energy.gov/ laboratories/slac-national-accelerator-laboratory/

[2] "SLAC renamed to SLAC Natl. Accelerator Laboratory" . *The Stanford Daily*. 16 October 2008. Archived from the original on 2012-01-29. Retrieved 2008-10-16.

[3] "Stanford Linear Accelerator Center renamed SLAC National Accelerator Laboratory" (Press release). SLAC National Accelerator Laboratory. 15 October 2008. Archived from the original on 2011-07-20. Retrieved 2011-07-20.

[4] Nobel Prize in Physics 1976. Half prize awarded to Burton Richter.

[5] Nobel Prize in Physics 1990 Award split between Jerome I. Friedman, Henry W. Kendall, and Richard E. Taylor.

[6] Nobel Prize in Physics 1995 Half prize awarded to Martin L. Perl.

[7] "Milestones:Stanford Linear Accelerator Center, 1962" . *IEEE Global History Network*. IEEE. Retrieved 3 August 2011.

[8] The Early World Wide Web at SLAC: Early Chronology and Documents

[9] "2006 Nobel Prize in Chemistry" . *SLAC Virtual Visitor Center*. Stanford University. n.d. Archived from the original on 5 August 2011. Retrieved 19 March 2015.

[10] A New Name for SLAC

[11] 23, 2009 - SLAC National Accelerator Laboratory to Receive $68.3 Million in Recovery Act Funding

[12] Saracevic, Alan T. "Silicon Valley: It's where brains meet bucks." *San Francisco Chronicle* 23 October 2005. p J2. Accessed 2005-10-24.

[13] Neal, R. B. (1968). "Chap. 5" . *The Stanford Two-Mile Accelerator* (PDF). New York, New York: W.A. Benjamin, Inc. p. 59. Retrieved 2010-09-17.

[14] Loew, G. A. (1984). "The SLAC Linear Collider and a few ideas on Future Linear Colliders" (PDF). *Proceedings of the 1984 Linear Accelerator Conference.*

[15] Rees, J. R. (1989). "The Stanford Linear Collider" . *Scientific American 1989* **261**: 36–43. See also a colleague's logbook at http://www.symmetrymagazine.org/cms/?pid= 1000294.

[16] Ken Baird, Measurements of A_{LR} and A_{lepton} from SLD http: //hepweb.rl.ac.uk/ichep98/talks_1/talk101.pdf

[17] http://www.slac.stanford.edu/gen/grad/GradHandbook/ slac.html

[18] Linac Coherent Light Source webpage

[19] "SOFT X-RAY MATERIALS SCIENCE (SXR) url=https://portal.slac.stanford.edu/sites/lcls_ public/Instruments/SXR/Pages/Specifications.aspx accessdate=2015-03-22" .

[20] Bostedt, C. et al. (2013). "Ultra-fast and ultra-intense x-ray sciences: First results from the Linac Coherent Light Source free-electron laser" . *Journal of Physics B* **46** (16): 164003. Bibcode:2013JPhB...46p4003B. doi:10.1088/0953-4075/46/16/164003.

[21] Rachel Ehrenberg, ScienceNews.org

[22] "LCLS-II Upgrade to Enable Pioneering Research in Many Fields" . *Cryogenic Society of America*. 8 July 2015. Retrieved 15 August 2015.

[23] FACET: SLAC's new user facility

[24] Stanford's SLAC Paleoparadoxia much thanks to Adele Panofsky, Dr. Panofsky's wife, for her reassembly of the bones of the Paleoparadoxia uncovered at SLAC.

[25] Bergmann, Uwe. "X-Ray Fluorescence Imaging of the Archimedes Palimpsest: A Technical Summary" (PDF). SLAC National Accelerator Laboratory. Retrieved 2009-10-04.

7.12.6 External links

- SLAC official webpage

  - SLAC Today, SLAC's online newspaper, published weekdays

  - *symmetry* magazine, SLAC's monthly particle physics magazine, with Fermilab

Coordinates: 37°24′53″N 122°13′18″W / 37.41472°N 122.22167°W

7.13   SPEAR

This article is about the particle accelerator. For other uses, see spear (disambiguation).

SPEAR (originally **Stanford Positron Electron Asymmetric Rings**, now simply a name) was a collider at the SLAC National Accelerator Laboratory. It began running in 1972, colliding electrons and positrons with an energy of 3 GeV. During the 1970s, experiments at the accelerator played a key role in particle physics research, including the discovery of the J/ψ meson (awarded the 1976 Nobel Prize in physics), many charmonium states, and the discovery of the tau (awarded the 1995 Nobel Prize in physics).

Today, SPEAR is used as a synchrotron radiation source for the Stanford Synchrotron Radiation Lightsource (SSRL). The latest major upgrade of the ring in that finished in 2004 rendered it the current name SPEAR3.

7.13.1   References

- Brief explanation of the acronym in SLACspeak
- 25th Anniversary Info from SLAC
- SPEAR history from CERN Courier

7.13.2   External links

- Official website
- SPEAR3 status

7.14   Sudbury Neutrino Observatory

The **Sudbury Neutrino Observatory** (**SNO**) is a neutrino observatory located 2,100 metres (6,800 ft) underground in Vale Inco's Creighton Mine in Sudbury, Ontario, Canada. The detector was designed to detect solar neutrinos through their interactions with a large tank of heavy water. The detector was turned on in May 1999, and was turned off on 28 November 2006. While new data is no longer being taken, the SNO collaboration will continue to analyze the data taken during that period for the next several years. The underground laboratory has been enlarged and continues to operate other experiments at SNOLAB. The SNO equipment itself is currently being refurbished for use in the SNO+ experiment.

7.14.1   Experimental motivation

The first measurements of the number of solar neutrinos reaching the earth were taken in the 1960s, and all experiments prior to SNO observed a third to a half fewer neutrinos than were predicted by the Standard Solar Model. As several experiments confirmed this deficit the effect became known as the solar neutrino problem. Over several decades many ideas were put forward to try to explain the effect, one of which was the hypothesis of neutrino oscillations. All of the solar neutrino detectors prior to SNO had been sensitive primarily or exclusively to electron neutrinos and yielded little to no information on muon neutrinos and tau neutrinos.

In 1984, Herb Chen of the University of California at Irvine first pointed out the advantages of using heavy water as a detector for solar neutrinos. Unlike previous detectors, using heavy water would make the detector sensitive to two reactions, one sensitive to all neutrino flavours, which would allow a detector to measure neutrino oscillations directly. The Creighton Mine in Sudbury, among the deepest in the world and accordingly low in background radiation, was quickly identified as an ideal place for Chen's proposed experiment to be built.

The SNO collaboration held its first meeting in 1984. At the time it competed with TRIUMF's KAON Factory proposal for federal funding, and the wide variety of universities backing SNO quickly led to it being selected for development. The official go-ahead was given in 1990.

The experiment observed the light produced by relativistic electrons in the water created by neutrino interactions. As relativistic electrons travel through a medium, they lose energy producing a cone of blue light through the Cherenkov effect, and it is this light that is directly detected.

7.14.2   Detector description

The SNO detector target consisted of 1,000 tonnes (1,102 short tons) of heavy water contained in a 6-metre-radius (20 ft) acrylic vessel. The detector cavity outside the vessel was filled with normal water to provide both buoyancy for the vessel and radiation shielding. The heavy water was viewed by approximately 9,600 photomultiplier tubes (PMTs) mounted on a geodesic sphere at a radius of about 850 centimetres (335 in). The cavity housing the detector is the largest man-made underground cavity in the world, requiring a variety of high-performance rock bolting techniques to prevent rock bursts.

The observatory is located at the end of a 1.5-kilometre-long (0.9 mi) drift, named the "SNO drift", isolating it from other mining operations. Along the drift are a number of operations and equipment rooms, all held in a clean room setting. Most of the facility is Class 3000 (fewer than 3,000 particles of 1 μm or larger per 1 m^3 of air) but the final cavity containing the detector is Class 1000.[*][1]

Charged current interaction

In the charged current interaction, a neutrino converts the neutron in a deuteron to a proton. The neutrino is absorbed in the reaction and an electron is produced. Solar neutrinos have energies smaller than the mass of muons and tau leptons, so only electron neutrinos can participate in this reaction. The emitted electron carries off most of the neutrino's energy, on the order of 5–15 MeV, and is detectable. The proton which is produced does not have enough energy to be detected easily. The electrons produced in this reaction are emitted in all directions, but there is a slight tendency for them to point back in the direction from which the neutrino came.

Neutral current interaction

In the neutral current interaction, a neutrino dissociates the deuteron, breaking it into its constituent neutron and proton. The neutrino continues on with slightly less energy, and all three neutrino flavours are equally likely to participate in this interaction. Heavy water has a small cross section for neutrons, and when neutrons capture on a deuterium nucleus a gamma ray (photon) with roughly 6 MeV of energy is produced. The direction of the gamma ray is completely uncorrelated with the direction of the neutrino. Some of the neutrons wander past the acrylic vessel into the light water, and since light water has a very large cross section for neutron capture these neutrons are captured very quickly. A gamma ray with roughly 2 MeV of energy is produced in this reaction, but because this is below the detector's energy threshold they are not observable. The gamma ray collides with an electron through Compton scattering and the accelerated electron can be detected through Cerenkov radiation.

Electron elastic scattering

In the elastic scattering interaction, a neutrino collides with an atomic electron and imparts some of its energy to the electron. All three neutrinos can participate in this interaction through the exchange of the neutral Z boson, and electron neutrinos can also participate with the exchange of a charged W boson. For this reason this interaction is dominated by electron neutrinos, and this is the channel through which the Super-Kamiokande (Super-K) detector can observe solar neutrinos. This interaction is the relativistic equivalent of billiards, and for this reason the electrons produced usually point in the direction that the neutrino was travelling (away from the sun). Because this interaction takes place on atomic electrons it occurs with the same rate in both the heavy and light water.

7.14.3 Experimental results and impact

On 18 June 2001, the first scientific results of SNO were published,[*][2][*][3] bringing the first clear evidence that neutrinos oscillate (i.e. that they can transmute into one another), as they travel in the sun. This oscillation in turn implies that neutrinos have non-zero masses. The total flux of all neutrino flavours measured by SNO agrees well with the theoretical prediction. Further measurements carried out by SNO have since confirmed and improved the precision of the original result.

Although Super-K had beaten SNO to the punch, having published evidence for neutrino oscillation as early as 1998, the Super-K results were not conclusive and did not specifically deal with solar neutrinos. SNO's results were the first to directly demonstrate oscillations in solar neutrinos. This was important to the standard solar model. The results of the experiment had a major impact on the field, as evidenced by the fact that two of the SNO papers have been cited over 1,500 times, and two others have been cited over 750 times.[*][4] In 2007, the Franklin Institute awarded the director of SNO Art McDonald with the Benjamin Franklin Medal in Physics.[*][5]

7.14.4 Other possible analyses

The SNO detector would have been capable of detecting a supernova within our galaxy if one had occurred while the detector was online. As neutrinos emitted by a supernova are released earlier than the photons, it is possible to alert the astronomical community before the supernova is visible. SNO was a founding member of the Supernova Early Warning System (SNEWS) with Super-Kamiokande and the Large Volume Detector. No such supernovas have yet been detected.

The SNO experiment was also able to observe atmospheric neutrinos produced by cosmic ray interactions in the atmosphere. Due to the limited size of the SNO detector in comparison with Super-K the low cosmic ray neutrino signal is not statistically significant at neutrino energies below 1 GeV.

7.14.5 Participating institutions

Large particle physics experiments require large collaborations. With approximately 100 collaborators, SNO was a rather small group compared to collider experiments. The participating institutions have included:

Canada

- Carleton University

- Laurentian University

- Queen's University – designed and built many calibration sources and the device for deploying sources

- TRIUMF

- University of British Columbia

- University of Guelph

Although no longer a collaborating institution, Chalk River Laboratories led the construction of the acrylic vessel that holds the heavy water, and Atomic Energy of Canada Limited was the source of the heavy water.

United Kingdom

- University of Oxford – developed much of the experiment's Monte Carlo analysis program (SNOMAN), and maintained the program

United States

- LBNL – Led the construction of the geodesic structure that holds the PMTs

- PNNL

- LANL

- University of Pennsylvania – designed and built the front end electronics and trigger

- University of Washington – designed and built proportional counter tubes for detection of neutrons in the third phase of the experiment

- Brookhaven National Laboratory

- University of Texas at Austin

- Massachusetts Institute of Technology

7.14.6 Honours and awards

- Asteroid 14724 SNO is named in honour of SNO.

- In November 2006, the entire SNO team was awarded the inaugural John C. Polanyi Award for "a recent outstanding advance in any field of the natural sciences or engineering" conducted in Canada.*[6]

7.14.7 See also

- SNOLAB – A permanent underground physics laboratory being built around SNO

- SNO+ – The successor of SNO

- Homestake experiment - predecessor experiment conducted 1970-1994 in a mine at Lead, South Dakota

7.14.8 References

[1] "The Sudbury Neutrino Observatory – Canada's eye on the universe". *CERN Courier*. CERN. 4 December 2001. Retrieved 2008-06-04.

[2] Ahmad, QR et al. (2001). "Measurement of the Rate of $v_e + d \rightarrow p + p + e^-$ Interactions Produced by ^{8}B Solar Neutrinos at the Sudbury Neutrino Observatory". *Physical Review Letters* **87** (7): 071301. arXiv:nucl-ex/0106015. Bibcode:2001PhRvL..87g1301A. doi:10.1103/PhysRevLett.87.071301.

[3] "Sudbury Neutrino Observatory First Scientific Results". 3 July 2001. Retrieved 2008-06-04.

[4] "SPIRES HEP Results". *SPIRES*. SLAC. Retrieved 2009-10-06.

[5] "Arthur B. McDonald, Ph.D.". *Franklin Laureate Database*. Franklin Institute. Retrieved 2008-06-04.

[6] "Past Winners – The Sudbury Neutrino Observatory". NSERC. 3 March 2008. Retrieved 2008-06-04.

7.14.9 External links

- SNO's official site

- Joshua Klein's *Introduction to SNO, Solar Neutrinos, and Penn at SNO*

- "Experiment Cave". *WIRED Science*. Episode 104. 2007-10-24. PBS.

- Written and Directed by David Sington (2006-02-21). "The Ghost Particle". *Nova*. Season 34. Episode 3306 (607). PBS.

Coordinates: 46°28′00″N 81°10′22″W / 46.46667°N 81.17278°W

7.15 Super-Kamiokande

"Super-K" redirects here. For other uses, see Super K (disambiguation).

Coordinates: 36°25′32.6″N 137°18′37.1″E / 36.425722°N 137.310306°E[*][1]

Super-Kamiokande (full name: **Super-Kamioka Neutrino Detection Experiment**, abbreviated to **Super-K** or **SK**) is a neutrino observatory located under Mount Kamioka near the city of Hida, Gifu Prefecture, Japan. The observatory was designed to search for proton decay, study solar and atmospheric neutrinos, and keep watch for supernovae in the Milky Way Galaxy.

7.15.1 Description

The Super-K is located 1,000 m (3,300 ft) underground in the Mozumi Mine in Hida's Kamioka area. It consists of a cylindrical stainless steel tank that is 41.4 m (136 ft) tall and 39.3 m (129 ft) in diameter holding 50,000 tons of ultra-pure water. The tank volume is divided by a stainless steel superstructure into an inner detector (ID) region that is 33.8 m (111 ft) in diameter and 36.2 m (119 ft) in height and outer detector (OD) which consists of the remaining tank volume. Mounted on the superstructure are 11,146 photomultiplier tubes (PMT) 50 cm (20 in) in diameter that face the ID and 1,885 20 cm (8 in) PMTs that face the OD. There is a Tyvek and blacksheet barrier attached to the superstructure that optically separates the ID and OD.

A neutrino interaction with the electrons or nuclei of water can produce a charged particle that moves faster than the speed of light in water (not to be confused with exceeding the speed of light in a vacuum). This creates a cone of light known as Cherenkov radiation, which is the optical equivalent to a sonic boom. The Cherenkov light is projected as a ring on the wall of the detector and recorded by the PMTs. Using the timing and charge information recorded by each PMT, the interaction vertex, ring direction and flavor of the incoming neutrino is determined. From the sharpness of the edge of the ring the type of particle can be inferred. The multiple scattering of electrons is large, so electromagnetic showers produce fuzzy rings. Highly relativistic muons, in contrast, travel almost straight through the detector and produce rings with sharp edges.

7.15.2 Detector

The Super-Kamiokande (SK) is a Cherenkov detector used to study neutrinos from different sources including the Sun, supernovae, the atmosphere, and accelerators for proton de-

cay. The experiment began in April 1996 and was shut down for maintenance in July 2001, a period known as "SK-I". Since an accident occurred during maintenance, the experiment resumed in October 2002 with only half of its original number of ID-PMTs. In order to prevent further accidents, all of the ID-PMTs were covered by fiber-reinforced plastic (FRP) with acrylic front windows. This phase from October 2002 to another closure for an entire reconstruction in October 2005 is called "SK-II". In July 2006, the experiment resumed with the full number of PMTs and stopped in September 2008 for electronics upgrades. This period was known as "SK3-III". The period after 2008 is known as "SK-IV". The phases and their main characteristics are summarised in table 1.[*][2]

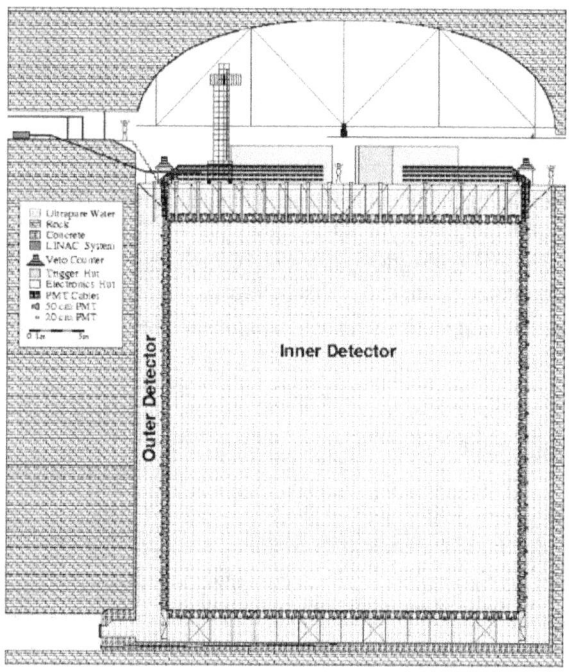

A cross-section of the Super-Kamiokande detector.

SK-IV upgrade

In the previous phases, the ID-PMTs processed signals by custom electronics modules called analog timing modules (ATMs). Charge-to-analog converters (QAC) and time-to-analog converters (TAC) are contained in these modules that had dynamic range from 0 to 450 pico Coulomb (pC) with 0.2 pC resolution for charge and from −300 to 1000 ns with 0.4 ns resolution for time. There were two pairs of QAC/TAC for each PMT input signal, this prevented dead time and allowed the readout of multiple sequential hits that may arise, e.g. from electrons that are decay products of stopping muons.[*][3]

The SK system was upgraded in September 2008 in order

to maintain the stability in the next decade and improve the throughput of the data acquisition systems, QTC-based electronics with Ethernet (QBEE).[*][4] The QBEE provides high-speed signal processing by combining pipelined components. These components are a newly developed custom charge-to-time converter (QTC) in the form of an application-specific integrated circuit (ASIC), a multi-hit time-to-digital converter (TDC), and field-programmable gate array (FPGA).[*][5] Each QTC input has three gain ranges – "Small", "Medium" and "Large" – the resolutions for each are shown in Table.[*][6]

For each range, analog to digital conversion is conducted separately, but the only range used is that with the highest resolution that is not being saturated. The overall charge dynamic range of the QTC is 0.2–2500 pC, five times larger than the old . The charge and timing resolution of the QBEE at the single photoelectron level is 0.1 photoelectrons and 0.3 ns respectively, both are better than the intrinsic resolution of the 20-in. PMTs used in SK. The QBEE achieves good charge linearity over a wide dynamic range. The integrated charge linearity of the electronics is better than 1%. The thresholds of the discriminators in the QTC are set to −0.69 mV (equivalent to 0.25 photoelectron, which is the same as for SK-III). This threshold was chosen to replicate the behavior of the detector during its previous ATM-based phases.[*][7]

Water tank

The outer shell of the water tank is a cylindrical stainless-steel tank with 39 m in diameter and 42 m in height. The tank is self-supporting, with concrete backfilled against the rough-hewn stone walls to counteract water pressure when the tank is filled. The capacity of the tank exceeds 50 ktons of water.[*][8]

PMTs and associate structure

The basic unit for the ID PMTs is a "supermodule", a frame which supports a 3×4 array of PMTs. Supermodule frames are 2.1 m in height, 2.8 m in width and 0.55 m in thickness. These frames are connected to each other in both the vertical and horizontal directions. Then the whole support structure is connected to the bottom of the tank and to the top structure. In addition to serving as rigid structural elements, supermodules simplified the initial assembly of the ID. Each supermodule was assembled on the tank floor and then hoisted into its final position. Thus the ID is in effect tiled with supermodules. During installation, ID PMTs were pre-assembled in units of three for easy installation. Each supermodule has two OD PMTs attached on its back side. The support structure for the bottom PMTs is

attached to the bottom of the stainless-steel tank by one vertical beam per supermodule frame. The support structure for the top of the tank is also used as the support structure for the top PMTs.

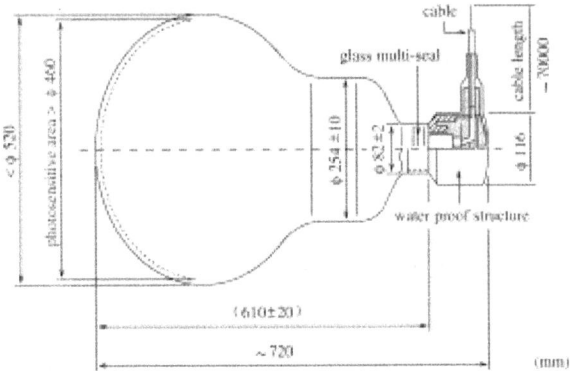

Schematic view of a 50 cm PMT.

Cables from each group of 3 PMTs are bundled together. All cables run up the outer surface of the PMT support structure, i.e., on the OD PMT plane, pass through cable ports at the top of the tank, and are then routed into the electronics huts.

The thickness of the OD varies slightly, but is on average about 2.6 m on top and bottom, and 2.7 m on the barrel wall, giving the OD a total mass of 18 ktons. OD PMTs were distributed with 302 on the top layer, 308 on the bottom, and 1275 on the barrel wall.

To protect against low energy background from radon decay products in the air, the roof of the cavity and the access tunnels were sealed with a coating called Mineguard® produced by Urylon in Canada. Mineguard® is a spray-applied polyurethane membrane developed for use as a rock support system and radon gas barrier in the mining industry.[*][9]

The average geomagnetic field is about 450 mG and is inclined by about 45° with respect to the horizon at the detector site. This presents a problem for the large and very sensitive PMTs which prefer a much lower ambient field. The strength and uniform direction of the geomagnetic field could systematically bias photoelectron trajectories and timing in the PMTs. To counteract this 26 sets of horizontal and vertical Helmholtz coils are arranged around the inner surfaces of the tank. With these in operation the average field in the detector is reduced to about 50 mG. The magnetic field at various PMT locations were measured before the tank was filled with water.[*][10]

A standard fiducial volume of approximately 22.5 ktons is defined as the region inside a surface drawn 2.00 m from the ID wall to minimize the anomalous response causing by natural radioactivity in the surrounding rock.

7.15.3 Monitoring system

Online monitoring system

An online monitor computer located in the control room reads data from the DAQ host computer via an FDDI link. It provides shift operators with a flexible tool for selecting event display features, makes online and recent-history histograms to monitor detector performance, and performs a variety of additional tasks needed to efficiently monitor status and diagnose detector and DAQ problems. Events in the data stream can be skimmed off and elementary analysis tools can be applied to check data quality during calibrations or after changes in hardware or online software.[11]

Realtime supernova monitor

To detect and identify such bursts as efficiently and promptly as possible Super-Kamiokande is equipped with an online supernova monitor system. About 10,000 total events are expected in Super-Kamiokande for a supernova explosion at the center of our Galaxy. Super-Kamiokande can measure a burst with no dead-time, up to 30,000 events within the first second of a burst. Theoretical calculations of supernova explosions suggest that neutrinos are emitted over a total time-scale of tens of seconds with about a half of them emitted during the first one or two seconds. The Super-K will search for event clusters in specified time windows of 0.5, 2 and 10 s.[12] Data are transmitted to real-time SN-watch analysis process every 2 min and analysis is completed typically in 1 min. When supernova (SN) event candidates are found, R_{mean} is calculated if the event multiplicity is larger than 16, where Rmean is defined as the average spatial distance between events, i.e.

$$R_{mean} = \frac{\sum_{i=1}^{N_{multi}-1} \sum_{j=i+1}^{N_{multi}} |r_i - r_j|}{N_{multi} C_2}$$

Neutrinos from supernovae interact with free protons, producing positrons which are distributed so uniformly in the detector that R_{mean} for SN events should be significantly larger than for ordinary spatial clusters of events. In the Super-Kamiokande detector, Rmean for uniformly distributed Monte Carlo events shows that no tail exists below R_{mean} [?]1000 cm. For the "alarm" class of burst, the events are required to have R_{mean} [?]900 cm for 25[?] N_{multi} [?]40 or R_{mean} [?]750 cm for N_{multi} >40. These thresholds were determined by extrapolation from SN1987A data.[13][14] The system will run special processes to check for spallation muons when burst candidates meeting "alarm" criteria and make a primarily decision for further process. If the burst candidate passes these checks, the data will be reanalyzed using an offline process and a final decision will be made within a few hours. During the Super-Kamiokande I running, this never occurred. One of the important capabilities for [Super-Kamiokande] is to reconstruct the direction to supernova. By neutrino–electron scattering, $v_x + e^- \rightarrow v_x + e^-$, a total of 100–150 events are expected in case of a supernova at the center of our Galaxy.[15] The direction to supernova can be measured with angular resolution

$$\delta\theta \sim \frac{30°}{\sqrt{N}}$$

where N is the number of events produced by the ν–e scattering. The angular resolution, therefore, can be as good as $\delta\theta \sim 3°$ for a supernova at the center of our Galaxy.[16] In this case, not only time profile and the energy spectrum of a neutrino burst, but also the information on direction of supernova can be provided.

Slow control monitor and offline process monitor

There is a process called the "slow control" monitor, as part of the online monitoring system, watches the status of the HV systems, the temperatures of electronics crates and the status of the compensating coils used to cancel the geomagnetic field. When any deviation from norms is detected, it will alert physicists to prompt to investigate, take appropriate action, or notify experts.[17]

To monitor and control the offline processes that analyze and transfer data, a set of software was sophisticatedly developed. This monitor allows non-expert shift physicists to identify and repair common problems to minimize down time, and the software package was a significant contribution to the smooth operation of the experiment and its overall high lifetime efficiency for data taking.[18]

7.15.4 Research

Solar neutrino

Also see: Neutrino Oscillation.

The energy of Sun becomes from the nuclear fusion in its core where a helium atom and an electron neutrino are generated by 4 protons. These neutrinos emitted from this reaction are called solar neutrinos. Photons, created by the nuclear fusion in the center of the Sun, take millions of years to reach the surface; on the other hand, solar neutrinos arrive at the earth in eight minutes due to their lack of interactions with matter. Hence, solar neutrinos make it possible for us to observe the inner Sun in "real-time" that takes millions of years for visible light.[19]

In 1999, the Super-Kamiokande detected strong evidence of neutrino oscillation that successful explain to the solar neutrino problem. The Sun and about 80% of the visible stars produce their energy by the conversion of hydrogen to helium via

$$2e^- + 4p \rightarrow {}^2He + 2v_e + 26.73 MeV$$

Consequently stars are source of neutrinos including our Sun. These neutrinos primarily come through the pp-chain in lower masses, and for cooler stars, primarily through CNO-chains of heavier masses.

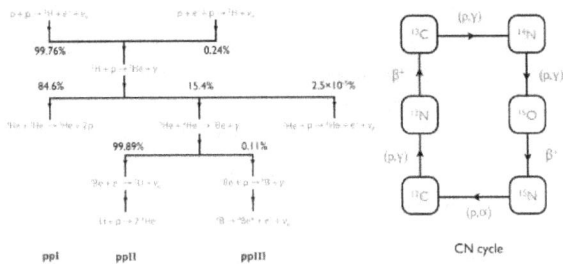

The left frame shows the three principal cycles comprising the pp chain (ppI, ppII, and ppIII), the neutrinos sources associated with these cycles. The right frame shows the CN I cycle.

In the early 1990s, particularly with the uncertainties that accompanied the initial results from Kamioka II and the Ga experiments, no individual experiment required a non-astrophysical solution of the solar neutrino problem. But in aggregate, the Cl, Kamioka II, and Ga experiments indicated a pattern of neutrino fluxes that was not compatible with any adjustment of the SSM. This in turn helped motivate a new generation of spectacularly capable active detectors. These experiments are Super-Kamiokande, the Sudbury Neutrino Observatory (SNO), and Borexino. Super-Kamiokande was able to detect elastic scattering (ES) events

$$v_x + e^- \rightarrow v_x + e^-$$

which, due to the charged-current contribution to v_e scattering, has a relative sensitivity to v_e s and heavy-flavor neutrinos of ~7:1.[20] Since the direction of the recoil electron is constrained to be very forward, the direction of the neutrinos are kept in the direction of recoil electrons. Here, $cos\theta_{Sun}$ is provided where θ_{Sun} is the angle between the direction of recoil electrons and the Sun's position. This shows that the ${}^8B$ solar neutrino flux can be calculated to be $2.40 \pm 0.03(stat.)^{+0.08}_{-0.07}(sys.) \times 10^6 cm^{-2}s^{-1}$. Comparing to the SSM, the ratio is $\frac{Data}{SSM_{BP98}} = 0.465 \pm 0.005(stat.)^{+0.015}_{-0.013}(sys.)$.[21] The result clearly indicates the deficit of solar neutrinos.

Atmospheric neutrino

Atmospheric neutrinos are secondary cosmic rays produced by the decay of particles resulting from interactions of primary cosmic rays (mostly proton) with Earth atmosphere. We classified the observed atmospheric neutrino data into four types. Fully contained (FC) events have all their tracks

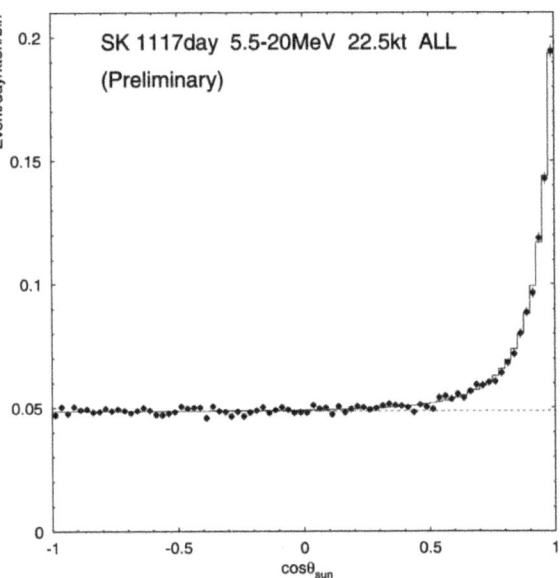

$cos\theta_{Sun}$ distribution above 5.5 MeV. The solid line indicates the best fit considering the flux as a free parameter.

in the inner detector while partially contained (PC) events have escaping tracks from the inner detector. Upward through-going muons (UTM) are produced in the rock beneath the detector and go through the inner detector. Upward stopping muons (USM) are also produced in the rock beneath the detector but stop in the inner detector.

The number of observed number of neutrinos is predicted uniformly regardless the zenith angle. However, Super-Kamiokande found that the number of upward going muon neutrinos (generated on the other side of the Earth) is half of the number of downward going muon neutrinos in 1998. This can be explained that neutrinos changes or oscillated into some other neutrinos that are not detected. This is called neutrino oscillation and this discovery indicates the finite mass of neutrinos and suggests to extend the Standard Model. Neutrinos oscillate in three flavors and all neutrinos have their rest mass. Later analysis in 2004 suggested a sinsinusoidal dependence of the event rate as a function of "Length/Energy" , which confirmed the neutrino oscillations.[22]

K2K Experiment

Main article: K2K experiment

The K2K experiment was a neutrino experiment from June 1999 to November 2004. This experiment was designed to verify oscillations observed by Super-Kamiokande through muon neutrinos. It gives first positive measurement of neutrino oscillations in conditions that both source and detector are under control. The Super-Kamiokande detector plays an important role in the experiment as the far detector. Later experiment T2K experiment continued as the second generation follow up to the K2K experiment.

T2K Experiment

Main article: T2K experiment

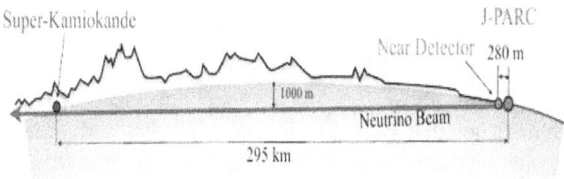

Passage of the muon neutrino beam from J-PARC to Super K

T2K (Tokai to Kamioka) experiment is a neutrino experiment collaborated by several countries including Japan, United States and others. The goal of T2K is to gain deeper understanding of parameters of neutrino oscillation. T2K has made a search for oscillations from muon neutrinos to electron neutrinos, and announced the first experimental indications for them in June 2011.[23] The Super-Kamiokande detector plays as the "far detector". The Super-K detector will record the Cerenkov radiation of muons and electrons created by interactions between high energy neutrinos and water.

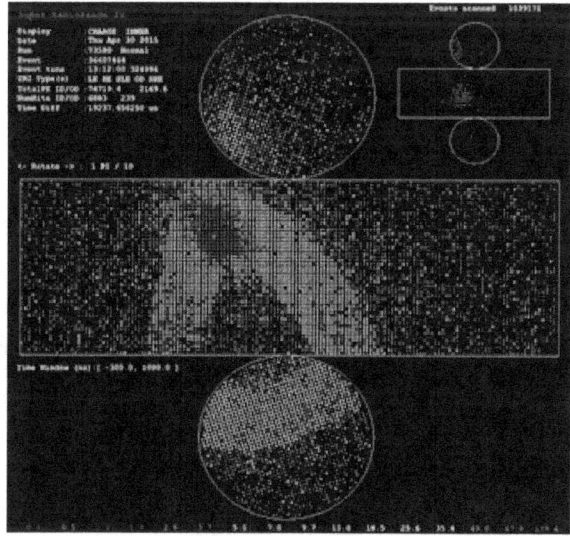

An almost-live event display from Super-K

Proton Decay

Proton is assumed to be absolutely stable in Standard Model. However, the Grand Unified Theories (GUTs) predict that can decay into lighter energetic charged particles such as electrons, muons, pions or others which can be observed. Kamiokande helps to rule out some of the theories. Super-Kamiokande is currently the largest detector for observation of proton decay.

7.15.5 Purification

Water purification system

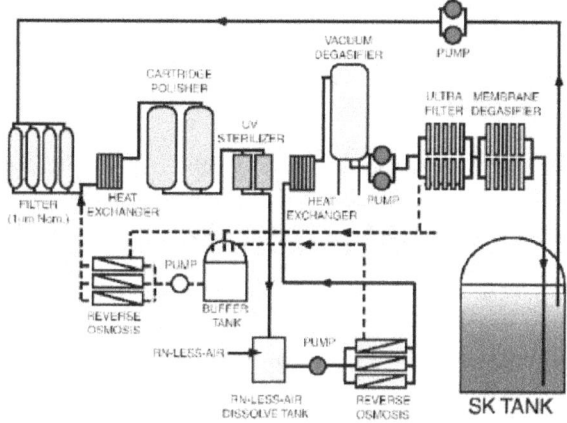

A schematic view of the water purification system.

The 50 tons pure water is continually reprocessed at rate about 30 tons/h in a close system since early 2002. Now, raw mine water is recycled through the first step (particle filters and RO) for some time before other processes, which involve expensive expendables, are imposed. Initially, water from the Super-Kamiokande tank is passed through nominal 1 μm mesh filters to remove dust and particles, which reduce the transparency of the water for Cherenkov photons, and provide a possible radon source inside the Super-Kamiokande detector. Heat exchanger is used to cool down the water in order to reduce the PMT dark noise level as well as suppress the growth of bacteria. Surviving bacteria are killed by UV sterilizer stage. A cartridge polisher (CP) eliminates heavy ions which also reduce water transparency and include radioactive species. The CP module increases the typical resistivity of recirculating water from 11MΩ cm to 18.24 MΩ cm, approaching chemical limit.[24] Originally, ion-exchanger (IE) was included in system, but it was removed when IE resin was found to be a significant radon source. The RO step that removes additional particulates, and the introduction of Rn-reduced air into the water that increases radon removal efficiency in the vacuum degasifier (VD) stage which follows were installed in 1999. After that, a VD removes dissolved gases in the water. These gases are dissolved in water with a serious background of events source for solar neutrinos in the MeV energy range and the dissolved oxygen encourages the growth of bacteria. The removing efficiency of removing is about 96%. Then, the ultra filter (UF) is introduced to remove particles whose minimum size corresponds to molecular weight approximately 10,000 (or about 10 nm diameter) thanks to hollow fiber membrane filters. Finally, a membrane degasifier (MD) removes radon dissolved in water, and the measured removal efficiency for radon is about 83%. The concentration of radon gases is miniaturized by realtime detecters. In June, 2001 typical radon concentrations in water coming into the purification system from the Super-Kamiokande tank were <2 mBq m*−3, and in water output by the system, 0.4±0.2 mBq m*−3.*[25]

Air purification system

Purified Air is supplied in the gap between the water surface and the top of the Super-Kamiokande tank. The air purification system contains three compressors, a buffer tank, dryers, filters, and activated charcoal filters. A total of 8 m^3 of activated charcoal is used. The last 50 L of charcoal is cooled to −40 °C to increase removal efficiency for radon. Typical flow rates, dew point, and residual radon concentration are 18 m^3/h, −65 °C (@+1 kg/cm^2), and a few mBq m*−3, respectively. Typical radon concentration in the dome air is measured to be 40 Bq m*−3. Radon levels in the mine tunnel air, near the tank cavity dome, typically reach 2000–3000 Bq m*−3 during the warm season, from May until October, while from November to April the radon level is approximately 100–300 Bq m*−3. This variation is due to the chimney effect in the ventilation pattern of the mine tunnel system; in cold seasons, fresh air flows into the Atotsu tunnel entrance that is a relatively short path through exposed rock before reaching the experimental area, while in the summer, air flows out the tunnel, drawing radon-rich air from deep within the mine past the experimental area.*[26]

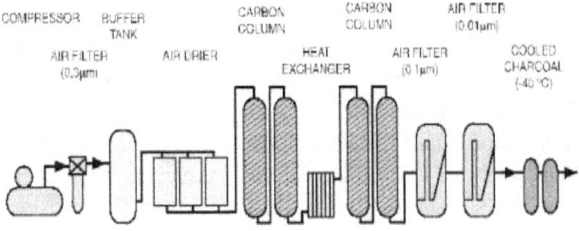

A schematic view of the air purification system.

In order to keep radon levels in the dome area and water purification system below 100 Bq m*−3, fresh air is continually pumped at approximately 10 m^3/min from outside the mine which generates a slight over-pressure in the Super-Kamiokande experimental area to minimize the entry of ambient mine air. A "Radon Hut" (Rn Hut) was constructed near the Atotsu tunnel entrance to house equipment for the dome air system: a 40 hp air pump with 10 m^3 min*−1 /15 PSI pump capacity, air dehumidifier, carbon filter tanks, and control electronics. In autumn 1997, an extended intake air pipe was installed at a location approximately 25 m above the Atotsu tunnel entrance. This low level satisfies that goals of air quality so that carbon filter regeneration operations would no longer be required.*[27]

7.15.6   Data processing

Offline data processing is produced both in Kamioka and United States.

In Kamioka

The offline data processing system is located in Kenkyuto and is connected to Super-Kamiokande detector with 4 km FDDI optical fiber link. On average, date flow from online system is 450 kbytes s*−1 on average, corresponding to 40 Gbytes day*−1 or 14 Tbytes yr*−1. Magnetic tapes are used in offline system to store data and most of the analysis is accomplished here. The offline processing system is designed platform-independent because different computer architectures are used for data analysis. Because of this, the data structures are based on ZEBRA bank system developed in CERN as well as the ZEBRA exchange system.*[28]

Event data from Super-Kamiokande online DAQ system basically contains a list of number of hit PMT, TDC and ADC counts, GPS time-stamps and other housekeeping data. For solar neutrino analysis, lowering the energy threshold is a constant goal, so it is a continual effort to improve the efficiency of reduction algorithms; however, changes in calibrations or reduction methods require reprocessing of earlier data. Typically, 10 Tbytes of raw data is processed every month so that a large amount of CPU power and high-speed I/O access to the raw data. In addition, extensive Monte Carlo simulation processing is also necessary.*[29]

Offline system was designed to meet demand of all these: tape storage of a large database (14 Tbytes yr−1), stable semi-realtime processing, nearly continuous re-processing and Monte Carlo simulation. The computer system consists of 3 major sub-systems: the data server, the CPU farm and the network at the end of Run I.*[30]

In US

A system dedicated to offsite offline data processing was set up at the Stony Brook University in Stony Brook, NY to process raw data sent from Kamioka. Most of the reformatted raw data is copied from system facility in Kamioka. At Stony Brook, a system was set up for analysis and further processing. At Stony Brook the raw data were processed with a multi-tape DLT drive. The first stage data reduction processes were done for the high energy analysis and for the low energy analysis. The data reduction for the high energy analysis was mainly for atmospheric neutrino events and proton decay search while the low energy analysis was mainly for the solar neutrino events. The reduced data for the high energy analysis was further filtered by other reduction processes and the resulting data were stored on disks. The reduced data for the low energy were stored on DLT tapes and sent to University of California, Irvine for further processing.

This offset analysis system continued for 3 years until their analysis chains were proved to produce equivalent results. Thus, in order to limited manpower, collaborations were concentrated to single combined analysis*[31]

7.15.7 History

Construction of the predecessor of the present Kamioka Observatory, the Institute for Cosmic Ray Research, University of Tokyo began in 1982 and was completed in April, 1983. The purpose of the observatory was to detect whether proton decay exists, one of the most fundamental questions of elementary particle physics.*[32]*[33]*[34]*[35]*[36]

The detector, named KamiokaNDE for Kamioka Nucleon Decay Experiment, was a tank 16.0 m (52 ft) in height and 15.6 m (51.2 ft) in width, containing 3,048 metric tons (3,000 tons) of pure water and about 1,000 photomultiplier tubes (PMTs) attached to its inner surface. The detector was upgraded, starting in 1985, to allow it to observe solar neutrinos. As a result, the detector (KamiokaNDE-II) had become sensitive enough to detect neutrinos from SN 1987A, a supernova which was observed in the Large Magellanic Cloud in February 1987, and to observe solar neutrinos in 1988. The ability of the Kamiokande experiment to observe the direction of electrons produced in solar neutrino interactions allowed experimenters to directly demonstrate for the first time that the sun was a source of neutrinos.

The Super-Kamiokande project was approved by the Japanese Ministry of Education, Science, Sports and Culture in 1991 for total funding of approximately $100 M. The US portion of the proposal, which was primarily to build the OD system, was approved by the US Department of Energy in 1993 for $3 M. In addition the US has also contributed about 2000 20 cm PMTs recycled from the IMB experiment.*[37]

Despite successes in neutrino astronomy and neutrino astrophysics, Kamiokande did not achieve its primary goal, the detection of proton decay. Higher sensitivity was also necessary to obtain high statistical confidence in its results. This led to the construction of Super-Kamiokande, with fifteen times the water and ten times as many PMTs as Kamiokande. Super-Kamiokande started operation in 1996.

The Super-Kamiokande Collaboration announced the first evidence of neutrino oscillation in 1998.*[38] This was the first experimental observation supporting the theory that the neutrino has non-zero mass, a possibility that theorists had speculated about for years.

On November 12, 2001, about 6,600 of the photomultiplier tubes (costing about $3000 each*[39]) in the Super-Kamiokande detector imploded, apparently in a chain reaction or cascading failure, as the shock wave from the concussion of each imploding tube cracked its neighbours. The detector was partially restored by redistributing the photomultiplier tubes which did not implode, and by adding protective acrylic shells that are hoped will prevent another chain reaction from recurring (Super-Kamiokande-II).

In July 2005, preparations began to restore the detector to its original form by reinstalling about 6,000 PMTs. The work was completed in June 2006, whereupon the detector was renamed Super-Kamiokande-III. This phase of the experiment collected data from October 2006 till August 2008. At that time, significant upgrades were made to the electronics. After the upgrade, the new phase of the experiment has been referred to as Super-Kamiokande-IV. SK-IV continues to run*, collecting data on various natural sources of neutrinos, as well as acting as the far detector for the Tokai-to-Kamioka (T2K) long baseline neutrino oscillation experiment.

7.15.8 Results

On February 23, 1987, a supernova explosion occurred in the Large Magellanic Cloud. From this explosion, the supernova neutrinos were detected for the first time. Kamiokande detected 11 events of these neutrinos. This observation confirmed that the theory of supernova explosion was correct and was the dawn of a new era in neutrino astronomy.*[40]

In 1998, Super-K found first strong evidence of neutrino oscillation from the observation of muon neutrinos changed into tau-neutrinos.[*][41]

SK has set limits on proton lifetime and other rare decays and neutrino properties. SK set a lower bound on protons decaying to kaons of 5.9×10^{33} yr[*][42]

7.15.9 In popular culture

Super-Kamiokande is the subject of German photographer Andreas Gursky's 2007 image, *Kamiokande*.[*][43] The detector was a topic in the television series *Cosmos: A Space-time Odyssey*.

7.15.10 See also

- Masatoshi Koshiba

- Yoji Totsuka

- Supernova 1987A

- Solar neutrino problem

- Sudbury Neutrino Observatory

- K2K experiment

- T2K experiment

7.15.11 References

[1] S. Fukuda et al. (April 2003), "The Super-Kamiokande detector", *Nuclear Instruments and Methods in Physics Research* **A501** (2–3): 418–462, Bibcode:2003NIMPA.501..418F, doi:10.1016/S0168-9002(03)00425-X

[2] K. Abe et al. (11 February 2014), "Calibration of the Super-Kamiokande detector", *Nuclear Instruments and Methods in Physics Research Section A: Accelerators, Spectrometers, Detectors and Associated Equipment* **737**: 253–272, arXiv:1307.0162, Bibcode:2014NIMPA.737..253A, doi:10.1016/j.nima.2013.11.081

[3] K. Abe et al. (11 February 2014), "Calibration of the Super-Kamiokande detector", *Nuclear Instruments and Methods in Physics Research Section A: Accelerators, Spectrometers, Detectors and Associated Equipment* **737**: 253–272, arXiv:1307.0162, Bibcode:2014NIMPA.737..253A, doi:10.1016/j.nima.2013.11.081

[4] S. Yamada; "et al," (2009), *IEEE Transactions on Nuclear Science*, NS-57: 248 Missing or empty |title= (help)

[5] H. Nishino et al. (2009), "High-speed charge-to-time converter ASIC for the Super-Kamiokande detector", *Nuclear Instruments and Methods A* **610**: 710, arXiv:0911.0986, Bibcode:2009NIMPA.610..710N, doi:10.1016/j.nima.2009.09.026

[6] K. Abe et al. (11 February 2014), "Calibration of the Super-Kamiokande detector", *Nuclear Instruments and Methods in Physics Research Section A: Accelerators, Spectrometers, Detectors and Associated Equipment* **737**: 253–272, arXiv:1307.0162, Bibcode:2014NIMPA.737..253A, doi:10.1016/j.nima.2013.11.081

[7] K. Abe et al. (11 February 2014), "Calibration of the Super-Kamiokande detector", *Nuclear Instruments and Methods in Physics Research Section A: Accelerators, Spectrometers, Detectors and Associated Equipment* **737**: 253–272, arXiv:1307.0162, Bibcode:2014NIMPA.737..253A, doi:10.1016/j.nima.2013.11.081

[8] S. Fukuda et al. (1 April 2003), "The Super-Kamiokande detector", *Nuclear Instruments and Methods A* **51**: 418–462, Bibcode:2003NIMPA.501..418F, doi:10.1016/S0168-9002(03)00425-X

[9] S. Fukuda et al. (1 April 2003), "The Super-Kamiokande detector", *Nuclear Instruments and Methods A* **51**: 418–462, Bibcode:2003NIMPA.501..418F, doi:10.1016/S0168-9002(03)00425-X

[10] S. Fukuda et al. (1 April 2003), "The Super-Kamiokande detector", *Nuclear Instruments and Methods A* **51**: 418–462, Bibcode:2003NIMPA.501..418F, doi:10.1016/S0168-9002(03)00425-X

[11] S. Fukuda et al. (1 April 2003), "The Super-Kamiokande detector", *Nuclear Instruments and Methods A* **51**: 418–462, Bibcode:2003NIMPA.501..418F, doi:10.1016/S0168-9002(03)00425-X

[12] S. Fukuda et al. (1 April 2003), "The Super-Kamiokande detector", *Nuclear Instruments and Methods A* **51**: 418–462, Bibcode:2003NIMPA.501..418F, doi:10.1016/S0168-9002(03)00425-X

[13] S. Fukuda et al. (1 April 2003), "The Super-Kamiokande detector", *Nuclear Instruments and Methods A* **51**: 418–462, Bibcode:2003NIMPA.501..418F, doi:10.1016/S0168-9002(03)00425-X

[14] Hirata, K et al. (6 April 1987), "Observation of a neutrino burst from the supernova SN1987A", *Phys. Rev. Lett.* (American Physical Society) **58** (14): 1490–1493, Bibcode:1987PhRvL..58.1490H, doi:10.1103/PhysRevLett.58.1490

[15] S. Fukuda et al. (1 April 2003), "The Super-Kamiokande detector", *Nuclear Instruments and Methods A* **51**: 418–462, Bibcode:2003NIMPA.501..418F, doi:10.1016/S0168-9002(03)00425-X

[16] S. Fukuda et al. (1 April 2003), "The Super-Kamiokande detector", *Nuclear Instruments and Methods A* **51**: 418–462, Bibcode:2003NIMPA.501..418F, doi:10.1016/S0168-9002(03)00425-X

[17] S. Fukuda et al. (1 April 2003), "The Super-Kamiokande detector", *Nuclear Instruments and Methods A* **51**: 418–462, Bibcode:2003NIMPA.501..418F, doi:10.1016/S0168-9002(03)00425-X

[18] S. Fukuda et al. (1 April 2003), "The Super-Kamiokande detector", *Nuclear Instruments and Methods A* **51**: 418–462, Bibcode:2003NIMPA.501..418F, doi:10.1016/S0168-9002(03)00425-X

[19] The official Super-Kamiokande home page/research

[20] A.B. Balantekin et al. (July 2013), "Neutrino oscillations", *Progress in Particle and Nuclear Physics* **71**: 150–161, arXiv:1303.2272, Bibcode:2013PrPNP..71..150B, doi:10.1016/j.ppnp.2013.03.007

[21] J.N Bahcall; S Basu; M.H Pinsonneault (1998), "How uncertain are solar neutrino predictions?", *Phys. Lett. B* **433**: 1–8, arXiv:astro-ph/9805135, Bibcode:1998PhLB..433....1B, doi:10.1016/S0370-2693(98)00657-1

[22] The Super-Kamiokande Homepage

[23] The official homepage of T2K experiment

[24] S. Fukuda et al. (1 April 2003), "The Super-Kamiokande detector", *Nuclear Instruments and Methods A* **51**: 418–462, Bibcode:2003NIMPA.501..418F, doi:10.1016/S0168-9002(03)00425-X

[25] S. Fukuda et al. (1 April 2003), "The Super-Kamiokande detector", *Nuclear Instruments and Methods A* **51**: 418–462, Bibcode:2003NIMPA.501..418F, doi:10.1016/S0168-9002(03)00425-X

[26] S. Fukuda et al. (1 April 2003), "The Super-Kamiokande detector", *Nuclear Instruments and Methods A* **51**: 418–462, Bibcode:2003NIMPA.501..418F, doi:10.1016/S0168-9002(03)00425-X

[27] S. Fukuda et al. (1 April 2003), "The Super-Kamiokande detector", *Nuclear Instruments and Methods A* **51**: 418–462, Bibcode:2003NIMPA.501..418F, doi:10.1016/S0168-9002(03)00425-X

[28] S. Fukuda et al. (1 April 2003), "The Super-Kamiokande detector", *Nuclear Instruments and Methods A* **51**: 418–462, Bibcode:2003NIMPA.501..418F, doi:10.1016/S0168-9002(03)00425-X

[29] S. Fukuda et al. (1 April 2003), "The Super-Kamiokande detector", *Nuclear Instruments and Methods A* **51**: 418–462, Bibcode:2003NIMPA.501..418F, doi:10.1016/S0168-9002(03)00425-X

[30] S. Fukuda et al. (1 April 2003), "The Super-Kamiokande detector", *Nuclear Instruments and Methods A* **51**: 418–462, Bibcode:2003NIMPA.501..418F, doi:10.1016/S0168-9002(03)00425-X

[31] S. Fukuda et al. (1 April 2003), "The Super-Kamiokande Detector", *Nuclear Instruments and Methods in Physics Research Section* **501**: 418–462

[32] The official Super-Kamiokande home page

[33] American Super-K home page

[34] Pictures and illustrations

[35] Official report on the accident (in PDF format)

[36] Logbook entry of first neutrinos seen at Super-K generated at KEK

[37] S. Fukuda et al. (1 April 2003), "The Super-Kamiokande detector", *Nuclear Instruments and Methods A* **51**: 418–462, Bibcode:2003NIMPA.501..418F, doi:10.1016/S0168-9002(03)00425-X

[38] Fukuda, Y. et al. (1998). "Evidence for oscillation of atmospheric neutrinos". *Physical Review Letters* **81** (8): 1562–1567. arXiv:hep-ex/9807003. Bibcode:1998PhRvL..81.1562F. doi:10.1103/PhysRevLett.81.1562.

[39] Accident grounds neutrino lab

[40] Neutrinos from SuperNova Burst

[41] Kearns; Kajita; Totsuka (August 1999), "Detecting Massive Neutrinos", *Scientific American*

[42] "Search for proton decay via $p \rightarrow \nu K\flat$ using 260 kiloton · year data of Super-Kamiokande". *PHYSICAL REVIEW D* 90, 072005. 14 Oct 2014.

[43] http://whitehotmagazine.com/articles/andreas-gursky-matthew-marks-gallery/493

7.16 Supernova Early Warning System

The **SuperNova Early Warning System (SNEWS)** is a network of neutrino detectors designed to give early warning to astronomers in the event of a supernova in our home galaxy or a nearby galaxy such as the Large Magellanic Cloud or the Canis Major Dwarf Galaxy. Enormous numbers of neutrinos are produced in the core of a red giant star as it collapses on itself. In the current model the neutrinos are emitted well before the light from the supernova peaks, so in principle neutrino detectors could give advance warning to astronomers that a supernova has occurred and may soon be visible. The neutrino pulse from supernova 1987A arrived 3 hours before the associated photons – but SNEWS was not yet active and it was not recognised as a supernova event until after the photons arrived.

The current members of SNEWS are Borexino, Daya Bay, KamLAND, IceCube, LVD, and Super-Kamiokande.[*][1] The Sudbury Neutrino Observatory is not currently active as it is being upgraded to its successor program SNO+. SNEWS began operation prior to 2004, with three members (Super-Kamiokande, LVD, and SNO). As of June 2013, SNEWS has not issued any SN alerts.

7.16.1   See also

- Near-Earth supernova

- History of supernova observation

- Timeline of white dwarfs, neutron stars, and supernovae

- Supernova nucleosynthesis

- Supernova neutrinos

7.16.2   References

[1] http://snews.bnl.gov/news.html

7.16.3   External links

- Official website

- Antonioli, P. et al. (2004). "SNEWS: the SuperNova Early Warning System". *New Journal of Physics* **6**: 114–114. arXiv:astro-ph/0406214. Bibcode:2004NJPh....6..114A. doi:10.1088/1367-2630/6/1/114.

- Francis Reddy, "Time for SNEWS", *Astronomy* 3 June 2005

- NOVA podcast about SNEWS (the same in MP3 format)

Chapter 8

Text and image sources, contributors, and licenses

8.1 Text

- **Lepton** *Source:* https://en.wikipedia.org/wiki/Lepton?oldid=679886360 *Contributors:* Bryan Derksen, Andre Engels, PierreAbbat, Ben-Zin~enwiki, Heron, Xavic69, Fruge~enwiki, Fwappler, Ahoerstemeier, Julesd, Glenn, Mxn, A5, Wikiborg, Dysprosia, Radiojon, Imc, Morwen, Fibonacci, Bcorr, Phil Boswell, Donarreiskoffer, Robbot, Merovingian, Wikibot, Giftlite, Smjg, DocWatson42, Harp, Herbee, Xerxes314, Sysin, Knutux, LiDaobing, LucasVB, ClockworkLunch, RetiredUser2, Icairns, Mike Rosoft, Chris j wood, Martinl~enwiki, Smalljim, Giraffedata, Jumbuck, RobPlatt, Neonumbers, Ahruman, Computerjoe, Simon M, Woohookitty, Mindmatrix, Rjwilmsi, Strait, Erkcan, FlaBot, Danny-Wilde, Mastorrent, Celebere, Peterl, YurikBot, Bambaiah, Jimp, Salsb, Spike Wilbury, Jaxl, SCZenz, DeadEyeArrow, Tetracube, Smoggyrob, Dmuth, Jaysbro, Sbyrnes321, That Guy, From That Show!, SmackBot, Bazza 7, KocjoBot~enwiki, Jrockley, Mom2jandk, Cool3, Hmains, Complexica, DHN-bot~enwiki, Mesons, Yevgeny Kats, TriTertButoxy, SashatoBot, Ouzo~enwiki, Happy-melon, Kurtan~enwiki, Myasuda, Cydebot, Meno25, Photocopier, Michael C Price, Casliber, Thijs!bot, Headbomb, Newton2, Mentifisto, Autotheist, Steveprutz, NeverWorker, NicoSan, MartinBot, Arjun01, HEL, J.delanoy, Numbo3, Gombang, Num1dgen, Ceoyoyo, VolkovBot, Macedonian, Mocirne, TXiKiBoT, Anonymous Dissident, Abdullais4u, Antixt, Jhb110, Thanatos666, AlleborgoBot, SieBot, ToePeu.bot, RadicalOne, Ngexpert7, Jacob.jose, Hamiltondaniel, TubularWorld, Muhends, ClueBot, ICAPTCHA, UniQue tree, Snigbrook, Fyyer, IceUnshattered, Cmj91uk, LieAfterLie, Manu-ve Pro Ski, TimothyRias, Addbot, Betterusername, AgadaUrbanit, Ehrenkater, OlEnglish, Zorrobot, Andy2308, Legobot, Luckas-bot, Ptbotgourou, Maxim Sabalyauskas, Planlips, JackieBot, Icalanise, Citation bot, ٤٢مدي.ع.د.ماح24, ArthurBot, Almabot, Omnipaedista, Alexeymorgunov, 老陳, Tormine, MathFacts, Citation bot 1, MastiBot, Earthandmoon, EmausBot, John of Reading, Az29, Galaktiker, StringTheory11, Quondum, Surajt88, I hate whitespace, ClueBot NG, Scimath Genius, Braincricket, Widr, Helpful Pixie Bot, Bibcode Bot, Tyler6360534, Katagun5, Melenc, Derek-Winters, Prasanna4s, Machosquirrel, Devinhorn, KasparBot and Anonymous: 149

- **Electron** *Source:* https://en.wikipedia.org/wiki/Electron?oldid=682603060 *Contributors:* AxelBoldt, CYD, Mav, Bryan Derksen, AstroNomer~enwiki, Ap, Ed Poor, Andre Engels, Ryrivard, William Avery, SimonP, Peterlin~enwiki, Heron, Camembert, Stevertigo, Bdesham, Patrick, D, JohnOwens, Michael Hardy, Tim Starling, Ixfd64, Fruge~enwiki, Arpingstone, PingPongBoy, Egil, NuclearWinner, Ahoerstemeier, Suisui, Jebba, JWSchmidt, Kingturtle, Aarchiba, Glenn, Scott, Kwekubo, Andres, Jordi Burguet Castell, Mxn, Agtx, Timwi, Wikiborg, Reddi, Rednblu, Markhurd, Maximus Rex, E23~enwiki, Omegatron, Secretlondon, Jusjih, BenRG, Jeffq, Donarreiskoffer, Gentgeen, Robbot, Sanders muc, Vespristiano, Merovingian, Pingveno, Blainster, Hadal, Wikibot, Wereon, Widsith, HaeB, Diberri, Dmn, Dina, Giftlite, Christopher Parham, Ferkelparade, Fastfission, Zigger, Herbee, Dissident, Xerxes314, Curps, Michael Devore, Bensaccount, Ssd, Gilgamesh~enwiki, Vadmium, Gdr, Knutux, Slowking Man, Yath, Gzuckier, Pcarbonn, Joizashmo, Karol Langner, Anythingyouwant, RetiredUser2, Bbbl67, Elroch, Icairns, JohnArmagh, JimQ, Mike Rosoft, Mindspillage, Patrick L. Goes, Discospinster, Brianhe, Rich Farmbrough, Guanabot, Hidaspal, Vsmith, Deh, Ardonik, Roybb95~enwiki, Xezbeth, Zazou, Mani1, SpookyMulder, Dmr2, ZeroOne, Kjoonlee, Goplat, Calair, Nabla, Brian0918, RJHall, Pt, Jaques O. Carvalho, El C, Huntster, Edward Z. Yang, Susvolans, Art LaPella, RoyBoy, ~K, Bobo192, Army1987, Asierra~enwiki, Flxmghvgvk, AtomicDragon, Evgeny, AllyUnion, Bert Hickman, Deryck Chan, PeterisP, Beetle B., Obradovic Goran, (aeropagitica), Pearle, Mpulier, HasharBot~enwiki, Confusedmiked, Mote, Jumbuck, Gary, ChristopherWillis, Ricky81682, Benjah-bmm27, Riana, AzaToth, DonJStevens, BernardH, Malo, David Hochron, Bart133, EagleFalconn, Schapel, Omphaloscope, RainbowOfLight, RichBlinne, H2g2bob, DV8 2XL, Gene Nygaard, Redvers, StuTheSheep, Linas, Mindmatrix, GrouchyDan, StradivariusTV, Uncle G, BillC, Kurzon, Jeff3000, HcorEric X, Eleassar777, Ozielke, Wayward, Palica, Omega21, FreplySpang, Enzo Aquarius, Rjwilmsi, Shaadow, Strait, Mike Peel, Chekaz, Bubba73, Dar-Ape, Yamamoto Ichiro, FlaBot, RobertG, Latka, DannyWilde, Nihiltres, RexNL, Kolbasz, Thecurran, Srleffler, Physchim62, Chobot, DVdm, Unclevortex, Eric B, YurikBot, Wavelength, RobotE, Bambaiah, AcidHelmNun, Jimp, Peter G Werner, Wolfmankurd, Wigie, Ventolin, JabberWok, SpuriousQ, Lucinos~enwiki, Akamad, Ori Livneh, Gaius Cornelius, Shaddack, Eleassar, Rsrikanth05, Salsb, Hawkeye7, Spike Wilbury, Jaxl, Welsh, DarthVader, Długosz, BirgitteSB, SCZenz, Retired username, Ravedave, PhilipO, Adam Rock, Mlouns, Chichui, BOT-Superzerocool, Gadget850, Bota47, Kkmurray, James Trotter~enwiki, Dna-webmaster, Ms2ger, Light current, Lycaon, Imaninjapirate, Josh3580, Kriscotta, JoanneB, Peyna, Lpm, JLaTondre, Heavy bolter, RG2, GrinBot~enwiki, Sbyrnes321, ChemGardener, Itub, SmackBot, Zazaban, Incnis Mrsi, KnowledgeOfSelf, Royalguard11, Melchoir, J.Sarfatti, KocjoBot~enwiki, Stepa, Pandion auk, Jrockley, JoeMarfice, ZerodEgo, Edgar181, Yamaguchi 先生, Skizzik, Dauto, JSpudeman, Kurykh, Rajeevmass~enwiki, Persian Poet Gal, Pieter Kuiper, Jprg1966, Acrinym, Miquonranger03, MalafayaBot, Droll, Complexica, DHN-bot~enwiki, Sbharris, RAlafriz, Vladis1av, Vanished User 0001, Darthgriz98, Voyajer, Addshore, Percommode, Krich, DavidStern, Theonlyedge, Nakon, Nrcprm2026, DMacks, Daniel.Cardenas, Zeamays, Jonnyapple, Sadi Carnot, Bdushaw, Wilt, TriTertButoxy, Chymicus, UberCryxic, Bagel7, Mattfont, Heimstern, Jaganath, Ocatecir, Mr. Lefty, Ckatz, 16@r, Omnedon, Owlbuster, Waggers, SandyGeorgia, Spiel496, Funnybunny, HappyVR, Iridescent, Newone, NativeForeigner, J Di, Amakuru, Tawkerbot2, Chetvorno, Thermochap, CmdrObot, Ale jrb, Megaboz, RedRollerskate, Ruslik0, MrZap, McVities, WMSwiki, Bakanov, RobertLovesPi, Equendil, Cydebot, Acelor, Reywas92, Cantras, Bvcrist, LouisBB, Travelbird, Llort, David edwards, Tawkerbot4, Christian75, Narayanese, Ssilvers, Thijs!bot, Epbr123, Mbell, Dougsim, Nonagonal Spider, Headbomb, Yzmo, Marek69, West Brom 4ever, Tellyaddict, Cool Blue, Greg L, Sean William, VictorP, KrakatoaKatie, AntiVandalBot, WinBot, Skymt, Voyaging, Opelio, Tyco.skinner, Gef756, Chill doubt, Naturalnumber, Gdo01, Spencer, Leuko, CosineKitty, J-stan, Smith Jones, Acroterion, Magioladitis, WolfmanSF, Bennybp, Bongwarrior, VoABot II, A4, Nyq, JNW, JamesBWatson, باسم, Drondent, Slartibartfast1992, Jackal irl, Animum, Dirac66, 28421u2232nfenfcenc, Hveziris, User A1, Maliz, PoliticalJunkie, DerHexer, GregU, PEBill, MartinBot, BetBot~enwiki, Mermaid from the Baltic Sea, WizendraW, Xantolus, Thereen, CommonsDelinker, AlexiusHoratius, J.delanoy, DrKiernan, Rgoodermote, Numbo3, Acalamari, TheChrisD, Dispenser, LordAnubisBOT, JayMars, Lathrop, AntiSpamBot, TomasBat, NewEnglandYankee, Nwbeeson, SmoothK, Sunderland06, MetsFan76, Joshmt, Cometstyles, STBotD, RB972, Treisijs, D-Kuru, Dineshextreeme, Martial75, CardinalDan, Idioma-bot, Sheliak, Bondslave777, FeralDruid, X!, VolkovBot, ABF, Thisisborin9, Jacroe, Ryan032, Philip Trueman, DoorsAjar, TXiKiBoT, GimmeBot, Kriak, Hqb, GDonato, Anonymous Dissident, Crohnie, Monkey Bounce, Voorlandt, Mr. Hallman, Michael H 34, TBond, Wikiisawesome, Suriel1981, Rbdebole, Graymornings, Synthebot, Enviroboy, Rurik3, Generalguy11, !dea4u, Insanity Incarnate, Ceranthor, Yoos~enwiki, AlleborgoBot, Kalivd, EmxBot, Neparis, Swimallday, Ponyo, EJF, SieBot, Graham Beards, Scarian, CircafuciX, BotMultichill, Jauerback, Dawn Bard, Joncam, Caltas, Sergeanthuggy, Bentogoa, RadicalOne, Arbor to SJ, Prestonmag, Thadaddy3233, Oxymoron83, Antonio Lopez, KPH2293, Lightmouse, WingkeeLEE, Ealdgyth, BenoniBot~enwiki, Stustjohn, Dabomb87, PlantTrees, Dolphin51, Nergaal, Tomdobb, Muhends, WikipedianMarlith, ClueBot, Trojancowboy, GorillaWarfare, Artichoker, PipepBot, UniQue tree, The Thing That Should Not Be, Hongthay, Unbuttered Parsnip, GreenSpigot, Liekmudkipz, Mild Bill Hiccup, Correcting nonesense, NovaDog, Blanchardb, Richerman, RandomTREES, Rotational, Piledhigheranddeeper, Inala, DragonBot, Almcaeobtac, Jusdafax, MEJG, Gtstricky, Rhododendrites, Brews ohare, NuclearWarfare, Lunchscale, Jotterbot, PhySusie, Tonyfey, Lkruijsw, Kaiba, SchreiberBike, Stepheng3, Thingg, Jamyricks, Aitias, Melibarr05, Kurtcobain321, Scalhotrod, Versus22, Johnuniq, MasterOfHisOwnDomain, DumZiBoT, TimothyRias, Sjodenenator, XLinkBot, Maky, Rror, Avoided, Mitch Ames, Ilikepie2221, WikHead, Mgaarafan, SkyLined, Addbot, Chizkiyahuavraham, AVand, Some jerk on the Internet, Hurleymann1, Uruk2008, DOI bot, Tcncv, Booba5, AkhtaBot, Jessepfrancis, Ronhjones, Jncraton, Moosehadley, CanadianLinuxUser, WFPM, LaaknorBot, Chamal N, CarsracBot, FiriBot, Omnipedian, LinkFA-Bot, Ehrenkater, Pnacitum, Tide rolls, Lightbot, Potekhin, UPS Truck Driver, VP-bot, Luckas-bot, Yobot, Nergality, Kan8eDie, THEN WHO WAS PHONE?, Eric-Wester, Tonyrex, AnomieBOT, Shootbamboo, DemocraticLuntz, Rubinbot, Götz, Jim1138, IRP, Piano non troppo, Icalanise, Kingpin13, Mydickishuge24, Materialscientist, The High Fin Sperm Whale, Citation bot, Neurolysis, ArthurBot, LovesMacs, Mrhellcool, Rightly, Xqbot, IrishChemistPride, IrishChemistPride2, GeometryGirl, Restu20, Srich32977, S0aasdf2sf, John5955, Alan8, ProtectionTaggingBot, Omnipaedista, RibotBOT, TonyHagale, Phillycheesesteaks, LyleHoward, A. di M., RyanOrdemann, Peter470, Thehelpfulbot, Al Wiseman, FrescoBot, Surv1v4l1st, Eadon-com, Paine Ellsworth, Tobby72, Gauravdce07, Steve Quinn, C.Bluck, Citation bot 1, MarB4, Galmicmi, Gil987, Pinethicket, HRoestBot, Voltron Hax, Raen79, Hoo man, Yos233, Allthingstoallpeople, MastiBot, Kuririmo, Noel Streatfield, Ezhuttukari, Swifterthenyou, Noisalt, Jujutacular, Euchanels, Lissajous, Dude1818, December21st2012Freak, IJBall, Jauhienij, Utility Monster, FoxBot, Sheogorath, Jdlawlis, Odatus, Bestcallumuk, Sampathsris, DARTH SIDIOUS 2, Mean as custard, RjwilmsiBot, TjBot, MinicheddarsandelephantsFTW, Benjadow, Mcmonsterbrothers, Priceracks, Csilcock, Sohaib360, Androstachys, Techhead7890, EmausBot, Optiguy54, GoingBatty, Jjasharpe, Pcorty, TuHan-Bot, Hhhippo, HiW-Bot, John Cline, Harddk, Fæ, Josve05a, StringTheory11, Wackywace, Quondum, GianniG46, Fizicist, Wayne Slam, Raynor42, Arnaugir, Jacksccsi, Brandmeister, Donner60, Negovori, RockMagnetist, Mni9791, ClueBot NG, HLachman, Hermajesty21, Jacobkh, Letoya123, Samsau ninjaguy, Ggonzalm, Moritz37, Braincricket, Helpful Pixie Bot, Geo7777, SzMithrandir, Bibcode Bot, Dfbowsmountainer, Ymblanter, Vagobot, Paolo Lipparini, Wzrd1, Lk00la1dl, JacobTrue, Socal212, Begman5, Mark Arsten, Cadiomals, Jikepaddy, Caterpillar111, Macymae, 06seagsa, BEEPTHENOOB, ItzzRevolution, Shawn Worthington Laser Plasma, Duxwing, Klilidiplomus, Uopchem251, Joe0x7F, BattyBot, Justincheng12345-bot, Cyberbot II, ChrisGualtieri, GoShow, Ankap~enwiki, Glenzo999, Barant2, BrightStarSky, Dexbot, Astromango2215, Webclient101, Mogism, 331dot, Spray787, Vanquisher.UA, Lugia2453, Kondormari, Reatlas, JellyBean4.1, Prof.Professer, The User 111, Bluemanyoung, Rohitgunturi, Ugog Nizdast, The Herald, Jwratner1, Zahid2233, MorshusApprentice, 2005-Fan, Phub Dorji, Epic Failure, Gindor, Ian98989898, Monkbot, SkateTier, Waldmannevan, Wiki1098, Wulfiedude14, Mario Castelán Castro, Fleivium, Crystallizedcarbon, Mcwikigeek, Jesus is the Light of my life, Acesoli, Soumilm, Lemmegetyou, Flying g shot, SirLagsalott, Tetra quark, Skipfortyfour, Stim 2.0, KasparBot, Fazbear7891 and Anonymous: 935

- **Positron** *Source:* https://en.wikipedia.org/wiki/Positron?oldid=679431995 *Contributors:* Bryan Derksen, Andre Engels, Hhanke, Peterlin~enwiki, Patrick, Looxix~enwiki, Hashar, Stismail, Pstudier, BenRG, Donarreiskoffer, Robbot, Merovingian, LGagnon, David Gerard, Decumanus, Giftlite, Jmnbpt, Xerxes314, Utcursch, Xmnemonic, Knutux, Icairns, Tumbarumba, Mike Rosoft, Vsmith, Jpk, Murtasa, Bender235, El C, Asierra~enwiki, La goutte de pluie, Tra, Atlant, Spangineer, Wtmitchell, SidP, RogerBarnett, Gene Nygaard, WojciechSwiderski~enwiki, UTSRelativity, Richard Arthur Norton (1958-), MartinSpacek, Jannex, Ma Baker, Robert K S, Palica, MassGalactusUniversum, V8rik, BD2412, Canderson7, Seraphimblade, Mike Peel, FlaBot, DannyWilde, Kolbasz, Fresheneesz, Tardis, Wrightbus, Chobot, YurikBot, Bambaiah, Jimp, Arado, Chuck Carroll, Bergsten, Damato, Hellbus, Astriks, Salsb, MidnightWolf, Vanished user kjdioejh329io3rksdkj, Complainer, Robertvan1, Howcheng, SCZenz, Zwobot, Scottfisher, Lt-wiki-bot, Arthur Rubin, Terbospeed, Vicarious, GrinBot~enwiki, SmackBot, Unyoyega, Edgar181, KingRaptor, Skizzik, Kmarinas86, Complexica, DHN-bot~enwiki, Sbharris, Rodri316, JorisvS, Mr. Vernon, Rock4arolla, MTSbot~enwiki, Iridescent, Michaelbusch, Darth Sader, CapitalR, CmdrObot, Eric, Cofax48, Yzphub, HalJor, Cydebot, Nick Y., Bvcrist, Yolocavo, HPaul, Quibik, Thijs!bot, Gamer007, Headbomb, Rlupsa, Davidhorman, Shadow Blaziken, Griba2010, Escarbot, Poshzombie, Harrylentil, JAnDbot, CosineKitty, Rob Mahurin, MegX, Recurring dreams, Mother.earth, Jetterman, Maliz, Gwern, Geboy, Boddey, Rustyfence, CliffC, Rigmahroll, CommonsDelinker, HEL, Solarswordsman, AntiSpamBot, Y2H, Sheliak, TreasuryTag, Philip Trueman, TXiKiBoT, DavidRThomas, Corticopia, Agradada, Corvus cornix, Lerdthenerd, Gabriel Vidal, Seraphita~enwiki, AlleborgoBot, SieBot, Winchelsea, Triwbe, RadicalOne, Flyer22, WingkeeLEE, BenoniBot~enwiki, Drgarden, ClueBot, Fyyer, Gabriel Vidal Álvarez, Piledhigheranddeeper, Maxtitan, Djr32, Tyler, EncyclopediaUpdaticus, DumZiBoT, AgnosticPreachersKid, TheRealVolucrix, SkyLined, Addbot, DOI bot, AkhtaBot, CanadianLinuxUser, Jim10701, AgadaUrbanit, Positroni, Tide rolls, Luckas-bot, Yobot, Allowgolf~enwiki, Götz, Jim1138, Taupositron, Materialscientist, Citation bot, Yathimc, Jakouso, Stephen.G.McAteer, FrescoBot, Paine Ellsworth, Dscraggs, Citation bot 1, I dream of horses, Tom.Reding, Wdanbae, RjwilmsiBot, Samdacruel, EmausBot, WikitanvirBot, Harddk, H3llBot, Quondum, I kabir, MGWeatherman08, Carmichael, Teapeat, ClueBot NG, Snotbot, Helpful Pixie Bot, Ieditpagesincorrectly, Bibcode Bot, Slaughter182, Sergeant Cribb, Drizzt182, Dexbot, Crocgandhi, Bigdumpy, Planetguy2345, KasparBot, Kobiej100, People73, Person420 and Anonymous: 148

- **Muon** *Source:* https://en.wikipedia.org/wiki/Muon?oldid=680138424 *Contributors:* AxelBoldt, The Epopt, CYD, Mav, Bryan Derksen, Zundark, Roadrunner, Bkellihan, Youandme, Tim Starling, EddEdmondson, Looxix~enwiki, Ahoerstemeier, Angela, Rob Hooft, Kbk, Donarreiskoffer, AlexPlank, Robbot, Merovingian, Rholton, Ojigiri~enwiki, Auric, Roscoe x, Bkell, Millosh, Wikibot, Ruakh, Diberri, Giftlite, Wizzy, Herbee, Xerxes314, Bodhitha, LiDaobing, Pcarbonn, DragonflySixtyseven, Deglr6328, Eb.hoop, Rich Farmbrough, Pjacobi, Vsmith, Mani 1, STGM, Kjoonlee, RJHall, Army1987, Danski14, Anthony Appleyard, RobPlatt, Keenan Pepper, RJFJR, Ceyockey, Falcorian, Woohookitty, Xinghuei, Bennetto, Graham87, Vanderdecken, Rjwilmsi, Strait, Mike Peel, Bubba73, DoubleBlue, Dougluce, FlaBot, Fivemack, DannyWilde, Sp00n, Lmatt, Goudzovski, Srleffler, Chobot, YurikBot, Wavelength, Bambaiah, Limulus, JabberWok, Hellbus, Salsb, SCZenz, Ravedave, Scottfisher, Tetracube, Lt-wiki-bot, E Wing, Roberto DR, CrniBombarder!!!, Sbyrnes321, Eog1916, SmackBot, Incnis Mrsi, Melchoir, Stifle, Gilliam, Dauto, Bluebot, Tigerhawkvok, Sbharris, Colonies Chris, Can't sleep, clown will eat me, Yevgeny Kats, Dane Sorensen, JorisvS, Mets501, JoeBot, CapitalR, SchmittM, Ruslik0, Ken Gallager, Rotiro, A876, Corpx, Thijs!bot, Headbomb, D.H, Bm gub, Andrew Carlssin, Spencer, Kariteh, Deflective, Belg4mit, Swpb, Mother.earth, Nono64, HEL, Hans Dunkelberg, 5Q5, Tarotcards, Coppertwig, Bermy88, Jarry1250, Thecinimod, Sheliak, Cuzkatzimhut, VolkovBot, Larryisgood, VasilievVV, TXiKiBoT, Anonymous Dissident, Mihaip, Graymornings, SalomonCeb, SieBot, Csmart287, Gerakibot, Statue2, Mhouston, StewartMH, ClueBot, Polyamorph, DnetSvg, Esbboston, Saritepe, Stefan Ritt, BarretB, Kajabla, Addbot, Roentgenium111, Toyokuni3, Download, Ehrenkater, Lightbot, Luckas-bot, Yobot, Evaders99, Kulmalukko, AnomieBOT, Icalanise, Kingpin13, Materialscientist, Citation bot, Kotika98, ArthurBot, Xqbot, Cjxc92, Gilo1969, Srich32977, Misterigloo, Kyng, A. di M., Paine Ellsworth, Citation bot 1, Citation bot 4, Rameshngbot, Isofox, Jetstoknowhere, TobeBot, Trappist the monk, Puzl bustr, RjwilmsiBot, EmausBot, John of Reading, WikitanvirBot, Dewritech, GoingBatty, Milledit, Naviguessor, StringTheory11, Medeis, Suslindisambiguator, Quondum, Timetraveler3.14, Layona1, Aerthis, Mikhail Ryazanov, Frietjes, CaroleHenson, Kebil, Bibcode Bot, Jesusmonkey, NotWith, BattyBot, Kisokj, Liam135, MuonRay, Tony Mach, Telfordbuck, Krotera, Ajdigregorio, Seoman2snowlock, Monkbot, Jromerofontalvo, KasparBot, Corrupt Titan, QzPhysics and Anonymous: 191

- **Tau (particle)** *Source:* https://en.wikipedia.org/wiki/Tau_(particle)?oldid=680139275 *Contributors:* Bryan Derksen, Iluvcapra, Ahoerstemeier, Bueller 007, Schneelocke, Dysprosia, Donarreiskoffer, Merovingian, Rorro, David19999, Millosh, Harp, Herbee, Codepoet, Xerxes314, Bodhitha, CryptoDerk, Icairns, Rich Farmbrough, Pjacobi, Martpol, Sunborn, Kjoonlee, El C, Reuben, JellyWorld, RobPlatt, RJFJR, Falcorian, Dmitry Brant, Christopher Thomas, Palica, Rjwilmsi, Strait, Mike Peel, FlaBot, DannyWilde, Goudzovski, Chobot, RobotE, Bambaiah, Acid-HelmNun, JabberWok, Eleassar, Salsb, SCZenz, Zwobot, Ospalh, PS2pcGAMER, Bota47, Someones life, Poulpy, Physicsdavid, Incnis Mrsi, Dauto, Pieter Kuiper, Loodog, JorisvS, MTSbot~enwiki, WISo, Q43, Thijs!bot, Headbomb, Davidhorman, Hcobb, Escarbot, RogueNinja, Yill577, Soulbot, Kostisl, STBotD, Sheliak, Joyko~enwiki, VolkovBot, Fences and windows, TXiKiBoT, Awl, Jba138, SieBot, OKBot, ImageRemovalBot, Plastikspork, Djr32, Alexbot, TimothyRias, Assosiation, BodhisattvaBot, SkyLined, J Hazard, Addbot, Eric Drexler, Ronhjones, ChenzwBot, Jklukas, Theozzfancometh, Skippy le Grand Gourou, Luckas-bot, Yobot, Grebaldar, AnomieBOT, Icalanise, Citation bot, Xqbot, Blennow, Franco3450, 老陳, Paine Ellsworth, Jonesey95, Three887, Plasticspork, 3ph, Miracle Pen, RjwilmsiBot, TjBot, Ripchip Bot, EmausBot, Dcirovic, Suslindisambiguator, Quondum, Rezabot, Helpful Pixie Bot, Bibcode Bot, BG19bot, Sudsguest, YFdyh-bot, Redcliffe maven, TwoTwoHello, Akro7, KasparBot, JPPepper, QzPhysics and Anonymous: 48

- **Neutrino** *Source:* https://en.wikipedia.org/wiki/Neutrino?oldid=681496713 *Contributors:* AxelBoldt, Chenyu, Bryan Derksen, Zundark, The Anome, Tarquin, Andre Engels, Xaonon, XJaM, William Avery, Roadrunner, DrBob, Heron, Cwitty, MimirZero, Spiff~enwiki, Edward, Patrick, Ken Arromdee, EddEdmondson, Ezra Wax, Gdarin, Meekohi, Bcrowell, Cyde, Arpingstone, Alfio, Looxix~enwiki, Strebe, JWSchmidt, Julesd, Glenn, Nikai, Andres, Evercat, Rob Hooft, TheSeez, Crissov, Wikiborg, Reddi, Lfh, Cos111, Tpbradbury, Fibonacci, Warofdreams, Twang, Donarreiskoffer, Drxenocide, Robbot, Findel, Zandperl, Nurg, Masao, Merovingian, Bobunf, Rursus, Meelar, Matty j, Intangir, Wikibot, Wereon, Duien, Jimduck, Bbx, David Gerard, Giftlite, Graeme Bartlett, DocWatson42, Laudaka, Mikez, Harp, Lethe, HangingCurve, Xerxes314, Anville, Dratman, Curps, Jorge Stolfi, Eequor, Mdob, Espetkov, LiDaobing, Elroch, Icairns, Doug Danner, Nickptar, Fg2, Lrenh, Deglr6328, Hmmm~enwiki, Mattman723, Helohe, Rich Farmbrough, Hydrox, Cacycle, Pjacobi, Vsmith, Dbachmann, Mani 1, Pavel Vozenilek, Ralfoide, Sunborn, Neko-chan, Kharhaz, RJHall, Charm, Haxwell, RoyBoy, Smalljim, Cje~enwiki, Viriditas, Cwolfsheep, Foobaz, I9Q79oL78KiL0QTFHgyc, La goutte de pluie, Thewayforward, Thuktun, Fleurot~enwiki, Quaoar, Alansohn, Anthony Appleyard, ChristopherWillis, Calton, Axl, Mac Davis, Hdeasy, RJFJR, Dirac1933, TenOfAllTrades, Vuo, Cmprince, Pauli133, Gene Nygaard, Lyuokdea, Flying fish, Richard Arthur Norton (1958-), Woohookitty, Swamp Ig, Insaneinside, Benhocking, Nakos2208~enwiki, GregorB, SDC, Joke137, Fxer, Palica, RedBLACKandBURN, Ashmoo, Graham87, Qwertyus, Raymond Hill, Drbogdan, Rjwilmsi, Coemgenus, Strait, John187, Staecker, Jmcc150, Salix alba, Mike Peel, Vegaswikian, Oblivious, Ligulem, R.e.b., Jehochman, The wub, FlaBot, Ian Pitchford, DannyWilde, Itinerant1, RexNL, Gurch, Kolbasz, Goudzovski, Sperxios, Scythe33, Smithbrenon, Chobot, Nagytibi, DVdm, YurikBot, Bambaiah, Vuvar1, Phmer, RussBot, Ohwilleke, Witan, Xihr, Bhny, Chris Capoccia, JabberWok, Gaius Cornelius, Salsb, Grafen, Długosz, Gillis, SCZenz, Ravedave, Abb3w, CecilWard, Santaduck, Bota47, Maunus, Dna-webmaster, Ms2ger, Rhynchosaur, DrWorm, Alias Flood, Ilmari Karonen, Nimbex, Phr en, Otto ter Haar, Fragman, That Guy, From That Show!, AndrewWTaylor, Palapa, Morgan wascko, SmackBot, Trainbrain27, Reedy, Tom Lougheed, Melchoir, The Monster, Arbe, Mscuthbert, BiT, Ohnoitsjamie, Dauto, Kmarinas86, GregRM, Decowski, Yurigerhard, DHNbot~enwiki, Sbharris, Colonies Chris, Hengsheng120, Nap~enwiki, Sergio.ballestrero, Милан Јелисавчић, Cophus, Mayrel, Wen D House, Engwar, Jdlambert, Webmaster Pete, TheMaster42, Pwjb, Akriasas, DenisRS, Rjn~enwiki, Kukini, Yevgeny Kats, Ged UK, Jjpcondor, Gobonobo, ThorAvaTahr, JorisvS, Makyen, Libera~enwiki, Aeluwas, Dicklyon, Mets501, MTSbot~enwiki, Galactor213, Fredil Yupigo, Masoninman, Newone, Richard75, Chalnoth, Mssgill, Valoem, Abeneal, Rszasz, CmdrObot, Calmargulis, Olaf Davis, Vyznev Xnebara, MrFizyx, Thubsch, Myasuda, Alton, Astralusenet, Icek~enwiki, Szdori~enwiki, Hyperdeath, Gogo Dodo, HPaul, RC Master, Q43, Michael C Price, Quibik, Christian75, DumbBOT, Joe Chick, Thijs!bot, Martin Hogbin, Naucer, Headbomb, Dtgriscom, WVhybrid, Esemono, James086, Second Quantization, Davidhorman, Weasel5i2, Jonny-mt, D.H, Greg L, Mentifisto, Luna Santin, Guy Macon, Seaphoto, Alphachimpbot, Astavats, Parande, DagosNavy, JAnDbot, Deflective, MER-C, CosineKitty, Savant13, Magioladitis, WolfmanSF, VoABot II, Nyq, Websterwebfoot, Bakken, DMcanada, Christoph Scholz~enwiki, Seleucus, Dirac66, Adrian J. Hunter, LorenzoB, NJR ZA, Khalid Mahmood, Squidonius, Pavel Jelínek, Gwern, Denis tarasov, Glrx, R'n'B, Fatka, Maurice Carbonaro, MrBell, Aqwis, Salih, Nalumc, Plasticup, Warut, Nwbeeson, Rosenknospe, Juliancolton, Mike Clough, Bonadea, Lseixas, Sheliak, Cuzkatzimhut, Jharris1993, VolkovBot, Camrn86, AlnoktaBOT, DrJohnPCostella, TXiKiBoT, The Original Wildbear, Nxavar, MinotAuruS, Cgr1123, Awl, Michael H 34, HannesHultgren~enwiki, SuperLonghorn, BotKung, SwordSmurf, Norbu19, Richwil, Tomaxer, The assassin 47, Morangm, AlleborgoBot, Thunderbird2, Angelastic, SieBot, Fredelige, PlanetStar, Zelab, Laoris, Ergateesuk, Stratman07, Jerryobject, RadicalOne, Aaarnooo, ScAvenger lv, John fromer, Thehotelambush, ShadowPhox, Ergo4sum, AWeishaupt, Jahilia, Extensive~enwiki, Sagredo, Martarius, ClueBot, Justin W Smith, Plastikspork, Apparentslug, Der Golem, Mild Bill Hiccup, Msgarrett, Hyh1048576, Mostargue, Paulcmnt, Ajoykt, Rhmtsang~enwiki, Xmantis, JayVora, Excirial, Kain Nihil, Wacko375, TonyBermanseder, Cenarium, Alastair301, Jwfvalle, Billrob458, Askahrc, Bouhadef, Termatt56, Johnuniq, PSimeon, Ost316, Nepenthes, Jprw, PL290, Jht4060, Buchler, MystBot, SkyLined, NCDane, Stephen Poppitt, Rical, Addbot, DOI bot, Download, Tide rolls, Lightbot, Taketa, SPat, Zorrobot, GDK, Gameseeker, Olsen-Fan, Legobot, Luckas-bot, Yobot, Bunnyhop11, Zagothal, Amble, Wikipedian Penguin, Azcolvin429, LibrarianofBabel, Dickdock, Robert Treat, AnomieBOT, Fatal!ty, Wrongfilter, Jim1138, Icalanise, Dakarateka, Mahmudmasri, Citation bot, Tano-kun, Vuerqex, GB fan, Xqbot, Blennow, JimVC3, Nickkid5, Cydelin, Srich32977, GrouchoBot, Abce2, Baba476, Backpackadam, QMarion II, 78.26, Aashaa, Ernsts, A. di M., Eldudarino, FrescoBot, LucienBOT, Paine Ellsworth, Tobby72, Ajgw56, WurzelT, Citation bot 1, Merongb10, RandomDSdevel, Pinethicket, HRoestBot, Rameshngbot, Tom.Reding, Swamper777, Phil John Hawkins, PRONIZ, Rknop, Jkforde, Nieuwenh, Puzl bustr, Tawe, Higgshunter, Beladee, Lotje, Persian knight shiraz, DrSinn, Miracle Pen, EngineerFromVega, RjwilmsiBot, MalapropX14, DexDor, Ripchip Bot, Phlegat, John of Reading, MindBlender, Architeuthidae, Racerx11, RA0808, Theonhwiki, 8digits, Themorrissey, NorthernRaven, Hhhippo, ZéroBot, ره‌ۆﻥﺵﻯﻥ ﺏﻩﺍﺭ, StringTheory11, Waperkins, Herp Derp, H3llBot, Quondum, UniversumExNihilo, Almatinez, Aschwole, Mayur, Kranix, Maschen, Eg-T2g, LarsJanZeeuwRules, LikeLakers2, Rocketrod1960, H1tchh1ker4, Mikhail Ryazanov, ClueBot NG, PeterKirk69, 4Jmaster, Gilderien, Navasj, Law of Entropy, Confuddledone, 336, Helpful Pixie Bot, Mightyname, Electriccatfish2, Kronn8, Curb Chain, Bibcode Bot, Tirebiter78, Rm1271, Neutral current, Stehgdop, Watson system, Koska One, Cs1791, Chewkaflax, Ddanndt, Johan.lundberg, Dragonami, Ownedroad9, Quickcrazy78, Rajibganguly01, Uioplk, Acalloni, Zedshort, TF SHaDowMAn, Yaroslav Nikitenko, Neutrinoread, Jakekong, Achowat, Pritombose, Ysawires, GodsAccident, Blakee911, Pieceofchit1, BattyBot, Layth888, Friedncrispy, Dja1979, Adyyy, Dexbot, Jdjwright, Mogism, Jaxcp3, Reatlas, Provacitu74, Coladar, CensoredScribe, Juan tanesia12, Rmohapat, Rmohapatra, Christophe1946, Klingerdinger, EWPage, Monkbot, Richard Henry Eckert, Jazzwhiz101, Crystallizedcarbon, Fimatic, PerpetuaLux, Isambard Kingdom, Lxplot, TheHecster, DN-boards1, Matan Kovac, KasparBot, Alarana, Saturn comes back around, CumbleSpuzz and Anonymous: 576

- **Electron neutrino** *Source:* https://en.wikipedia.org/wiki/Electron_neutrino?oldid=674917553 *Contributors:* Bryan Derksen, Bobrayner, Rjwilmsi, Strait, Bgwhite, Jeffhoy, Lockesdonkey, Dna-webmaster, GDallimore, Thijs!bot, Headbomb, Oreo Priest, Magioladitis, Maurice Carbonaro, Nwbeeson, Ggenellina, FourteenDays, SieBot, SkyLined, Addbot, LaaknorBot, PieterJanR, Luckas-bot, Rubinbot, Citation bot, ArthurBot, Xqbot, Carlog3, Paine Ellsworth, Citation bot 1, TjBot, EmausBot, John of Reading, Optiguy54, TuHan-Bot, JSquish, Quondum, Kasirbot, Helpful Pixie Bot, Bibcode Bot, Love's Labour Lost, Tpaine krk, Svebert, Makecat-bot, DD4235, Scipsycho and Anonymous: 10

- **Muon neutrino** *Source:* https://en.wikipedia.org/wiki/Muon_neutrino?oldid=680138955 *Contributors:* Bryan Derksen, The Anome, Twang, Goudzovski, Eleassar, Dna-webmaster, SmackBot, Ruslik0, Thijs!bot, Headbomb, Magioladitis, Ggenellina, FourteenDays, SieBot, Thesavagenorwegian, Muhends, Ajoykt, SkyLined, Addbot, PieterJanR, Luckas-bot, AnomieBOT, Rubinbot, Icalanise, ArthurBot, RibotBOT, Paine Ellsworth, Citation bot 1, TjBot, Mithril, EmausBot, ZéroBot, StringTheory11, Quondum, Kasirbot, Bibcode Bot, Sunitharay, Artdk, DaveW51, QzPhysics and Anonymous: 7

- **Tau neutrino** *Source:* https://en.wikipedia.org/wiki/Tau_neutrino?oldid=644666618 *Contributors:* B.d.mills, Kfitzner, Rjwilmsi, Salsb, TriTert-Butoxy, Newone, Thijs!bot, Headbomb, Fences and windows, Ggenellina, FourteenDays, SieBot, SkyLined, Addbot, Ronhjones, Luckas-bot, Rubinbot, JackieBot, Icalanise, Citation bot, ArthurBot, Carlog3, LucienBOT, Paine Ellsworth, Citation bot 1, TjBot, EmausBot, K6ka, Joe Gazz84, ZéroBot, Kasirbot, Bibcode Bot and Anonymous: 5

- **Neutrino detector** *Source:* https://en.wikipedia.org/wiki/Neutrino_detector?oldid=663331886 *Contributors:* Rursus, Giftlite, Leonard G., DragonflySixtyseven, Fg2, Pjacobi, ArnoldReinhold, RJHall, A2Kafir, Ferrierd, Bjones, MiG, Strait, Mike Peel, Goudzovski, YurikBot, Phr en, SmackBot, Rspanton, MTSbot~enwiki, Thijs!bot, Naucer, Superhilac, Headbomb, Escarbot, R'n'B, CommonsDelinker, GenghisDon, Fountains of Bryn Mawr, Idioma-bot, Sheliak, Edewolf, SkyLined, Addbot, Ozzy1948, Skippy le Grand Gourou, Snaily, Luckas-bot, ArthurBot, Xqbot, Tomwsulcer, Trafford09, Mnmngb, Erik9bot, Tom.Reding, A930913, ClueBot NG, Helpful Pixie Bot, BattyBot and Anonymous: 21

- **Neutrino oscillation** *Source:* https://en.wikipedia.org/wiki/Neutrino_oscillation?oldid=678950918 *Contributors:* Edward, Michael Hardy, SebastianHelm, Taxman, BenRG, Robbot, Nurg, Giftlite, Jmnbpt, Xerxes314, ConradPino, Tubedogg, B.d.mills, Lazarus666, Mike Rosoft, Rich Farmbrough, ESkog, Cedders, Worldtraveller, Keenan Pepper, Bsadowski1, Falcorian, Flying fish, Linas, JFG, Jugger90, Plrk, Yurik, Rjwilmsi, Strait, Mike Peel, Dudegalea, Erkcan, Itinerant1, Goudzovski, Chobot, Hairy Dude, Kordas, Aaronwinborn, Salsb, Jamesg, Thiseye, Santaduck, Gadget850, Banus, Tosus, Teply, GrinBot~enwiki, Eatcacti, MacsBug, SmackBot, Haymaker, Arbe, Stepa, Saros136, Bluebot, Timothy Clemans, QFT, DMacks, Q9a, Mets501, IRevLinas, Kurtan~enwiki, Harold f, MrFizyx, Rotiro, Michael C Price, Hugozam, Headbomb, D.H, Credema, Yellowdesk, Spartaz, DanPMK, Leyo, Choihei, Higgsino, Lseixas, Improve~enwiki, EverGreg, Jasondet, Paulfharrison, Aardvarkleg, Al Leween, Von Crayola, Mild Bill Hiccup, Hyh1048576, Alexbot, Jwfvalle, SchreiberBike, Asf107, SkyLined, Luismarques83, Addbot, Zorrobot, Luckas-bot, Yobot, Wireader, AnomieBOT, Citation bot, Blennow, Mnmngb, Paine Ellsworth, Ysyoon, Jonesey95, Pmokeefe, RjwilmsiBot, CaptRik, Hhhippo, Arbnos, Cymru.lass, Timetraveler3.14, Xronon, Gareth Griffith-Jones, Ben morphett, Mightyname, Bibcode Bot, BG19bot, Dwightboone, MskKrieger, Arcandam, Zanpan, Klingerdinger, Drscientific, Sircier, Soham92 and Anonymous: 101

- **Pontecorvo–Maki–Nakagawa–Sakata matrix** *Source:* https://en.wikipedia.org/wiki/Pontecorvo%E2%80%93Maki%E2%80%93Nakagawa%E2%80%93Sakata_matrix?oldid=677906442 *Contributors:* Julesd, Jordi Burguet Castell, Giftlite, Bender235, Strait, Goudzovski, Ohwilleke, Leo C Stein, Hydraton31, Headbomb, R'n'B, Cuzkatzimhut, TXiKiBoT, FourteenDays, Jasondet, Copyeditor42, Addbot, Debresser, Luckas-bot, Yobot, Citation bot, Blennow, Ace111, A. di M., D'ohBot, Puzl bustr, John of Reading, ZéroBot, ClueBot NG, Bibcode Bot, Franzl aus tirol, Foreveriii, Danholly and Anonymous: 15

- **Solar neutrino problem** *Source:* https://en.wikipedia.org/wiki/Solar_neutrino_problem?oldid=678435470 *Contributors:* AxelBoldt, Bryan Derksen, Zundark, The Anome, AstroNomer~enwiki, Malcolm Farmer, Chrislintott, Roadrunner, Maury Markowitz, Heron, Boud, JohnOwens, Ken Arromdee, Michael Hardy, Cyde, Ahoerstemeier, William M. Connolley, Notheruser, JWSchmidt, Wolfstu, Loren Rosen, Timwi, Reddi, Taxman, Phys, Thue, Bearcat, Robbot, Peak, Mlaine, Davidl9999, Hadal, Diberri, Stirling Newberry, Giftlite, Harp, Marcika, Wikibob, Jorge Stolfi, Gzornenplatz, Chowbok, Kaldari, DragonflySixtyseven, AmarChandra, Neutrality, IcycleMort, JTN, NeilTarrant, Worldtraveller, Viriditas, Jason One, Enirac Sum, MattHaffner, Axeman89, Flying fish, Dzordzm, GregorB, Chinacat, Tevatron~enwiki, Graham87, Strait, Mike Peel, Goudzovski, Gepay, Physicsdavid, Phr en, Kgf0, Solarkennedy, SmackBot, Eloy, Agradman, Dicklyon, Topace10, Ruslik0, Korander, Cyhawk, Boardhead, Synergy, Headbomb, FST777, Fuchsias, BSVulturis, Charliet, Turtlens, Pere prlpz, Maurice Carbonaro, CaptinJohn, ^demonBot2, Zentau7, Bsartre, ImageRemovalBot, VsBot, Five-toed-sloth, DumZiBoT, MystBot, Addbot, Fluffernutter, KamikazeBot, Citation bot, Br77rino, Mnmngb, Trebauchet1986, Tom.Reding, IVAN3MAN, Alph Bot, Chuck Kincy, ZéroBot, ClueBot NG, Oklahoma3477, Bibcode Bot, BG19bot, Petermahlzahn, Aaaaaaum, Gshahali, Soham92, MinorStoop and Anonymous: 53

- **Sterile neutrino** *Source:* https://en.wikipedia.org/wiki/Sterile_neutrino?oldid=679151361 *Contributors:* Rursus, Giftlite, Xerxes314, Robert Brockway, Jkl, Rich Farmbrough, Pjacobi, I9Q79oL78KiL0QTFHgyc, Cgmusselman, Count Iblis, Ceyockey, GregorB, Rjwilmsi, Strait, Goudzovski, Tinlad, SCZenz, Reyk, Nickst, Fredvanner, Colonies Chris, V1adis1av, QFT, Mayrel, JorisvS, Dan Gluck, Foice, Michael C Price, Thijs!bot, Headbomb, Qwerty Binary, Wdtaylor1066, DanPMK, .anacondabot, Jayanthtn, Ccrummer, R'n'B, Mikek999, Fordi, Rod57, Sheliak, VolkovBot, Paradoctor, Wing gundam, BartekChom, Wwheaton, Alexbot, Jwfvalle, DumZiBoT, BodhisattvaBot, MystBot, SkyLined, Addbot, Eric Drexler, Yobot, Romul~enwiki, Stuffed cat, Citation bot, Omnipaedista, Ernsts, Paine Ellsworth, DrilBot, Jonesey95, Tom.Reding, Michael9422, JLincoln, Bj norge, Dskrvk, EmausBot, ZéroBot, StringTheory11, Timetraveler3.14, Maschen, Bibcode Bot, BG19bot, Chris-Gualtieri, Egofofo, AloisKabelschacht, Ajbilan, Monkbot, ⊠□, Lxplot and Anonymous: 34

- **Lepton number** *Source:* https://en.wikipedia.org/wiki/Lepton_number?oldid=678945693 *Contributors:* Xavic69, Alfio, Phys, Robbot, RScheiber, Jag123, Flying fish, SeventyThree, Yurik, Bambaiah, JabberWok, That Guy, From That Show!, SmackBot, Incnis Mrsi, Fuhghettaboutit, Doug Bell, Khazar, Man pl, Blinking Spirit, Cydebot, Thijs!bot, Headbomb, WolfmanSF, Leyo, FaTTshady74, A4bot, Murkee, Gerakibot, Peachypoh, Addbot, Peti610botH, Yobot, Amirobot, AnomieBOT, Unara, Citation bot, GrouchoBot, 靖天子, HRoestBot, EmausBot, Timetraveler3.14, ChuispastonBot, Bibcode Bot, BG19bot, Monkbot, MarkovianStumble and Anonymous: 24

- **Chirality (physics)** *Source:* https://en.wikipedia.org/wiki/Chirality_(physics)?oldid=681061355 *Contributors:* XJaM, Patrick, Michael Hardy, Ettlz, Dcljr, Smack, Phys, Jeffq, Nickptar, .:Ajvol:., Deadworm222, Linas, Nneonneo, Erkcan, Itinerant1, Goudzovski, YurikBot, Bambaiah, TobiS, Larsobrien, SmackBot, Tom Lougheed, Hmains, Bluebot, Radagast83, Acdx, Yevgeny Kats, Aiwendil42, WhiteHatLurker, CmdrObot, Avillia, Michael C Price, Difty, BetacommandBot, Headbomb, Roggg, B-80, Icep, Stannered, Alphachimpbot, CosineKitty, Arch dude, Frankie816, R sirahata, Leftfoot69, HEL, DrKiernan, Cuzkatzimhut, VolkovBot, PhysPhD, Paolo.dL, BartekChom, ClueBot, CristianCantoro, Bob108, SchreiberBike, RP459, Addbot, Jkasd, Drkarat, Yobot, Synchronism, Choij, Citation bot, ArthurBot, Ernsts, FrescoBot, Bj norge, Hanswehrli, Maschen, Donner60, RockMagnetist, Preon, Luizpuodzius, Vkpd11, Brieuc varphi, ChrisGualtieri, BronzeRatio, Linoush, YiFeiBot and Anonymous: 39

- **Helicity (particle physics)** *Source:* https://en.wikipedia.org/wiki/Helicity_(particle_physics)?oldid=680910551 *Contributors:* Stevertigo, Michael Hardy, Phys, Robbot, Fropuff, Lumidek, Rich Farmbrough, Dmr2, DannyWilde, Chobot, YurikBot, RussBot, Conscious, Archelon, SmackBot, Unint, Loodog, Kurtan~enwiki, Rumsey, Barticus88, Headbomb, WVhybrid, Davidhorman, Stannered, David Eppstein, Infovarius, HEL, Cuzkatzimhut, Ryan032, Red Act, PhysPhD, OKBot, Muhends, PipepBot, DumZiBoT, Addbot, Luckas-bot, Yobot, Niout, Jim1138, ArthurBot, Alfa137, EmausBot, WikitanvirBot, ZéroBot, Maschen, RockMagnetist, Tommy.Hudec, Lxplot and Anonymous: 11

- **Michel parameters** *Source:* https://en.wikipedia.org/wiki/Michel_parameters?oldid=599591714 *Contributors:* TakuyaMurata, Erkcan, Goudzovski, Dialectric, Headbomb, Coppertwig, Why Not A Duck, Citation bot 1, Trappist the monk, Scheeler, Bibcode Bot and Anonymous: 1

- **Weak isospin** *Source:* https://en.wikipedia.org/wiki/Weak_isospin?oldid=679239808 *Contributors:* Xavic69, Charles Matthews, Giftlite, RScheiber, Hidaspal, Ian Pitchford, Chobot, Roboto de Ajvol, YurikBot, Bambaiah, Jimp, RussBot, Paul D. Anderson, Jheriko, Bbabba, Cydebot, Michael C Price, Headbomb, Igodard, Tokei-so, Andre.holzner, Pamputt, AlleborgoBot, Muhends, L.smithfield, Tvine, Addbot, Icalanise, ArthurBot, Ernsts, Puzl bustr, John of Reading and Anonymous: 17

- **Weak hypercharge** *Source:* https://en.wikipedia.org/wiki/Weak_hypercharge?oldid=679239006 *Contributors:* Xavic69, Pjacobi, MeltBanana, David Schaich, BD2412, Chobot, Roboto de Ajvol, YurikBot, Bambaiah, Paul D. Anderson, Incnis Mrsi, Dauto, Michael C Price, Headbomb, Andre.holzner, HEL, Anonymous Dissident, Pamputt, Antixt, Tvine, MystBot, Addbot, Luckas-bot, Yobot, Götz, Ernsts, Slightsmile, QuantumSquirrel, ResidentAnthropologist, Helpful Pixie Bot, Bambi12~enwiki, EzPz4 and Anonymous: 20

- **B − L** *Source:* https://en.wikipedia.org/wiki/B_%E2%88%92_L?oldid=621966745 *Contributors:* Phys, Herbee, Xerxes314, Jag123, Starwed, Mike Peel, Bgwhite, Roboto de Ajvol, RussBot, Conscious, Gaius Cornelius, SmackBot, QFT, Doug Bell, Cydebot, Michael C Price, Headbomb, Maliz, Andre.holzner, Pamputt, ClueBot, Auntof6, MystBot, SkyLined, Addbot, Mjamja, Luckas-bot, Yobot, Ptbotgourou, Ernsts, A. di M., Erik9bot and Anonymous: 7

- **X (charge)** *Source:* https://en.wikipedia.org/wiki/X_(charge)?oldid=605281425 *Contributors:* XJaM, BD2412, SmackBot, Michael C Price, Headbomb, Addbot, Xqbot, Ernsts, A. di M. and Anonymous: 1

- **Weak interaction** *Source:* https://en.wikipedia.org/wiki/Weak_interaction?oldid=679571094 *Contributors:* AxelBoldt, Chenyu, Sodium, Bryan Derksen, Tarquin, AstroNomer~enwiki, Andre Engels, XJaM, Heron, JohnOwens, Gdarin, Delirium, Andrewa, Andres, Emperorbma, Timwi, Fibonacci, Phys, Phil Boswell, Lowellian, Mayooranathan, Tobias Bergemann, Giftlite, Sj, Herbee, Xerxes314, Jcobb, Mckaysalisbury, Munkee, Toby Woodwark, Bbbl67, Icairns, AmarChandra, Lumidek, Jørgen Friis Bak, Discospinster, ArnoldReinhold, Roybb95~enwiki, Gianluigi, Joanjoc~enwiki, Shanes, AJP, AtomicDragon, Danski14, Alansohn, Arthena, Axl, SidneySM, Hwefhasvs, DV8 2XL, Nightstallion, Kazvorpal, Linas, StradivariusTV, Benbest, Bbatsell, Palica, Tevatron~enwiki, Graham87, BD2412, Ketiltrout, Rjwilmsi, Strait, Erkcan, The wub, FlaBot, Naraht, Itinerant1, Srleffler, Chobot, Krishnavedala, YurikBot, Borgx, Bambaiah, Hairy Dude, Jimp, Sillybilly, Conscious, Epolk, JabberWok, Gaius Cornelius, Shaddack, SCZenz, Irishguy, Shimei, Willtron, RG2, Phr en, That Guy, From That Show!, Luk, SmackBot, David Kernow, Tom Lougheed, WookieInHeat, Dauto, Chris the speller, Philosopher, Moshe Constantine Hassan Al-Silverburg, Complexica, DHN-bot~enwiki, Zirconscot, BIL, Wen D House, "alyosha", Maxwahrhaftig, Akriasas, Vina-iwbot~enwiki, Bdushaw, TTE, SashatoBot, Fontenello, Herr apa, Condem, Tony Fox, MottyGlix, JRSpriggs, Heartofgoldfish, Calmargulis, Green caterpillar, Joelholdsworth, Cydebot, Michael C Price, Mtpaley, Thijs!bot, ChKa, Kichwa Tembo, Headbomb, Hcobb, Icep, Escarbot, AntiVandalBot, Jimeree, Steelpillow, JAnDbot, Magioladitis, Swpb, بسام, Wormcast, DAGwyn, Giggy, Khalid Mahmood, Gah4, Tarotcards, 2help, Lighted Match, DorganBot, Halmstad, Idiomabot, VolkovBot, Jcuadros, Hilarious Bookbinder, TXiKiBoT, Rei-bot, CaptinJohn, Awl, Shenanegins, BotKung, Wingedsubmariner, Antixt, Xxxlilbritxxx, Ptrslv72, Monty845, AlleborgoBot, SieBot, Paolo.dL, Skyentist, Ptr123, ClueBot, Bondchic007, SuperHamster, Erudecorp, Rotational, Jackey0105, Alexbot, Cenarium, Zomno, Zahnrad, He6kd, TimothyRias, InternetMeme, Timo Metzemakers, Stephen Poppitt, Addbot, Some jerk on the Internet, Markdman, ChenzwBot, Ehrenkater, Tide rolls, Luckas-bot, Yobot, Les boys, Kilom691, THEN WHO WAS PHONE?, Rifter0x0000, Duping Man, Dickdock, Magog the Ogre, AnomieBOT, Materialscientist, Citation bot, Quebec99, Kreigiron, Xqbot, Drilnoth, BurntSynapse, GrouchoBot, Omnipaedista, RibotBOT, Workanode, Jaz1305, Mnmngb, Dave3457, FrescoBot, Charles.walker, LucienBOT, Ionutzmovie, Grandiose, Pinethicket, Boulaur, Rameshngbot, RedBot, 23790AD, Tea with toast, Jauhienij, FoxBot, Earthandmoon, RjwilmsiBot, Itamarhason, Newty23125, EmausBot, WikitanvirBot, GA bot, GoingBatty, Splibubay, StringTheory11, Braswiki, Git2010, Wayne Slam, Jsayre64, Maschen, ChuispastonBot, ClueBot NG, VinculumMan, Physics is all gnomes, Fjpyanez, Mouse20080706, Helpful Pixie

Bot, Geo7777, Bibcode Bot, Junaid2754, Bolatbek, Phbarnacle, Neutral current, Glevum, Idenshi, Marioedesouza, Dexbot, Spray787, Reatlas, CsDix, Jamesmcmahon0, Ihatedirac2k13, Kharkiv07, Jwratner1, YimmyYohnson, Monkbot, BalderdashVonDrivel, ASCarretero, Malerisch, Lachlan Newland, Tetra quark, KasparBot and Anonymous: 155

- **Antiparticle** *Source:* https://en.wikipedia.org/wiki/Antiparticle?oldid=680600374 *Contributors:* AxelBoldt, CYD, Mav, Bryan Derksen, Andre Engels, Josh Grosse, Stevertigo, Mrwojo, Patrick, RTC, Paddu, CesarB, Nikai, Nikola Smolenski, Charles Matthews, The Anomebot, Wik, Omegatron, Bevo, Altenmann, Merovingian, Intangir, Wikibot, Martinwguy, Giftlite, Bogdanb, Harp, BenFrantzDale, Herbee, Spencer195, Fleminra, Jason Quinn, Zeimusu, Mako098765, Karol Langner, Mike Rosoft, Helohe, Rich Farmbrough, Guanabot, Pjacobi, Guanabot2, Mr. Billion, Joanjoc~enwiki, Kghose, Cmdrjameson, Giraffedata, Matt McIrvin, HasharBot~enwiki, Pediddle, Deror avi, Woohookitty, Mindmatrix, Wdyoung, GregorB, SeventyThree, Justin Ormont, Palica, Marudubshinki, Tevatron~enwiki, Rjwilmsi, Ae77, MZMcBride, KaiMartin, FlaBot, Krackpipe, Commander Nemet, Roboto de Ajvol, YurikBot, Borgx, Bambaiah, Zhaladshar, Spike Wilbury, Bota47, Terbospeed, Mkossick, Tim314, タ チ コ マ robot, SmackBot, FocalPoint, Alsandro, Srnec, Dauto, Octahedron80, Drphilharmonic, Marcus Brute, Vinaiwbot~enwiki, Jake-helliwell, Grumpyyoungman01, Newone, Mellery, Van helsing, Tim1988, Myasuda, Gogo Dodo, Goldencako, Thijs!bot, Headbomb, Tyco.skinner, JAnDbot, Steveprutz, Ferritecore, Jpod2, Singularity, Dbiel, TomasBat, Eternalmatt, Joshmt, DorganBot, Cuckooman4, VolkovBot, TXiKiBoT, Red Act, Anonymous Dissident, AlleborgoBot, SieBot, Likebox, RadicalOne, Flyer22, KoenDelaere, Thomega, RW Marloe, BrightRoundCircle, Davidmosen, Jacob.jose, Anyeverybody, ClueBot, Diagramma Della Verita, Alexbot, Eeekster, Rishi.bedi, SilvonenBot, NellieBly, Lilaspastia, SkyLined, AkhtaBot, CarsracBot, Lightbot, Legobot, Luckas-bot, Yobot, Planlips, Csmallw, AnomieBOT, Citation bot, Vuerqex, ArthurBot, Xqbot, Omnipaedista, RibotBOT, Muhwang, EmausBot, John of Reading, L Kensington, Benazhack, ClueBot NG, Geekingreen, Mesoderm, Bibcode Bot, B wik, Mark Arsten, Rm1271, Penguinstorm300, Robotsheepboy, YFdyh-bot, 77Mike77, संजीव कुमार, Dert567, Monkbot, Nazo!nin, KasparBot and Anonymous: 100

- **Beta decay** *Source:* https://en.wikipedia.org/wiki/Beta_decay?oldid=682280135 *Contributors:* AxelBoldt, Chenyu, Trelvis, Mav, Peterlin~enwiki, Ellywa, Andrewa, Cyan, Mxn, Robertb-dc, Shizhao, Pstudier, Jusjih, Twang, Donarreiskoffer, Robbot, Enochlau, Giftlite, Donvinzk, Harp, Herbee, Xerxes314, Radius, Mdob, Antandrus, Icairns, Ukexpat, Jørgen Friis Bak, Discospinster, Guanabot, Vsmith, Roo72, Gianluigi, Joanjoc~enwiki, Neilrieck, Bobo192, Army1987, Drw25, Nk, Haham hanuka, AjAldous, Wtmitchell, Saga City, Falcorian, Flying fish, Eleassar777, Gimboid13, Graham87, Rjwilmsi, FlaBot, Ground Zero, Itinerant1, Goudzovski, Chobot, AllyD, Bgwhite, Roboto de Ajvol, YurikBot, JWB, Jimp, JabberWok, Romanc19s, Spike Wilbury, Johantheghost, Reyk, Geoffrey.landis, JLaTondre, MacsBug, SmackBot, Incnis Mrsi, Ixtli, Gilliam, Skizzik, Dauto, Chris the speller, Octahedron80, Yurigerhard, Sbharris, Audriusa, Tsca.bot, V1adis1av, Decltype, Akulkis, Hgilbert, Polonium, Adj08, JorisvS, Steipe, Mets501, Tuttt, Happy-melon, Civil Engineer III, Timrem, CRGreathouse, JohnCD, Rwflammang, Joelholdsworth, WeggeBot, Kanags, HPaul, Neil9999, Barticus88, Headbomb, Escarbot, AntiVandalBot, Edokter, LibLord, Salgueiro~enwiki, MSBOT, .anacondabot, VoABot II, SHCarter, Pixel ;-), Geekmansworld, Kevinmon, Johnbibby, Dirac66, Edward321, Geboy, Andre.holzner, Catmoongirl, Happyfacesrock, It Is Me Here, Rominandreu, Mcat2, KylieTastic, DorganBot, Y2H, Sheliak, Club house, Milesisgreat, VolkovBot, Hqb, JhsBot, BotKung, Pishogue, FMasic, Cnilep, Tresiden, PlanetStar, Jasondet, Paolo.dL, Arjen Dijksman, Sean.hoyland, Rjc34, Muhends, Sidhu ghanta, Loren.wilton, ClueBot, R000t, Maxtitan, EhJJ, Mikaey, SoxBot III, Directormq, RP459, SkyLined, Addbot, Yakiv Gluck, DOI bot, Man utd suger, Ayrenz, Mdnahas, Zorrobot, Spacy73, Skippy le Grand Gourou, आशीष भटनागर, Luckas-bot, Yobot, Fraggle81, TaBOTzerem, THEN WHO WAS PHONE?, Azylber, Kulmalukko, AnomieBOT, SamuraiBot, Citation bot, Xqbot, FrescoBot, Tekmeme, Jonesey95, Achim1999, Minivip, ApusChin, Double sharp, Bj norge, Andrea105, John of Reading, Acather96, Dewritech, GoingBatty, JSquish, Δ, Coasterlover1994, Illinikiwi, ClueBot NG, Movses-bot, Widr, Meea, MerllwBot, Bibcode Bot, BG19bot, ElphiBot, Onewhohelps, Cadiomals, Ragnarstroberg, Glevum, Currb, CeraBot, Idenshi, Goyala1, Stigmatella aurantiaca, ChrisGualtieri, Pvoytas, Monkbot, Raytuzio, Hyperclassic, Scipsycho, KasparBot and Anonymous: 174

- **Double beta decay** *Source:* https://en.wikipedia.org/wiki/Double_beta_decay?oldid=679399596 *Contributors:* Bryan Derksen, Donarreiskoffer, Ruakh, Giftlite, Jason Quinn, Rich Farmbrough, Dnwq, CDN99, Kjkolb, Jumbuck, RJFJR, Flying fish, Yurik, Strait, Ian Pitchford, Goudzovski, Chobot, Bgwhite, YurikBot, JabberWok, Tomvds, Nekura, Itub, MacsBug, Stepa, V1adis1av, QFT, JorisvS, Beetstra, Ruslik0, WISo, Headbomb, QuantumEngineer, Magioladitis, EvilFred, Amcarroll32, J.delanoy, AstroHurricane001, Red Harvest, Rominandreu, Sheliak, Rambatino, AlexDenney, Pamputt, Antixt, Northfox, SieBot, Jasondet, Aardvarkleg, Emk, Jwfvalle, Muro Bot, Stepheng3, Addbot, Particleguy, Spacy73, Yobot, Dreamer08, AnomieBOT, Rubinbot, PianoDan, Carturo222, Brandon5485, Mikespedia, Double sharp, Higgshunter, Inthedryer, John of Reading, Dewritech, MGMPhysics, Timetraveler3.14, KlapdorKleingrothaus, Bibcode Bot, BG19bot, Dja1979, Illia Connell, Dexbot, Fun4free2, Sheyga, Lxplot and Anonymous: 50

- **Fermion** *Source:* https://en.wikipedia.org/wiki/Fermion?oldid=682673767 *Contributors:* AxelBoldt, Chenyu, Derek Ross, CYD, Mav, Bryan Derksen, The Anome, Ben-Zin~enwiki, Alan Peakall, Dominus, Dcljr, Looxix~enwiki, Glenn, Nikai, Andres, Wikiborg, David Latapie, Phys, Bevo, Stormie, Olathe, Donarreiskoffer, Robbot, Merovingian, Rorro, Wikibot, HaeB, Giftlite, Fropuff, Xerxes314, Vivektewary, JoJan, Karol Langner, Tothebarricades.tk, Icairns, Hidaspal, Vsmith, Laurascudder, Lysdexia, Ashlux, Graham87, Magister Mathematicae, Kbdank71, Syndicate, Strait, Protez, Drrngrvy, FlaBot, Srleffler, Chobot, YurikBot, RobotE, Jimp, Bhny, Captaindan, SpuriousQ, Salsb, Lomn, Enormousdude, CharlesHBennett, Federalist51, Tom Lougheed, Unyoyega, Jrockley, MK8, BabuBhatt, Complexica, Zachorious, Shalom Yechiel, QFT, Garry Denke, Daniel.Cardenas, SashatoBot, Flipperinu, Dan Gluck, LearningKnight, Happy-melon, Paulfriedman7, Cydebot, Meno25, Zalgo, Thijs!bot, Mbell, Headbomb, Nick Number, Orionus, Shlomi Hillel, CosineKitty, NE2, Mwarren us, ZPM, Vanished user ty12kl89jq10, Joshua Davis, R'n'B, Tensegrity, Rod57, Dgiraffes, Alpvax, VolkovBot, TXiKiBoT, Red Act, Anonymous Dissident, Abdullais4u, בל יכול, Tanhueiming, Antixt, Haiviet~enwiki, EmxBot, Kbrose, SieBot, Likebox, Jojalozzo, Dhatfield, Oxymoron83, TubularWorld, ClueBot, Seervoitek, Rodhullandemu, Jorisverbiest, Feebas factor, ChandlerMapBot, Nilradical, Wikeepedian, Stephen Poppitt, Addbot, Vectorboson, Luckas-bot, Yobot, Planlips, Dickdock, AnomieBOT, Icalanise, Materialscientist, Xqbot, Br77rino, Balaonair, 老陳, Paine Ellsworth, Blackoutjack, Kikeku, Rameshngbot, Tom.Reding, RedBot, Alarichus, Michael9422, Silicon-28, TjBot, EmausBot, WikitanvirBot, Quazar121, Solomonfromfinland, JSquish, Fimin, Quondum, AManWithNoPlan, EdoBot, ClueBot NG, PBot1, EthanChant, Bibcode Bot, BG19bot, Petermahlzahn, KingKhan85, ChrisGualtieri, BoethiusUK, DerekWinters, Tentinator, JNrgbKLM, Mohit rajpal, KasparBot, Jiswin1992 and Anonymous: 120

- **Flavour (particle physics)** *Source:* https://en.wikipedia.org/wiki/Flavour_(particle_physics)?oldid=681888935 *Contributors:* Schewek, Michael Hardy, Nurg, Xerxes314, Varlaam, Andycjp, R. fiend, DragonflySixtyseven, CALR, STGM, Andrew Gray, Knowledge Seeker, Egg, Alai, Sylvain Mielot, Linas, Mindmatrix, SpNeo, Drrngrvy, YurikBot, Bambaiah, Hairy Dude, NTBot~enwiki, Bhny, Cossy, Długosz, SCZenz, Nick, Karl Andrews, SmackBot, Incnis Mrsi, Dauto, Doug Bell, Zero sharp, Ompty, BFD1, Ruslik0, Cydebot, Hydraton31, Xxanthippe, Michael C Price, Thijs!bot, Headbomb, FelixP~enwiki, Rompe, Hayesgm, Knotwork, CosineKitty, Robin S, Askielboe, Yonidebot, Choihei,

I310342~enwiki, Thecinimod, VolkovBot, A4bot, Kresadlo, Maxim, Odellus, Ptrslv72, SieBot, VVVBot, The Stickler, Muhends, PixelBot, Jtle515, Count Truthstein, DumZiBoT, MystBot, SkyLined, Addbot, ZeroOmega, SpBot, Ehrenkater, HerculeBot, Luckas-bot, Ptbotgourou, Magog the Ogre, Icalanise, Omnipaedista, Citation bot 1, Xtermin8R645, B2NVB2, Jrobbinz123, 777sms, Bizzurp, EmausBot, VinculumMan, AvocatoBot, Drift chambers, Skynden, Isambard Kingdom and Anonymous: 42

- **Generation (particle physics)** *Source:* https://en.wikipedia.org/wiki/Generation_(particle_physics)?oldid=672806813 *Contributors:* Selket, Arkuat, Aleron235, Giftlite, Harp, Xerxes314, Keith Edkins, Discospinster, FT2, Army1987, John Vandenberg, PaulHanson, DannyWilde, Fosnez, Vossman, Mushin, Conscious, Bhny, FFLaguna, SmackBot, InverseHypercube, Anastrophe, Dauto, Jmnbatista, Titus III, Aeluwas, SchmittM, Vyznev Xnebara, Q43, Thijs!bot, Headbomb, Nick Number, JAnDbot, Kborland, Choihei, TomasBat, Idioma-bot, VolkovBot, Nx-avar, SieBot, Muhends, TimothyRias, MystBot, Addbot, Luckas-bot, HieronymousCrowley, Icalanise, Citation bot, Xqbot, LucienBOT, JIROT, Akesich, Afteread, Bookalign, Barak90, ZéroBot, ChuispastonBot, Bibcode Bot, Blaspie55, Prokaryotes, Isambard Kingdom and Anonymous: 24

- **Leptoquark** *Source:* https://en.wikipedia.org/wiki/Leptoquark?oldid=674819966 *Contributors:* Bcorr, Herbee, Xerxes314, Pjacobi, David Schaich, Bender235, Erkcan, Timboe, Hairy Dude, Conscious, PJTraill, Colonies Chris, Mesons, Linus M., Headbomb, Sanitycult, Calwiki, Waltoncats, Addbot, Mjamja, Wireader, Citation bot, Ernsts, Citation bot 1, Slightsmile, ResidentAnthropologist, Bibcode Bot and Anonymous: 9

- **Koide formula** *Source:* https://en.wikipedia.org/wiki/Koide_formula?oldid=667222329 *Contributors:* XJaM, Wikiborg, BenRG, Giftlite, Mporter, Herbee, Jason Quinn, Arivero, Pavel Vozenilek, Amaurea, Hairy Dude, Jimp, Hmains, Bluebot, Vladislav, Kendrick7, CmdrObot, Headbomb, Autotheist, Mu7, Trusilver, Potatoswatter, Amgasiago, Samdhatte, MystBot, Addbot, Yobot, AnomieBOT, Icalanise, Citation bot, A. di M., Citation bot 1, RedBot, Trappist the monk, EmausBot, Suslindisambiguator, Bibcode Bot, Ervin Goldfain, IluvatarBot, Kennethaw88, Mfb and Anonymous: 24

- **Majorana fermion** *Source:* https://en.wikipedia.org/wiki/Majorana_fermion?oldid=680786745 *Contributors:* Pablo Mayrgundter, Jerzy, BenRG, Finlay McWalter, Lumos3, Phil Boswell, Giftlite, Dmmaus, Chris Howard, ArnoldReinhold, Bender235, Sburke, Rjwilmsi, Bubba73, Hairy Dude, Chris Capoccia, Buster79, SCZenz, Larsobrien, Thnidu, Teply, Incnis Mrsi, Modest Genius, JorisvS, Brienanni, Tmangray, Vttoth, Cydebot, Ntsimp, Difluoroethene, Quibik, Whatever1111, Headbomb, Davidhorman, Tjmayerinsf, Bpmullins, R sirahata, HEL, MistyMorn, Ljgua124, Satani, Pamputt, Afernand74, Coinmanj, Jwfvalle, Another Believer, Dthomsen8, Addbot, Roentgenium111, GDK, Luckas-bot, Yobot, Ptbotgourou, Amirobot, AnomieBOT, Citation bot, Obersachsebot, MIRROR, Omnipaedista, The Interior, Astiburg, Nicolas Perrault III, Paine Ellsworth, Abductive, Tom.Reding, Amonet, Trappist the monk, RRBiswas, RobinPolt, JSquish, Quondum, AlbertusmagnusOP, Brandmeister, KarlsenBot, Claradea, ClaudeDes, Editør, Tyzoid, Bibcode Bot, BG19bot, Ymblanter, Moguns, 理想主⊠, YFdyh-bot, M Krikke, Neutrinomajorana, MrCondense, Anton.akhmerov, Kowtje, Kolen Cheung, Anrnusna, Alien Putsch resistant, IRW0, ScrapIronIV, DrKitts and Anonymous: 55

- **Mikheyev–Smirnov–Wolfenstein effect** *Source:* https://en.wikipedia.org/wiki/Mikheyev%E2%80%93Smirnov%E2%80%93Wolfenstein_effect?oldid=662909591 *Contributors:* Mjb, Bbx, Cmapm, Flying fish, GregorB, Strait, Jehochman, Goudzovski, Conscious, Larsobrien, Stepa, Bluebot, Sbharris, Dicklyon, WISo, Michael C Price, Headbomb, Maliz, STBot, HEL, TXiKiBoT, Tomaxer, Legoktm, Paulfharrison, Jht4060, Addbot, Luckas-bot, Yobot, Citation bot, Blennow, EmausBot, John of Reading, ZéroBot, Kasirbot, Bibcode Bot and Anonymous: 24

- **Quark** *Source:* https://en.wikipedia.org/wiki/Quark?oldid=681925889 *Contributors:* AxelBoldt, Derek Ross, Vicki Rosenzweig, Mav, Bryan Derksen, The Anome, Gareth Owen, Andre Engels, PierreAbbat, Peterlin~enwiki, Ben-Zin~enwiki, Zoe, Heron, Montrealais, Hfastedge, Edward, Dante Alighieri, Ixfd64, CesarB, Card~enwiki, NuclearWinner, Looxix~enwiki, Ahoerstemeier, Elliot100, Docu, J-Wiki, Nanobug, Aarchiba, Julesd, Glenn, Schneelocke, Jengod, A5, Timwi, Dysprosia, DJ Clayworth, Phys, Ed g2s, Bevo, Olathe, MD87, Jni, Phil Boswell, Sjorford, Donarreiskoffer, Robbot, Sanders muc, Moncrief, Merovingian, PxT, Texture, Bkell, UtherSRG, Widsith, Ancheta Wis, Giftlite, ShaunMacPherson, Harp, Nunh-huh, Lupin, Herbee, Leflyman, Monedula, 0x6D667061, Xerxes314, Anville, Hoho~enwiki, Alison, Beardo, Moogle10000, Wronkiew, Jackol, Bobblewik, Bodhitha, Piotrus, Kaldari, Elroch, Icairns, Zfr, TonyW, Ukexpat, BrianWilloughby, Grunt, O'Dea, Jiy, Discospinster, Rich Farmbrough, Guanabot, T Long, Vsmith, Saintswithin, SocratesJedi, Mani1, Bender235, Lancer, RJHall, Mr. Billion, El C, Kwamikagami, Laurascudder, Susvolans, Triona, Axezz, Bobo192, Army1987, C S, Ziggurat, Rangelov, Matt McIrvin, Jojit fb, Nk, Pentalis, Obradovic Goran, Fwb22, Lysdexia, Benjonson, Alansohn, Gary, Gintautasm, Guy Harris, Keenan Pepper, MonkeyFoo, Lectonar, Mac Davis, Wdfarmer, Snowolf, Schapel, Knowledge Seeker, Evil Monkey, VivaEmilyDavies, CloudNine, Kusma, Kazvorpal, Kay Dekker, Crosbiesmith, Mogigoma, Linas, Mindmatrix, JarlaxleArtemis, ScottDavis, LOL, Wdyoung, Before My Ken, Tylerni7, Jwanders, Dataphiliac, AndriyK, Noetica, Wayward, Wisq, Palica, Marudubshinki, Calréfa Wéná, GSlicer, Graham87, Deltabeignet, Kbdank71, Yurik, Crzrussian, Rjwilmsi, Bremen, Marasama, SpNeo, Mike Peel, Bubba73, DoubleBlue, Matt Deres, Yamamoto Ichiro, Algebra, Dsnow75, RobertG, Nihiltres, Jeff02, RexNL, TeaDrinker, Chobot, DVdm, Jpacold, Gwernol, Elfguy, Roboto de Ajvol, YurikBot, Wavelength, Bambaiah, Sceptre, Hairy Dude, Jimp, Phantomsteve, TheDoober, Dobromila, JabberWok, CambridgeBayWeather, Chaos, Salsb, Wimt, Ugur Basak, NawlinWiki, Spike Wilbury, Bossrat, SCZenz, Randolf Richardson, Danlaycock, Tony1, DRosenbach, Robertbyrne, Dna-webmaster, WAS 4.250, Closedmouth, Pietdesomere, Heathhunnicutt, Kevin, Banus, RG2, Kamickalo, That Guy, From That Show!, Veinor, MacsBug, SmackBot, Aigarius, BBandHB, Incnis Mrsi, InverseHypercube, C.Fred, Bazza 7, Ikip, Anastrophe, Jrockley, Eskimbot, AnOddName, Jonathan Karlsson, Edgar181, Gilliam, Dauto, NickGarvey, Vvarkey, Bluebot, KaragouniS, Keegan, Dahn, Bigfun, Miquonranger03, OrangeDog, Silly rabbit, Metacomet, Tripledot, Nbarth, DHN-bot~enwiki, Sbharris, Colonies Chris, Hallenrm, Scwlong, Gsp8181, Can't sleep, clown will eat me, Mallorn, Jeff DLB, TKD, Addshore, Mqjjb30e, Cybercobra, Khukri, B jonas, Jdlambert, Lpgeffen, Nrcprm2026, Akriasas, Zadignose, Jóna Þórunn, Bdushaw, Beyazid, TriTertButoxy, SashatoBot, SciBrad, Doug Bell, Soap, Richard L. Peterson, John, Mgiganteus1, SpyMagician, Edconrad, Loadmaster, 2T, Waggers, SandyGeorgia, Ravi12346, Dbzfrk15146, Peyre, Newone, GDallimore, Happy-melon, Majora4, Chovain, Tawkerbot2, Cryptic C62, JForget, Vaughan Pratt, Hello789, ZICO, SUPRATIM DEY, Ruslik0, CuriousEric, Paulfriedman7, Logical2u, Myasuda, RoddyYoung, Typewritten, Cydebot, Abeg92, Mike Christie, Grahamec, Gogo Dodo, Jayen466, 879(CoDe), Michael C Price, Tawkerbot4, Ameliorate!, Akcarver, Gimmetrow, SallyScot, Casliber, Thijs!bot, Epbr123, NeoPhyteRep, LeBofSportif, Markus Pössel, Anupam, Sopranosmob781, Headbomb, Marek69, John254, KJBurns, MichaelMaggs, Escarbot, Eleuther, Ice Ardor, Aadal, AntiVandalBot, SmokeyTheCat, Tyco.skinner, Exteray, RobJ1981, Rsocol, Ke garne, Deflective, Husond, MER-C, CosineKitty, Andonic, East718, Pkoppenb, DanPMK, Magioladitis, WolfmanSF, Thasaidon, Bongwarrior, VoABot II, باسم, Inertiatic076, Kevinmon, Christoph Scholz~enwiki, Aka042, Giggy, Tanvirzaman, Johnbibby, Cyktsui, ArchStanton69, Ace42, Allstarecho, Shijualex, DerHexer, Elandra, Denis tarasov, MartinBot, Poeloq, Dorvaq, CommonsDelinker, HEL, J.delanoy, Nev1, Ops101ex, DrKiernan, Hgpot, Ferdyshenko, Jigesh, DJ1AM, Tarotcards, Coppertwig,

TomasBat, Nikbuz, SJP, FJPB, Vainamainien, Tiggydong, Robprain, Sheliak, Cuzkatzimhut, Lights, X!, VolkovBot, CWii, ABF, John Darrow, Holme053, Nousernamesleft, Ryan032, GimmeBot, Davehi1, A4bot, Captain Courageous, Guillaume2303, Anonymous Dissident, Drestros power, Qxz, Anna Lincoln, Eldaran~enwiki, Leafyplant, Don4of4, PaulTanenbaum, Abdullais4u, Jbryancoop, Mbalelo, Gilisa, Eubulides, Chronitis, Seresin, Dustybunny, Insanity Incarnate, Upquark, Edge1212, Ollieho, AOEU Warrior, SieBot, Graham Beards, WereSpielChequers, Csmart287, Guguma5, Winchelsea, Jbmurray, Caltas, Vanished User 8a9b4725f8376, Keilana, Bentogoa, Aillema, RadicalOne, Arbor to SJ, Elcobbola, Physics one, Dhatfield, RSStockdale, Son of the right hand, Ngexpert5, Ngexpert6, Ngexpert7, Psycherevolt, Sean.hoyland, Mygerardromance, Dabomb87, Nergaal, Muhends, Romit3, SallyForth123, Atif.t2, ClueBot, The Thing That Should Not Be, Wwheaton, Xeno malleus, Harland1, Piledhigheranddeeper, Maxtitan, DragonBot, Glopso, Choonkiat.lee, Himynameisdumb, Worth my salt, Arthur Quark, Estirabot, Brews ohare, Jotterbot, PhySusie, Brianboulton, Dekisugi, ANOMALY-117, Sallicio, Yomangan, Jtle515, Katanada, DumZiBoT, TimothyRias, XLinkBot, Vayalir, Oldnoah, Saintlucifer2008, Nathanwesley3, Dragonfiremage, Devilist666, Mancune2001, Jbeans, WikiDao, SkyLined, Truthnlove, Airplaneman, Eklipse, Addbot, Eric Drexler, AVand, Some jerk on the Internet, Captain-tucker, Giants2008, Iceblock, Ronhjones, Quarksci, Mseanbrown, Looie496, LaaknorBot, Peti610botH, AgadaUrbanit, Tide rolls, Vicki breazeale, Gail, Extruder~enwiki, Abduallah mohammed, Dealer77, Luckas-bot, Yobot, Fraggle81, Cflm001, Legobot II, Amble, Mmxx, Superpenguin1984, Worm That Turned, The Vector Kid, Planlips, Fangfyre, TestEditBot, Azcolvin429, Vroo, Synchronism, Bility, Orion11M87, AnomieBOT, Xi rho, Rubinbot, Jim1138, Bookaneer, Yotcmdr, Crystal whacker, Sonic h, Materialscientist, Citation bot, Pitke, Vuerqex, Bci2, ArthurBot, LilHelpa, Xqbot, Jeffrey Mall, AbigailAbernathy, Srich32977, Alex2510, Almabot, Uscbino, Pmlineditor, RibotBOT, Shmomuffin, Gunjan verma81, Chotarocket, Ernsts, Renverse, A. di M., Weekendpartier, FrescoBot, Paine Ellsworth, DelphinidaeZeta, Steve Quinn, Citation bot 1, AstaBOTh15, Pinethicket, Jonesey95, Calmer Waters, Skyerise, Pmokeefe, Jschnur, Searsshoesales, Jrobbinz123, Lissajous, Turian, Lando Calrissian, Wotnow, Ansumang, Reaper Eternal, 564dude, Jackvancs, Bobotast, MINTOPOINT, TjBot, DexDor, Антон Глінисты, Daggersteel10, Chiechiecheist, EmausBot, John of Reading, WikitanvirBot, Duskbrood, FergalG, Slightsmile, Barak90, Wikipelli, TheLemon1234, Manofgrass, Brazmyth, H3llBot, Stoneymufc29, GeorgeBarnick, Brandmeister, Ego White Tray, TYelliot, ClueBot NG, Gilderien, A520, Cheeseequalsyum, Timothy jordan, 123Hedgehog456, Maplelanefarm, 336, Helpful Pixie Bot, Jeffreyts11, 123456789malm, Bibcode Bot, BG19bot, Hurricanefan25, MusikAnimal, Davidiad, MosquitoBird11, Mydogpwnsall, MrBill3, Njavallil, Glacialfox, Walterpfeifer, Thebannana, CE9958, Marioedesouza, Mediran, Dexbot, Rishab021, Cjean42, Sriharsh1234, Sam boron100, Wankybanky, Wikitroll12345, RojoEsLardo, Jwratner1, NottNott, Saebre, JNrgbKLM, KheltonHeadley, AspaasBekkelund, HectorCabreraJr, Hazinho93, Quadrupedi, QuantumMatt101, Philipphilip0001, Monkbot, RiderDB, Egfraley, Tetra quark, Weed305, KasparBot and Anonymous: 705

- **Quark–lepton complementarity** *Source:* https://en.wikipedia.org/wiki/Quark%E2%80%93lepton_complementarity?oldid=667466538 *Contributors:* Chuunen Baka, RJFJR, Stephenb, SmackBot, BryanG, JorisvS, Cydebot, BetacommandBot, Koeplinger, Headbomb, Mentifisto, Yahel Guhan, STBotT, Ergo leu, Paulfharrison, BartekChom, TubularWorld, Yobot, Omnipaedista, Citation bot 1, Bibcode Bot, BattyBot and Anonymous: 8

- **Standard Model** *Source:* https://en.wikipedia.org/wiki/Standard_Model?oldid=679087631 *Contributors:* AxelBoldt, Derek Ross, CYD, Bryan Derksen, The Anome, Ed Poor, Andre Engels, Roadrunner, David spector, Isis~enwiki, Youandme, Ram-Man, Stevertigo, Edward, Patrick, Boud, Michael Hardy, SebastianHelm, Looxix~enwiki, Julesd, Glenn, AugPi, Mxn, Raven in Orbit, Reddi, Phr, Tpbradbury, Populus, Haoherb428, Phys, Floydian, Bevo, Pierre Boreal, AnonMoos, BenRG, Jeffq, Dmytro, Drxenocide, Robbot, Nurg, Securiger, Texture, Roscoe x, Fuelbottle, Mattflaschen, Tobias Bergemann, Alan Liefting, Ancheta Wis, Giftlite, Dbenbenn, Harp, Herbee, Monedula, LeYaYa, Xerxes314, Dratman, Alison, JeffBobFrank, Dmmaus, Pharotic, Brockert, Bodhitha, Andycjp, Sonjaaa, HorsePunchKid, APH, Icairns, AmarChandra, Gscshoyru, Kate, Arivero, FT2, Rama, Vsmith, David Schaich, Xezbeth, D-Notice, Dfan, Bender235, Pt, El C, Laurascudder, Shanes, Drhex, Fogger~enwiki, Brim, Rbj, Jeodesic, Jumbuck, Alansohn, Gary, ChristopherWillis, Guy Harris, Axl, Sligocki, Kocio, Stillnotelf, Alinor, Wtmitchell, Egg, TenOfAllTrades, H2g2bob, Killing Vector, Linas, Mindmatrix, Benbest, Dodiad, Mpatel, Faethon, TPickup, Faethon34, Palica, Dysepsion, Faethon36, Qwertyca, Drbogdan, Rjwilmsi, Zbxgscqf, Macumba, Strangethingintheland, Dstudent, R.e.b., Bubba73, Drrngrvy, Agasicles, FlaBot, Naraht, Agasides, DannyWilde, Dave1g, Itinerant1, Gparker, Jrtayloriv, Goudzovski, Chobot, Bgwhite, FrankTobia, YurikBot, Bambaiah, Ohwilleke, VoxMoose, Bhny, JabberWok, Bovineone, Krbabu, SCZenz, JulesH, Davemck, Lomn, E2mb0t~enwiki, Dna-webmaster, Jrf, Dv82matt, Tetracube, Hirak 99, Arthur Rubin, Netrapt, JLaTondre, Caco de vidro, RG2, GrinBot~enwiki, That Guy, From That Show!, Hal peridol, SmackBot, YellowMonkey, Tom Lougheed, Melchoir, Bazza 7, KocjoBot~enwiki, Jagged 85, Thunderboltz, Setanta747 (locked), Skizzik, Dauto, Chris the speller, Bluebot, TimBentley, Sirex98, Silly rabbit, Complexica, Metacomet, DHN-bot~enwiki, MovGP0, QFT, Kittybrewster, Addshore, Jmnbatista, Cybercobra, Jgwacker, BullRangifer, Soarhead77, Daniel.Cardenas, Yevgeny Kats, Byelf2007, TriTertButoxy, Craig Bolon, Ajnosek, Ekjon Lok, Bjankuloski06, Tarcieri, Waggers, JarahE, Michaelbusch, Lottamiata, Newone, Twas Now, IanOfNorwich, Srain, Patrickwooldridge, J Milburn, Mosaffa, Gatortpk, Vessels42, Geremia, Van helsing, Harrigan, Phatom87, Cydebot, David edwards, Verdy p, Michael C Price, Xantharius, Crum375, JamesAM, Thijs!bot, Epbr123, Headbomb, Phy1729, Stannered, Tariqhada, Seaphoto, Orionus, Voyaging, Gnixon, Jbaranao, Jrw@pobox.com, Len Raymond, Narssarssuaq, Bakken, CattleGirl, Davidoaf, Vanished user ty12kl89jq10, Lvwarren, Taborgate, Leyo, HEL, J.delanoy, Hans Dunkelberg, Stephanwehner, Wbellido, Aoosten, Jacksonwalters, The Transliterator, DadaNeem, Student7, Joshmt, WJBscribe, Jozwolf, Hexane2000, BernardZ, Awren, Sheliak, Physicist brazuca, Schucker, Goop Goop, Fences and windows, Dextrose, Mcewan, Swamy g, TXiKiBoT, Sharikkamur, Thrawn562, Voorlandt, Escalona, Setreset, PDFbot, Pleroma, UnitedStatesian, Piyush Sriva, Kacser, Billinghurst, Francis Flinch, Moose-32, Ptrslv72, David Barnard, SieBot, ShiftFn, Robdunst, Jim E. Black, SheepNotGoats, Gerakibot, Nozzer42, Mr swordfish, Wing gundam, Bamkin, Likebox, Arthur Smart, HungarianBarbarian, Commutator, KathrynLybarger, Iomesus, C0nanPayne, Crazz bug 5, ClueBot, Superwj5, Wwheaton, Garyzx, SuperHamster, Elsweyn, Maldmac, DragonBot, Djr32, Diagramma Della Verita, Nymf, Eeekster, Brews ohare, NuclearWarfare, PhySusie, Ordovico, Mastertek, DumZiBoT, BodhisattvaBot, Guarracino, Mitch Ames, Truthnlove, Stephen Poppitt, Tayste, Addbot, Deepmath, Eric Drexler, DWHalliday, Mjamja, Leszek Jańczuk, NjardarBot, Mwoldin, Bassbonerocks, Barak Sh, AgadaUrbanit, Lightbot, Smeagol 17, Abjiklam, Ve744, Luckas-bot, Yobot, Orion11M87, AnomieBOT, JackieBot, Icalanise, Citation bot, ArthurBot, Northryde, LilHelpa, Xqbot, Sionus, Professor J Lawrence, Tomwsulcer, Edsegal, GrouchoBot, Trongphu, QMarion II, Ernsts, A. di M., Bytbox, FrescoBot, Paine Ellsworth, Aliotra, Steve Quinn, Citation bot 1, Rameshngbot, MJ94, RedBot, Masti-Bot, Aknochel, Sijothankam, Puzl bustr, Beta Orionis, Physics therapist, Bj norge, Innotata, Jesse V., RjwilmsiBot, Mathewsyriac, Afteread, EmausBot, Bookalign, WikitanvirBot, Wilhelm-physiker, Bdijkstra, DerNeedle, Kenmint, Dbraize, Tanner Swett, HeptishHotik, مه, منٛہین ٻہاۀ, Suslindisambiguator, Quondum, Webbeh, UniversumExNihilo, Vanished user fijw983kjaslkekfhj45, Maschen, RockMagnetist, Stormymountain, Ζeτα ζ, Whoop whoop pull up, Isocliff, ClueBot NG, Smtchahal, Snotbot, Tonypak, O.Koslowski, CharleyQuinton, Dsperlich, Theopolisme, ZakMarksbury, Helpful Pixie Bot, Bibcode Bot, BG19bot, Tirebiter78, AvocatoBot, Lukys~enwiki, Stapletongrey, Ownedroad9, Chip123456, ChrisGualtieri, Khazar2, Billyfesh399, Rhlozier, JYBot, Dexbot, Doom636, Rongended, Cerabot~enwiki, CuriousMind01, Cjean42, Jayanta mallick, Joeinwiki, Kowtje, JPaestpreornJeolhlna, Eyesnore, Euan Richard, Nigstomper, Particle physicist, Prokaryotes, Jernahthern, Ginsuloft, Dimension10, JNrgbKLM, Krabaey, 1codesterS, FelixRosch, Monkbot, Delbert7, BradNorton1979, Lathamboyle, Tetra quark, KasparBot, Buckbill10 and Anonymous: 357

- **List of particles** *Source:* https://en.wikipedia.org/wiki/List_of_particles?oldid=682746251 *Contributors:* AxelBoldt, Danny, Rmhermen, Stevertigo, Bdesham, Ahoerstemeier, Stan Shebs, Docu, Salsa Shark, Nikai, Evercat, Schneelocke, Charles Matthews, Jitse Niesen, CBDunkerson, Bevo, Raul654, Donarreiskoffer, Robbot, Sanders muc, Merovingian, Pengo, Giftlite, Herbee, Xerxes314, Dratman, Jeremy Henty, Alensha, Bodhitha, Physicist, Hayne, Quadell, RetiredUser2, Mysidia, Icairns, Asbestos, D6, Urvabara, Discospinster, Rich Farmbrough, FT2, Qutezuce, ArnoldReinhold, Neko-chan, El C, Laurascudder, Susvolans, EmilJ, Physicistjedi, Minghong, Gbrandt, Eddideigel, Axl, Mac Davis, David Ko, Radical Mallard, RJFJR, Count Iblis, Dirac1933, TenOfAllTrades, LFaraone, Oleg Alexandrov, Linas, JarlaxleArtemis, Duncan.france, GregorB, Cedrus-Libani, Karam.Anthony.K, Palica, Rjwilmsi, Zbxgscqf, JLM~enwiki, Strait, Ems57fcva, Krash, Dan Guan, DannyWilde, Lmatt, Goudzovski, Chobot, YurikBot, Bambaiah, Vuvar1, Madkayaker, Hydrargyrum, Presscorr, Chaos, Salsb, Tavilis, SCZenz, Lexicon, TUSHANT JHA, Dna-webmaster, Tomvds, Poulpy, Cstmoore, TLSuda, NeilN, MacsBug, Tom Lougheed, McGeddon, Bazza 7, WookieInHeat, Derdeib, Yamaguchi 先生, Betacommand, Bluebot, Master of Puppets, DHN-bot~enwiki, Raistuumum, Juancnuno, Kittybrewster, Acepectif, Ligulembot, TriTertButoxy, ArglebargleIV, Khazar, John, FrozenMan, JorisvS, 041744, Dr Greg, Slakr, Mets501, Scorpion0422, Cbuckley, Iridescent, TwistOfCain, Happy-melon, JRSpriggs, Flickboy, Van helsing, Lithium6, Neelix, Rotiro, Cydebot, Quibik, Christian75, Omicronpersei8, Thijs!bot, Qwyrxian, TauLibrus, Headbomb, Inner Earth, 49, Guptasuneet, Scottmsg, WinBot, Elmoosecapitan, Tyco.skinner, AubreyEllenShomo, Arch dude, Johnman239, Mwarren us, TheEditrix2, CalamusFortis, MartinBot, Sadisticsuburbanite, Bissinger, Anaxial, CommonsDelinker, Maurice Carbonaro, Zojj, OliverHarris, Joshmt, Adanadhel, Lseixas, Graphite Elbow, VolkovBot, Jmrowland, Quilbert, Anonymous Dissident, Dstary, Escalona, JPMasseo, Figureskatingfan, Inx272, Meters, Antixt, Hamish a e fowler, GoddersUK, Bluetryst, SieBot, Ishvara7, WereSpielChequers, Audrius u, VovanA, Paolo.dL, RSStockdale, Anchor Link Bot, StewartMH, Explicit, ClueBot, Unbuttered Parsnip, Nolimitownass, DragonBot, Atomic7732, TimothyRias, SkyLined, Addbot, DOI bot, Jojhutton, Favonian, LinkFA-Bot, OlEnglish, Teles, Legobot, Luckas-bot, Yobot, Dov Henis, Azcolvin429, AnomieBOT, Götz, Icalanise, Flewis, Materialscientist, OllieFury, Vuerqex, ArthurBot, Vulcan Hephaestus, Blennow, Reality006, Coretheapple, Jcimorra, RibotBOT, Ernsts, A. di M., Axelfoley12, Zosterops, FrescoBot, Paine Ellsworth, Citation bot 1, JIK1975, Tom.Reding, Diffequa, WikitanvirBot, Racerx11, 112358sam, Aegnor.erar, Hops Splurt, HESUPERMAN, Hhhippo, AvicBot, JSquish, StringTheory11, Waperkins, Bamyers99, Suslindisambiguator, L Kensington, DennisIsMe, RockMagnetist, ClueBot NG, Snotbot, Primergrey, Vio45lin, Widr, MsFionnuala, Oklahoma3477, Bibcode Bot, CityOfSilver, Cap'n G, BML0309, Dan653, Twocount, Penguinstorm300, Dexbot, LightandDark2000, Ohiggy, TwoTwoHello, Andyhowlett, Printersmoke, Orion 2013, ARUNEEK, Seino van Breugel, AspaasBekkelund, TheMagikCow, Vyom27, ParkersComments, Selva Ganapathy and Anonymous: 290

- **Timeline of particle discoveries** *Source:* https://en.wikipedia.org/wiki/Timeline_of_particle_discoveries?oldid=679101637 *Contributors:* Rmhermen, Tempshill, Harp, Xerxes314, Bodhitha, Perey, Discospinster, Cmdrjameson, JohnAlbertRigali, Crosbiesmith, Rjwilmsi, Strait, Bubba73, Goudzovski, David H Braun (1964), Yamara, SCZenz, Tony1, GrinBot~enwiki, Attilios, SmackBot, Onionmon, Dl2000, JeffW, Lottamiata, Newone, Jhlawr, Headbomb, D.H, Pkoppenb, JNW, Joshmt, Chronitis, SieBot, A. Carty, BartekChom, Muhends, Wprlh, Addbot, DOI bot, MizzoulaB, Luckas-bot, Icalanise, Pepo13, Citation bot, Gogiva, Xqbot, Omnipaedista, Ulm, Carlog3, Citation bot 1, Wbm1058, Bibcode Bot, Dexbot, Trinitresque, Makecat-bot, JanJaeken, ElŞahin, Revolution1221, Monkbot and Anonymous: 24

- **Carl David Anderson** *Source:* https://en.wikipedia.org/wiki/Carl_David_Anderson?oldid=673049798 *Contributors:* XJaM, Michael Hardy, Mic, Ixfd64, Salsa Shark, Maximus Rex, Diberri, Giftlite, Everyking, JillandJack, ChicXulub, Mako098765, D6, TheBlueWizard, Eb.hoop, Guanabot, Aris Katsaris, Roo72, Djordjes, CanisRufus, Brim, Jumbuck, Ksnow, KingTT, Dirac1933, Adrian.benko, Richard Arthur Norton (1958-), Woohookitty, Rocastelo, Betsythedevine, Emerson7, Tevatron~enwiki, Jebur~enwiki, Kbdank71, Wikix, MZMcBride, XLerate, Stilgar135, DannyWilde, Kolbasz, Srleffler, YurikBot, Conscious, Salsb, NawlinWiki, Bota47, Mike Dillon, LeonardoRob0t, Garion96, GrinBot~enwiki, SmackBot, CRKingston, ZerodEgo, Brooklynl, Y e l m, WestA, JackO'Lantern, Michael David, John, Beetstra, Dicklyon, Elb2000, GiantSnowman, Kedar63, T-W, Drinibot, Cydebot, HPaul, Thijs!bot, Headbomb, RobotG, Hannes Eder, JAnDbot, Gcm, Duendeverde, Joshmt, VolkovBot, Jimmyeatskids, GcSwRhIc, SieBot, Gborchardt, Arjen Dijksman, OKBot, Kumioko (renamed), Wuhwuzdat, Muhends, RS1900, Joao Xavier, Californicator, Masterpiece2000, DragonBot, Cardinalem, DumZiBoT, Addbot, Numbo3-bot, Lightbot, Zorrobot, Luckas-bot, Yobot, Götz, Citation bot, Xqbot, Davshul, Der Blaue Reiter, Topherwhelan, Citation bot 1, Fat&Happy, RedBot, Full-date unlinking bot, TobeBot, RjwilmsiBot, EmausBot, Amanda.nelson12, Lemeza Kosugi, Captain Assassin!, Orange Suede Sofa, ClueBot NG, Nasmem, Bibcode Bot, AvocatoBot, Purdygb, VIAFbot, Churn and change, ArmbrustBot, Jonarnold1985, Alyssachristina, KasparBot and Anonymous: 28

- **Clyde Cowan** *Source:* https://en.wikipedia.org/wiki/Clyde_Cowan?oldid=668049228 *Contributors:* Magnus Manske, XJaM, Rbrwr, JWSchmidt, Charles Matthews, Raul654, Owen, Ancheta Wis, Harp, Everyking, Anárion, Kate, D6, Rich Farmbrough, Bender235, TheParanoidOne, Ropcat, Goudzovski, Jaraalbe, RussBot, Salsb, ENDelt260, SmackBot, Grey Shadow, Hmains, Chris the speller, EncMstr, Ser Amantio di Nicolao, BrownHairedGirl, Nobunaga24, Astuishin, Kriston, Briancua, Bhaverchuck, Cydebot, Thijs!bot, Mojo Hand, Arch dude, Grimlock, CommonsDelinker, Johnpacklambert, BOTijo, Kumioko (renamed), Gabodon, Alexbot, SchreiberBike, Addbot, Lightbot, Luckas-bot, Uh matt lame~enwiki, Full-date unlinking bot, EmausBot, John of Reading, Delagyela, ZéroBot, Suslindisambiguator, Bibcode Bot, EWPage, KasparBot and Anonymous: 15

- **Raymond Davis, Jr.** *Source:* https://en.wikipedia.org/wiki/Raymond_Davis%2C_Jr.?oldid=679032747 *Contributors:* AxelBoldt, Rsabbatini, Giftlite, Capitalistroadster, Gamaliel, Jdavidb, PDH, Howardjp, Ukexpat, Klemen Kocjancic, Deglr6328, Clubjuggle, D6, Rich Farmbrough, Laurascudder, Shanes, Rje, Alansohn, Snowolf, Ksnow, Bbsrock, SteinbDJ, RyanGerbil10, Etacar11, Emerson7, Bunchofgrapes, Wikix, Strait, The wub, Bhadani, FlaBot, Srleffler, CJLL Wright, Adoniscik, Whosasking, Algebraist, Kummi, YurikBot, RussBot, Donald Albury, LeonardoRob0t, Garion96, Le Hibou~enwiki, SmackBot, Cassandro, Brossow, Bluebot, AdamSmithee, Timothy Clemans, Y e l m, Jiminy pop, Andrei Stroe, Michael David, John, EricR, HennessyC, Cydebot, Customcabf100, Ning-ning, Headbomb, WVhybrid, Bunzil, RobotG, JAnDbot, Gcm, Felix116, Waacstats, Strikehold, HOT L Baltimore, CommonsDelinker, Lseixas, Thismightbezach, GrahamHardy, Idioma-bot, VolkovBot, TXiKiBoT, Jimmyeatskids, ElinorD, GcSwRhIc, SieBot, Kaspo, PolarBot, Kumioko, Dabomb87, Someone111111, All Hallow's Wraith, Joao Xavier, Pointillist, Masterpiece2000, PixelBot, Eustress, Cardinalem, Qwfp, Rror, Good Olfactory, Kbdankbot, Addbot, DOI bot, Luckas-bot, Yobot, Amirobot, NLWASTI, AnomieBOT, Materialscientist, ArthurBot, Xqbot, Davshul, Mnmngb, Citation bot 1, Apwj5060, Tom.Reding, MJ94, Charlesmartinsmith, TobeBot, El Mayimbe, RjwilmsiBot, Lemeza Kosugi, Fæ, H3llBot, Suslindisambiguator, Bibcode Bot, Physicsch, CountryRadio, Monkbot, Jonarnold1985, Hiram Abiff, KasparBot and Anonymous: 27

ChrisGualtieri, VIAFbot, Bibliophilen, Pinocchio3000, Arosariorivera, PADDTWILL007, KasparBot, Csumstudent and Anonymous: 80

- **Louis Michel (physicist)** *Source:* https://en.wikipedia.org/wiki/Louis_Michel_(physicist)?oldid=659833659 *Contributors:* Gene Nygaard, Goudzovski, Ser Amantio di Nicolao, Optimale, Eastfrisian, Myasuda, Gonzo fan2007, Headbomb, Waacstats, CommonsDelinker, Victor Blacus, Maurice Carbonaro, Cuzkatzimhut, @pple, Monegasque, Alexbot, Giuseppegaeta, Addbot, Citation bot, Omnipaedista, Full-date unlinking bot, RjwilmsiBot, ZéroBot, Suslindisambiguator, Balm oral winds or sand ring ham, KLBot2, Bibcode Bot, VIAFbot, KasparBot and Anonymous: 1

- **Wolfgang Pauli** *Source:* https://en.wikipedia.org/wiki/Wolfgang_Pauli?oldid=679117483 *Contributors:* Magnus Manske, CYD, Zundark, The Anome, Little guru, Maria Renee Jenkins, Peterlin~enwiki, Ben-Zin~enwiki, KF, Stevertigo, Michael Hardy, Liftarn, Mic, Kalki, AlexR, Looxix~enwiki, Ahoerstemeier, Docu, Salsa Shark, Glenn, Ciphergoth, Smack, Vanished user 5zariu3jisj0j4irj, Bemoeial, Wik, Maximus Rex, Jeffq, Robbot, Fredrik, AdamReed, Goethean, Diderot, Timrollpickering, Modeha, HaeB, Lzur, Cyrius, Giftlite, Harp, Ryz, Curps, Waltpohl, Mdob, ConradPino, Phe, Gunnar Larsson, PDH, Pmanderson, Edsanville, Abdull, Trevor MacInnis, Sparky2002b, Chris Howard, D6, Rich Farmbrough, Guanabot, Pjacobi, Vsmith, Aris Katsaris, Djordjes, Rgdboer, Hayabusa future, Apyule, Physicistjedi, Lokifer, Alansohn, TheParanoidOne, Plumbago, Snowolf, Aranae, Ksnow, Wtmitchell, Dirac1933, Crosbiesmith, PoccilScript, Kzollman, GregorB, Isnow, Emerson7, Rjwilmsi, Koavf, Sdornan, DrTorstenHenning, Brighterorange, Erkcan, The wub, Ttwaring, Titoxd, FlaBot, Nystrxz, Themanwithoutapast, KFP, JohnMarkStrain, Srleffler, Valentinian, DaGizza, Jaraalbe, GangofOne, Bgwhite, YurikBot, Pippo2001, Neilbeach, Salsb, Thane, Wiki alf, Bachrach44, LaszloWalrus, Gerhard51, RUL3R, Bota47, TransUtopian, Curpsbot-unicodify, Garion96, Victor falk, SmackBot, Roger Hui, Lestrade, KnowledgeOfSelf, KocjoBot~enwiki, Eskimbot, MalafayaBot, Josteinn, DHN-bot~enwiki, Sbharris, Newport, Voyajer, GRuban, WestA, Khoikhoi, Wen D House, Khukri, Deknyff, Shamir1, Bidabadi~enwiki, Sadi Carnot, Lucretius~enwiki, SashatoBot, Ser Amantio di Nicolao, John, Syrcatbot, Hillelg, Rglovejoy, Meco, ShakingSpirit, ChazYork, Newone, Hawkestone, Twas Now, Nádvorník, Courcelles, Paulmlieberman, Calmargulis, Banedon, Dgw, Cydebot, Gogo Dodo, Khatru2, MWaller, Michael C Price, My Flatley, DumbBOT, Storeye, Headbomb, Wildthing61476, BehnamFarid, Bunzil, CharlotteWebb, AntiVandalBot, RobotG, Lmaltier, Gioto, Tillman, Storkk, JAnDbot, NBeale, Husond, MER-C, Dmerthe, Chickyfuzz123, Seifer1886, FerranJorba, VoABot II, Misheu, Generalstudent, Here2fixCategorizations, Midgrid, Tanvirzaman, Bellbird, Trioculite, Dirac66, LorenzoB, Duendeverde, Murraypaul, MartinBot, TheEgyptian, CommonsDelinker, Johnpacklambert, Wikipediausernumber1, HEL, J.delanoy, Rrostrom, Numbo3, SureFire, SuperGirl, OfficeGirl, Cpiral, LordAnubisBOT, Petersec, Naniwako, R613vlu, DadaNeem, RB972, Inwind, Idioma-bot, Deor, VolkovBot, ABF, Jeff G., Soliloquial, TXiKiBoT, A4bot, Wassermann~enwiki, Uannis~enwiki, CoolKid1993, Spinningspark, AlleborgoBot, Resurgent insurgent, SieBot, BotMultichill, Utternutter, LKNUTZ, Vojvodaen, Sean.hoyland, Dabomb87, Wahrmund, RS1900, ClueBot, PipepBot, All Hallow's Wraith, QueenAdelaide, Shinigami27, Masterpiece2000, Alexbot, Surfrider310, Thingg, Cardinalem, Alphatronic, Egmontaz, Darkicebot, Bletchley, Zacharie Grossen, Good Olfactory, Darktaco, Addbot, Solartime, Blethering Scot, ChenzwBot, Zanestuchman, Tide rolls, Lightbot, Zorrobot, Luckas-bot, Yobot, Bunnyhop11, Sprachpfleger, TheWindyCity, AnakngAraw, AnomieBOT, Steamturn, Jim1138, Ularevalo98, Ninahexan, Materialscientist, Citation bot, Xqbot, Davshul, AbigailAbernathy, Usedfan19, Omnipaedista, MadGeographer, Erik9, Cmaric, FrescoBot, Ironboy11, Citation bot 1, Merongb10, Full-date unlinking bot, PrinceRegentLuitpold, TobeBot, Trappist the monk, Bronescu, Jamietw, Tbhotch, TjBot, CerberusAlpha, Sprout333, Racerx11, T-dawg003, TeleComNasSprVen, Hhhippo, JSquish, Jordancelticsfan, Ὁ οἶστρος, Suslindisambiguator, Phruizler, Staszek Lem, Maschen, Donner60, DASHBotAV, Rocketrod1960, ClueBot NG, Bakrnl, 123Hedgehog456, Masssly, Antiqueight, Asalrifai, Helpful Pixie Bot, MusikAnimal, Fudgeballboy, Snow Blizzard, Monster9999, Brad7777, Anbu121, Ajaxfiore, Ninmacer20, The Elixir Of Life, Southbigsean0300, Dexbot, VIAFbot, BreakfastJr, Ihatepauldirac, Ihatedirac2k13, Crispulop, Monochrome Monitor, JaconaFrere, TheEpTic, Eddiewardos, Brandon axe, Monkbot, Internucleotide, Michael Dominik Fischer, Epigogue, Pablovegan98, NilubonT, KasparBot, Wikitidy and Anonymous: 226

- **Martin Lewis Perl** *Source:* https://en.wikipedia.org/wiki/Martin_Lewis_Perl?oldid=682649183 *Contributors:* Danny, Rmhermen, Altenmann, Ancheta Wis, Phe, Icairns, Klemen Kocjancic, D6, Srbauer, Nk, TheParanoidOne, Snowolf, Ksnow, Gene Nygaard, RyanGerbil10, Emerson7, Jack Cox, BD2412, MZMcBride, FlaBot, DannyWilde, Srleffler, Valentinian, YurikBot, NTBot~enwiki, Takwish, Salsb, Equilibrial, GrinBot~enwiki, SmackBot, PrimeHunter, Ser Amantio di Nicolao, John, Eastfrisian, Kriston, HennessyC, Justinian3, Cydebot, Hydraton31, LeeSawyer, Thijs!bot, Headbomb, RobotG, Prolog, Gcm, Postcard Cathy, MER-C, Magioladitis, Connormah, Waacstats, Chemical Engineer, Ceancata, Plindenbaum, VolkovBot, JamieXO, TXiKiBoT, Jimmyeatskids, Broadbot, SieBot, Arjen Dijksman, Ufinne, All Hallow's Wraith, Nsk92, Frontierblog, Neuber hia, Joao Xavier, Masterpiece2000, Djr32, Alexbot, Cardinalem, Athrion, WikHead, Kbdankbot, Addbot, Jacopo Werther, Lightbot, OlEnglish, Luckas-bot, Yobot, Amirobot, NLWASTI, AnomieBOT, Comparativist1, Albatross48, ArthurBot, LilHelpa, Xqbot, Davshul, Omnipaedista, Mnmngb, FrescoBot, Ysyoon, W E Hill, TobeBot, RjwilmsiBot, EmausBot, ZéroBot, Suslindisambiguator, Kelainoss, BG19bot, Taief.shahed, Jimbo113453, Pinocchio3000, Zibetta, Hiram Abiff, KasparBot, Mermaidawesome and Anonymous: 31

- **Bruno Pontecorvo** *Source:* https://en.wikipedia.org/wiki/Bruno_Pontecorvo?oldid=681458345 *Contributors:* Panairjdde~enwiki, Xerxes314, Line, Ukexpat, Rich Farmbrough, Pissipo, Balubino, Bill Thayer, Bobo192, Flying fish, NightOnEarth, Strait, FlaBot, Ground Zero, Musical Linguist, Goudzovski, EamonnPKeane, RussBot, Nobs01, Gaius Cornelius, AlexeiK, DonaldDuck, Rjensen, Dormidondt, Renata3, Stumps, Blablabla, Attilios, SmackBot, Pavlovič, CRKingston, Stepa, Kintetsubuffalo, Hmains, Bluebot, OrphanBot, Threeafterthree, JJstroker, MilborneOne, Alessandro57, Cydebot, Tec15, SeNeKa, Headbomb, MarkV, WolfmanSF, Bulbeck, Serguei S. Dukachev, Hugo999, Rei-bot, N-HH, Kumioko (renamed), RS1900, Martarius, Fasettle, Alexbot, Aecharri, Burubuz~enwiki, Assosiation, Addbot, DOI bot, LaaknorBot, Lightbot, Zorrobot, PieterJanR, Luckas-bot, Yobot, Mauro Lanari, Materialscientist, USConsLib, Enok, Jsmith1000, Omnipaedista, Ace111, PasswordUsername, RjwilmsiBot, WikitanvirBot, Avdonina, AvicBot, ZéroBot, JeanneMish, Givenunion, Helpful Pixie Bot, BendelacBOT, VIAFbot, Lola Rennt, Tatianyc, Theblabe, FFranzinetti, Monkbot, KasparBot and Anonymous: 40

- **Frederick Reines** *Source:* https://en.wikipedia.org/wiki/Frederick_Reines?oldid=675418591 *Contributors:* Magnus Manske, Mav, DW, Mic, Rednblu, Maximus Rex, Raul654, DHN, Ancheta Wis, Giftlite, Harp, Everyking, ChicXulub, Phe, PDH, Deglr6328, D6, Aris Katsaris, Too Old, Srbauer, Nk, Alansohn, Ksnow, Bbsrock, Dirac1933, Gene Nygaard, RyanGerbil10, Carcharoth, Twthmoses, Emerson7, Rjwilmsi, Nightscream, Koavf, Strait, G Clark, Srleffler, RussBot, Hawkeye7, Gadget850, Nikkimaria, LeonardoRob0t, GrinBot~enwiki, SmackBot, KocjoBot~enwiki, Gaff, PrimeHunter, Colonies Chris, Rlevse, Vladislav, Υ e l m, RFD, G716, Andrei Stroe, Ser Amantio di Nicolao, Margoz, Kriston, HennessyC, CmdrObot, Drinibot, WeggeBot, Cydebot, Gogo Dodo, Khatru2, Thijs!bot, Headbomb, RobotG, Gcm, The Transhumanist, Waacstats, Cardamon, David Eppstein, Ceancata, HOT L Baltimore, CommonsDelinker, GMA Divo, DadaNeem, Plindenbaum, VolkovBot, TXiKiBoT, Jimmyeatskids, Sintaku, Blandis, Gilisa, Ponyo, SieBot, Moletrouser, SalineBrain, Binksternet, All Hallow's Wraith, Joao Xavier, Masterpiece2000, Alexbot, Eustress, Cardinalem, MystBot, Addbot, Pyfan, LaaknorBot, Lightbot, Innapoy, Legobot, Luckas-bot, Yobot, Amirobot, NLWASTI, AnomieBOT, ArthurBot, Xqbot, Davshul, J JMesserly, GrouchoBot, Anotherclown, Mnmngb, FrescoBot, Jonesey95, Fat&Happy,

TobeBot, RjwilmsiBot, WildBot, GoingBatty, AvicBot, Lemeza Kosugi, JeanneMish, Wkroppjr, ClueBot NG, Helpful Pixie Bot, Bibcode Bot, CarloMartinelli, Physicsch, VIAFbot, Bigsean12001, PeterTAnteater, Jonarnold1985, KasparBot, Ceannlann gorm and Anonymous: 29

- **Shoichi Sakata** *Source:* https://en.wikipedia.org/wiki/Shoichi_Sakata?oldid=682509448 *Contributors:* Jason Quinn, SmackBot, Kurtan~enwiki, Headbomb, Magioladitis, Waacstats, TXiKiBoT, Enkyo2, Rotational, Addbot, SpBot, Zorrobot, Yobot, AnomieBOT, Jeni, GrouchoBot, Omnipaedista, Claudiodib, Diannaa, MegaSloth, RjwilmsiBot, ZéroBot, Snotbot, VIAFbot, Kounosuke7japan, KasparBot and Anonymous: 6

- **Melvin Schwartz** *Source:* https://en.wikipedia.org/wiki/Melvin_Schwartz?oldid=660413543 *Contributors:* Jokestress, Klemen Kocjancic, D6, Freakofnurture, Grechelonsurge, El C, Ben davison, Ksnow, Jack Cox, MZMcBride, FlaBot, Srleffler, Valentinian, YurikBot, Equilibrial, DAJF, SmackBot, Roger Hui, Ekamaloff~enwiki, Nixeagle, G716, Andrei Stroe, Michael David, Ser Amantio di Nicolao, John, Dicklyon, HennessyC, Cydebot, BushidoWarlord, Lugnuts, Thijs!bot, Ginosal, Headbomb, RobotG, Mallomar, JAnDbot, Gcm, Waacstats, N.Nahber, Dbodell, Tvoz, Warrickball, Plindenbaum, VolkovBot, TXiKiBoT, Antoni Barau, Cheesebox, SieBot, Arjen Dijksman, RS1900, All Hallow's Wraith, Joao Xavier, Excirial, Cardinalem, RogDel, Good Olfactory, Addbot, Ginosbot, Numbo3-bot, Lightbot, Zorrobot, Luckas-bot, Yobot, Amirobot, NLWASTI, Việt Chi, ArthurBot, Xqbot, Davshul, GrouchoBot, Mnmngb, El Mayimbe, RjwilmsiBot, Alph Bot, EmausBot, ZéroBot, Suslindisambiguator, Jonke20, Bibcode Bot, JYBot, VIAFbot, KasparBot and Anonymous: 25

- **Jack Steinberger** *Source:* https://en.wikipedia.org/wiki/Jack_Steinberger?oldid=678932576 *Contributors:* Mic, Ahoerstemeier, RedWolf, Pibwl, Fastfission, Everyking, OldakQuill, PDH, D6, Rich Farmbrough, Aris Katsaris, CanisRufus, El C, Ctrl build, NickCatal, Alansohn, TheParanoidOne, Cdc, A Kit, Ksnow, Ceyockey, Siafu, Emerson7, Kbdank71, Kane5187, Rjwilmsi, Koavf, The jt, FlaBot, Srleffler, Valentinian, YurikBot, Wavelength, NTBot~enwiki, Leuliett, Bota47, Zargulon, Attilios, SmackBot, Nixeagle, Khukri, G716, StN, Andrei Stroe, Ser Amantio di Nicolao, John, Dicklyon, Cheez277, SchmittM, HennessyC, Drinibot, Cydebot, Odie5533, Thijs!bot, Headbomb, Second Quantization, Darev, RobotG, Wizmo, Gcm, MER-C, Magioladitis, Waacstats, Ceancata, CommonsDelinker, Plindenbaum, Alffsteinberger, VolkovBot, TXiKiBoT, Jimmyeatskids, AlleborgoBot, Ponyo, SieBot, Monegasque, Arjen Dijksman, Hello71, Fratrep, RS1900, All Hallow's Wraith, Joao Xavier, Cardinalem, Zwinglisjubilee, Zacharie Grossen, Addbot, Jacopo Werther, Wulf Isebrand, Numbo3-bot, Lightbot, Luckas-bot, Yobot, Bunnyhop11, Amirobot, NLWASTI, KamikazeBot, Ulric1313, Citation bot, Davshul, Omnipaedista, Mnmngb, Togodumnus, W-C, Ironboy11, D'ohBot, Citation bot 1, Jonesey95, Thinking of England, El Mayimbe, RjwilmsiBot, EmausBot, Suslindisambiguator, Jonke20, Usctommytrojan, Gelfandn, Bibcode Bot, Dazzab555, Ninmacer20, VIAFbot, Janx8686, Bibliophilen, Sayitclearly, Monkbot, Jonarnold1985, Ellipapa, KasparBot, Csumstudent and Anonymous: 25

- **J. J. Thomson** *Source:* https://en.wikipedia.org/wiki/J._J._Thomson?oldid=682729002 *Contributors:* General Wesc, Bryan Derksen, Andre Engels, Youssefsan, XJaM, Fcueto, Deb, Heron, D, Booyabazooka, Fred Bauder, BrianHansen~enwiki, Mic, Bcrowell, Delirium, Ahoerstemeier, Nikai, Sethmahoney, John K, EdH, Charles Matthews, Reddi, Stone, Maximus Rex, Grendelkhan, Lord Emsworth, Penfold, Secretlondon, Proteus, Twang, Robbot, Fredrik, Schutz, Flauto Dolce, Wereon, Raeky, Diberri, Jeremiah, Ancheta Wis, Giftlite, Tom harrison, Everyking, Mpntod, Dratman, Curps, Mboverload, Jackol, Golbez, Sesel, ChicXulub, Gadfium, Utcursch, Alexf, Antandrus, Mr impossible, Am088, PDH, Jossi, 1297, DragonflySixtyseven, Necrothesp, Icairns, Muijz, Deglr6328, Adashiel, ELApro, Mike Rosoft, Kmccoy, D6, TheBlueWizard, CALR, DanielCD, Ultratomio, Cruvers, RossPatterson, Discospinster, Eel, KarlaQat, Vsmith, YUL89YYZ, Aris Katsaris, Upi, Gianluigi, Bender235, ESkog, Djordjes, Mashford, RJHall, Karmafist, Tirdun, Bluap, Deanos, Wareh, Bobo192, Whosyourjudas, Meggar, Robotje, K0hlrabi, AtomicDragon, Pharos, Merope, Ranveig, Jumbuck, Alansohn, Ben davison, Atlant, Andrewpmk, Craigy144, Echuck215, Lightdarkness, Fritzpoll, Malo, Phyllis1753, Bart133, Snowolf, Ksnow, Wtmitchell, Bbsrock, Wtshymanski, AnIco, Woohookitty, TigerShark, Etacar11, Kurzon, Benbest, ^demon, WadeSimMiser, JeremyA, Jeff3000, Thruston, GregorB, Mathewtse, MarcoTolo, Emerson7, Rnt20, Deltabeignet, MC MasterChef, Kbdank71, Zzedar, Rjwilmsi, Mayumashu, Angusmclellan, Koavf, Vary, Salix alba, GreetingsEarthling, Crazynas, Bubba73, Ttwaring, Yamamoto Ichiro, FlaBot, Windchaser, Nivix, RexNL, Gurch, Kolbasz, Srleffler, Valentinian, Chobot, DVdm, Antiuser, Bgwhite, NSR, Gwernol, Cornellrockey, UkPaolo, The Rambling Man, YurikBot, Sceptre, Jimp, Phantomsteve, RussBot, Gaius Cornelius, NawlinWiki, Wiki alf, Journalist, Dhollm, Cholmes75, Moe Epsilon, Mikeblas, Aaron Schulz, Michael Drew, Kkmurray, Alpha 4615, Wknight94, Zzuuzz, Theda, KGasso, Tevildo, CWenger, HereToHelp, Mais oui!, Allens, Katieh5584, Kungfuadam, GrinBot~enwiki, SmackBot, KnowledgeOfSelf, Piccadilly, Salmyxn, Jab843, Canthusus, Colonel Tom, Edgar181, Aksi great, Gilliam, Steverich, Ohnoitsjamie, Rmosler2100, Famouslongago, Keegan, Ishango, Stubblyhead, EncMstr, RayAYang, Josteinn, Xchbla423, Can't sleep, clown will eat me, MyNameIsVlad, Nick Levine, Vladis1av, Awh, OrphanBot, Avb, Rrburke, Mr.Z-man, The tooth, Fullstop, Ian01, Kntrabssi, Dreadstar, BinaryTed, DMacks, Bidabadi~enwiki, Sadi Carnot, Ohconfucius, SashatoBot, Silvem, John, Crazyviolinist, Chodorkovskiy, JorisvS, Hemmingsen, PseudoSudo, Les.hopper, Stwalkerster, Dicklyon, Waggers, Dhp1080, Ryulong, Thrindel, Dl2000, Emx~enwiki, Iridescent, Theone00, Wjejsckenewr, J Di, Cbrown1023, CapitalR, Courcelles, Bertport, Tawkerbot2, T-W, Nkuzmik, Chetvorno, Owen214, Xcentaur, JForget, Jorcoga, Ninetyone, KyraVixen, Jackietang33, JohnCD, Drinibot, Cooljeanius, KnightLago, Nbach, Stevo1000, Dgw, WeggeBot, Moreschi, Chicheley, Cydebot, Jackyd101, Gogo Dodo, Corpx, Chasingsol, Alucard (Dr.), Shirulashem, Christian75, DumbBOT, Chrislk02, Phydend, Sharonlees, JodyB, Cubfanpgh, Storeye, Seicer, Thijs!bot, Epbr123, Daa89563, Hacky, HappyInGeneral, Sagaciousuk, Andyjsmith, Dougsim, Headbomb, Marek69, Marziepants, John254, Ronald W Wise, Phoe, Bunzil, CTZMSC3, Escarbot, The Lord of Time, AntiVandalBot, RobotG, Majorly, Seaphoto, Prolog, Mary Mark Ockerbloom, Bakabaka, Dylan Lake, L0b0t, DShamen, Fireice, Myanw, Golgofrinchian, Tigga, Husond, DuncanHill, Gcm, Pfzngn, Dsp13, Hut 8.5, PhilKnight, LittleOldMe, .anacondabot, Paulmulcahy1982, Connormah, Bongwarrior, VoABot II, Bothar, Vanish2, JNW, Midgrid, Bubba hotep, Dirac66, Allstarecho, David Eppstein, Xargon666x6, Turpificatus, Just James, Glen, DerHexer, Patstuart, Kayau, Seba5618, Flaming Ferrari, Youkai no unmei, MartinBot, Poeloq, Matt-rex, R'n'B, CommonsDelinker, VirtualDelight, Llllllou, SSSidhu, J.delanoy, Mattgk, Trusilver, SureFire, SuperGirl, Discott, Rhinestone K, Uncle Dick, Rdkng478, Boltslt21, Davidprior, Acalamari, Salih, Turtlebean2, McSly, Jeepday, Flobthelog, AKucia, Quarma, Floaterfluss, Andraaide, NewEnglandYankee, Fountains of Bryn Mawr, SJP, Gregfitzy, Toon05, KylieTastic, Kvdveer, Hstarkey, Secleinteer, Scewing, Wikieditor06, VolkovBot, CWii, The Duke of Waltham, Jeff G., Eje004, AlnoktaBOT, Philip Trueman, TXiKiBoT, Oshwah, Jomasecu, Jimmyeatskids, GDonato, Miranda, Warlord dehacker, GcSwRhIc, Sintaku, Seraphim, SelketBot, Jackfork, LeaveSleaves, ^demonBot2, Raymondwinn, Duncan.Hull, Madhero88, Gillyweed, Synthebot, CoolKid1993, Rhopkins8, Cnilep, Insanity Incarnate, Dessymona, Deconstructhis, Aqwfyj, Mehmet Karatay, EJF, SieBot, Tiddly Tom, Caltas, Cwkmail, RJaguar3, Yintan, Tataryn, Mothmolevna, LeadSongDog, Arda Xi, Keilana, Bentogoa, Toddst1, Flyer22, Prof .Woodruff, Exert, 2hiyup2, Arjen Dijksman, Prestonmag, Summeree, Oxymoron83, Bagatelle, KoshVorlon, Oculi, JackSchmidt, Joshii, Macy, Sunrise, OKBot, Latics, Mygerardromance, Firefly322, WikiLaurent, Liddy01, Escape Orbit, AutoFire, Muhends, RS1900, Beeblebrox, ClueBot, Dcaps, Binksternet, PipepBot, Foxj, The Thing That Should Not Be, Lawrence Cohen, Parkjunwung, Cygnis insignis, R000t, Arakunem, Drmies, Kattttttt, Phahn7, CounterVandalismBot, Xenon54, Nrucker05, GregVolk, Mspraveen, Masterpiece2000, Seanwal111111, DragonBot, Veryprettyfish, Robert Skyhawk, Excirial, Alexbot, Jusdafax,

Samasamas1, Bow chika wow wow, Christine1107, Monobi, Master10060, SpikeToronto, NuclearWarfare, Jotterbot, Razorflame, Arun cbe 888, Mikaey, ChrisHamburg, Kakofonous, Tincalf65, Cardinalem, Aitias, AlphaSpartan117, Walshie112233, Ghaytorade, Versus22, NERIC-Security, Vanished User 1004, Bletchley, Spitfire, Gnowor, MessinaRagazza, Rror, Integralolrivative, Avoided, Splunge launcher, Skarebo, NellieBly, Meeeeoooow, Alexius08, Khilsati, Good Olfactory, Thatguyflint, Kbdankbot, Addbot, Narayansg, Some jerk on the Internet, Guoguo12, Ronhjones, Movingboxes, MalachiK, Ironholds, Mr. Wheely Guy, Fluffernutter, Richmond96, FiriBot, Favonian, Wiemis, 5 albert square, Itfc+canes=me, Ehrenkater, Tide rolls, Nibraas, AlexJFox, Jedishive, Legobot, Luckas-bot, Yobot, Ptbotgourou, Senator Palpatine, Archon-Magnus, THEN WHO WAS PHONE?, JEms123, Gunnar Hendrich, Annette46, Eric-Wester, Tempodivalse, AnomieBOT, DemocraticLuntz, A More Perfect Onion, NathanoNL, Steamturn, Jim1138, IRP, Piano non troppo, Ularevalo98, Khcf6971, AdjustShift, Aditya, Waterden, Flewis, Materialscientist, Xcesspower, Citation bot, Bob Burkhardt, Xqbot, Jayarathina, Zad68, Cureden, The sock that should not be, Melmann, Addihockey10, Capricorn42, Nasnema, Gensanders, Davshul, Jmundo, Abce2, Omnipaedista, 10petersons, B-rad35, Prunesqualer, RibotBOT, SassoBot, Kieryh, Nedim Ardoğa, Mjkkdd, Polargeo, Shadowjams, A. di M., SD5, Green Cardamom, Paine Ellsworth, M0000p, VS6507, MichealH, Delphinus1997, Wifione, Jamesooders, A little insignificant, Atlantia, Citation bot 1, Biker Biker, I dream of horses, Per Ardua, ImageTagBot, MJ94, Supreme Deliciousness, A8UDI, Serols, Turian, Reconsider the static, Jauhienij, Chaugen1, ActivExpression, Tjlafave, Trappist the monk, Dinamik-bot, Oracleofottawa, Vrenator, MrX, Leondumontfollower, MajorStovall, Xdmndx, Jeffrd10, Linguisticgeek, Diannaa, Weedwhacker128, ThinkEnemies, Tbhotch, Reach Out to the Truth, Minimac, Killer00112, DARTH SIDIOUS 2, Rasberybunn, Twonernator, Sciencenerd998, Marco Gentili, Peaceworld111, Wikislemur, NerdyScienceDude, Sticky taint, EmausBot, Acather96, WikitanvirBot, Gfoley4, Stief, Heracles31, Super48paul, RA0808, RenamedUser01302013, Tommy2010, Wikipelli, K6ka, Yankee02, Anirudh Emani, Lamb99, Mark Purdy, Hhhippo, JSquish, ZéroBot, Lemeza Kosugi, Kinimv, Fæ, Gokuluday, A930913, H3llBot, Suslindisambiguator, Alyff rs, Fizicist, Wayne Slam, Tolly4bolly, Wagino 20100516, Cyberdog958, L Kensington, Donner60, Sailsbystars, Autoerrant, ChuispastonBot, RockMagnetist, Matthewrbowker, Peter Karlsen, GrayFullbuster, TYelliot, DASHBotAV, Ely411, Taliesin717, Xanchester, ClueBot NG, Elodzinski, Fiveninefive, Crtcollector, Ijf3, O.Koslowski, Conservikid, Widr, XXXXXsciecneboy, Helpful Pixie Bot, HMSSolent, Sharingandid, Lowercase sigmabot, Aldeyain, BG19bot, MusikAnimal, GKFX, Onewhohelps, Rm1271, ERJANIK, Aranea Mortem, IiJohnny xD, Joshem1995, Mathetudes, Snow Blizzard, Pkenny2332, The Traditionalist, Glacialfox, Buechlein, Anbu121, Roopraiash, Absconditus, Tutelary, Ben524524, Cyberbot II, Kiera5678, MadGuy7023, JYBot, Dexbot, Mogism, Thepasta, Ash-wapati, HelicopterLlama, Lugia2453, VIAFbot, Frosty, Sriharsh1234, Telfordbuck, Champrevis, Awsome117, Byrdtrack990, Mr.scienceman, Epicgenius, Herbert the III, Red-eyed demon, AmericanLemming, Nonsenseferret, Tentinator, ArmbrustBot, Ugog Nizdast, The Herald, My name is not dave, Ginsuloft, Kalem5678, EricEnfermeroMobile, Dredge16, Starmeow, JaconaFrere, Surcoe, TheEpTic, Austinsagan, Monkbot, Jonarnold1985, Vigina slap, Benniijones, KH-1, Anirudhan rajagopalan, Carolineholla, Jimbofarn, FourViolas, UofM1234321, Govindaharihari, Rkgchpc, KasparBot, JesseXD, John Dalton69, Leawesomepotato, Dawingwingherro, 57yo and Anonymous: 1428

- **Lincoln Wolfenstein** *Source:* https://en.wikipedia.org/wiki/Lincoln_Wolfenstein?oldid=666933763 *Contributors:* David Schaich, Bender235, Cmapm, Flying fish, Linas, Ruud Koot, GregorB, Rjwilmsi, Goudzovski, Jaraalbe, RussBot, CRKingston, Stepa, Ephraim33, Fuhghettaboutit, Ser Amantio di Nicolao, Headbomb, Connormah, Waacstats, Tommieboi, SieBot, Rosiestep, Sean.hoyland, DumZiBoT, Addbot, Lightbot, Zorrobot, Materialscientist, Omnipaedista, DefaultsortBot, RjwilmsiBot, ZéroBot, Asi013, Tannerscheeler, VIAFbot, KasparBot, Zip of king, Pstcrn and Anonymous: 12

- **Cowan–Reines neutrino experiment** *Source:* https://en.wikipedia.org/wiki/Cowan%E2%80%93Reines_neutrino_experiment?oldid=663825642 *Contributors:* Michael Hardy, Jengod, Maximus Rex, Raul654, Twang, Harp, Bobblewik, Nickptar, Xeroc, Gene Nygaard, Rjwilmsi, Strait, Zagloba~enwiki, Goudzovski, Whosasking, Spike Wilbury, Phr en, Eloy, JoshuaZ, Headbomb, CosineKitty, Grimlock, LorenzoB, STBot, Plasticup, The Original Wildbear, Kaspar.jan, BOTarate, Johnuniq, TimothyRias, SkyLined, Addbot, Lightbot, Ettrig, Luckas-bot, Yobot, Ptbotgourou, Citation bot, Xqbot, Whatsoevernever, Mnmngb, Lester Welch, EmausBot, Salvoaiola, Helpful Pixie Bot, Curb Chain, Bibcode Bot, Idenshi, ChrisGualtieri, EWPage and Anonymous: 14

- **DESY** *Source:* https://en.wikipedia.org/wiki/DESY?oldid=682709168 *Contributors:* The Anome, Schneelocke, Robbot, Altenmann, Gnomon Kelemen, Chowbok, Ookami~enwiki, Rich Farmbrough, Bcat, Markussep, Cmdrjameson, Gbrandt, Tpikonen, Lectonar, M3tainfo, Dirac1933, Ceyockey, Linas, Sympleko, Rjwilmsi, HappyCamper, Erkcan, Gurch, Spudbeach, Howcheng, Multichill, SCZenz, Voidxor, Cambion, Andreask~enwiki, KnowledgeOfSelf, Bggoldie~enwiki, Chris the speller, Bluebot, Sergio.ballestrero, BNutzer, Vriullop, JorisvS, Green Giant, Dnheff, CmdrObot, Headbomb, Leolaursen, Ultracobalt, Dmellor, The Anomebot2, Connor Behan, Uvainio, Keith D, Tikiwont, Jotempe, DorganBot, Amikake3, Jferrando, Rei-bot, Broadbot, SwordSmurf, Novadeath69, StewartMH, Mild Bill Hiccup, Polyamorph, Thgoiter, Alexbot, Addbot, Nz26, Sebastian scha., Yobot, Amirobot, AnomieBOT, Yevin, Omnipaedista, Ace111, Nameless23, FrescoBot, Ironboy11, 4bpp, EmausBot, Wikfr, Khazar2, Tony Mach, Mfb, Trackteur, KasparBot, Chrisemblhh and Anonymous: 46

- **DONUT** *Source:* https://en.wikipedia.org/wiki/DONUT?oldid=660311327 *Contributors:* Rjwilmsi, Strait, SCZenz, Emilio floris, Stepa, Yamaguchi 先生, LogicalFrank, KnightLago, Headbomb, Animum, Seba5618, STBot, Thrawn562, Pennstatephil, Eidetic Man, ClueBot, WriterListener, Alexbot, Jesseflr09, Wertystorm, Weirdo813, XLinkBot, Addbot, Willking1979, DOI bot, Citation bot, MIRROR, Amaury, W-C, LucienBOT, Waitwat?, Citation bot 1, Wikipelli, ClueBot NG, Nnnn443, Bibcode Bot, Mrchip124 and Anonymous: 15

- **Fermilab** *Source:* https://en.wikipedia.org/wiki/Fermilab?oldid=681315607 *Contributors:* Mav, Shsilver, Rmhermen, Minesweeper, Ahoerstemeier, Docu, Rob Hooft, Bevo, Robbot, Goethean, Gnomon Kelemen, Wikibot, Tom harrison, Fastfission, Xerxes314, Jason Quinn, Bodhitha, Andycjp, H Padleckas, Crawdad, Thorwald, N328KF, O'Dea, Adambondy, Paul August, Brian0918, Ylee, Arancaytar, Dralwik, PaulHanson, Andrew Gray, Clay1039, Ayeroxor, SidP, Freshraisin, Dirac1933, Linas, Riffsyphon1024, Jpers36, Acone, Jugger90, Tevatron~enwiki, Reisio, Saperaud~enwiki, Rjwilmsi, Jivecat, Strait, Vegaswikian, Erkcan, The wub, Ttwaring, Bob Wiyadabebe-Iytsaboi, Matt Deres, NekoDaemon, Chobot, Blando728, Kummi, YurikBot, RussBot, KevinCuddeback, Bovineone, SCZenz, Voidxor, Scottfisher, JasonAD, Besselfunctions, ThunderBird, Curpsbot-unicodify, Ybbor, GrinBot~enwiki, Timothyarnold85, Morgan wascko, SmackBot, Clockhappy, TestPilot, Stepa, Liaocyed, Frumpet, Nil Einne, HeartofaDog, Chris the speller, DroEsperanto, Dual Freq, Sct72, Backspace, Jumping cheese, Savidan, Kevlar67, Pulu, Ser Amantio di Nicolao, DavidGC, Soap, Mcshadypl, DavidBailey, Jaganath, Bucksburg, Simkiott, Rob Shanahan, PRRfan, IceHunter, CzarB, CapitalR, CmdrObot, Itomchandler, N2e, Ken Gallager, Studiousstud, Cydebot, Archange56, Thijs!bot, Ucanlookitup, Headbomb, Rjshade, EdJohnston, Neatpete86, AntiVandalBot, Yellowdesk, SamIAmNot, Ingolfson, Barek, Albany NY, Z22, Magioladitis, Jllm06, Swpb, MMD61764, The Anomebot2, MrWarMage, Speedracer0883, STBot, Emsox, CommonsDelinker, HEL, DandyDan2007, Maurice Carbonaro, Athaenara, Fastspinecho, Plasticup, DadaNeem, DorganBot, VIOLENTRULER, JeffreyRMiles, Funandtrvl, Master z0b, VolkovBot, Rodstur, TheQuandry, DavidBrahm, Mbehnkeil, TXiKiBoT, PKDASD, Vegeta206, Hughey, Hmwith, SieBot, Sonicology, WereSpielChequers, Jauerback, The Parsnip!, Yintan, Ketone16, Flyer22, Lightmouse, Anchor Link Bot, Martarius, Sfan00 IMG, ClueBot, Frmorrison, Blanchardb,

DragonBot, Diagramma Della Verita, ResidueOfDesign, Beamjockey, Kakofonous, Joe N, Addbot, Aaronjhill, Mjamja, AkhtaBot, Zahd, Asippel89, FermilabUser, Write-out, Lightbot, Igor26, Luckas-bot, Yobot, Naudefjbot~enwiki, AnomieBOT, Archon 2488, AdjustShift, Citation bot, LouriePieterse, Multixfer, Shadowjams, Spellage, Steve Quinn, Jschnur, Kalmbach, Higgshunter, 777sms, DrCrisp, RjwilmsiBot, Emaus-Bot, RA0808, Solomonfromfinland, Gayshark, H3llBot, Hotzemoerkerk, ClueBot NG, Markthoms1, Frietjes, Helpful Pixie Bot, Tirebiter78, Sunshine Warrior04, MeanTuring, BonifaceFR, BattyBot, Mogism, Kennethaw88, Ramendoctor, Nikevcowsky, KasparBot and Anonymous: 136

- **Homestake experiment** *Source:* https://en.wikipedia.org/wiki/Homestake_experiment?oldid=652066409 *Contributors:* William Avery, Ubiquity, Smalljim, Flying fish, Oldelpaso, Strait, Mike Peel, Whosasking, Bluebot, V1adis1av, Mvsrhollywood, Jaeger5432, JohnCD, Reywas92, Customcabf100, Headbomb, The Anomebot2, Addbot, Lightbot, Citation bot, Scorbatt, GrouchoBot, Mnmngb, LucienBOT, Citation bot 1, Tom.Reding, Coronium, RjwilmsiBot, ClueBot NG, BatesIsBack, Bibcode Bot, Monkbot, Kankerhomo and Anonymous: 9

- **Kamioka Observatory** *Source:* https://en.wikipedia.org/wiki/Kamioka_Observatory?oldid=677943356 *Contributors:* RTC, Palfrey, Carlj7, Fukumoto, Andycjp, David Schaich, AnyFile, Flying fish, Rjwilmsi, Tim!, Strait, Mike Peel, Vegaswikian, Freshgavin, Mtu, SmackBot, Telescope, Kurtan~enwiki, CmdrObot, No1lakersfan, Safalra, Quibik, Uddzislaw, After Midnight, Headbomb, RobDe68, Douggers, Jllm06, The Anomebot2, BigrTex, Tikiwont, Lightmouse, Bojechko, Vendeka, Uzdzislaw, Addbot, SpBot, Lightbot, Luckas-bot, Yobot, Shinkansen Fan, Trinitrix, BsBsBs, Galoubet, Citation bot, Scorbatt, Tom.Reding, Beberger, Trappist the monk, H3llBot, Brycehughes, Mariguld, BG19bot, ImJimHill, Praxiphenes, LukasMatt, Deadin1984, The Quixotic Potato and Anonymous: 15

- **Large Electron–Positron Collider** *Source:* https://en.wikipedia.org/wiki/Large_Electron%E2%80%93Positron_Collider?oldid=676806098 *Contributors:* Michael Hardy, Looxix~enwiki, J'raxis, Ehn, BenRG, Sanders muc, ElBenevolente, Mattflaschen, Harp, Bodhitha, Gadfium, Lumidek, Laurascudder, Apyule, Ranveig, Zyqqh, Barrkel, Dirac1933, Gortu, Gene Nygaard, Lyuokdea, Dan100, Linas, Duncan.france, Strait, Goudzovski, Common Man, YurikBot, Arado, Spike Wilbury, Gary84, SCZenz, FlyingPenguins, Light current, SmackBot, Shmoo~enwiki, Philc 0780, Gakhandal, TheMaster42, Ryan Roos, LeoNomis, Mets501, JarahE, WISo, Headbomb, Boru318, Escarbot, Taborgate, Pagw, Yonidebot, STBotD, Kenneth M Burke, Idioma-bot, Themel, AlleborgoBot, Tombomp, Jedgilbert, AWeishaupt, Ctxppc, Muhends, Wwheaton, Alexbot, Gdlong, DumZiBoT, Addbot, Mortense, LaaknorBot, Lightbot, כיב מלמד, Legobot, Luckas-bot, Yobot, Rubinbot, Citation bot, Xqbot, SassoBot, RedBot, EmausBot, WikitanvirBot, Quantanew, Ethaniel, Quondum, Amdyrowlands, ChiZeroOne, Deer*lake, Wisconsinbadger, Bibcode Bot, Bm gub2, BG19bot, CERNwebeditors, Tony Mach, Bibliophilen, AlbertAndTheLion, Ellipapa, Csumstudent and Anonymous: 48

- **Lawrence Berkeley National Laboratory** *Source:* https://en.wikipedia.org/wiki/Lawrence_Berkeley_National_Laboratory?oldid=681266550 *Contributors:* Mav, Ed Poor, Menchi, Minesweeper, Jiang, Jengod, Stone, Wik, Dragons flight, Frish, Eman, Postdlf, DocWatson42, Mintleaf~enwiki, Fastfission, Bobblewik, Geni, Mako098765, Klemen Kocjancic, D6, Rich Farmbrough, AlanBarrett, Bender235, Duk, Hashar-Bot~enwiki, PaulHanson, Sbeath, TommyBoy, Middenface, Falcorian, Lensovet, Jarwulf, Mendaliv, Rjwilmsi, Tim!, NeonMerlin, NekoDaemon, ChongDae, Floriang, Ragesoss, Arcimpulse, Scottfisher, Mjsabby, DaveOinSF, Curpsbot-unicodify, Garion96, Sardanaphalus, Melchoir, Lagringa, Gyrobo, Cybercobra, Pulu, Bejnar, Andrei Stroe, Glacier109, Chymicus, Codepro, Tmangray, Chetvorno, CmdrObot, MrFizyx, N2e, Location, Trojan2006, Cydebot, MC10, Dynaflow, Headbomb, Trevyn, WVhybrid, Cutlassdude70, Yill577, Magioladitis, Panser Born, Rettetast, CommonsDelinker, Johnpacklambert, Tikiwont, BenTels, Flatterworld, VolkovBot, Mercurywoodrose, AlleborgoBot, Phe-bot, Ketone16, Lightmouse, Bgordski, OKBot, Rosiestep, HairyWombat, DragonBot, Sun Creator, Stepheng3, Qwfp, AgnosticPreachersKid, Deltawk, Northwesterner1, Good Olfactory, Addbot, AkhtaBot, LaaknorBot, SPat, Zorrobot, Legobot, Luckas-bot, Yobot, Fossett&Elvis, ArthurBot, Spesh531, FrescoBot, Saehrimnir, Fat otter, Redrose64, Skyerise, TobeBot, Cfsgfds, EmausBot, Djembayz, Lab-comm, KarlsenBot, DianeChojnowski, Strike Eagle, BG19bot, Compfreak7, Berklabsci, Cosmopolite1, E8xE8, Adele0622, Prisencolin, Unician, KasparBot, TonyaKayPetty and Anonymous: 43

- **Lawrence Livermore National Laboratory** *Source:* https://en.wikipedia.org/wiki/Lawrence_Livermore_National_Laboratory?oldid=681915095 *Contributors:* Trelvis, Bryan Derksen, Maury Markowitz, Edward, RTC, Michael Hardy, Oliver Pereira, Jiang, Fullerton, Wnissen, Pizza Puzzle, Hashar, Jengod, WhisperToMe, Gentgeen, Roscoe x, Bkell, DocWatson42, Christopher Parham, Greyengine5, Fastfission, Edcolins, Gadfium, Dubaduba~enwiki, Bumm13, Neutrality, Hellisp, Avatar, Kate, Rich Farmbrough, Guanabot, User2004, RoyBoy, TMC1982, Kjkolb, Jumbuck, PaulHanson, Apoc2400, Sligocki, Rwendland, DV8 2XL, Crosbiesmith, Stemonitis, Alvis, JonBirge, Jarwulf, Emerson7, BD2412, Kbdank71, Ketiltrout, Saperaud~enwiki, Rjwilmsi, ElKevbo, Awotter, Bmpower, NekoDaemon, ChongDae, Error9900, Simesa, YurikBot, RussBot, AVM, Anders.Warga, RadioFan, Hydrargyrum, ArséniureDeGallium, Davemck, Scottfisher, TimK MSI, Mjsabby, WAS 4.250, Georgewilliamherbert, Mike Dillon, DaveOinSF, Mike Selinker, Reject, Curpsbot-unicodify, Junglecat, AtomCrusher, SmackBot, Henriok, Agateller, EncMstr, Letdorf, Can't sleep, clown will eat me, Backspace, WestA, Cybercobra, Nutschig, RandomP, Pulu, Jeremyb, Tim Ross, Glacier109, Will Beback, Lambiam, John, Loodog, Mbeychok, Eleveneleven, Nobunaga24, Feureau, Mets501, Zepheus, Tmangray, Natrajdr, CapitalR, Eastlaw, Triumph Sisyphus, Van helsing, N2e, MaxEnt, Trojan2006, Cydebot, ChardingLLNL, Jedonnelley, Skittleys, Nick Ottery, Dynaflow, Asiaticus, Thijs!bot, Headbomb, WVhybrid, Missvain, WillMak050389, Mrshaba, CosineKitty, .anacondabot, ARHAP-STF, Magioladitis, Enoent, Charlesreid1, Artlondon, Cgingold, Beagel, DerHexer, DGG, Johnpacklambert, Tikiwont, Nwbeeson, Heyitspeter, Hugo999, VolkovBot, Shortride, Johnfos, Jgui~enwiki, TXiKiBoT, Mercurywoodrose, JhsBot, Luuva, Chueleven, Cmcnicoll, 4wajzkd02, I Like Cheeseburgers, Ketone16, Lisatwo, Lightmouse, JL-Bot, EmanWilm, Martarius, ClueBot, Djr32, Two Hearted River, XLinkBot, Addbot, DOI bot, Jojhutton, Grafikwerice, Dougbateman, Lightbot, Luckas-bot, Yobot, Amirobot, AnomieBOT, Piano non troppo, Hg6996, Trentonx, Nedim Ardoğa, Spesh531, Spellage, Ass mcgee, Mills20, DrilBot, Metricmike, Julien1978, TobeBot, JnRouvignac, Mramz88, El Mayimbe, DARTH SIDIOUS 2, Look2See1, H3llBot, Wikfr, Dannyauble, ClueBot NG, Matthiaspaul, BrekekekexKoaxKoax, ServiceAT, Abdulazeez,ani, John222222, Frze, LAgirl5252, Pcalbright, 1abqdad, Ema--or, Lseaveratnif, Cyberbot II, Gooball3man, ChrisGualtieri, Tmieieima, Cactus83, ScotXW, Jason Carpenter13, Meresarb, Kwpavan, KasparBot and Anonymous: 83

- **MINOS** *Source:* https://en.wikipedia.org/wiki/MINOS?oldid=678948067 *Contributors:* Jdpipe, Nine Tail Fox, Wikiborg, Pigsonthewing, Nurg, Awolf002, Bodhitha, Frau Holle, Flying fish, Graham87, Rjwilmsi, Strait, Mike Peel, Erkcan, Goudzovski, Tone, Johndarrington, PM Poon, Scottfisher, Crumley, Clockhappy, Benjaminevans82, Bluebot, Sergio.ballestrero, V1adis1av, Headbomb, D.H, Hartnell, Z22, LorenzoB, Oren0, STBot, Ajoykt, Addbot, Luckas-bot, Yobot, Ptbotgourou, Citation bot, Scorbatt, MIRROR, GrouchoBot, Spellage, Citation bot 1, Claudiodib, Jonesey95, Tom.Reding, Puddingrice, Trappist the monk, Bibcode Bot, BG19bot, Quickcrazy78, BattyBot, StarryGrandma, Dja1979, Mogism, Monkbot and Anonymous: 24

- **MiniBooNE** *Source:* https://en.wikipedia.org/wiki/MiniBooNE?oldid=645099398 *Contributors:* Northgrove, Mako098765, Pjacobi, Scentoni, Benhocking, Rjwilmsi, Strait, Mike Peel, Xaque, Rekleov, Phmer, Morgan wascko, SmackBot, Arbe, V1adis1av, Dc3~enwiki, Michael C Price,

Headbomb, Rod57, Fratrep, MystBot, Addbot, DOI bot, Lightbot, Luckas-bot, Citation bot, Scorbatt, MIRROR, Citation bot 1, Tom.Reding, Full-date unlinking bot, Bibcode Bot, BattyBot and Anonymous: 15

- **SLAC National Accelerator Laboratory** *Source:* https://en.wikipedia.org/wiki/SLAC_National_Accelerator_Laboratory?oldid=680114102 *Contributors:* Bryan Derksen, Edward, Minesweeper, Andrewa, Peter Kaminski, Mxn, Jengod, Nv8200pa, Ed g2s, Finlay McWalter, Sanders muc, Voyager640, DocWatson42, Dinomite, Fastfission, Peruvianllama, Leonard G., Xinoph, Wronkiew, Bobblewik, Neilc, Geni, CryptoDerk, Beland, Chrisn4255, Icairns, Jewbacca, Klemen Kocjancic, Squash, ChrisRuvolo, Rich Farmbrough, Art LaPella, Spoon!, Janna Isabot, Mike Schwartz, Dreish, Viriditas, Jag123, Rainer Bielefeld~enwiki, SPUI, Natelipkowitz, Tom Yates, Msh210, PaulHanson, Aranae, Wtshymanski, RJFJR, Dirac1933, Linas, RHaworth, Justinlebar, David Levy, Saperaud~enwiki, Erkcan, DannyWilde, NekoDaemon, Chobot, YurikBot, Darkstar949, JabberWok, Hydrargyrum, Rsrikanth05, SCZenz, Dionea, Scottfisher, TimK MSI, Karit~enwiki, Arthur Rubin, Brianlucas, ArielGold, Curpsbot-unicodify, Physicsdavid, GrinBot~enwiki, Kgf0, SmackBot, Hydrogen Iodide, Stepa, Miquonranger03, RadWorkerII, Colonies Chris, Annoyedgrunt, Cybercobra, PetesGuide, Akriasas, Pulu, Evil genius, Disavian, Iridescent, Astrobayes, Jerry-va, DanHickstein, CmdrObot, Raysonho, Old Guard, Johnlogic, Myasuda, Cydebot, Mblumber, JFreeman, Thijs!bot, Epbr123, Wikid77, Headbomb, WillMak050389, Thadius856, Qwerty Binary, Z22, Magioladitis, The Anomebot2, Wormcast, Hallonsten, JaGa, Jvimal, Racepacket, J.delanoy, Athaenara, BrokenSphere, Aboutmovies, Belovedfreak, Minesweeper.007, Joshmt, S, Philip Trueman, The Original Wildbear, Anonymous Dissident, SieBot, Umrguy42, Oxymoron83, Kqxr, Stepheng3, DumZiBoT, PSimeon, J Hazard, Addbot, Download, زرش, User0529, Luckas-bot, Yobot, Nallimbot, Neutrinoless, Orion11M87, AnomieBOT, Archon 2488, Galoubet, İncelemeelemani, Citation bot, Cnwilliams, S4wilson, Iluvkitties991, DrCrisp, John of Reading, Look2See1, Profrocshae, Andyfreeberg, Helpful Pixie Bot, Bibcode Bot, PearlSt82, Edgitar86, Comfr, Darvii, Sonĝanto, This Name is Ironic, Phbuck, Maderthaner, Monkbot, MistyEye, Jtinsman, KasparBot and Anonymous: 56

- **SPEAR** *Source:* https://en.wikipedia.org/wiki/SPEAR?oldid=664792927 *Contributors:* SarekOfVulcan, Strait, SCZenz, Voidxor, Od Mishehu, Backspace, Happy-melon, Headbomb, The Anomebot2, Hallonsten, Lightbot, Dexbot and Anonymous: 5

- **Sudbury Neutrino Observatory** *Source:* https://en.wikipedia.org/wiki/Sudbury_Neutrino_Observatory?oldid=670843307 *Contributors:* Bryan Derksen, Maury Markowitz, Montrealais, Looix~enwiki, Taxman, Bearcat, Robbot, Astronautics~enwiki, Hadal, Wikibob, Joseph Dwayne, Qui1che, Rich Farmbrough, Kmccarty, Shanes, CDN99, Duk, Jag123, Hooperbloob, Fleurot~enwiki, Msh210, RoySmith, Gene Nygaard, Ceyockey, Falcorian, Flying fish, Linas, Mindmatrix, Kralizec!, PeregrineAY, Tevatron~enwiki, Rnt20, Graham87, Rjwilmsi, Tim!, Strait, Hunor, DVdm, NSR, RussBot, Hellbus, Varnav, Golfcam, Flup, Cstaffa, SmackBot, CyclePat, Stepa, Bluebot, EncMstr, Sbharris, OrphanBot, Jiminy pop, NeilFraser, Euchiasmus, M@sk, ShakingSpirit, Korandder, Safalra, Rotiro, Cydebot, Ntsimp, Boardhead, WISo, Headbomb, DMcanada, The Anomebot2, CommonsDelinker, HEL, Tikiwont, Blotto adrift, DonovanHawkins, Wilmot1, Kelapstick, Judge Nutmeg, Don4of4, Jakowa~enwiki, Neparis, Android Mouse Bot 3, ImageRemovalBot, Jht4060, Addbot, DOI bot, Turtlejuice, LaaknorBot, Lightbot, Luckas-bot, Archon 2488, Icalanise, Xqbot, Scorbatt, Paine Ellsworth, StephenFuqua, Tom.Reding, EmausBot, WikitanvirBot, Arbnos, Timetraveler3.14, EdoBot, Widr, Bibcode Bot, BHBrunt, DrVentureWasRight, Monkbot, Vieque, Csumstudent and Anonymous: 63

- **Super-Kamiokande** *Source:* https://en.wikipedia.org/wiki/Super-Kamiokande?oldid=682631815 *Contributors:* Brion VIBBER, Zundark, The Anome, Css, Ed Poor, Maury Markowitz, MimirZero, Hfastedge, Tzaquiel, Alfio, Looix~enwiki, Wik, Astronautics~enwiki, Harp, Fukumoto, Dissident, Everyking, Hoho~enwiki, Niteowlneils, Matt Crypto, Bobblewik, Gyrofrog, Andycjp, Quadell, Kusunose, Neutrality, TomRitchford, Rich Farmbrough, Spoon!, Hooperbloob, Jeltz, Dental, Alfvaen, Sudachi, Jackhynes, Flying fish, Linas, BillC, Tevatron~enwiki, Graham87, Ketiltrout, Wikix, Rjwilmsi, Strait, Pitan, Mike Peel, Tintazul, Alex Kapranoff, Goudzovski, WikiWikiPhil, YurikBot, Osomec, Neilbeach, Baru~enwiki, Mysid, Kkmurray, Physicsdavid, GrinBot~enwiki, KnightRider~enwiki, N3v, Chris the speller, Cattus, Noir~enwiki, Cybercobra, Eitan Freedenberg, Kurtan~enwiki, A876, Was a bee, Michael C Price, Markluffel, ZeZo, Thijs!bot, Headbomb, Douggers, Bm gub, Boffob, Tikiwont, Sheliak, Zentau7, JukoFF, Neparis, MissMJ, Boneyard90, Addbot, OlEnglish, Luckas-bot, Amirobot, Citation bot, S7evyn, Trafford09, Mnmngb, Spellage, FrescoBot, Paine Ellsworth, Discovery1998, Trebauchet1986, Jonesey95, Tom.Reding, RjwilmsiBot, Mmm333k, ZéroBot, Arbnos, Timetraveler3.14, Bibcode Bot, BG19bot, Yowanvista, ImJimHill, Epicgenius, Spluzer, Monkbot, Li722 and Anonymous: 68

- **Supernova Early Warning System** *Source:* https://en.wikipedia.org/wiki/Supernova_Early_Warning_System?oldid=679090183 *Contributors:* Anville, Pjacobi, BRW, Falcorian, Flying fish, Daniel Case, Strait, Chobot, Tone, Gaius Cornelius, Eleassar, ERcheck, Modest Genius, Cybercobra, JoshuaZ, PuerExMachina, WISo, Headbomb, Dtgriscom, Swpb, Gwern, Maurice Carbonaro, Rod57, Wingedsubmariner, Scog, Addbot, Yobot, AnomieBOT, Citation bot, Xymostech, Nacen, ZéroBot, Bibcode Bot, Yaroslav Nikitenko, Monkbot, Jimwade562 and Anonymous: 7

8.2 Images

- **File:1e0657_scale.jpg** *Source:* https://upload.wikimedia.org/wikipedia/commons/a/a8/1e0657_scale.jpg *License:* Public domain *Contributors:* Chandra X-Ray Observatory: 1E 0657-56 *Original artist:* NASA/CXC/M. Weiss

- **File:A_Super-Kamiokande_Realtime_Monitor_result.gif** *Source:* https://upload.wikimedia.org/wikipedia/en/6/69/A_Super-Kamiokande_Realtime_Monitor_result.gif *License:* Fair use *Contributors:* http://t2k-experiment.org/t2k/ *Original artist:* T2K experiment homepage

- **File:A_cross-section_of_the_Super-Kamiokande_detector.jpg** *Source:* https://upload.wikimedia.org/wikipedia/en/e/e4/A_cross-section_of_the_Super-Kamiokande_detector.jpg *License:* Fair use *Contributors:* http://www.sciencedirect.com/science/article/pii/S016890020300425X *Original artist:* S. Fukuda

- **File:AirShower.svg** *Source:* https://upload.wikimedia.org/wikipedia/commons/2/2c/AirShower.svg *License:* CC BY 3.0 *Contributors:* originally from nl.wikipedia; description page is/was here. *Original artist:* Mpfiz

- **File:Air_Purification_system_of_Super-Kamiokande.jpg** *Source:* https://upload.wikimedia.org/wikipedia/en/1/18/Air_Purification_system_of_Super-Kamiokande.jpg *License:* Fair use *Contributors:* http://www.sciencedirect.com/science/article/pii/S016890020300425X *Original artist:* S. Fuduka

- **File:Alpha_1_racetrack,_Uranium_235_electromagnetic_separation_plant,_Manhattan_Project,_Y-12_Oak_Ridge.jpg** *Source:* https://upload.wikimedia.org/wikipedia/commons/c/ce/Alpha_1_racetrack%2C_Uranium_235_electromagnetic_separation_

- **File:Oscillations_electron_short.svg** *Source:* https://upload.wikimedia.org/wikipedia/en/3/3e/Oscillations_electron_short.svg *License:* CC0 *Contributors:* ? *Original artist:* ?

- **File:Oscillations_muon_long.svg** *Source:* https://upload.wikimedia.org/wikipedia/en/1/12/Oscillations_muon_long.svg *License:* CC0 *Contributors:* ? *Original artist:* ?

- **File:Oscillations_muon_short.svg** *Source:* https://upload.wikimedia.org/wikipedia/en/1/1f/Oscillations_muon_short.svg *License:* CC0 *Contributors:* ? *Original artist:* ?

- **File:Oscillations_tau_long.svg** *Source:* https://upload.wikimedia.org/wikipedia/en/2/28/Oscillations_tau_long.svg *License:* CC0 *Contributors:* ? *Original artist:* ?

- **File:Oscillations_tau_short.svg** *Source:* https://upload.wikimedia.org/wikipedia/en/4/4b/Oscillations_tau_short.svg *License:* CC0 *Contributors:* ? *Original artist:* ?

- **File:Oscillations_two_neutrino.svg** *Source:* https://upload.wikimedia.org/wikipedia/en/8/86/Oscillations_two_neutrino.svg *License:* CC0 *Contributors:* ? *Original artist:* ?

- **File:PETRA_III_Max_von_Laue_Hall_2014.png** *Source:* https://upload.wikimedia.org/wikipedia/commons/a/ac/PETRA_III_Max_von_Laue_Hall_2014.png *License:* CC0 *Contributors:* Own work *Original artist:* Uvainio

- **File:PETRA_III_Max_von_Laue_Hall_inside_2014.png** *Source:* https://upload.wikimedia.org/wikipedia/commons/e/e1/PETRA_III_Max_von_Laue_Hall_inside_2014.png *License:* CC0 *Contributors:* Own work *Original artist:* Uvainio

- **File:PLOS_logo_2012.svg** *Source:* https://upload.wikimedia.org/wikipedia/commons/b/b0/PLOS_logo_2012.svg *License:* CC BY-SA 3.0 *Contributors:* Public Library of Science (PLOS) *Original artist:* PLOS

- **File:P_vip.svg** *Source:* https://upload.wikimedia.org/wikipedia/en/6/69/P_vip.svg *License:* PD *Contributors:* ? *Original artist:* ?

- **File:Pairproduction.png** *Source:* https://upload.wikimedia.org/wikipedia/commons/8/84/Pairproduction.png *License:* CC-BY-SA-3.0 *Contributors:* Transferred from en.wikipedia *Original artist:* Original uploader was Davidhorman at en.wikipedia. Later version(s) were uploaded by Falcorian at en.wikipedia.

- **File:Particles_and_antiparticles.svg** *Source:* https://upload.wikimedia.org/wikipedia/commons/c/cd/Particles_and_antiparticles.svg *License:* CC BY-SA 3.0 *Contributors:* This vector image was created with Inkscape. *Original artist:* Anynobody

- **File:Pauli.jpg** *Source:* https://upload.wikimedia.org/wikipedia/commons/4/43/Pauli.jpg *License:* Public domain *Contributors:* http://nobelprize.org/nobel_prizes/physics/laureates/1945/pauli-bio.html *Original artist:* Nobel foundation

- **File:Photo_of_the_Week-_An_Incredible_Journey_--_Transporting_a_600-ton_Magnet_(9324124048).jpg** *Source:* https://upload.wikimedia.org/wikipedia/commons/c/cc/Photo_of_the_Week-_An_Incredible_Journey_--_Transporting_a_600-ton_Magnet_%289324124048%29.jpg *License:* Public domain *Contributors:* Photo of the Week: An Incredible Journey -- Transporting a 600-ton Magnet *Original artist:* ENERGY.GOV

- **File:PiPlus_muon_decay.svg** *Source:* https://upload.wikimedia.org/wikipedia/commons/6/69/PiPlus_muon_decay.svg *License:* CC0 *Contributors:* Own work *Original artist:* Krishnavedala

- **File:Portal-puzzle.svg** *Source:* https://upload.wikimedia.org/wikipedia/en/f/fd/Portal-puzzle.svg *License:* Public domain *Contributors:* ? *Original artist:* ?

- **File:PositronDiscovery.jpg** *Source:* https://upload.wikimedia.org/wikipedia/commons/6/69/PositronDiscovery.jpg *License:* Public domain *Contributors:* Anderson, Carl D. (1933). "The Positive Electron". *Physical Review* **43** (6): 491–494. DOI:10.1103/PhysRev.43.491. *Original artist:* Carl D. Anderson (1905–1991)

- **File:Pp-chain_and_CNO_chain.jpg** *Source:* https://upload.wikimedia.org/wikipedia/en/8/87/Pp-chain_and_CNO_chain.jpg *License:* Fair use *Contributors:* http://www.sciencedirect.com.ezproxy.lib.purdue.edu/science/article/pii/S0146641013000264 *Original artist:* A.B. Balantekin

- **File:Proton_proton_cycle.svg** *Source:* https://upload.wikimedia.org/wikipedia/commons/a/ac/Proton_proton_cycle.svg *License:* CC BY 2.5 *Contributors:* file:Proton proton cycle.png *Original artist:* Dorottya Szam

- **File:QCDphasediagram.svg** *Source:* https://upload.wikimedia.org/wikipedia/commons/b/bc/QCDphasediagram.svg *License:* CC BY-SA 3.0 *Contributors:* Own work *Original artist:* TimothyRias

- **File:Quark_masses_as_balls.svg** *Source:* https://upload.wikimedia.org/wikipedia/commons/b/b5/Quark_masses_as_balls.svg *License:* CC BY-SA 3.0 *Contributors:* Own work *Original artist:* Incnis Mrsi

- **File:Quark_structure_proton.svg** *Source:* https://upload.wikimedia.org/wikipedia/commons/9/92/Quark_structure_proton.svg *License:* CC BY-SA 2.5 *Contributors:* Own work *Original artist:* Arpad Horvath

- **File:Quark_weak_interactions.svg** *Source:* https://upload.wikimedia.org/wikipedia/commons/6/66/Quark_weak_interactions.svg *License:* Public domain *Contributors:* Derivative work, from public down work uploaded to en.wikipedia. original *Original artist:*

- Original work: [1]

- **File:Question_book-new.svg** *Source:* https://upload.wikimedia.org/wikipedia/en/9/99/Question_book-new.svg *License:* Cc-by-sa-3.0 *Contributors:*

  Created from scratch in Adobe Illustrator. Based on Image:Question book.png created by User:Equazcion *Original artist:* Tkgd2007

- **File:RaE1.jpg** *Source:* https://upload.wikimedia.org/wikipedia/commons/2/2b/RaE1.jpg *License:* CC BY-SA 4.0 *Contributors:* Own work *Original artist:* HPaul

- **File:Radioactive.svg** *Source:* https://upload.wikimedia.org/wikipedia/commons/b/b5/Radioactive.svg *License:* Public domain *Contributors:* Created by Cary Bass using Adobe Illustrator on January 19, 2006. *Original artist:* Cary Bass

- **File:Raymond_Davis,_Jr_&_GW_Bush.jpg** *Source:* https://upload.wikimedia.org/wikipedia/commons/8/8c/Raymond_Davis%2C_Jr_%26_GW_Bush.jpg *License:* Public domain *Contributors:* [1] [2] *Original artist:* National Science Foundation

- **File:Red_pog.svg** *Source:* https://upload.wikimedia.org/wikipedia/en/0/0c/Red_pog.svg *License:* Public domain *Contributors:* ? *Original artist:* ?

- **File:Right_left_helicity.svg** *Source:* https://upload.wikimedia.org/wikipedia/commons/a/a9/Right_left_helicity.svg *License:* Public domain *Contributors:* en:Image:Right left helicity.jpg *Original artist:* en:User;HEL, User:Stannered

- **File:SF_From_Marin_Highlands3.jpg** *Source:* https://upload.wikimedia.org/wikipedia/commons/d/da/SF_From_Marin_Highlands3.jpg *License:* Public domain *Contributors:* en:User:Paul.h *Original artist:* en:User:Paul.h

- **File:SLAC_Entrance.jpg** *Source:* https://upload.wikimedia.org/wikipedia/commons/7/7c/SLAC_Entrance.jpg *License:* CC BY 3.0 *Contributors:* Own work *Original artist:* Jvimal

- **File:SLAC_LogoSD.png** *Source:* https://upload.wikimedia.org/wikipedia/en/a/aa/SLAC_LogoSD.png *License:* Fair use *Contributors:* http://www-group.slac.stanford.edu/com/images/slac_logos_2012branding/SLAC_LogoSD.png *Original artist:* ?

- **File:SLAC_detector.jpg** *Source:* https://upload.wikimedia.org/wikipedia/commons/0/0a/SLAC_detector.jpg *License:* CC-BY-SA-3.0 *Contributors:* Own work *Original artist:* Justin Lebar

- **File:SLAC_long_view.jpg** *Source:* https://upload.wikimedia.org/wikipedia/commons/8/87/SLAC_long_view.jpg *License:* CC-BY-SA-3.0 *Contributors:* Own work *Original artist:* Justin Lebar

- **File:SLAC_pit_and_detector.jpg** *Source:* https://upload.wikimedia.org/wikipedia/commons/a/aa/SLAC_pit_and_detector.jpg *License:* CC-BY-SA-3.0 *Contributors:* Own work *Original artist:* Justin Lebar

- **File:SLAC_tunnel_2.jpg** *Source:* https://upload.wikimedia.org/wikipedia/commons/0/0d/SLAC_tunnel_2.jpg *License:* Public domain *Contributors:* http://energy.gov/articles/labchat-particle-accelerators-lasers-and-discovery-science-may-17-1pm-est *Original artist:* Brad Plummer

- **File:Schwebungsfall.svg** *Source:* https://upload.wikimedia.org/wikipedia/commons/1/1a/Schwebungsfall.svg *License:* Public domain *Contributors:* Own work *Original artist:* Daniel Schaal Farbing

- **File:Science.jpg** *Source:* https://upload.wikimedia.org/wikipedia/commons/5/54/Science.jpg *License:* Public domain *Contributors:* ? *Original artist:* ?

- **File:Scientist.svg** *Source:* https://upload.wikimedia.org/wikipedia/commons/0/03/Scientist.svg *License:* CC-BY-SA-3.0 *Contributors:* Own work *Original artist:* Viktorvoigt

- **File:Standard_Model_Feynman_Diagram_Vertices.png** *Source:* https://upload.wikimedia.org/wikipedia/commons/7/75/Standard_Model_Feynman_Diagram_Vertices.png *License:* CC BY-SA 3.0 *Contributors:* I made it in Adobe Illustrator *Original artist:* Garyzx

- **File:Standard_Model_of_Elementary_Particles.svg** *Source:* https://upload.wikimedia.org/wikipedia/commons/0/00/Standard_Model_of_Elementary_Particles.svg *License:* CC BY 3.0 *Contributors:* Own work by uploader, PBS NOVA [1], Fermilab, Office of Science, United States Department of Energy, Particle Data Group *Original artist:* MissMJ

- **File:Stanford-linear-accelerator-usgs-ortho-kaminski-5900.jpg** *Source:* https://upload.wikimedia.org/wikipedia/commons/8/8a/Stanford-linear-accelerator-usgs-ortho-kaminski-5900.jpg *License:* Public domain *Contributors:* United States Geological Survey *Original artist:* Peter Kaminski

- **File:Strong_force_charges.svg** *Source:* https://upload.wikimedia.org/wikipedia/commons/b/b6/Strong_force_charges.svg *License:* CC BY-SA 3.0 *Contributors:* Own work, Created from Garret Lisi's Elementary Particle Explorer *Original artist:* Cjean42

- **File:Stylised_Lithium_Atom.svg** *Source:* https://upload.wikimedia.org/wikipedia/commons/e/e1/Stylised_Lithium_Atom.svg *License:* CC-BY-SA-3.0 *Contributors:* based off of Image:Stylised Lithium Atom.png by Halfdan. *Original artist:* SVG by Indolences. Recoloring and ironing out some glitches done by Rainer Klute.

- **File:Supernova-1987a.jpg** *Source:* https://upload.wikimedia.org/wikipedia/commons/4/43/Supernova-1987a.jpg *License:* CC BY 3.0 *Contributors:* Supernova 1987A: Halo for a Vanished Star, Mosaic of Supernova 1987A. *Original artist:* First image: Dr. Christopher Burrows, ESA/STScI and NASA; Second image: Hubble Heritage team.

- **File:Supernova_SN1987A_in_the_Large_Magellanic_Cloud_-_GPN-2000-000948.jpg** *Source:* https://upload.wikimedia.org/wikipedia/commons/2/2a/Supernova_SN1987A_in_the_Large_Magellanic_Cloud_-_GPN-2000-000948.jpg *License:* Public domain *Contributors:* Great Images in NASA Description *Original artist:* NASA, The Hubble Heritage Team, STScI, AURA

- **File:Symbol_book_class2.svg** *Source:* https://upload.wikimedia.org/wikipedia/commons/8/89/Symbol_book_class2.svg *License:* CC BY-SA 2.5 *Contributors:* Mad by Lokal_Profil by combining: *Original artist:* Lokal_Profil

- **File:T2K_experiment_beam.png** *Source:* https://upload.wikimedia.org/wikipedia/en/f/f2/T2K_experiment_beam.png *License:* Fair use *Contributors:* http://t2k-experiment.org/t2k/ *Original artist:* T2K experiment

- **File:Table_isotopes_en.svg** *Source:* https://upload.wikimedia.org/wikipedia/commons/c/c4/Table_isotopes_en.svg *License:* CC BY-SA 3.0 *Contributors:*

- Table_isotopes.svg *Original artist:* Table_isotopes.svg: Napy1kenobi

- **File:Tau_lepton.svg** *Source:* https://upload.wikimedia.org/wikipedia/commons/f/f8/Tau_lepton.svg *License:* CC BY 3.0 *Contributors:* File:Standard Model of Elementary Particles.svg *Original artist:* user:MissMJ

- **File:Tau_neutrino.svg** *Source:* https://upload.wikimedia.org/wikipedia/commons/a/ac/Tau_neutrino.svg *License:* CC BY 3.0 *Contributors:* File:Standard Model of Elementary Particles.svg *Original artist:* user:MissMJ

8.3 Content license